D0812190

# FIBER OPTICS HANDBOOK

## Optical and Electro-Optical Engineering Series

**Robert E. Fischer and Warren J. Smith, *Series Editors***

### Published

NISHIHARA, HARUNA, SUHARA • *Optical Integrated Circuits*
RANCOURT • *Optical Thin Films Users' Handbook*
THELEN • *Design of Optical Interference Coatings*

### Forthcoming Volumes

FANTONE • *Optical Fabrication*
FISCHER • *Optical Design*
GILMORE • *Expert Vision Systems*
JACOBSON • *Thin Film Deposition*
JOHNSON • *Infrared System Design*
ROSS • *Laser Communications*
STOVER • *Optical Scattering*

### Other Published Books of Interest

CSELT • *Optical Fibre Communication*
HECHT • *The Laser Guidebook*
HEWLETT-PACKARD • *Optoelectronics Applications Manual*
KAO • *Optical Fiber Systems*
KEISER • *Optical Fiber Communications*
MAC LEOD • *Thin Film Optical Filters*
MARSHALL • *Free Electron Lasers*
OPTICAL SOCIETY OF AMERICA • *Handbook of Optics*
SMITH • *Modern Optical Engineering*
TEXAS INSTRUMENTS • *Optoelectronics*
WYATT • *Radiometric System Design*

# FIBER OPTICS HANDBOOK

## For Engineers and Scientists

**Frederick C. Allard** Editor
Naval Underwater Systems Center
Member, Optical Society of America



**McGRAW-HILL PUBLISHING COMPANY**

New York St. Louis San Francisco Auckland Bogotá
Caracas Hamburg Lisbon London Madrid Mexico
Milan Montreal New Delhi Oklahoma City
Paris San Juan São Paulo Singapore
Sydney Tokyo Toronto

Library of Congress Cataloging-in-Publication Data

Allard, Frederick C.
Fiber optics handbook: for engineers and scientists/Frederick C. Allard.
p. cm.—(Optical and electro-optical engineering series)
ISBN 0-07-001013-7
1. Optical communications. 2. Fiber optics. I. Title.
II. Series.
TK5103.59.A44 1989
621.382'75—dc20

89-37274
CIP



234567890 DOC/DOC 9543210

ISBN 0-07-001013-7

*The editors for this book were Daniel Gonneau and Steven Wagman, the designer was Naomi Auerbach, and the production supervisor was Suzanne W. Babeuf. This book was set in Times Roman. It was composed by the McGraw-Hill Publishing Company Professional and Reference Division composition unit.*

*Printed and bound by R. R. Donnelley & Sons Company.*

*To my wife, Suzanne*

# CONTENTS

# CONTRIBUTORS

**James C. Daly** *Department of Electrical Engineering, University of Rhode Island, Kingston, RI 02881-0805* (CHAP. 7)

**Anthony Fascia** *Department of Electrical Engineering, University of Rhode Island, Kingston, RI 02881-0805* (CHAP. 7)

**Felix Kapron** *Bell Communications Research, Morristown, NJ 07960-1910* (CHAP. 4)

**Kenneth Li** *PCO, Inc., Chatsworth, CA 91311-6289* (CHAPS. 5 AND 6)

**Peirangelo Morra** *Centro Studi e Laboratori Telecomunicazioni S.p.A., 274-10148 Torino, Italy* (CHAP. 3)

**G. D. Pitt** *Renishaw Transducer Systems Ltd., Wotton-Under-Edge, Gloucestershire GL 12 7DN, England* (CHAP. 8)

**David E. Quinn** *Corning Glass Works, Corning, NY 14831* (CHAP. 1)

**Murray Ramsay** *STC Technology, Ltd., Harlow, Essex CM 17 9NA, England* (CHAP. 2)

**Harish R. D. Sunak** *Department of Electrical Engineering, University of Rhode Island, Kingston, RI 02881-0805* (CHAP. 9)

**Emilio Vezzoni** *Centro Studi e Laboratori Telecomunicazioni S.p.A., 274-10148 Torino, Italy* (CHAP. 3)

**Paul Kit Lai Yu** *Department of Electrical Engineering and Computer Sciences, University of California, San Diego, La Jolla, CA 92093* (CHAPS. 5 AND 6)

# PREFACE

This Handbook provides engineers, scientists, and technical managers the basis for applying fiber optics to specific problems in information transfer. It provides the broad coverage and extensive bibliography required of a reference work by those already involved in fiber optics, yet it is written in an expository style to aid an entry-level reader. It should also be useful—by virtue of the numerous examples cited—to the experienced technician who is transferring from wire transmission to fiber optics and, as well, to the engineering student interested in the relative merits of specific components and techniques.

This Handbook comprehensively reviews the elements of applied fiber-optic information transfer. It includes definitions, operational principles, and comparative performance information on candidate components for fiber optics systems. Functional relationships among elements within fiber-optic systems are described and illustrated with examples relating to the transfer of digital and analog data and the transfer of other information, such as from sensors. The effects of various environmental parameters are explained, as well as other engineering-related factors, such as aging processes. Specific, standardized test methods are described.

The applications focus is maintained throughout the Handbook, resulting in a definitive guide that—in a readily accessible format—

- provides engineering information on fiber-optic components and techniques
- establishes the basis for selecting individual components and methods of operation by means of comparative evaluation
- guides the reader in applying specific information by means of specific engineering examples

The Handbook begins with a chapter describing the operating principles and various implementations of fiber-optic waveguides. Chapter 2 delineates the principles underlying cabling of optical fibers and includes examples of cables for specific applications.

Chapter 3 provides a comprehensive treatment of fiber-optic connectors, splices, and couplers, including a virtual catalog of techniques. Chapter 4 surveys the more important standardized fiber-optic test procedures.

Chapters 5 and 6 review optical source and detector technologies, respectively, for application to fiber-optic systems, while Chap. 7 describes the encoding strategies used to impress information on an optical carrier.

Chapter 8 comprehensively describes various fiber-optic sensors, and features extensive and specific quantitative and qualitative performance comparisons. Chapter 9—the closing chapter—provides a transmission system perspective, using numerous design examples of both analog and digital transmission systems.

## ACKNOWLEDGMENTS

Some of the authors' organizations are gratefully acknowledged for their extensive graphics contributions. Worthy of special mention in this regard are Corning Glass Works (Chap. 1), STC Technology, Ltd. (Chaps. 2 and 8), and CSELT (Chap. 3).

Several individuals provided special assistance to the editor during the preparation of the handbook. These include: Thomas J. Gryk, Naval Underwater Systems Center; Gregory A. Allard, Carnegie Mellon University; Gregory Majewski, Naval Underwater Systems Center; Frederic L. Allard, University of Bridgeport; David A. McQueeney, Naval Underwater Systems Center; and Professor James C. Daly, University of Rhode Island.

*Frederick C. Allard*

# FIBER OPTICS HANDBOOK

# CHAPTER 1
# OPTICAL FIBERS

**David E. Quinn**
*Corning Incorporated*

## *1.1 INTRODUCTION*

Research and development efforts over the last 20 years have led to commercial realization of low-cost, low-loss optical fibers. Their use as a preferred transmission medium in current communication systems is because fiber-optic systems offer greater information-carrying capacity over longer repeaterless distances at costs lower than conventional copper-wire systems. In addition to the initial reduced system costs, system capacity can be improved without adding fibers by upgrading the optoelectronic devices in the system as needed. For example, lasers that can be modulated at faster rates or that emit light over a narrow wavelength range can dramatically increase the information-carrying capacity, or bandwidth, of an optical-fiber system.

Along with the ability to be upgraded, current optical fibers have other beneficial properties that make them desirable as a transmission medium. For example, their high information-carrying capacity means that typical optical-fiber cables are only a fraction of the size of comparable copper wire cables. This makes them attractive for use in areas of limited space, such as in congested underground conduits in major cities or large office buildings. Also, since optical fibers typically are made from glass, a dielectric material, they are relatively unaffected by electromagnetic radiation. Therefore, they often can be installed in locations where copper wire would require additional shielding or protection to maintain system quality. This has enabled optical transmission systems to run with electric power lines where issues of geographic topology and access rights-of-way already have been resolved.

## *1.2 PROPAGATION THEORY*

This chapter begins with a discussion of the principles involved with the propagation of light through the fiber structure. (Generating the light signal, coupling it into the fiber, and detecting it as it exits the fiber will be discussed in other chapters of this handbook.)

An *optical fiber* is a structure that is designed to guide light over a distance, or path, that is not necessarily straight. Optical fibers accomplish light confinement by the total internal reflection of the light that is coupled into the end of the fiber. While typical optical fibers for telecommunication or data transmission are made from high-purity glass in a round fiber geometry, optical fibers can be fabricated from plastics, single crystals drawn into fibers, or hollow tubes filled with an appropriate fluid. Also, the fiber does not necessarily need to be circular in geometry. Planar, or flat slab, guides are used commercially for a variety of applications such as splitting a light signal into multiple signals or specific power levels, i.e., in optical couplers.

The concept of light propagation in an optical fiber can be discussed from two different perspectives, consistent with the notion of light having a *dual* nature: light is a discrete particle and it is also an electromagnetic wave that is influenced by the propagation medium. These two concepts of light behavior allow one to consider two different approaches to light propagation in an optical fiber.

The first perspective—that of ray theory—can provide an intuitive feel for the use of optical fibers in terms of light containment and light pulse spreading. This approach is based on the concept that a light particle has a discrete, defined trajectory. The second point of view—that of wave theory—is less intuitive, using electromagnetic wave behavior to explain the behavior of light within an optical fiber. Wave theory is useful in defining absorption, attenuation, and dispersion. Both approaches are presented in this chapter, and notable differences are given. Additional information is contained in the references, should the reader want more detailed information.

### 1.2.1 Ray Diagrams

*Internal Reflection.* Typical optical fibers are solid cylinders. A simplified, two-dimensional description, also applicable to planar waveguides, suffices to explain optical propagation in fibers. Figure 1.1 shows possible meridional ray paths in a step-index fiber in which the core region has a refractive index $n_1$ and the cladding region has a refractive index $n_2$ lower than $n_1$. The incoming ray strikes the core-air boundary at angle $\theta$ to the fiber axis. The air has a refractive index $n_0$,

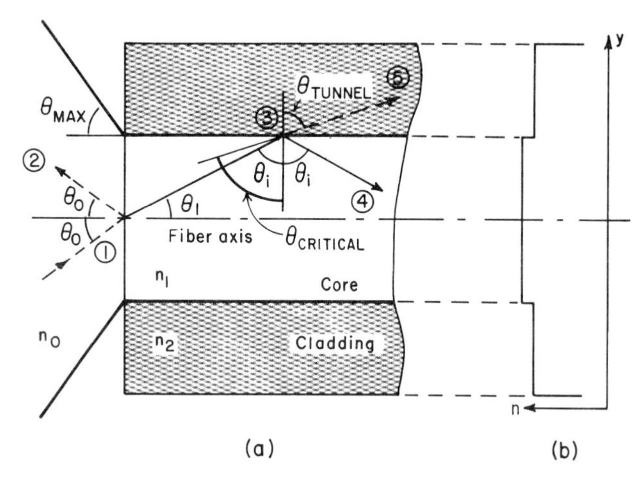


FIGURE 1.1 (*a*) Meridional ray paths in a step-index fiber; (*b*) step-index profile.

less than that of either the core or cladding. This ray is, in part, reflected from the core interface and, in part, transmitted into the core. The transmitted ray is refracted at the core surface and continues at an angle $\theta_1$ to the centerline of the fiber. The transmitted ray strikes the core-cladding interface at point 3 and is totally internally reflected to point 4, if $\theta_i$ is greater than a critical angle $\theta_{critical}$.

The thick lines to the left in Fig. 1.1 represent the maximum angle $\theta_{max}$ at which entering rays undergo total internal reflection at the core-cladding interface. If an entering ray strikes the core-cladding interface at an angle smaller than $\theta_{critical}$, it will be only partially reflected at the cladding, giving rise to a refracted ray in the cladding (5 in Fig. 1.1). The refracted ray initially propagates at an angle $\theta_{tunnel}$ within the cladding. This ray is only loosely bound within the fiber structure and eventually escapes from the fiber.

The ray angles in Fig. 1.1 are related through the laws of reflection and refraction. In the case of reflection, the angle of incidence equals the angle of reflection:

$$\theta_{incidence} = \theta_{reflection} \tag{1.1}$$

Refraction is expressed by Snell's law; i.e., the product of the refractive index and the sine of the ray angle in a given medium is equal to the product of the refractive index and the sine of the ray angle in any other medium within the system.

$$n_0 \sin \theta_0 = n_1 \sin \theta_1 = n_2 \sin \theta_2 \tag{1.2}$$

where $n_0$ = refractive index of the medium outside the fiber
$n_1$ = refractive index of the core region
$n_2$ = refractive index of the cladding region
$\theta_1$ = ray angle in the medium having a refractive index $n_1$ (the ray angle is referenced to the normal at the surface of the boundary)

From Snell's law, the sine of the critical angle is determined by setting $\theta_2$ equal to $\pi/2$ and $\theta$ equal to $\theta_{critical}$:

$$\sin \theta_{critical} = \frac{n_2}{n_1} \tag{1.3}$$

From this relationship, the numerical aperture (NA) can be defined as

$$\text{NA} = \sin \theta_{max} \tag{1.4}$$

where

$$\sin \theta_{max} = (n_1^2 - n_2^2)^{1/2} \tag{1.5}$$

Referring to Fig. 1.1, a ray entering at an angle $\theta_{max}$ will strike the core-cladding interface at the critical angle $\theta_{critical}$.

Therefore, the numerical aperture defines an acceptance cone within which all bound rays are contained. It is a convenient measure of the ability of an optical fiber to capture light from a wide-angle source such as a light-emitting diode

(LED). Generally, a higher NA indicates a higher source-to-fiber coupling efficiency.

In typical glass optical fibers fabricated from silica with an assumed cladding refractive index of 1.458 and a core index elevation of 0.01, an acceptance angle of 12° is obtained. The numerical aperture of such a fiber is 0.2. The upper limit of NA for commercially available glass optical fibers is nominally 0.29. Plastic fibers can have an NA greater than 0.5.

***Optical Path Length.*** As shown in Fig. 1.2, a serious problem can arise from the multiplicity of ray angles accepted by a typical step-index optical fiber. Rays entering at smaller angles take more direct paths than larger ones, the shortest path

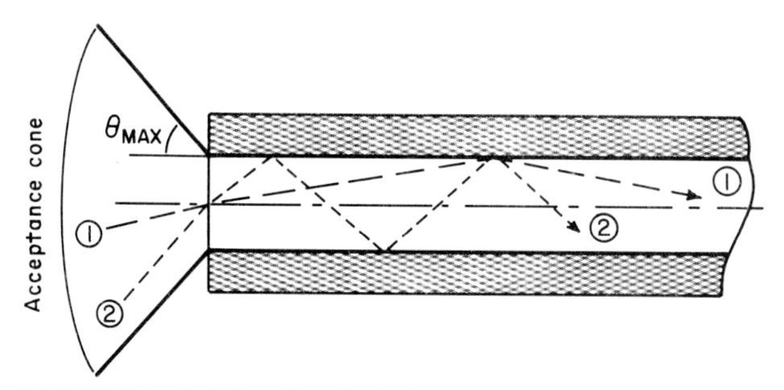


**FIGURE 1.2** Path length variation in a step-index fiber for different ray angles.

being the centerline. This difference in path length results in different arrival times for different rays at the end of the fiber. In Fig. 1.2, ray 1 arrives ahead of ray 2. The difference in arrival times for various rays results in light-pulse spreading.

A typical optical source emits rays at various angles. These rays will be accepted into a circular fiber if they are within the fiber's numerical aperture, just defined. Since the light is accepted at various ray angles, the different rays will have different path lengths down the fiber. Since all rays will travel at the same speed in this example, some rays, those with the shortest path length, or those accepted normal to the fiber end face, will reach the other end of the fiber first. Those accepted at slightly lower angles to the normal will arrive next, etc., until those introduced at an angle just below $\theta_{max}$ will exit the end of the fiber last.

Therefore, the instantaneous introduction of a light pulse at one end of the fiber gives a multiplicity of ray arrival times at the exit end, leading to pulse broadening, which increases as the length of fiber increases.

If the input signal is degraded far enough, it may fall below the detectable limit of the receiver. Also, if signals traverse a sufficiently long path, individual pulses may become indiscernible as they overlap one another. When discussed in terms of the electromagnetic-wave nature of light, this phenomenon of pulse spreading by virtue of different optical path lengths is called *intermodal dispersion.* Intermodal dispersion is commonly referred to simply as *modal dispersion* and will be discussed in greater detail later in this chapter.

Not only do optical sources produce rays over a range of angles, but they also produce rays over a range of wavelengths. As the refractive index of the glass fiber is not constant with respect to wavelength, the light pulse spreads because

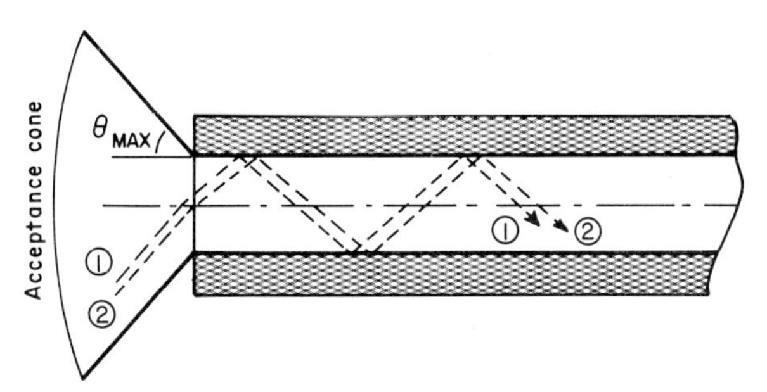


FIGURE 1.3 Chromatic dispersion caused by different wavelengths in light source.

different wavelengths travel at different speeds, even if they have the same ray path. This is referred to as *chromatic dispersion* and is illustrated in Fig. 1.3.

Pulse spreading as a function of geometrical path differences can be controlled by altering the refractive index of the core. Typically, a profile is chosen such that the refractive index is highest at the center of the core, with a parabolic fall-off in refractive index occurring as a function of core radius, Fig. 1.4*b*. In this graded-index fiber, the ray increases in speed as it moves farther away from the center of the fiber because of the progressively lower refractive index. Thus, low-angle rays, such as 1 in Fig. 1.4*a*, have a shorter geometrical path length, but move more slowly through the higher refractive index medium than do the high-angle rays, such as 3. A parabolic refractive index profile tends to compensate for the different optical path lengths of various rays by causing the rays to propagate at different speeds. Therefore, the rays, regardless of path length, propagate at approximately the same net velocity. This balancing reduces pulse spreading due to modal dispersion, each trajectory being thought of as a mode of propagation, and thereby provides for an optical fiber with greater information-carrying capacity.

As discussed previously, optical fibers used in transmission are three-dimensional, and therefore reflection occurs at a curved interface—not at a flat boundary. The complication of the reflection from this curved surface causes the loss of some rays that, by previous discussion, would be expected to be internally reflected or tightly bound to the fiber. Also, in three dimensions, light is not only introduced at an angle $\theta$ but also at an angle $\phi$, giving rise to the propagation of helical or skewed rays that travel off the center axis of the fiber. The path length differences of skew rays also can be compensated for by the grading of the core refractive index profile.

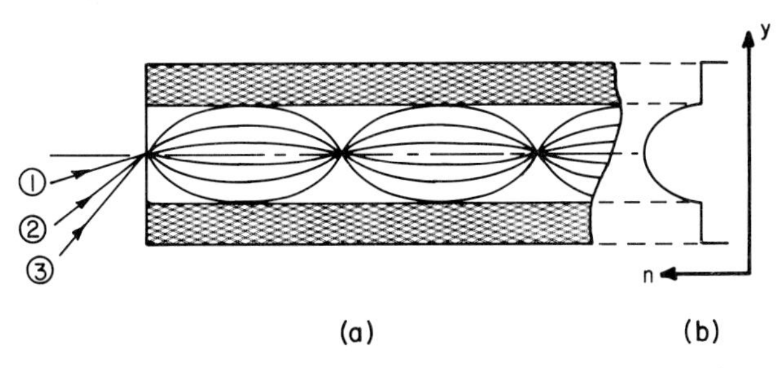


FIGURE 1.4 (*a*) Ray propagation in a graded-index fiber; (*b*) graded-index profile.

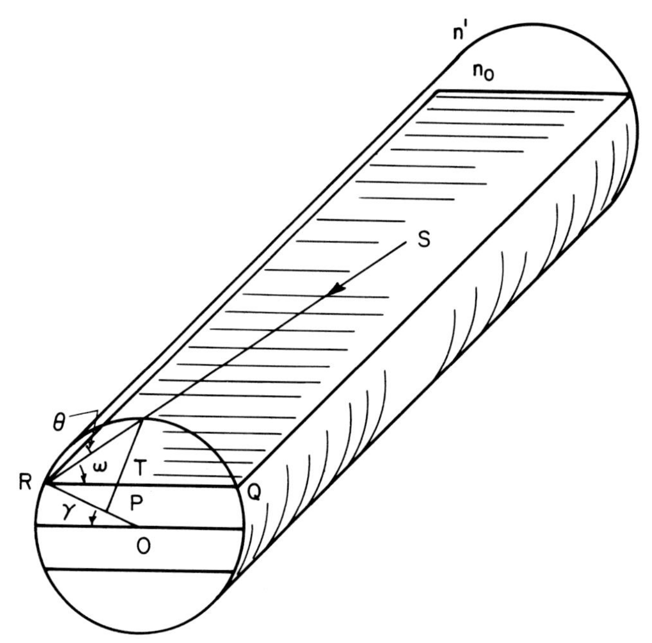


FIGURE 1.5 Skew-ray propagation in a step-index fiber. *(Reproduced from Ref. 1, with permission.)*

From simple geometric considerations, this skew or general ray propagation has been shown to be similar to meridional ray travel. From the work of Kapany,[1] the extrapolation of the two-dimensional case to the three-dimensional case again makes use of the discrete-particle concept of light.

In Fig. 1.5, the trajectory of a skew ray *SR* (entering from the right in the figure) is traced down the fiber. Here the ray enters at some skew angle and intersects the cylinder wall at *R*. For a given ray *SR*, intersecting the cylinder wall at this point, a plane containing the ray and a line parallel to the cylinder axis is defined. *RO* defines the normal to the surface at that point of intersection. The angle $\omega$ formed by *SR* and *RO* defines the trapping of the ray into the cylinder. Given *RT* as the projection of the unit length of the ray on the normal and $RP = RT \cos \gamma$ as the projection of *RT* on the normal,

$$\cos \omega = \sin \theta \cos \gamma \tag{1.6}$$

The values of $\gamma$ and $\theta$ are unique for any ray at any point in the cylinder. The numerical aperture for any fiber now becomes

$$\text{NA} = n \sin \theta \cos \gamma \leq (n_1^2 - n_2^2)^{1/2} \tag{1.7}$$

### 1.2.2 Electromagnetic Theory

***Propagation Modes.*** Ray diagrams are somewhat limited in describing some important fiber properties, such as propagation modes, referred to earlier as different ray trajectories, arising from the electromagnetic wave nature of light.

A limitation in the ray approach is that rays cannot actually be localized to a well-defined path. Also, the ray approach allows a continuum of possible rays to propagate between the normal to the fiber face and the critical angle. The latter condition is not observed in optical fibers because the oscillations or modes of light in the fiber are discrete, not continuous.

Mode theory can be developed using plane waves, where a plane wave is described by its direction—a vector perpendicular to the wavefronts (see Fig. 1.6), its amplitude, and its wavelength. In a dielectric medium, the last can be expressed as

$$\lambda = \frac{c}{fn} \tag{1.8}$$

where $c$ = speed of light in vacuum
$n$ = refractive index
$f$ = frequency

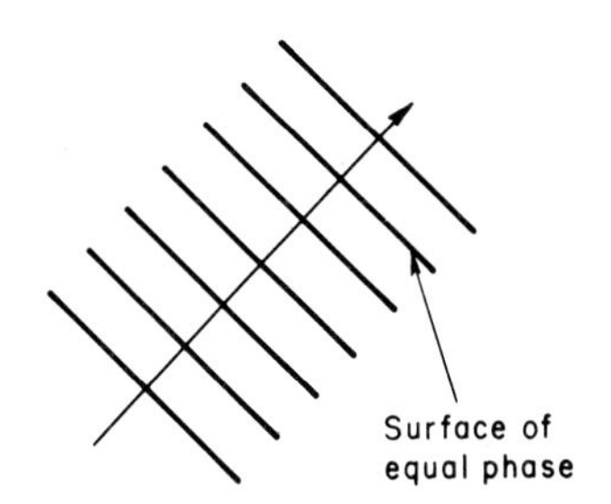


FIGURE 1.6 Plane-wave propagation.

From the previous discussions, the ray incident on the fiber end is "accepted" into the fiber at some angle with respect to the normal to the dielectric interface. As the ray, now considered as a wavefront, reflects at the upper boundary, a second, downward-propagating wave $B$ is created, similar to the reflection in Fig. 1.1. $B$ in turn reflects off the lower boundary to create a second reflected wave $C$. Wave theory requires that the original wave $A$ and reflected wave $C$ be in phase so that the entire collection of original and reflected waves can propagate. If the propagating wavefronts are not in phase, they decay as they traverse the fiber, as various in-phase and out-of-phase fronts destructively interfere.

By following the wavefronts in Fig. 1.7 as they traverse from point 1 to point 2, it can be observed that the distance traversed is $2H/\cos\theta$. The phase accumulated along the path is this distance times $2\pi/\lambda$. The total phase accumulated is equal to

$$\phi_1 = \frac{4\pi H}{\lambda \cos\theta} - 2\phi_R \tag{1.9}$$

where $\phi_R$ is an extra phase shift due to the reflection at each boundary.

If we do not follow the reflected ray down the path length but instead follow it directly down a straight line from 1 to 2, the number of wavefronts passed is the distance

$$\frac{2H \tan\theta \cdot \sin\theta}{\lambda}, \tag{1.10}$$

Thus, we do not accumulate wavefronts as quickly when we do not move parallel to the wave direction. The accumulated phase along this path is

$$\phi_2 = \frac{4\phi H \sin\theta \cdot \tan\theta}{\lambda} \tag{1.11}$$

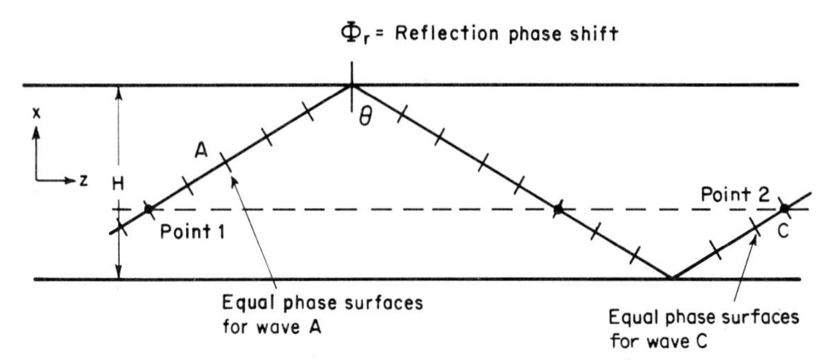


FIGURE 1.7 Plane-wave reflection.

For the original wave $A$ and the reflected wave $C$ to be phase-matched, they can differ only by an integer multiple of a wave period; i.e., the accumulated phases $\phi_1$ and $\phi_2$ can differ only by an integer multiple of $2\pi$.

Therefore,

$$\frac{4\pi H}{\lambda} \cos \theta - 2\phi_R = 2M\pi \tag{1.12}$$

where $M$ is a positive integer ($M = 0, 1, 2, \ldots$). Equation (1.12) is not satisfied for general values of $\theta$. For a specific value of integer $M$, at most a single value of $\theta$ will solve the equation; the collection of solutions for $M = 0, 1, 2, \ldots$ are the discrete ray angles that are allowed by ray theory.

Equation (1.12) also develops other important concepts of waveguide propagation. First, there are a finite number of modes that can propagate in a fiber. This is based on the fact that $M$ cannot become so large that $\cos \theta > 1$, since this condition cannot be satisfied by any angle.

Second, if the fiber structure is large so that $H/\lambda \gg 1$, there are many solutions, with very small spacing between the allowed $\theta$ values. In the limit of a very large fiber, the ray theory approach is regained: A large number of closely spaced modes are approximated very well by a continuous family of rays.

Also, as the fields of the plane waves repeat themselves along the $z$ axis at a distance of $\lambda/\sin \theta$, a periodicity along $z$ with a spatial frequency of $\beta = 2\pi \sin \theta/\lambda$ is seen. The quantity $\beta$ is defined as the axial propagation constant.

As the wavelength $\lambda$ varies, the values of $\cos \theta$ that satisfy the mode equation change, and so the propagation constant must change as well. This variation of $\beta$ with $\lambda$ is called the dispersion related to the mode. For a given mode $M$, $\lambda$ typically can be made large enough so that no $\cos \theta$ will satisfy the modal equation. The mode is then said to be *cut off*, and the wavelength at which the modal solution disappears is called the *cutoff wavelength* for the mode.

It also can be shown from the form of the phase shifts $\Phi_R$ that the modal equations always have solutions for any wavelength for the $M = 0$ mode. This is referred to as the fundamental mode of the waveguide, and it is never cut off. Optical fibers typically have fundamental modes also, and that wavelength below which the next higher mode can begin to propagate is called the cutoff wavelength for the fiber. When an optical fiber is operated above cutoff, only the fundamental mode can propagate; below cutoff, the waveguide is multimoded.

Like ray theory as it is developed from the two- to three-dimensional case, the wave approach also takes on added complexity. However, from the treatment just presented, a given ray trajectory continues to be thought of as the direction of a local plane wave. As before, this wave and the waves produced by subsequent reflections must interfere constructively, i.e, phase-match, for the wave to propagate. As before, only certain ray trajectories will satisfy the modal equation.

Beyond the intuitive approach to developing lightwave propagation, rigorous mathematical descriptions exist.

Lightwave propagation using local plane waves has been described and predicted by Maxwell's equations, based on his work in electromagnetic theory.[2] In Maxwell's treatise, the electromagnetic wave is described as having two components which are orthogonal. These are $\mathbf{E}(x, y, z)$, the electric field, and $\mathbf{H}(x, y, z)$, the magnetic field. The change in refractive index is assumed to be zero along the fiber length (i.e., $dn/dz = 0$).

Given the relationship of **E** and **H** to the parameters **D** and **B**, the electric flux density and the magnetic flux density, respectively, the following set of conditions can be written:

$$\nabla \times \mathbf{E} = -\frac{\partial \mathbf{B}}{\partial t} \quad \text{and} \quad \nabla \times \mathbf{H} = \frac{\partial \mathbf{D}}{\partial t} \tag{1.13}$$

$$\nabla \cdot \mathbf{D} = 0 \quad \text{and} \quad \nabla \cdot \mathbf{B} = 0 \tag{1.14}$$

where the curl and divergence of the vector fields are taken. Substituting

$$\mathbf{D} = \varepsilon \mathbf{E} \quad \text{and} \quad \mathbf{B} = \mu \mathbf{H} \tag{1.15}$$

and taking the curl of the resultant equation yields

$$\nabla \times (\nabla \times \mathbf{E}) = -\mu\varepsilon \frac{\partial^2 \mathbf{E}}{\partial t^2} \tag{1.16}$$

$$\nabla \times (\nabla \times \mathbf{H}) - \nabla(\ln \varepsilon) \times \nabla \times \mathbf{H}) = -\mu\varepsilon \frac{\partial^2 \mathbf{H}}{\partial t^2} \tag{1.17}$$

where $\varepsilon$ = dielectric permittivity and $\mu$ = magnetic permeability. Equations (1.16) and (1.17), in turn, result in the wave equations

$$\nabla^2 + \nabla(\mathbf{E} \cdot \nabla \ln \varepsilon) = \mu\varepsilon \frac{\partial^2 \mathbf{E}}{\partial t^2} \tag{1.18}$$

$$\nabla^2 \mathbf{H} + \nabla(\ln \varepsilon) \times (\nabla \times \mathbf{H}) = \mu\varepsilon \frac{\partial^2 \mathbf{H}}{\partial t^2} \tag{1.19}$$

where $\nabla^2$ = laplacian operator.

Given that the refractive index is constant along the fiber length,

$$E(x, y, z) = E(xy)e^{i(\beta z - \omega t)} \tag{1.20}$$

$$H(x, y, z) = H(xy)e^{i(\beta z - \omega t)} \tag{1.21}$$

with $\beta$ defined as the propagation constant and $\omega$ = angular frequency ($=2\pi f$).

In cylindrically symmetric waveguides, the fields are written

$$E = E(r, \phi)e^{-i(\omega t - \beta z)}e^{im\theta} \tag{1.22}$$

$$H = H(r, \phi)e^{-i(\omega t - \beta z)}e^{im\theta} \tag{1.23}$$

where $\omega$ = angular frequency = $2\pi f$, $\beta$ = a propagation constant, and $m$ = azimuthal mode number.

Given the constraints imposed by a finite area of refractive index difference, such that $n(r) = [\varepsilon_r(r)]^{1/2}$ and with the constraint that the propagating field be finite on axis and zero at infinity, these equations take on eigenvalue solutions. These solutions yield distinct and noncontinuous values for $\beta$. It is these distinct solutions that give rise to the propagating electromagnetic field patterns which are called *modes*.[3]

*Normalized Frequency.* The degree to which an electromagnetic wave is bound to the fiber is described by the fiber's normalized frequency $V$, which is defined as

$$V = ka(n_{\text{core}}^2 - n_{\text{clad}}^2)^{1/2} \tag{1.24}$$

where $k = 2\pi/\lambda$ and $V$ is a measure of both the width of the guiding region and its depth in terms of index contrast.

As the refractive indexes of the core, $n_{\text{core}}$, and cladding, $n_{\text{clad}}$, are known values (as is $a$, the core diameter), the $V$ value can be routinely calculated for fibers. This $V$ value will describe the number of modes propagating for a given fiber geometry. A normalized propagation constant $b$ is also defined[4] as

$$b = \frac{(\beta/k) - n_0}{n_1 - n_0} \tag{1.25}$$

where $\beta$ = propagation constant
$n_0$ = cladding index
$n_1$ = core index

Figure 1.8 shows a plot of $b$ vs. the normalized frequency $V$ for a step-index profile fiber. As can be seen, as $V$ increases, more modes are supported in the optical fiber. Further, as $V$ increases, $\beta$ increases to the limiting value of $\beta = n_1 k$ as $\beta$ becomes 1. As $V$ decreases, $\beta$ decreases for the mode to the lower limit value of $\beta = n_2 k$ as $b$ becomes 0.

The number of guided modes is important, as it is the modes, or electromagnetic patterns, that transfer the optical power. The higher-order modes are those that are most lossy and, thus, are responsible for power loss as the propagating light is influenced by bends, etc.

It is noted from Eq. (1.24) that $V$ is dependent on $k$ (which is equal to $2\pi/\lambda$), and thus $V$ is wavelength-dependent.

Complications in mode propagation arise as the fiber profile deviates from the ideal step geometry. Approximations to modal propagation in a parabolic index profile are achievable by using the Wentzel-Kramers-Brillouin (WKB) approximation. For graded index profiles described by the form

$$n^2(r) = n_1^2\left[1 - 2\Delta\left(\frac{r}{a}\right)^\alpha\right] \tag{1.26}$$

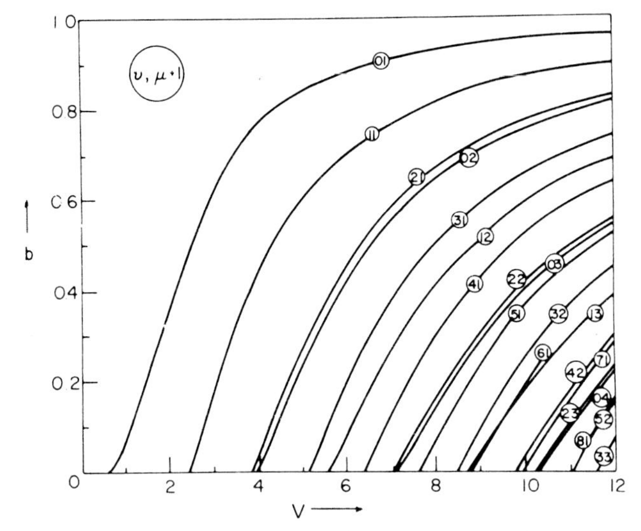


**FIGURE 1.8** Plot of the propagation constant $b$ vs. $V$ value for lower-order modes. *(Reproduced from D. Gloge, "Weakly Guiding Fibers," Applied Optics, vol. 10, no. 10, pp. 2252–2258, 1971.)*

where $r <$ core radius

$n(r) = n_{\text{clad}}$ for cladding material

$\alpha$ = profile exponent

$\Delta = (n_{\text{core}}^2 - n_{\text{clad}}^2)/2n_{\text{clad}}^2$

the propagation constant can now be written as

$$\beta_n = nk\left[1 - 2\Delta\left(\frac{m}{N}\right)^{\alpha/(\alpha+2)}\right]^{1/2} \tag{1.27}$$

where $N$ represents the total number of modes, described as

$$N = \left(\frac{\alpha}{\alpha + 2}\right)n_1^2k^2a^2\Delta \tag{1.28}$$

Again, it is seen that the number of supported modes increases directly with core size $a$ and index elevation $\Delta$.

Typical profiles are shown in Fig. 1.9 for $\alpha$ values ranging from 1 to infinity.[5]

## 1.3 FIBER TYPES

Optical fibers are usually classified into two types, *single-mode* and *multimode.* These fiber types are so named by the number of modes propagating at the wavelength of operation. Table 1.1 shows common dopants added to silica glass for

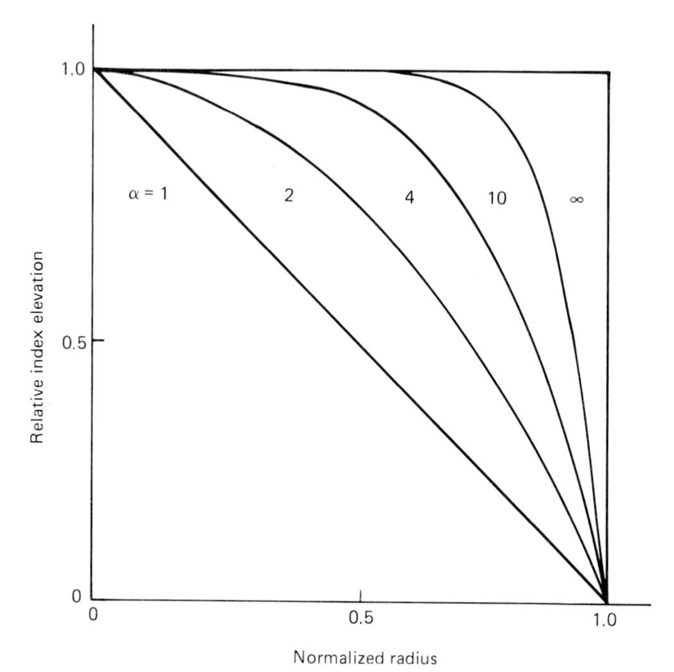


**FIGURE 1.9** Core refractive index profiles for various values of α. *(Reproduced from Ref. 5., with permission. Copyright 1973, AT&T.)*

**TABLE 1.1** Effects of Dopants in Silica

| | Index $n$ | Expansion $\alpha$ | Viscosity $\eta$ | Stability | Durability | Disperson $\lambda_0$ | Raleigh scattering | Infrared loss | Cost |
|---|---|---|---|---|---|---|---|---|---|
| $GeO_2$ | ↑ | ↑ | ↓ | ○ | ↓ | ↑ | ↑ ↑ | ○ | ↑ |
| $P_2O_5$ | ↑ | ↑ | ↓ ↓ | ○ | ↓ ↓ | ○ | ↑ | ↑ | ○ |
| $B_2O_3$ | ↓ | ↑ | ↓ ↓ | ○ | ↓ ↓ | ○ | ? | ↑ | ○ |
| F | ↓ ↓ | ○ | ↓ | ○ | ↓ ↓ | ○ | ? | ○ | ○ |
| $TiO_2$ | ↑ ↑ | ↓ | ↓ | ↓ | ↑ | ↑ | ? | ○ | ○ |
| $Al_2O_3$ | ↑ | ↑ | ↓ | ↓ ↓ | ? | ↑ | ? | ○ | ○ |

optical fibers and their effects on physical properties such as refractive index. Dopants, therefore, affect *material dispersion,* the tendency to broaden a light pulse. Material dispersion results from the interaction of light, which is composed of different wavelengths traveling at different speeds, with the medium. It should be noted that such fiber dopants as boron oxide ($B_2O_3$) and fluorine lower the refractive index of silica and thus are utilized as cladding dopants—not core dopants—to maintain the waveguide refractive index profile discussed previously. The relative effect of various dopants on the refractive index of silica as a function of concentration is seen in Fig. 1.10.

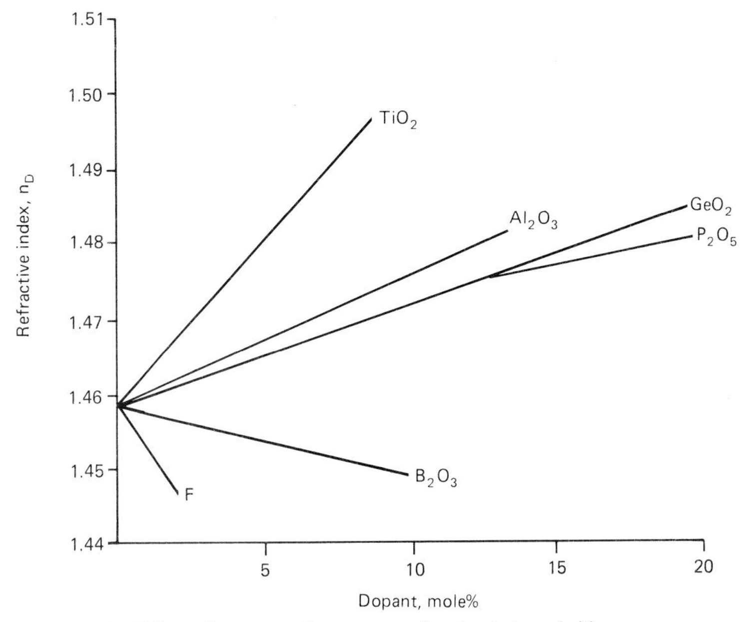


**FIGURE 1.10** Effect of common dopants on refractive index of silica.

### 1.3.1 Multimode Fiber

Multimode fiber is the fiber type in which more than one mode is propagating at the system operating wavelength. Currently, multimode fibers, ranging from those with as few as two modes to those with over a hundred modes, are usable in specific commercial applications. Typical multimode-fiber applications include telecommunication, with bandwidths of 1 to 2 GHz, premises wiring, with bandwidths of 500 to 1000 MHz, and links where power throughput is a greater need than bandwidth and, as a result, bandwidths of 50 to 100 MHz are sufficient.

Although multimode-fiber systems do not have the information-carrying capacity of single-mode fibers, they do offer several key advantages for specific systems. For example, the larger core diameters result in easier splicing of fibers, as core-to-core alignment is less critical. For this same reason, connectors used for the termination of fiber ends to various devices are less expensive. Also, given these larger cores, higher numerical apertures, and typically shorter link distances, multimode systems can use less expensive light sources such as light-emitting diodes (as opposed to lasers).

***Multimode Fiber Bandwidth.*** Multimode fibers have numerical apertures that typically range from 0.2 to 0.29 and have core sizes that range from 35 to 100 μm (Table 1.2). Core sizes larger than 100 μm typically are available in the form of plastic-clad silica or plastic-clad, plastic-core fibers. In these fibers, the plastic coating forms the cladding material, and thus are step-index fibers in design.

The tradeoff in multimode fibers is one of numerical aperture (which determines the amount of light coupled into the fiber) for bandwidth (the highest num-

**TABLE 1.2** Characteristics of Various Multimode Fibers

| Core/cladding diameters, μm | Attenuation, dB/km at 1300 nm | NA | Bandwidth, MHz · km at 1300 nm |
|---|---|---|---|
| 100/140 | 1.5 to 2 | 0.29 | 100–500 |
| 85/125 | 0.7 | 0.26 | 200–400 |
| 62.5/125 | 0.6 | 0.275 | 200–1200 |
| 50/125 | 0.6 | 0.23 | 400–1500 |
| 50/125 | 0.5 | 0.2 | 400–2500 |

ber of pulses per unit time that can be distinguished at the receiver end of the fiber). The higher the numerical aperture and the larger the core size, the higher the number of modes supported and the greater the need to "tune" the profile to minimize modal dispersion and, therefore, pulse spreading. This tuning is expressed in terms of refractive index as a function of core radius in Eq. (1.26). An $\alpha$ of 2 represents a parabolic-index profile, an $\alpha$ of 1 indicates a triangular profile, and an $\alpha$ of $\infty$ indicates a step-index profile. In multimode fibers, minimal pulse broadening is achieved at $\alpha = 2 - (12\Delta/5)$, or slightly over 2 (depending on core dopant material and wavelength of fiber operation). Olshansky[6] has calculated the optimal $\alpha$ as a function of wavelength for various core compositions, as shown in Fig. 1.11.

Beyond the discussion of fiber bandwidth for a given fiber, the question arises as to the bandwidth obtained when two fibers of different bandwidths are joined in a system. Unfortunately, the resultant bandwidth is not a simple expression of the bandwidth of the two fibers involved. Instead, the resultant bandwidth is a complex function of the power distribution in the joined fibers. For example, join multimode fiber 1 to multimode fiber 2. Normally these fibers are characterized for bandwidth at time of manufacture by injecting a light pulse down the fibers and measuring the signal degradation. In this example, the bandwidth performance of fiber 2 would be different from that characterized at the time of manufacture, since the input signal to fiber 2 is now the modified output signal of fiber 1, which is different from the signal used to characterize fiber 2 originally. A predictive technique is necessary to estimate the bandwidth of a multimode system composed of concatenated dissimilar fibers.

For the convenience of fiber users, the ability to predict the bandwidth of the system is defined in terms of the gamma parameter.[7] Assuming accurate core-to-core alignment to prevent perturbations of the modes as they are transferred from fiber to fiber, the bandwidth-length exponent $\gamma$ is dependent on the correlation between the fiber mode group delays[8] in the system. This correlation is based on probability theory and is an expression of the randomness of the variations occurring down the fiber length. Thus, the fluctuations that have no spatial or systematic dependence are referred to as *uncorrelated,* and those with some degree of common frequency are referred to as *correlated.* This can be expressed as the variance of the energy arrival time $\sigma^2$ through the concatenated fibers:

$$\sigma^2 = A^2 + B^2 \tag{1.29}$$

where $A^2$ describes the individual-fiber pulse broadening and $B^2$ describes the fiber-to-fiber interaction. The gamma parameter is defined from

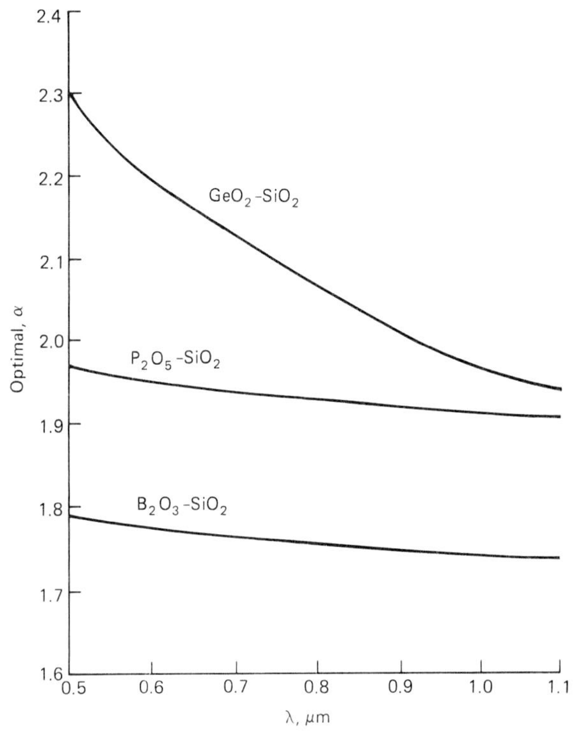


**FIGURE 1.11** Optimal α for a given core composition vs. wavelength. (*Adapted from H. M. Presby and I. P. Kaminow, "Binary Silica Optical Fibers: Refractive Index and Profile Dispersion Measurements," Applied Optics, vol. 15, no. 12, 1976, pp. 3029–3036, with permission.*)

$$\sigma^2 = \sigma_0^2 \left(\frac{L}{L_0}\right)^{2\gamma} \tag{1.30}$$

where $\gamma$ = pulse broadening of the link
$\sigma_0$ = average pulse broadening of each fiber in the link
$L_0$ = length of the fiber
$L$ = link length

Solving for $\gamma$ yields

$$\gamma = \frac{1}{2}\left[1 + \frac{\ln\,(1 + B^2/A^2)}{\ln N}\right] \tag{1.31}$$

where $N$ is the number of equal-length fibers in the link. For uncorrelated mode group delay times, $B = 0$, giving $\gamma = 0.5$. For correlated mode group delay times, $B = A(N - 1)$, and $\gamma = 1$, with $A$ and $B$ determined from differential mode delay data.[9]

*Multimode Fiber Losses.* After the bandwidth of multimode fibers, optical loss must also be considered. As the dopant level of the fiber is increased to raise the NA, the optical loss also increases as *Rayleigh scattering* (loss due to localized fluctuations in refractive index and density) removes light from the fiber. The typical losses, measured in decibels per kilometer, associated with various NA fibers are shown in Table 1.2.

In addition to this intrinsic loss, multimode fibers exhibit differing sensitivities to external influences, depending on the design. For example, the degree of light confinement varies with numerical aperture. Therefore, some designs are more sensitive to bending, which can cause the loss of power by the stripping of higher-order modes (i.e., rays traveling at steep angles in the core). Also, in system use, fibers with larger cores are more easily spliced together, since their losses are more insensitive to lateral offset of the cores or end separation because of the physically wider power distribution in the fiber.

The fiber chosen for any given link is that which best fits the system loss budget for a given cost, while maintaining the specified transmission rate. Fiber with 100-μm core and 140-μm cladding diameters is easily spliced because of the large core, but typically has lower bandwidth owing to the larger number of modes whose path lengths must be equalized. On the other hand, 62.5-μm-core fiber, with a 125-μm cladding diameter, may have higher bandwidth, but may also have higher loss, because of an increased sensitivity to bending losses as a result of a lower NA.

### 1.3.2 Single-Mode Fiber

Single-mode fiber typically is fabricated from the same materials and by the same process as multimode fiber. However, the difference in core size and dopant level between the two types of fiber results in different operating characteristics. With their small core size (typically 8 μm) and their low dopant level (typically 0.3 to 0.4 percent index elevation over the cladding index), single-mode fibers are defined by the normalized frequency parameter $V$, where $V \leq 2.405$. The $V$ value is a dimensionless number which relates the propagation of light down the fiber core to that propagating in the fiber cladding, in the terms of the expression

$$V = (n_1^2 - n_0^2)^{1/2} ka \tag{1.32}$$

Therefore, with $V \leq 2.405$, the fiber will allow only the fundamental mode to propagate down the fiber as the higher mode becomes lossy and is lost through the cladding.

Actually, whereas fibers with $V$ values between 0 and 2.405 contain only a single mode, they differ greatly with regard to system usefulness, since low $V$ value fibers can be extremely lossy owing to their poor light confinement at fiber bends.

This is illustrated in Fig. 1.12, which shows that portion of the power that propagates in the cladding as a function of $V$ value. From the graph, it is seen that the bimodal optical fiber (which supports the $LP_{11}$ and $LP_{01}$ modes) becomes single-mode in nature at a $V$ value of 2.405 because the $LP_{11}$ mode becomes radiative (i.e., $P_{clad}/P = 1$). At this $V$ value, the remaining mode, $LP_{01}$, is tightly bound with approximately 20 percent of the transmitted power propagating in the cladding material. As $V$ approaches a value of 1 or below, the majority of the power now is carried in the cladding material, where it is not tightly bound. This loosely contained power is more susceptible to loss, given such perturbations as bending of the fiber.

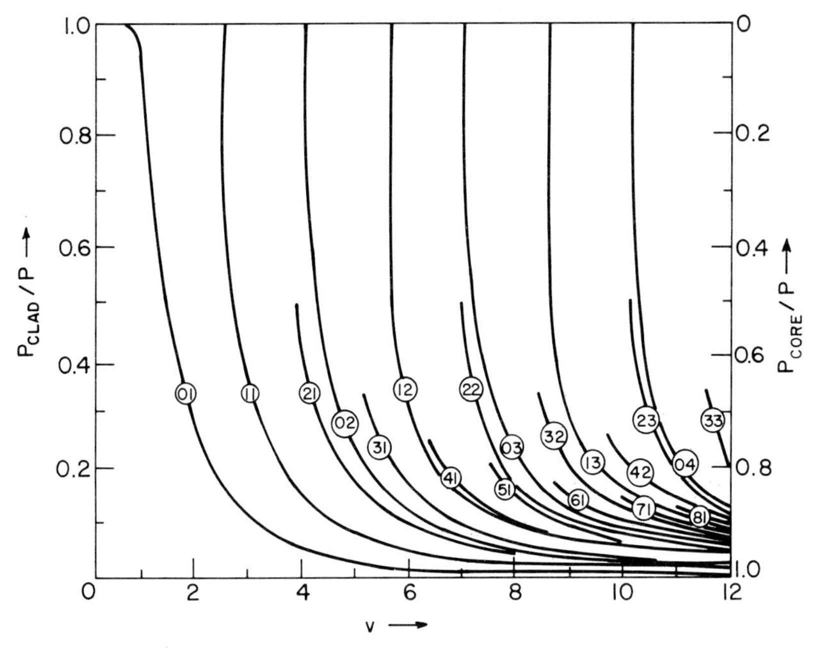


**FIGURE 1.12** Fraction of power-propagating in cladding vs. $V$ value, with mode numbers indicated. *(Reproduced from D. Gloge, "Weakly Guiding Fibers," Applied Optics, vol. 10, no. 10, 1971, pp. 2252–2258, with permisison.)*

***Single-Mode Fiber Bandwidth.*** Since only one mode propagates in a single-mode fiber, its bandwidth typically is not limited by fiber profile design and is therefore rarely mentioned. Rather, single-mode fiber information-carrying capacity is discussed in terms of dispersion. Dispersion, or pulse spreading, in single-mode fiber is largely a function of the spectral width of the optical source. Since light sources emit over a range of wavelengths, a light pulse will spread as the different wavelengths travel at different speeds down the fiber. This phenomenon of chromatic dispersion is several orders of magnitude smaller than that of modal dispersion found in multimode fiber. Single-mode fibers thus have bandwidths that typically are several orders of magnitude greater than those of multimode fibers. Also, while both multimode and single-mode fiber systems can be improved in performance by upgrading the optoelectronics, such as by using high-speed, narrow-spectral-width sources, a larger improvement can be obtained from single-mode fiber systems. This is seen in Fig. 1.13, which shows the improvement in bit rate or transmission capacity of typical single-mode fiber at 1310 nm as a result of such a change in system electronics. An increase of 2 orders of magnitude can be obtained by changing from a light-emitting diode to a distributed-feedback laser. It should be noted that this figure also illustrates the decrease in information capacity with wavelength, caused by chromatic dispersion.

***Single-Mode Fiber Loss.*** The reduced dopant level of single-mode fibers results in lower attenuation than in multimode fibers. Because typical dopant levels are only one-third those of multimode fibers, single-mode fiber attenuations are typically less than half those of multimode fibers, depending on the wavelength of

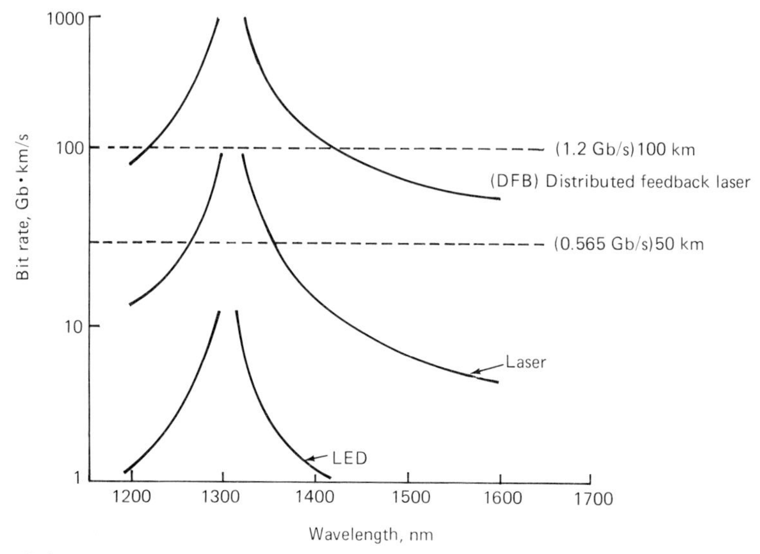


**FIGURE 1.13** Transmission capacity of single-mode fiber with various optical sources.

operation. This lower attenuation often allows for doubling the span length between repeaters or eliminating repeaters altogether.

The losses associated with bending the fiber are largely dependent on the operating characteristics of the fiber, such as operating wavelength. As discussed previously, there is a wavelength at which a single-mode fiber makes a transition from having two modes propagating to having only the fundamental mode—the second mode becoming lossy or radiative. This condition, referred to as the single-mode cutoff, occurs at a specific wavelength called the *single-mode cutoff wavelength.* This wavelength is defined as

$$\lambda_c = \left(\frac{2\pi a}{2.405}\right)(n_1^2 - n_2^2)^{1/2} \tag{1.33}$$

with all terms previously defined earlier in the chapter. As the operating wavelength moves to wavelengths longer than the cutoff wavelength, the fundamental mode also becomes increasingly radiative. The degree or speed at which the fundamental mode is lost depends on the environment of the fiber. Single-mode fibers operating at wavelengths longer than the cutoff wavelength become sensitive to power loss caused by bending the fiber, since the light is not tightly confined to the core.

The immediate implication is that the fiber core size is not synonymous with the size of the power distribution in the fiber. As a fiber becomes single-mode, the power distribution in the fiber, which is near gaussian in shape for single-mode, step-index profiles, is larger than the fiber core size, as measured by dopant index elevation, as shown in Fig. 1.14. This occurs because a fraction of the power propagates in the cladding of the single-mode fiber. This power distribution is referred to as the *fiber mode field* and has been correlated with the bend sensitivity and splicing sensitivity of single-mode fibers. Larger mode fields are

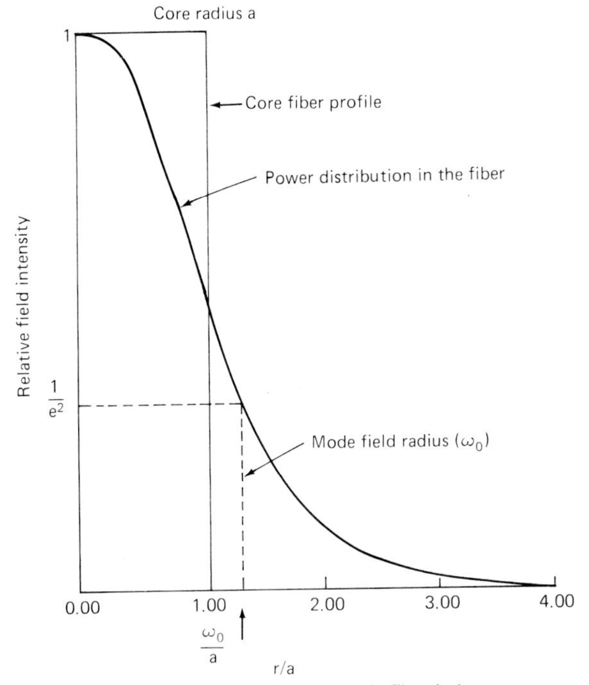


**FIGURE 1.14** Comparison of single-mode fiber index profile and optical field intensity distribution.

less sensitive to lateral offset during splicing, but more sensitive to losses incurred by bending during either installation or the cabling process.

The field intensity can be modified to optimize bending performance and minimize splice loss by modifying the refractive index of the core. Figure 1.15 shows that, by deviating from a step profile to one having a cladding index adjacent to the core below that of the remaining cladding material, or by segmenting the core index profile, the field intensity can be altered. Recent work[10] has shown that altering the mode field is achievable without compromising other aspects of fiber performance, e.g., attenuation.

## 1.4 OPTICAL LOSS

In the discussion so far, reference has been made to the loss of power both by intrinsic means, such as dopant level, and extrinsic means, such as bending. In a bulk sample, the loss can be broken down into components, such that

$$A_\lambda + R_\lambda + T_\lambda = 1 \tag{1.34}$$

where $A_\lambda$, $R_\lambda$, and $T_\lambda$ are fractions of power absorbed, reflected, and transmitted, respectively.

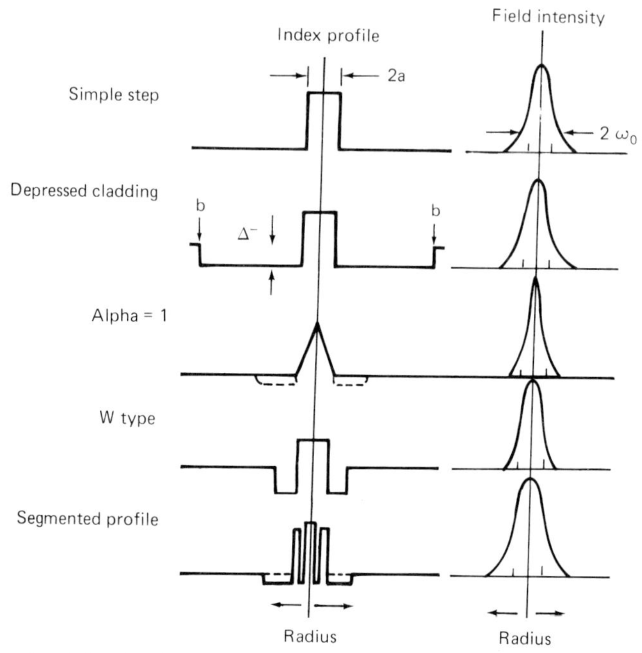


**FIGURE 1.15** Fiber core index profile vs. mode field diameter.

The transmitted component of the incident light power can be characterized further in terms of length and the absorption coefficient of the material:

$$T = T_0 e^{-\alpha x} \tag{1.35}$$

where $T$ = power transmitted over a fiber of length $x$
$T_0$ = initial coupled optical power
$\alpha$ = absorption coefficient of the material
$x$ = length or distance traversed in the material

This relationship is known as Lambert's law.

The loss of power in the fiber through absorption, scattering, etc., is referred to as its *attenuation*, which is expressed in terms of loss per unit of length by the formula

$$\text{Attenuation} = -\left(\frac{10}{L}\right)\log_{10}\left(\frac{\text{power out}}{\text{power in}}\right) \tag{1.36}$$

where, if $L$ is expressed in kilometers, the loss is defined in units of decibels per kilometer (dB/km). Since the attenuation relationship is a log relation, a 90 percent transmission corresponds to a 0.5-dB loss, 50 percent transmission to a 3-dB loss, and 10 percent transmission to a 10-dB loss. As seen previously, typical single-mode glass fiber attenuations are on the order of 0.35 to 0.40 dB/km at 1310 nm.

### 1.4.1 Rayleigh Scattering

Scattering loss is the loss associated with the interaction of the light with density fluctuations in the fiber. If the size of the defect site is less than one-tenth of the wavelength of the light, the phenomenon is termed *Rayleigh scattering.* This is the main loss mechanism in current fibers operating in the 800- to 1600-nm wavelength range. Defects of this size are associated with localized refractive index fluctuations in the glass matrix. The origin of these fluctuations is composition variations, or density variations of the same composition, that are frozen into the glass structure as it is cooled from the molten state. The defect so created forms an electric dipole when exposed to the electromagnetic light wave, and a secondary wave will be emitted. Rayleigh scattering in fiber can be expressed as

$$\gamma_{RS} = \left(\frac{8\pi^3}{3\lambda^4}\right)(\delta n^2)^2\,\delta V \tag{1.37}$$

where $(\delta n^2)^2$ is the square of the mean-squared fluctuations in refractive index and $\delta V$ is the volume associated with this index difference. If the glass composition tends to have a larger number and/or size of defects, its intrinsic loss will be higher. When optical fibers contain multiple dopants $(\delta n^2)^2$ becomes

$$(\delta n^2)^2 = \left(\frac{\delta n^2}{\delta p}\right)(\delta \rho)^2 + \sum_i \left(\frac{\delta n^2}{\delta c_i}\right)^2 (\delta C_i)^2 \tag{1.38}$$

where $\delta p$ is the density fluctuation and $\delta c_i$ is the concentration fluctuation of the $i$th component. In commercial optical-fiber compositions, it has been observed that the replacement of germanium dioxide ($GeO_2$) by phosphorus pentoxide ($P_2O_5$) results in reduced Rayleigh scattering.[11] A reduction in scattering also can be achieved by replacing boron oxide ($B_2O_3$), with $P_2O_5$,[12] although such reductions in scattering come at the expense of other parameters, such as numerical aperture.

An important characteristic of Rayleigh scattering is that it is inversely proportional to the fourth power of the wavelength, i.e., $\gamma_{RS} \propto 1/\lambda^4$. From a fiber-use standpoint, this means that systems designed for operation at longer wavelengths have lower intrinsic loss. Given a similar loss budget, system length can be increased by a factor of 3 to 4 by shifting the operating wavelength from 850 to 1310 nm, and by another factor of 2 by shifting from 1310 to 1550 nm, ignoring dispersion considerations for the moment.

For sites where the defects are larger than $\lambda/10$ (but on the order of $\lambda$), the scattering mechanism is referred to as *Mie scattering.*[13] These large defect sites are produced by inhomogeneities in the fiber and are associated with incomplete mixing of waveguide dopant components (i.e., phase separation) or defects formed in the fabrication process (such as gaseous voids or particulate inclusions). These large defects physically scatter light out of the fiber core. This type of scattering is easily characterized, as it is typically a forward scattering, where the loss now has an angular dependence. Mie scattering rarely is seen in commercially available silica-based fibers because of the high level of manufacturing expertise.

Forward light scattering (Raman scattering) and backward scattering (Brillouin scattering) are two additional scattering phenomena that can be seen in optical materials. Raman scattering is caused by molecular vibration of phonons in the glass matrix. The degree to which this type of scattering is seen therefore is dependent on the temperature of the material.

Brillouin scattering is induced by acoustic waves as opposed to thermal phonons. Brillouin scattering, like Raman scattering, is a nonlinear optical effect, since the frequency of the scattered light varies with the wavelength and frequency of the thermal or acoustic phonon. Since the frequency shift for the acoustic phonon is a maximum in the back direction, that is, maximum interference is seen as the forward and backward traveling waves interfere, Brillouin scattering typically is viewed as a backscatter phenomenon.

The importance of stimulated Raman scattering and stimulated Brillouin scattering is that they can be the limiting factor in high-power system designs. In submarine applications and certain cable television transmission schemes, system designers are trying to increase unrepeated transmission length or improve signal-to-noise ratio by using a higher-power light source in the system. However, if the transmitted power in the system is high enough—typically 10 mW to stimulate Brillouin scattering and 1 W to stimulate Raman scattering—system performance can be degraded, rather than improved, because loss is increased by the increased power. Fortunately, Brillouin scattering loss can be decreased by using a light source with a broad spectral width. A broad spectral width source decreases the light-material interaction responsible for generating the acoustic wave (the interaction is highly wavelength dependent). Raman scattering loss is not affected by spectral source width but fortunately requires at least an order of magnitude more power for onset.

### 1.4.2 Absorption

Absorption is a second major loss mechanism in optical fibers. Absorption can be intrinsic to the material of fabrication, or it can be induced by the presence of external contaminants incorporated during processing.

For optical fibers based on the silicate glass forming system, major absorptions are determined by the band structure of the glass. In pure silicon dioxide ($SiO_2$), the absorption observed is that occurring as the valence band electrons are excited to the conduction band. This high energy transition arises from incident light at short wavelengths, i.e., in the ultraviolet region.

***Intrinsic Absorption.*** The low intrinsic material absorption of the silicate glass system is the main reason for its use in commercial manufacture of low-loss glass fibers. In doped silicate fibers, the regions, or windows, of operation (Fig. 1.16) typically range from 850 to 1550 nm—between two intrinsic absorption edges. These two edges bracket the operating region in the ultraviolet (uv) region (160 to 400 nm) and the infrared region (beyond 2100 nm) because of two different absorption mechanisms, electronic excitation at short wavelengths and lattice vibration excitations at long wavelengths.

The uv absorption edge, known as the *Urbach edge,* is characterized by a steep increase in loss which is expressed[14] as

$$\gamma_{uv} = Ce^{E/E_0} \tag{1.39}$$

where $E$ is the photon energy and $C$ and $E_0$ are empirical constants.

This expression shows that the transmission roll-off into the forbidden gap in the $SiO_2$ band structure is exponential with photon energy. Thus the more ener-

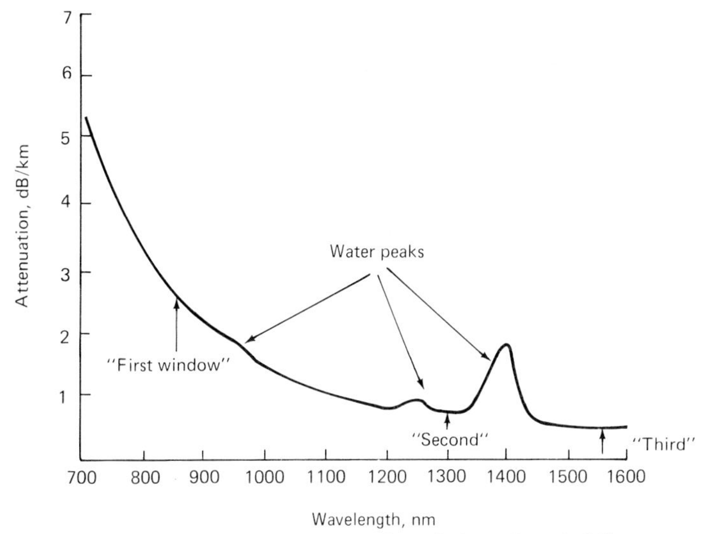


**FIGURE 1.16** Attenuation vs. wavelength for a single-mode optical fiber.

getic, shorter-wavelength light produces a steep and high loss edge in the uv. With the addition of dopants to modify the refractive index, the uv edge is shifted as the bandgap is modified. $P_2O_5$ and $GeO_2$ additions are known to shift the edge to longer wavelengths.

This shift is attributed to defects in the structure which require less energetic radiation to cause transfer to the conduction band. The defect identified with germanium-doped fibers is $Ge^{2+}$, which is a reduced form of $Ge^{4+}$—the typical oxidation state associated with germania silica glasses.

Long-wavelength transmission in silicate fibers is limited by an infrared (ir) absorption edge that exists beyond 2000 nm, beyond the wavelengths of current commercial interest. The intrinsic absorption of this edge is associated with the glass matrix. In silica glass, this absorption is caused by the vibration of the Si-O bond. In the 1000- to 1200-$cm^{-1}$ (8300- to 10,000-nm) region, the vibration is associated with the stretching of the Si-O-Si bonds. Bands in the 450- to 850-$cm^{-1}$ (12,000- to 22,000-nm) range are induced by the bending of the Si-O-Si bonds. The fundamental absorption of the Si-O bond has been shown to exist at approximately 8 μm, although overtones exist that lower this edge to approximately 3 μm. In general form this absorption can be written as

$$\gamma_{ir} = C_{ir} e^{-E/E_0} \tag{1.40}$$

with $C$ and $E_0$ again being empirical constants.

While core dopant materials typically do not affect the position of the ir edge, $B_2O_3$ is an exception. As a dopant, $B_2O_3$ lowers the refractive index of the glass below that of pure silica, but shifts the ir edge to shorter wavelengths. For this reason, commercial fibers for use in the 1550-nm window contain no $B_2O_3$. This *boron edge* is so severe that $B_2O_3$ now is excluded from fibers operating at 1300 nm. Figure 1.17 shows the attenuations in fibers that arise from the scattering and absorption mechanisms just discussed.

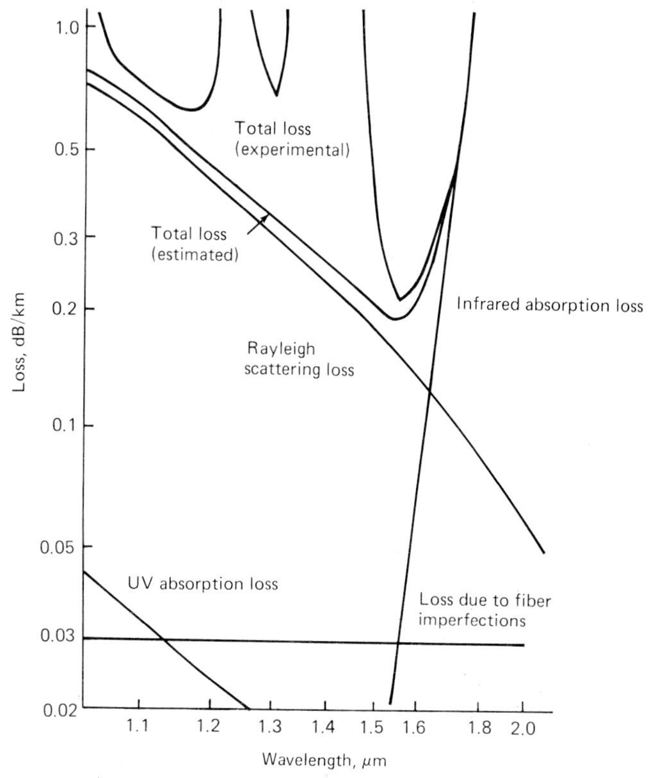


**FIGURE 1.17** Components of optical loss in optical fibers. *(Reproduced from T. Miya et al., "Ultimate Low-Loss Single-Mode Fibre at 1.55 μm," Electron. Lett., vol. 15, 1979, p. 106, with permission.)*

***Impurity Absorption.*** While bonding and structural absorptions associated with the materials of fabrication can alter a fiber's spectral attenuation, so, too, can impurities.

Trace metal impurities introduced into the fiber at any fabrication step will alter the band structure of the glass and thus alter its transmission characteristics. In transition metals, the absorption arises from promoting 3d electrons in the partially filled inner electron shells. This absorption is therefore an electronic, not a vibrational, absorption. The exact shape of the absorption is defined by the localized splitting or bonding of the particular impurity species. Typically, transition metals, such as iron (Fe), nickel (Ni), and chromium (Cr), can produce measurable absorptions at the parts per billion (ppb) level of impurity. Figure 1.18 shows the shape of the absorption curves associated with various transition metal ions such as $Fe^{2+}$, $Cr^{3+}$, and $Cu^{2+}$. The degree to which these impurities affect a system depends on the concentration of impurity, the wavelength of the absorption, and the transmission wavelength being used.

Water also can be considered an impurity in the optical fiber structure. Because most fibers are formed by the sintering of a porous preform of some nature, the opportunity for the entrapment of water by either chemical or mechanical means is great. When incorporated into the glass matrix, the water forms a silicon-hydroxyl (Si-OH) bond in the silicate system and does not exist as molec-

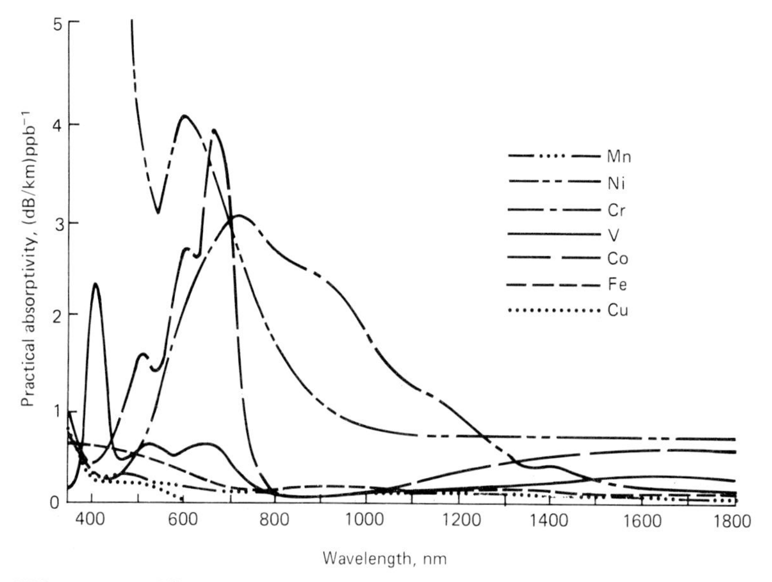


**FIGURE 1.18** Effect of impurities in silica on spectral absorptivity in optical fibers.

ular water. This bond has its fundamental absorption at 2700 nm, with overtones or harmonics that reside near the wavelength regions of commercial interest. These overtones exist at 1383 nm (between the 1310- and 1550-nm windows), 1250 nm (near the 1310-nm operating window), and 950 nm (between the 850- and 1310-nm windows). The determination of the amount of water present in a fiber is based on the conversion that 33 ppb of water will produce a 1.0-dB/km increase in attenuation at 1383 nm. At the 950-nm overtone, a 1.0-dB/km increase is seen for every 1 ppm of water. This absorption is narrow in character; it takes several hundred parts per billion of water to produce a measurable increase in the attenuation at 1310 nm.

***Induced Absorption.*** Absorption in glass fibers can be induced by the environment. The two major environmentally induced defects of current interest are those caused by altering the bond structure during manufacture (by means other than contamination) and those caused by ionizing radiation. The former defect is termed the *draw-induced defect* and shall be discussed first.

Most glass fiber is produced by heating a glass boule to temperatures in the 2000°C region, and pulling or drawing a glass fiber from the molten region. The molten glass experiences a draw tension. Once at the desired diameter, it is rapidly quenched so that a coating can be applied to facilitate handling of the fiber without breakage. A draw defect is a broken bond that is formed at the elevated temperature and is frozen into the glass as it is cooled from the molten state. This broken bond, actually a broken silicon-oxygen (Si-O) bond, absorbs at the 630-nm wavelength, as shown in Fig. 1.19. This loss typically can be removed by heat treating the fiber or by optimizing the fiber-drawing parameters to reduce its presence to nonmeasurable levels. Although the defect absorbs at 630 nm, the wavelength of a helium-neon laser, few fibers are used commercially at this wavelength.

Optical loss increases can also occur after the fiber has been fabricated by exposing the fiber to hydrogen. These loss increases have been observed in fiber

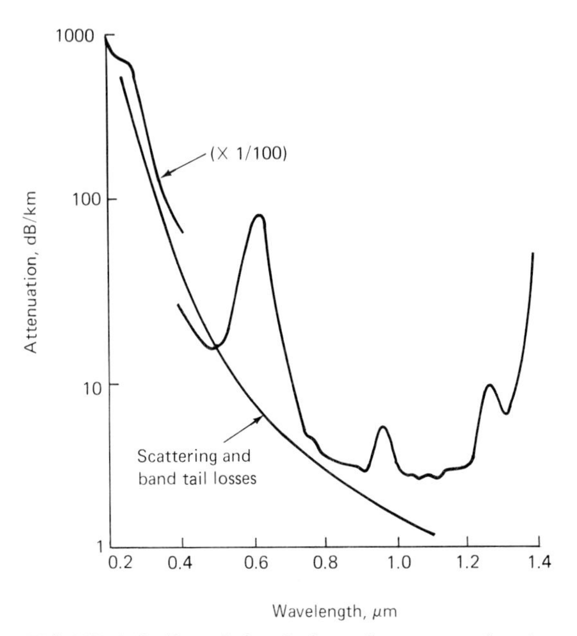


**FIGURE 1.19** Draw-induced absorption vs. wavelength. *(Reproduced from P. Kaiser, "Spectral Losses of Unclad Fibers Made from High-Grade Vitreous Silica," Appl. Phys. Lett., vol. 23, no. 1, 1973, p. 45., with permission.)*

transmission systems in the field as hydrogen is evolved in the cable, either from the organic filling compounds used in the cables or from galvanic release of hydrogen from the cable structure. The hydrogen released from the cable causes two types of attenuation increase in the fiber: reversible and permanent.

The reversible attenuation increase is associated with molecular hydrogen ($H_2$) which diffuses into the glass matrix. Once in the glass it resides in the interstitial spaces in the glass matrix. Because this is a diffusion-controlled process which is dependent on concentration (or $H_2$ pressure), temperature, and time, the hydrogen can diffuse out of the glass as the $H_2$ concentration decreases with time or temperature.

From the work of Rush,[15] Fig. 1.20 shows the increase in attenuation associated with the presence of molecular hydrogen. The absorption at long wavelengths, beyond 1600 nm, is a result of the fundamental vibration of molecular hydrogen in silica occurring at 2400 nm. The peak at 1.24 μm is the first overtone of this fundamental vibration. Overall, the increase in loss at 1310 nm is roughly half the loss increase at 1550 nm, making this absorption far more important to those systems operating in the third window, 1550 nm. Also, while those loss increases are shown for $H_2$ at 0.74 atm, typical cables evolve $10^{-3}$ to $10^{-4}$ atm of hydrogen.[15]

As the solubility of hydrogen is linear with pressure for pressures less than 1 atm, the increase in loss in a cabled fiber is very small and reversible in all instances.

An irreversible loss induced by hydrogen can also be seen under some conditions. This permanent loss increase involves the chemical bonding of the diffused hydrogen with defect sites in the glass. Figure 1.21 shows the loss increase in a

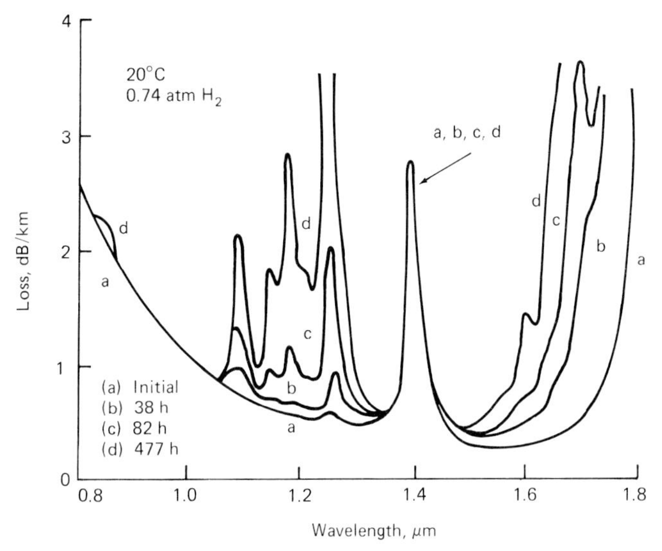


**FIGURE 1.20** Loss increases with time due to the absorption of $H_2$ in single-mode fiber. *(Reproduced from Ref. 15, with permission.)*

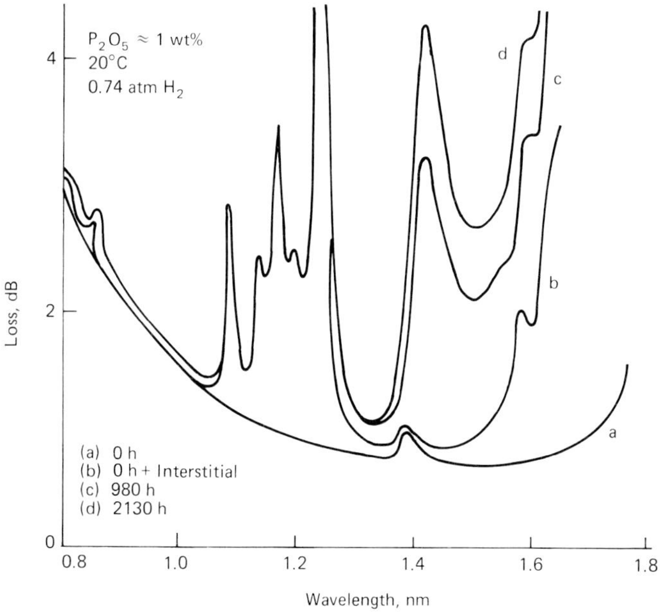


**FIGURE 1.21** Loss increase with time due to the absorption of $H_2$ in multimode fiber. *(Reproduced from Ref. 15, with permission.)*

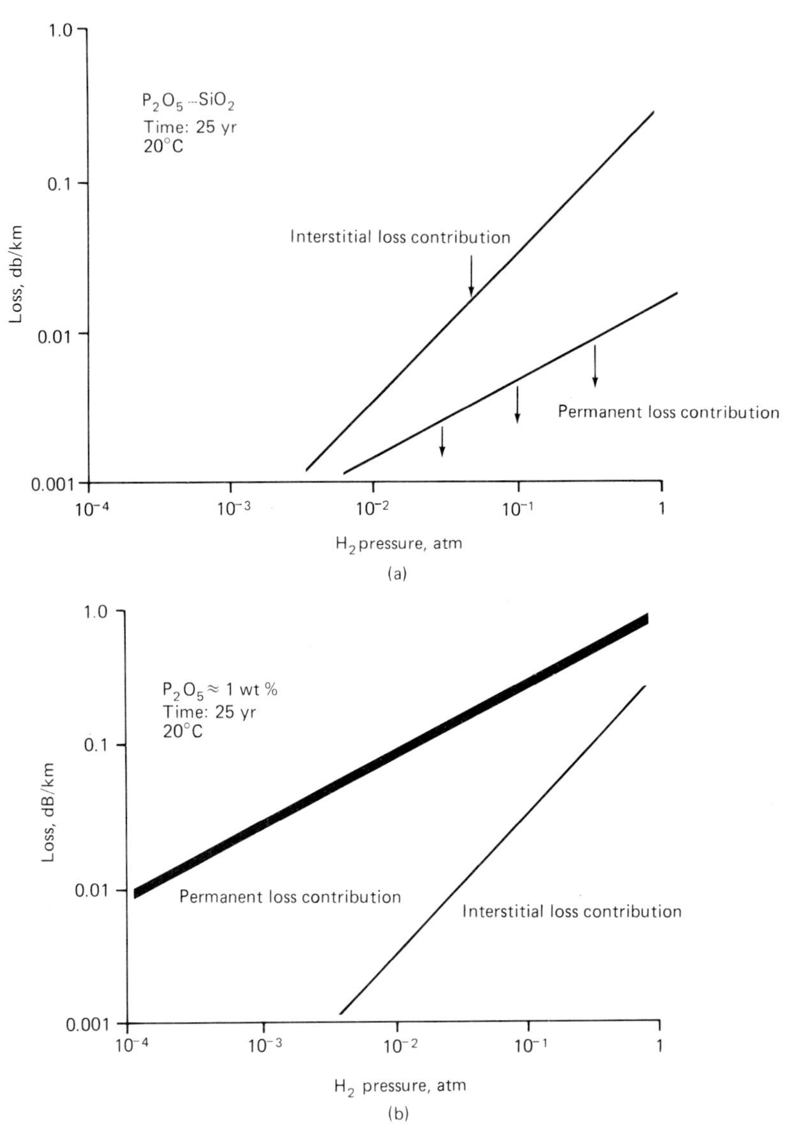


**FIGURE 1.22** Predicted loss increase due to $H_2$ (at 1.3 μm). (*a*) Single-mode fiber; (*b*) multimode fiber. (*Reproduced from Ref. 15, with permission.*)

multimode fiber for this irreversible loss increase, which occurs over longer times than the reversible loss just discussed. In this figure it is seen that a long-wavelength peak at 1600 nm develops, as well as a peak at 1400 nm. These peaks are associated with P-OH and Ge-OH bond formation.

This chemical bonding shows a dependence on the amount of phosphorus in the multimode fiber, with higher losses reported for higher-phosphorus-content multimode fibers. As a result, the majority of commercially available multimode fibers now contain minimal amounts of phosphorus (which is added to facilitate manufacturing for a variety of reasons). Overall, single-mode fibers that are of lower germania content and typically phosphorus-free (compared to multimode fiber) are far less sensitive to this absorption loss. Predicted loss increases for single-mode and multimode fiber are shown in Fig. 1.22*a* and *b* for a range of hydrogen pressures.

Optical absorption also can be caused by defects produced by ionizing radiation. Fibers that are resistant to radiation-induced optical loss are important in such applications as monitoring of nuclear reactors and special military uses. While extensive testing has been conducted, there does not exist a comprehensive defect model that explains the radiation response of optical fibers. This is partly because of the variety of characteristics tested: maximum loss, recovery rate after initial exposure, loss as a function of dose rate, and loss as a function of temperature. The variables of fiber composition and testing wavelength also have been shown to be important, as is the type of radiation, e.g., gamma rays vs. x-rays.

Figure 1.23 shows the generalized response of a glass fiber to radiation exposure, including the maximum absorption after exposure, and followed by a slow temperature-dependent recovery to a condition of permanent loss change. Overall, irradiation of a glassy material by high-energy particles has two effects on the medium: (1) ionization of certain structures in the medium by electronic excitation and (2) direct displacement of the atomic structure by elastic scattering of the structure. Since electronic defects are created in the form of electrons or holes, the band structure of the material is modified. Materials whose band structure already is modified by the presence of such defects as water or draw-induced defects respond differently to irradiation. This suggests that the best radiation re-

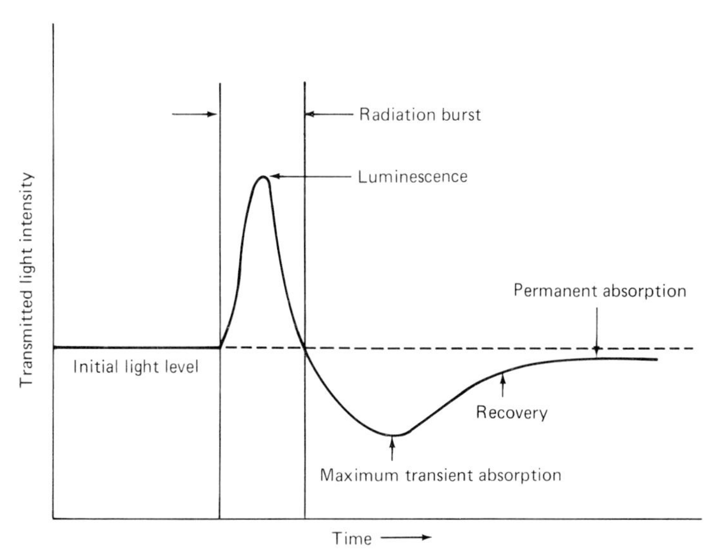


**FIGURE 1.23** Radiation-induced loss in optical fibers as a function of time.

sistance would be obtained with pure $SiO_2$ fibers that are processed to avoid the formation of defects. To date, conclusive evidence of this performance has not been observed.

Given the variety of test conditions, some basic generalizations on radiation performance can be made. First, as seen in Fig. 1.24, the addition of phosphorus to the fiber increases radiation-induced loss, as evidenced by phosphorus-containing fibers showing higher elevated loss. However, adding phosphorus to germania-silica has been seen to improve the low-temperature recovery over simple germania-silica fibers. Second, in germania-silica fibers, the induced loss is high in the uv and decreases monotonically with increasing wavelength. Finally, the radiation response in terms of induced loss varies with germania content. High-NA multimode fibers show higher induced loss than lower, germania-content, single-mode fibers. This is believed to be caused by the higher number

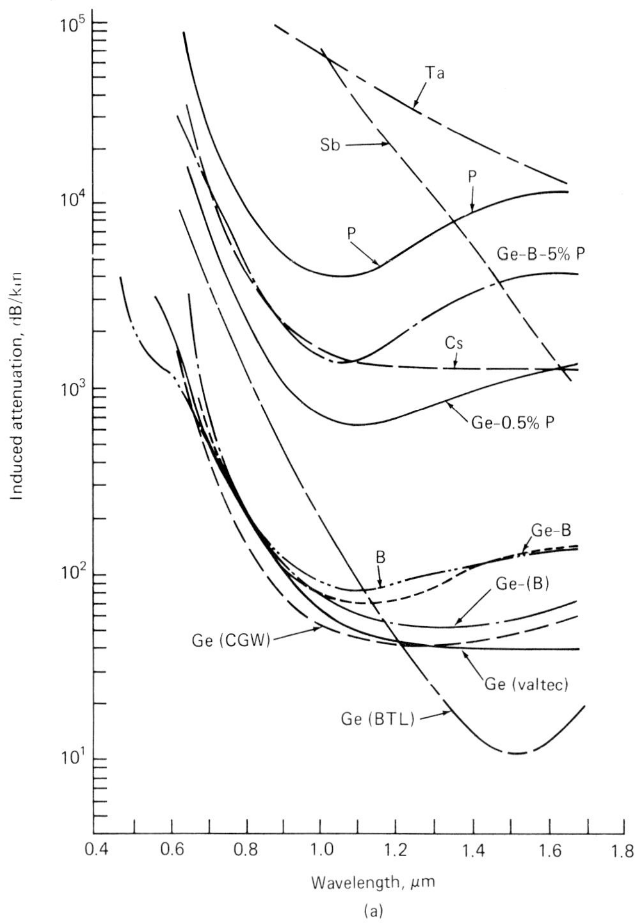


**FIGURE 1.24** Radiation-induced attenuation for optical fibers of various compositions. (*a*) Doped silica core fibers; (*b*) silica core fibers. (*Reproduced from Ref. 17, with permission.*)

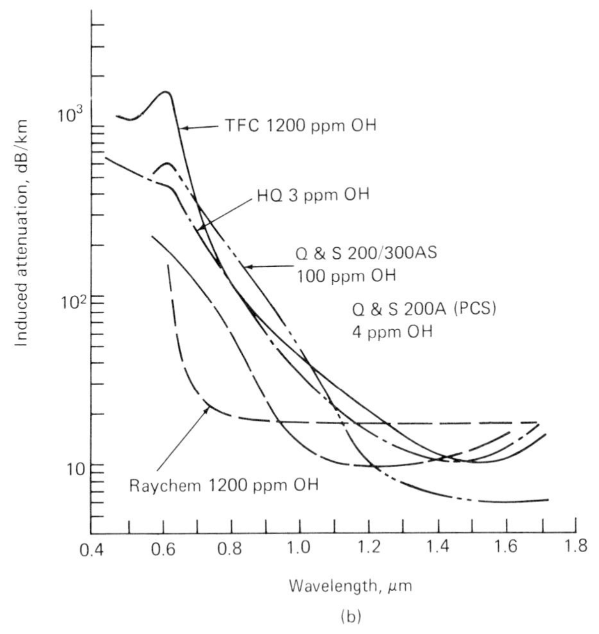


**FIGURE 1.24** *(Continued)*

of unmixed Ge atoms or Ge density fluctuations in the more heavily doped fibers. However, it must be noted that the different fiber types frequently experience different processing conditions, such as draw temperature, which may influence radiation performance.

Work in recent years to reduce radiation sensitivity has centered around incorporating rare or trace dopants to the optical fiber.[16] Dopants investigated have included cerium, cesium, antimony, and others. The purpose of these dopants was not to remove those defects responsible for the attenuation increase, but to induce a controllable defect site. That is, any defect induced upon irradiation would react with the compositionally induced defect and absorb outside the wavelength of interest. To date, these efforts have been unsuccessful; the induced loss of fibers containing these dopants remains high.

Along with the intentional co-doping of the core region, studies have been conducted to determine the role of impurities such as hydroxyl content. While water levels have been shown to be important in some studies,[17] other studies[18] have shown it to have no impact on the radiation response of the fiber. This conflict may be based, in part, on the fact that different studies may very well have used fibers that again were fabricated by different processing techniques. Therefore, there may have been different initial defect structures in the fibers studied.

The temperature dependence of the radiation response of optical fibers is somewhat easier to explain, but is difficult to control or predict. Whereas the temperature response—the ability to resist induced loss or the ability to recover after exposure—may vary from fiber to fiber and from study to study, low-temperature loss typically is worse than elevated temperature loss. From simple band structure considerations, this trend can be rationalized by the fact that carrier mobility is higher at elevated temperatures. The loss at elevated temperatures therefore would be lower, since the higher thermal activation would more readily accommodate electron-hole recombination, thereby removing the absorbing species.

This defect model is supported by the experimental data.[19] As the shallow traps requiring the least activation energy for depopulation would be removed first, the deeper remaining traps would require more energy for removal. If produced by incident light, these defects would be expected to lie at the shorter, more energetic portion of the spectrum, such as the uv. This band structure concept also is compatible with the experimental evidence that annealing an irradiated fiber at elevated temperatures improves recovery. Also, successful photobleaching[20]—the removal of radiation-induced loss by illuminating the core with light of a specific wavelength—would support the preceding argument. The effect of photobleaching has been seen to be most effective in pure $SiO_2$ fibers, although this phenomenon has been observed to some extent in all fibers and with a variety of light sources.

## 1.5 BENDING LOSS

Loss of light from a fiber can also be induced by bending the fiber. This is a very important loss mechanism, as the presence or absence of this loss mechanism is determined by the fiber user. Improper cabling may produce small systematic perturbations to the fiber, causing an elevation of the initial loss. These losses, caused by small-amplitude (nanometer), high-spatial-frequency (millimeter) perturbations are called *microbending losses.* Even if cabled correctly, the fiber can be installed in tight-diameter bends (e.g., centimeters in radius) that also will raise the attenuation. This large-diameter bend loss is referred to as *macrobend loss.*

### 1.5.1 Microbend Loss

Microbend losses are those associated with small perturbations of the fiber, induced by such factors as uneven coating application or cabling-induced stresses. The result of the perturbations is to cause the coupling of propagating modes in the fiber by changing the optical path length. This destabilization of the modal distribution causes lower-order modes to couple to higher-order modes that are lossy in nature. Although treated in detail by others,[21] multimode fiber loss from random bends can be calculated from the basic relationship[22]

$$M_L = N\langle h\rangle^2\left(\frac{a^4}{b^6\Delta^3}\right)\left(\frac{E}{E_F}\right)^{3/2} \tag{1.41}$$

where $M_L$ = multimode loss
$N$ = number of bumps
$h$ = height of the bump per unit length
$b$ = fiber diameter
$a$ = core radius
$E_F$ = elastic modulus of the fiber
$E$ = elastic modulus of the surrounding medium
$\Delta$ = index difference between core and cladding

From this empirical relationship, it is seen that increasing the refractive index of the core decreases the fiber's sensitivity to bending loss. Increasing overall fiber diameter also decreases sensitivity, whereas increasing core diameter increases sensitivity, since the fiber will have a greater modal volume and tend toward more lossy modes. The physical basis for the loss in the fiber is that a bend will change the optical path length of the fiber. The light propagating at the inside of the bend will travel a shorter distance than that traveling on the outside of the bend. To maintain coherence, the mode phase velocity must increase. But when a fiber bend is below the critical radius, this propagation velocity will exceed the speed of light and some of the light within the fiber is converted to higher-order modes and becomes radiative. The loss of these higher-order modes causes a gradual increase in attenuation.

Figure 1.25 shows the nature of the loss induced by bending. In multimode fiber, the overall attenuation is shifted higher as higher-order modes are lost. The bend loss associated with multimode fiber is therefore wavelength-independent. The single-mode fiber loss resulting from macrobending shows a severe increase in attenuation at long wavelengths from the unperturbed state, while the microbend loss of this fiber exhibits a more wavelength-independent loss. This difference in loss character is based on the fact that the microbend loss affects all wavelengths as the light encounters the perturbations. The degree or extent of interaction depends on the size of the perturbation and the wavelength of light. This description also shows the importance of the fiber's surrounding medium. A low-modulus coating generally will improve the insensitivity to random bends. However, this low-modulus coating may not produce sufficient mechanical protection for industrial handling, and a secondary, higher-modulus coating often is applied. If a secondary coating is not applied, the lower-modulus coating typically is applied to a greater thickness, i.e., several hundred micrometers. Coatings are further discussed in Chap. 2.

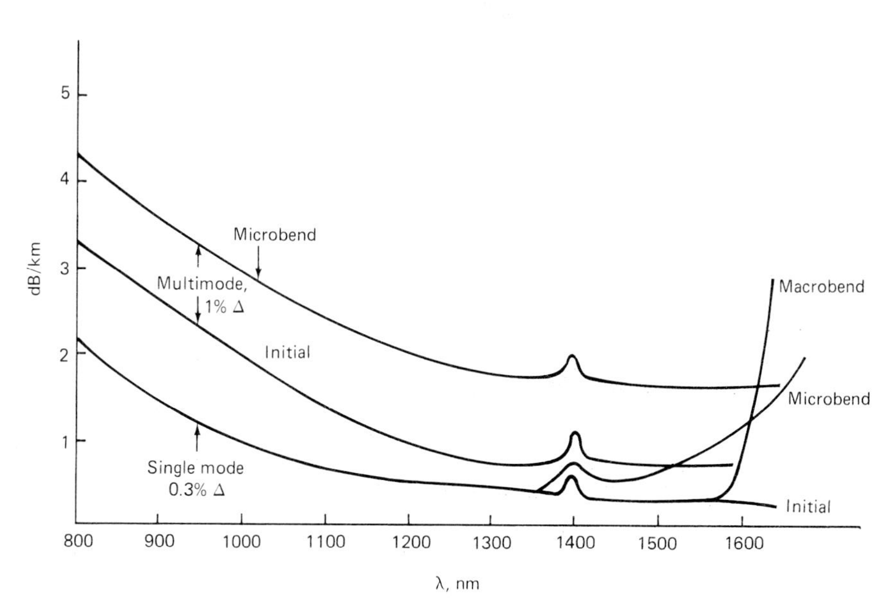


**FIGURE 1.25** Bend-induced loss elevation of optical fibers.

### 1.5.2 Macrobend Loss

*Macrobend losses* are those losses observed when a fiber or cable is bent to a radius of several centimeters or less. These severe bends can be introduced during installation, for example, when the cable is placed in junction boxes or splice trays.

Empirically, macrobend loss can be expressed as[23]

$$\gamma_{\text{bend}} = 10 \log \frac{\alpha + 2}{(2\alpha)(a/R\Delta)} \tag{1.42}$$

where $\Delta$ = index elevation
$R$ = bend radius
$a$ = core radius
$\alpha$ = profile parameter

The trend of reduced bend sensitivity with decreasing core size and increasing dopant level is consistent with a more rigorous treatment of macrobending than is given here.

In single-mode fiber, the presence of a macrobending-induced loss is easily identified. Referred to as the *bend edge,* the condition whereby the primary mode becomes radiative results in an attenuation edge where the fiber no longer transmits. This is explained by the fact that, after single-mode fiber cutoff is reached, the mode field expands in size with increasing wavelength in a predictable manner. As the bend radius is decreased, a radius is reached at which the primary mode is lost at a given wavelength. As the bend radius is further decreased, the bend edge shifts to still shorter wavelengths. Thus, in single-mode fibers, the loss is wavelength-dependent.

Figure 1.26 shows the bending sensitivity of single-mode fibers. Optimized for operation at the 1550-nm, low-loss window, a single-mode fiber must be designed to be resistant to bend loss to maintain its low loss at the 1550-nm wavelength. As the fiber is bent to an indicated radius, the elevation of the 1550-nm attenuation is measured. This measurement is done at different single-mode cutoff wavelengths $\lambda_c$ and with fibers of different zero dispersion wavelength $\lambda_0$. As the bend sensitivity changes with the degree of power confinement, which is determined by the difference between the operating wavelength and single-mode cutoff wavelength. As the difference between these two parameters is decreased, and as the single-mode cutoff wavelength is shifted toward longer wavelengths, the optical loss caused by fiber bends decreases.

## 1.6 FIBER DISPERSION

Fibers produced for commercial systems can be limited in use by one of two major phenomena: attenuation and dispersion. An attenuation limit is reached when the transmitted signal degrades below the usable detection limit of the optical receiver. If sufficient signal is available, the fiber dispersion determines the ability to distinguish the individual output pulses and tends to be the limiting factor for use. While modal dispersion is of primary concern with multimode fibers, as discussed previously, two other forms of dispersion exist: material dispersion and waveguide dispersion.

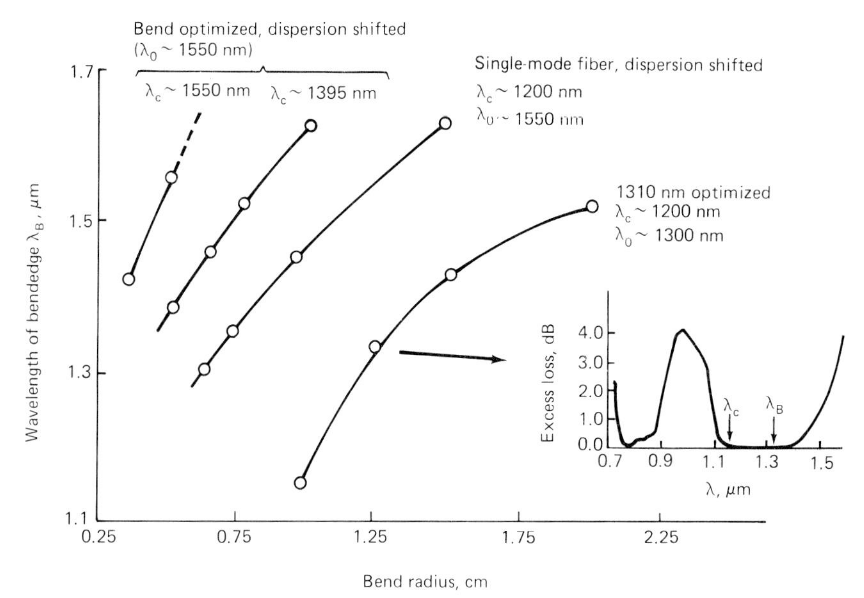


**FIGURE 1.26** Bend sensitivity of different single-mode fiber designs.

### 1.6.1 Material Dispersion

Material dispersion is the phenomenon by which a light pulse is spread out by the wavelength-dependent interaction of light with the material medium. The degree of this dispersion is a function of the *source spectral width,* the range of optical frequencies propagating in the medium, and the material whose atomic structure and index will influence the light propagation. Typically, the speed of light is defined as that speed observed in a vacuum. In a solid material this speed, or phase velocity of the propagating wavefront, is reduced to

$$v_{\text{phase}} = \frac{c}{n} \quad \text{at } \lambda \tag{1.43}$$

where $c$ = speed of light in vacuum and $n$ = refractive index.

The change in speed occurs because the atomic structure assumes a dipole nature as it experiences the electromagnetic field induced by the propagating light. The dispersion of the input signal can be written as

$$\tau_m = \left(\frac{L}{c}\right)\lambda^2 \frac{\delta\lambda}{\lambda}\left(\frac{d^2 n}{d\lambda^2}\right) \tag{1.44}$$

where $c$ = speed of light in vacuum
$\delta\lambda/\lambda$ = relative spectral width of the source
$L$ = length traversed
$n$ = refractive index

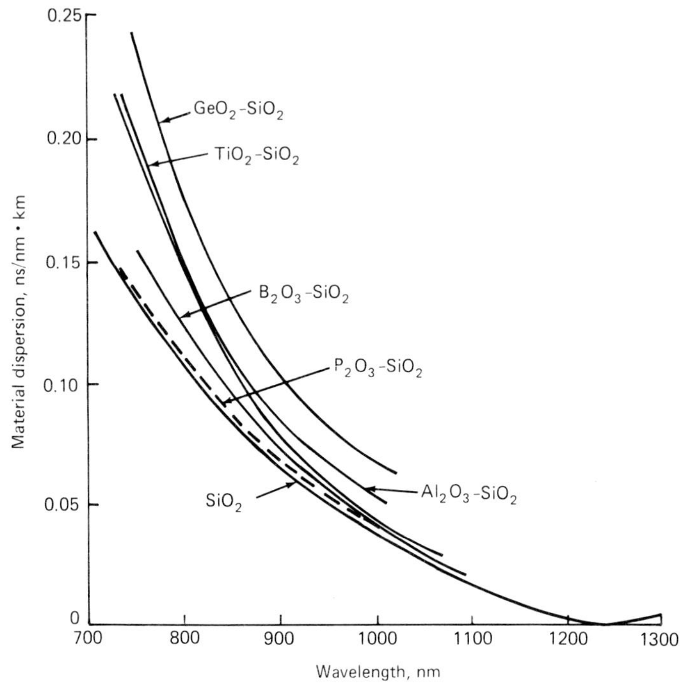


**FIGURE 1.27** Material dispersion as a function of wavelength for various glass fiber compositions. *(Reproduced from Ref. 23, with permission.)*

Dispersion is typically expressed in units of time delay per nanometer of spectral wavelength per unit distance, for example, ps/(nm · km).

Figure 1.27 shows the material dispersion of fibers of several different compositions.

Figure 1.28 shows the change in material dispersion of a single-mode fiber as the $GeO_2$ dopant concentration is changed in the fiber. It is seen that pure $SiO_2$ has a "zero" dispersion at about 1280 nm, whereas typical single-mode fibers with a 0.3 to 0.4 percent $\Delta$ have zero chromatic dispersion at approximately 1310 nm.

### 1.6.2 Waveguide Dispersion

Waveguide dispersion occurs because the propagation of the light in the core and cladding is different.

From the work of Dyott and Stern,[24] this wavelength-sensitive and core-position-sensitive dispersion is expressed as

$$\frac{dt_w}{d\lambda} = \left(-\frac{L}{2\pi c}\right)V^2\left(\frac{\partial^2\beta}{\partial V^2}\right) \tag{1.45}$$

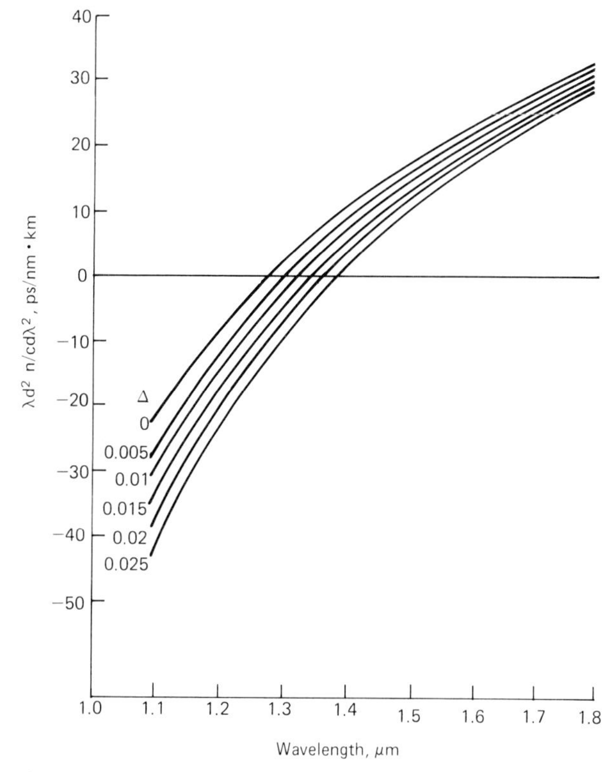


**FIGURE 1.28** Material dispersion vs. wavelength and index change based on $GeO_2$.

where $dt_w$ = the time delay induced by waveguide dispersion
$V$ = normalized frequency parameter
$\beta$ = propagation constant
$L$ = length

For a given small refractive index change of 0.1 percent, the waveguide dispersion is 1 to 2 orders of magnitude smaller than the material dispersion. However, it should be noted that the two different dispersions are of different signs. Thus, the dispersion associated with the material of fabrication and the dispersion associated with the waveguide structure can be balanced at a zero-dispersion wavelength where maximum data transmission is achieved, as shown in Fig. 1.29. It is the summation of these dispersion phenomena that dictates the use of those fabrication materials discussed previously. Dopants such as germanium and phosphorus can be added in the correct amounts, such that low-loss fibers can be produced that do not have high dispersion at wavelengths where lasers or light-emitting diodes (LEDs) are commercially available.

It is this balance that underlies the design of single-mode fibers for 1310 nm operation, where material doping is limited to 8 to 10 percent (by weight) germanium dioxide, with the corresponding appropriate single-mode cutoff wavelength, which is a factor in the waveguide dispersion. For single-mode fibers operating at

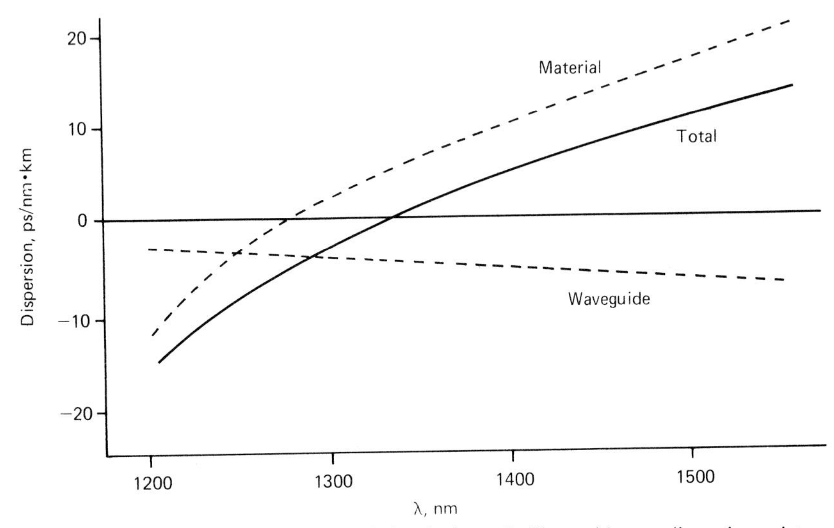


**FIGURE 1.29** Dispersion vs. wavelength for single-mode fiber, with zero-dispersion point shown.

the longer wavelengths, i.e., 1550 nm, the zero-dispersion wavelength can be shifted to produce fibers of high information capacity. This shifting is done by modifying the refractive index profile from the typical step index to an index profile that is more triangular in shape, with a higher on-axis dopant level, as shown in Fig. 1.30. A multiple-index core profile can flatten this dispersion over a range

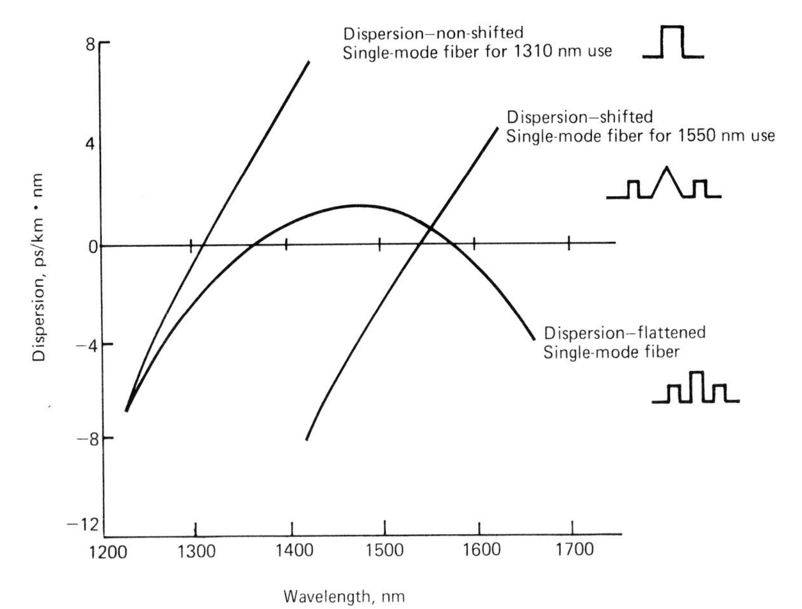


**FIGURE 1.30** Spectral dispersion for various single-mode fiber profiles.

of wavelengths, allowing wavelength-division multiplexing (WDM), for increased system capacity.

## 1.7 FIBER STRENGTH

The intended application of a fiber determines the strength requirement for it. While virtually all glass fiber is supplied in coated form, field use typically requires the fiber to be in some additionally reinforced form, such as cable.

The critical area in determining the strength of the fiber is the outside surface. It is this glass surface that has small flaws that determine the mechanical behavior of the fiber. The existence and growth of these flaws determine the stress level at which fiber fails.

### 1.7.1 Theoretical Fiber Strength

The major difference between a glass fiber and the bulk glass from which it is produced is the high surface-area-to-volume ratio of the fiber. The high surface area increases the likelihood of surface flaws, which can initiate failure.

From glass structure considerations, a variety of defects can be present that will initiate failure. For example, glass is composed of tetrahedral units that share corners, but not edges, when bonded to form the matrix. Beyond this short-range bonding order, unit cells are random in orientation. A silicate glass structure therefore can become stressed if impurities such as cations (other than Si) are present. Cation impurities will cause weak bond formations. Also, such impurities can reside in the open areas of the matrix and induce a physical stress in the structure by their large size. An ideal, flawless glass has a theoretical strength equal to the energy necessary to break the glass-forming bond. This ideal strength is defined as[25]

$$\sigma^2 = \frac{2\gamma E}{8d} \tag{1.46}$$

where $\gamma$ = surface energy of the material
$E$ = Young's modulus
$d$ = bond distance

For the silicon-oxygen bond, which is one of the strongest existing bonds, theoretical strengths on the order of 2 to 3 $\times$ $10^6$ psi (14 to 21 $GN/m^2$) have been calculated. These theoretical strengths are in good agreement with measurements made by Griffith[26] taken in the 1920s.

As one moves away from the ideal, flawless glass piece, assumptions must be made. These assumptions are that the flaws in the glass will be surface cracks, present in some number, at random orientation, with a distribution of crack depths. To establish a relationship between strength and flaw size, we start with the well-known fracture mechanics relationship[27]

$$K_I = Y\sigma\sqrt{a} \tag{1.47}$$

where $K_I$ is known as the stress intensity factor that relates an externally applied tensile stress $\sigma$ to a flaw of depth $a$. $Y$ is a constant that accounts for the mode of loading and the crack geometry. From Eq. (1.47), the stress-intensity factor can be increased by either increasing the applied stress or the depth of the flaw. If the applied stress for a given flaw depth is increased sufficiently, a critical stress-intensity factor is reached, $K_I = K_{Ic}$, and failure occurs. This critical stress-intensity factor is a measure of a material's resistance to fracture and is a material constant.

For the condition of failure, Eq. (1.47) can be rearranged:

$$\sigma_f = \frac{K_{Ic}}{Y\sqrt{a}} \tag{1.48}$$

where $\sigma_f$ is the stress at failure or the failure stress of the material. Equation (1.48) gives a general relation between strength and flaw size in which the strength is inversely proportional to the square root of the flaw depth, or the larger the flaw, the lower the strength. Figure 1.31 shows the fiber strength as a function of the depth.

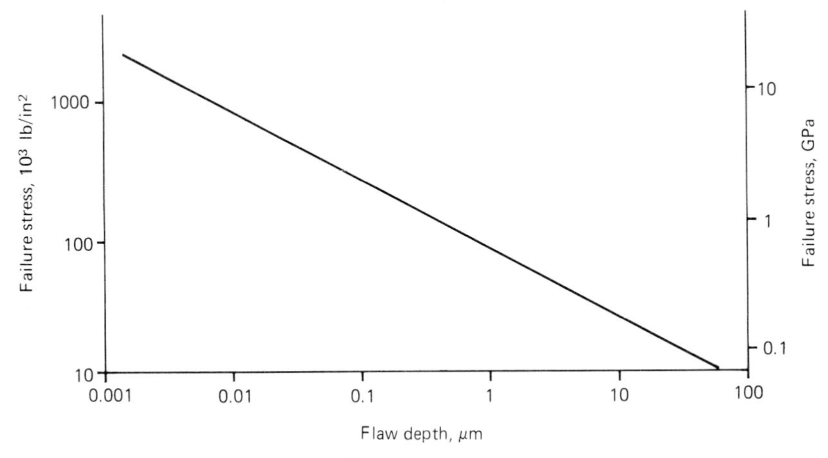


FIGURE 1.31 Failure stress vs. flaw size.

Recently, strengths near $10^6$ psi (about 7 $GN/m^2$) have been reported for short lengths of optical glass fibers. This corresponds to a flaw size of approximately 0.003 μm. However, the existence of flaws depends on the amount of surface area tested, which, in turn, depends on the length of fiber tested. A longer fiber will have a higher probability for a larger flaw. This flaw-size distribution requires that all fiber be tested at some minimum load to ensure a minimum fiber strength over the entire length.

### 1.7.2 Fiber Strength Distribution

Flaws are distributed over the fiber surface and vary in size. To ensure a minimum strength, a fiber is usually dynamically loaded over its length to remove weak sections. Known as *proof testing,* a long length of fiber is loaded to typically 50,000 psi (345 $MN/m^2$) over consecutive short lengths. This removes flaws larger than approximately 2.3 μm and yields fiber lengths of known minimum strength. Thus, a continuous reel of manufactured fiber may result in several reels of shorter-length fiber after proof testing, when low-strength areas are removed from the original reel.

The strength of the fiber can be characterized further by approaching the fiber strength from a statistical perspective. The strength distribution of optical fiber can be conveniently displayed by plotting data according to a two-parameter Weibull distribution,[28] which is given in its linear form by

$$\ln \ln \left(\frac{1}{1 - F}\right) = m \ln \sigma_f - m \ln \sigma_o \tag{1.49}$$

where $F$ is the cumulative failure probability and $m$ and $\sigma_o$ are the Weibull slope and scale parameters, respectively. The Weibull modulus $m$ is a measure of scatter in the strength and is inversely related to standard deviation.

Strength data are obtained by placing fiber samples of specified length in a given environment, usually air, and loading the samples at a predetermined rate. Samples 10 to 20 m long usually are used for testing. Data on several kilometers of fiber are necessary to determine strength for long-fiber applications. Short samples, 1 m for example, would require many thousands of tests to ensure a sufficient sampling of the fiber. Strain rates of 0.04 mm/min are typical for this testing.

After the failure strength of the test fibers has been recorded, the data are plotted. This is done first by ranking the data from weakest to strongest and then assigning a failure probability $F$ to each strength value using the equation[29]

$$F = \frac{R - 0.3}{J + 0.4} \tag{1.50}$$

where $R$ is the rank from 1 to the maximum number of specimens tested, $J$. When a line is fitted through the strength data with Eq. (1.49), strength $\sigma_f$ is the independent variable and the failure probability is the dependent variable.

Figure 1.32 represents a Weibull distribution of strength data from standard single-mode fiber using 20-m gauge lengths in air at a strain rate of 0.04/min. A linear fit will not represent the data in Fig. 1.32 because there are distinct regions to the data. Region I represents the high-strength, low-variability region which corresponds to the intrinsic-type flaws ($m \approx 15$ to 30). Region II is a lower-strength, higher-variability region ($m \approx 2$ to 4), corresponding to flaws most likely to be induced during manufacturing and handling. Region III represents the region of the distribution truncated by the 50,000 psi proof stress or screening of the fiber. By eliminating the weak flaws, proof testing establishes a minimum strength and gives a region of low variability near the proof stress with a Weibull slope approaching that of region I.

The various methods of strength-testing optical fibers has been discussed in detail.[7] For example, strength data from 1-m gauge lengths are not sufficient for long-length telecommunications applications, since the probability of finding an extrinsic-type flaw is small unless many thousands of tests are performed. For long-length applications, data on kilometers of fiber are necessary. Figure 1.33 shows the difference in a Weibull plot where, by using longer test fibers and thus sampling a longer fiber length, the presence of lower stress breaks was observed.

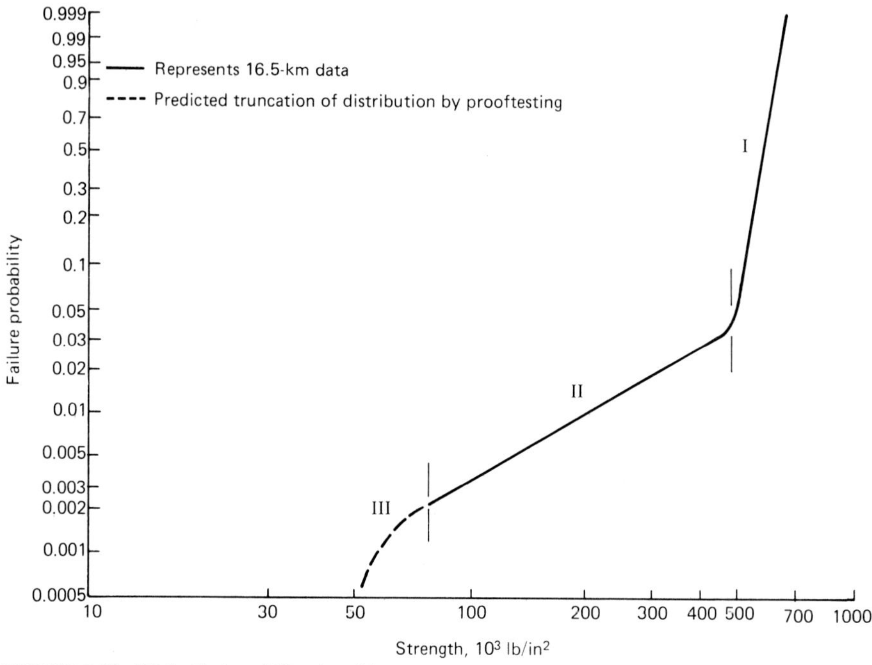


**FIGURE 1.32** Weibull plot of fiber breaking stress.

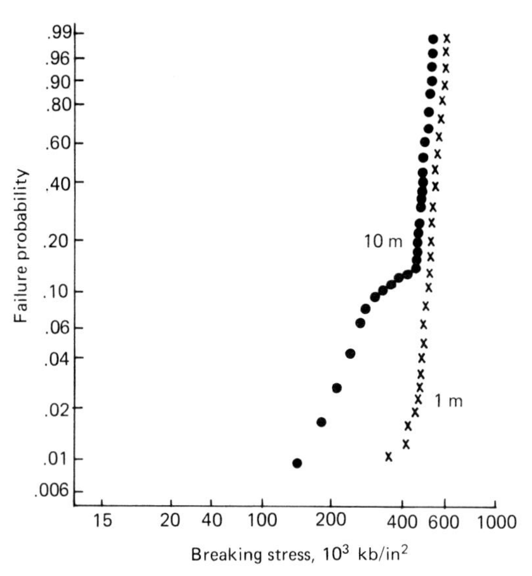


**FIGURE 1.33** Fiber failure probability vs. breaking stress for 1-m and 10-m gauge lengths.

### 1.7.3 Fatigue

Beyond the mechanical failure that occurs during dynamic loading, it is well known that optical glass fibers can exhibit delayed failure when under sufficient stress in a humid environment. Failure under these conditions results from growth of small flaws on the fiber surface; eventually the flaws reach the critical dimensions for failure. This phenomenon of subcritical crack growth is commonly referred to as *fatigue* and has been described on a molecular scale by Michaleske and Freiman[30] as "a specific chemical reaction between strained bonds in vitreous silica and water, which can be used to explain environmental enhanced crack growth."

In this model, water breaks a stressed silicon-oxygen bond at the crack tip and thereby extends the crack. This process proceeds by the chemistry[30]

$$\begin{array}{c} | \quad\quad | \quad\quad\quad\quad\quad\quad | \quad\quad\quad | \\ \text{—Si— O —Si—} + H_2O \leftrightarrow \text{—SiOH} + \text{—SiOH} \\ | \quad\quad | \quad\quad\quad\quad\quad\quad | \quad\quad\quad | \end{array} \tag{1.51}$$

in which the bridging oxygen bond in the silica matrix is broken and the hydroxyl ion is incorporated into the structure.

At low stress, the $OH^-$ bonding is uniform and can result in tip "blunting," where the crack does not propagate. At higher stresses the bonding becomes more rapid and the resultant corrosion causes the crack to propagate. In an effort to model fatigue in brittle materials such as glass, researchers have devised several theories to express the crack growth phenomenon in terms of the applied stress. One such theory is known as the "power law crack velocity model," where crack velocity $v$ is related to the stress intensity factor $K_I$ by[31]

$$v = AK_I^n \tag{1.52}$$

where $A$ and $n$ are crack growth parameters. The stress intensity factor $K_I$ is related to the flaw depth $a$ and the applied stress $\sigma$ by

$$K_I = Y\sigma\sqrt{a} \tag{1.53}$$

where $Y$ depends on the flaw geometry. Substituting Eq. (1.53) into Eq. (1.52), one can express the crack velocity in terms of the applied stress,

$$v = \frac{da}{dt} = A(Y\sigma\sqrt{a})^n \tag{1.54}$$

Equation (1.54) can be rewritten in an integral form as

$$\int_{a_0}^{a} \sqrt{a}^n da = \int_0^t AY^n\sigma^n \, dt \tag{1.55}$$

where $a_0$ is the initial crack length, $a$ is the crack length at failure, and $t_f$ is the time to failure. Equation (1.54) now can be integrated for either a static or dynamic stress history. For the static case,

$$\sigma = \text{constant} = \sigma_a \tag{1.56}$$

and when substituted into Eq. (1.55) yields the following relationship between time to failure and a constant applied stress:

$$t_f = BS_i^{n-2}\sigma^{-n} \tag{1.57}$$

where $S_i$ is the strength in the absence of fatigue, or *inert strength,* and $B = 2K_{Ic}^{2-n}/AY^2(n - 2)$.

In the case of dynamic fatigue, a stress applied to the fiber is increased linearly until failure occurs; the failure stress is recorded as the strength. One can imagine that the longer the test or the slower the stress rate, the lower the strength since more time is allowed for crack growth. The relationship between the stress rate $\sigma$ and the strength $\sigma_f$ is derived by integrating Eq. (1.55) with

$$\sigma_f = \dot{\sigma}t \tag{1.58}$$

and is given by

$$\begin{aligned} \sigma_f^{n+1} &= B(n - 2)S_i\sigma \\ &= B(n + 1)S_i^{n-2}\dot{\sigma} \end{aligned} \tag{1.59}$$

Equation (1.59) can be linearized by taking the logarithm of both sides, which gives

$$\ln \sigma_f = \left(\frac{1}{n + 1}\right) \ln [B(n + 1)S_i^{n-2}] + \left(\frac{1}{n + 1}\right) \ln \dot{\sigma} \tag{1.60}$$

where $1/(n + 1)$ is the slope of the $\ln \sigma_f$ vs. $\ln \sigma$ line. It is in this form that strength data for a range of stressing rates can be plotted.

***n Parameter.*** From Eq. (1.60), the parameter $n$ is determined by fitting the slope of the line to dynamic fatigue data, as shown in Fig. 1.34. This parameter is particularly important in that it gives a measure of the material's resistance to subcritical-crack growth. That is, the higher the value for $n$, the less sensitive the strength is to changes in the stress rate. This is shown in Fig. 1.34, where the change in strength for an order-of-magnitude change in the stress rate is given for various $n$ values. Note that, since $n$ is inversely related to the slope of the $\ln \sigma_f$ vs. $\ln \dot{\sigma}$ curve, the fatigue resistance does not increase significantly for $n$ values above 100. Therefore a material with an $n$ value of 200 is, for practical purposes, no less resistant to fatigue than one with an $n$ of 600. On the other hand, a significant difference in fatigue resistance exists for $n$ values between 21 and 30. For standard silica fiber with $n$ approximately equal to 20, proof testing or dynamic loading of the fiber typically is done at stress levels 3 to 4 times larger than the intended-use stress to assure longer-term fiber reliability.

***Mechanisms of Strength Improvement.*** Fiber strength can be improved by reducing the Young's modulus to decrease the stress at the surface, balancing some of the stress by imparting a compressive layer of alternative glass composition to

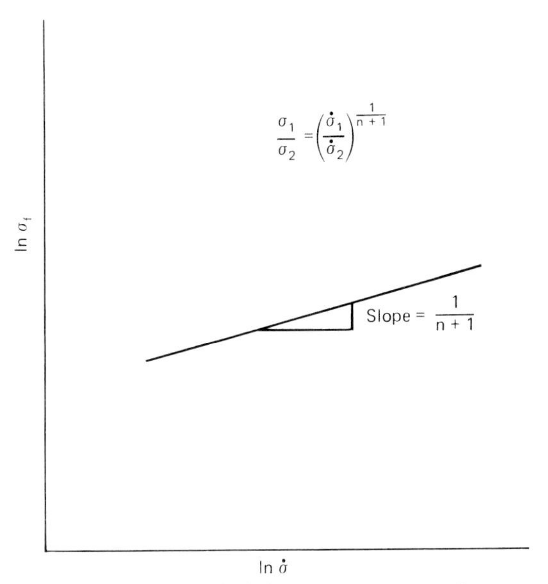


**FIGURE 1.34** Dynamic fatigue measurements of glass fiber.

the surface[32] and by applying a thin layer of carbon to the glass surface to limit the amount of water which reaches the crack tip.[33]

## 1.8 SPECIAL FIBERS

Depending on the application of the fiber, it may be desirable to depart from the circularly symmetric, high-quality glass fiber discussed thus far. Fibers have been fabricated from various materials—sometimes with highly specific constructions—to achieve a particular functionality for special applications.

### 1.8.1 Polarization-Maintaining Fibers

In previous discussions of single-mode fibers, it was assumed that the propagating mode propagated uniformly down the fiber. However, light propagates within single-mode fibers in orthogonal modes—the $HE_{11}^x$ and $HE_{11}^y$ modes. The orthogonally polarized modes propagate with their respective propagation constants $\beta_x$ and $\beta_y$. The difference in propagation of the two modes is given in a concise description of polarization properties of optical fibers by Kaminow,[34] and will be described here. Given a constant index along the fiber length in the $z$ direction, the modal birefringence $B$ can be written as

$$B = (\beta_x - \beta_y)\left(\frac{2\pi}{\lambda}\right) \tag{1.61}$$

where $\lambda$ = optical path length. $B$ represents the difference in effective indexes for the polarized modes. Given a perfect optical fiber with no variation in refractive index in the $x$ and $y$ directions and with perfectly circular geometry, the modal birefringence introduced into the fiber can be maintained along its length. For a given polarization introduced into the fiber at an angle $\theta$ with respect to the $x$ axis, the light will have different polarizations as a function of path length, expressed as a phase shift:

$$\Phi_{(X)} = (\beta_x - \beta_y)z \tag{1.62}$$

The path length corresponding to that distance over which the original input polarization again is achieved is defined as the fiber beat length of

$$\Phi(L) = 2\pi \qquad \text{and} \qquad L = \frac{\lambda}{\beta} \tag{1.63}$$

While a perfectly constructed fiber (i.e., one free of index variations, possessing perfect circular symmetry, and free from random bends or twists) can maintain the original input polarization, departures from perfection will cause coupling of the two orthogonal modes. Complete coupling from one polarized mode to the other can occur if a perturbation is such that

$$|\beta_x - \beta_y| = k + \Delta k \tag{1.64}$$

where $k = 2\pi/\Lambda$ is the spatial frequency of the perturbation in the $z$ direction
$\Delta k = \pi/l$
$l$ = length of the fiber

To realize the benefits of a polarization-maintaining fiber—for example, making physical measurements by the interference of two coherent beams—special single-mode fibers are fabricated so that the modal birefringence is maximized. This is done by maximizing the geometrical portion of the birefringence (for example, by using a noncircular core shape), or by maximizing the material birefringence (by induced strain in the fiber).

While the material contribution to the birefringence depends on the material and location, the geometric contribution, for small birefringence, can be approximated by

$$G \cong C\left(\frac{\lambda}{d}\right)e^2\Delta^{3/2} \tag{1.65}$$

where $d$ = mean core diameter
$\Delta = \Delta n/n$ = core/cladding index ratio
$e = [1 - (dy/dx)^2]^{1/2}$ with $dx$ and $dy$ referring to the minor and major core diameters, respectively
$C$ = constant dependent on fiber $V$ value

For large birefringence, $G$ becomes $C'(\Delta n)^2$, with $C'$ dependent on eccentricity and effective $V$ value. The strain birefringence induced from the materials of fabrication is taken as an average of the material birefringence over the core and cladding.

Polarization-maintaining fibers are utilized when the difference in the propa-

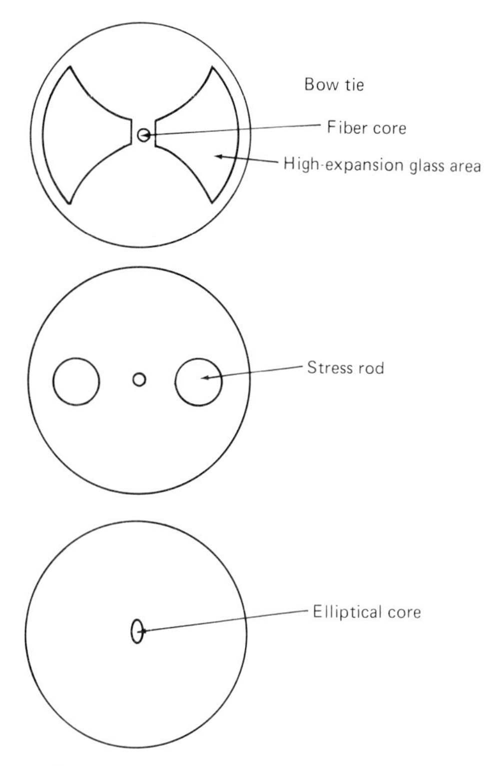


**FIGURE 1.35** Fiber designs for polarization maintaining fiber.

gation of the two orthogonal modes offers benefits. In fiber sensors, for example, the difference in propagation of one of the orthogonal polarizations is subjected to a parameter being measured (such as magnetic field or pressure) and is compared to another fiber leg serving as a reference. In heterodyne communications systems, signals are transmitted separately on each of the orthogonal modes, so that information-carrying capacity is increased.

The fiber designs used to increase $\beta$ in fiber by geometric or material means are shown in Fig. 1.35. Typical beat lengths achieved in these fiber designs, fabricated by a variety of techniques, are on the order of fractional to several millimeters.

### 1.8.2 Plastic Fibers

Fibers fabricated from plastic have been in commercial use longer than glass fibers. This commercial use was driven by the ability to produce plastic fibers of large NA (0.5 or greater), with large core sizes (greater than 1000 μm) and at low cost (cents per meter), although such inexpensive plastic fibers have high loss (thousands of decibels per kilometer).

Limitations of plastic-clad, plastic-core fibers are numerous. Although they can be fabricated with high numerical apertures, the plastic fabrication process does not routinely allow for the grading of the core index profile. Therefore, these fibers exhibit high modal dispersion, giving low data transmission rates (a few megabits per second over a few hundred meters). High modal dispersion dominates as the limiting factor in transmission rate, even though plastics also have a high material dispersion (300 to 700 ns/km). The high attenuation associated with plastic fibers also limits their typical working distances to less than 1 km. Figure 1.36 shows an attenuation vs. wavelength plot for polymethylmethacrylate-core plastic fiber. Large absorption bands from the carbon-hydrogen (C-H) bond vibration are present, with many overtones in the spectra. Operating windows are in the shorter wavelength region of 500 to 700 nm, where optical sources such as LEDs are available. Therefore, plastic fiber systems can be made of low-cost components.

Characterizing radiation response for plastic fibers is more complicated than for glass fibers because of the larger number of materials that can be used to fabricate plastic fibers. The overall tendency of plastic fibers is to liberate hydrogen by breaking C-H bonds. The liberated hydrogen then recombines in the polymer structure to form compositions other than the original composition. The same kind of recombination is also often observed if a C-C bond is broken. Since the new compositions vary drastically in properties from the material of choice, refractive index perturbations usually arise, giving high scattering loss.

Thermal degradation of a plastic fiber is also a concern. Changes in polymer structure can occur in most commonly used plastics at temperatures of 80 to 100°C, such that mechanical reliability, as well as optical loss, may be a concern. However, applications requiring short fibers, high NA, tight bend diameters, or

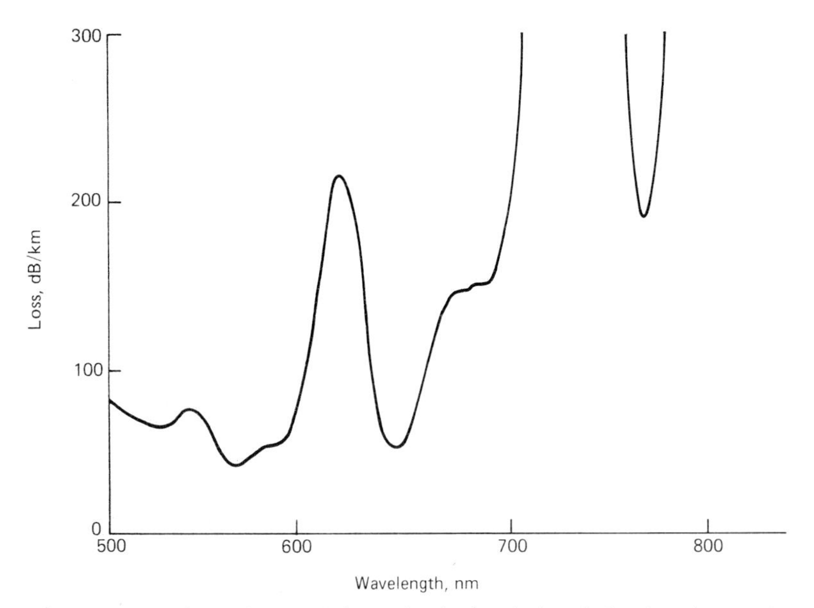


**FIGURE 1.36** Spectral attenuation of plastic (polymethylmethacrylate-core) fiber. (*Reproduced from T. Kanio et al., Low Loss Poly (Methyl Methacrylate -d5) Core Optical Fibers," App. Phys. Lett., vol. 41, no. 9, 1982, pp. 802–804, with permission.*)

inexpensive system components (such as connectors) provide opportunities to use plastic fibers.

Plastic-clad silica fibers are an alternative to all-glass or all-plastic fibers. Attenuations of plastic-clad silica fibers are typically 5 to 10 dB/km at the 800- to 1310-nm transmission window. This loss is higher than for all-glass fibers and is caused by the high-loss plastic cladding, which carries part of the optical power. Absorptions in plastic-clad silica fiber are similar to those in glass fiber (from $OH^-$, $Fe^{2+}$, etc.), although higher-NA fibers are possible in plastic-clad silica when low-index plastic is the cladding material. However, plastic-clad silica fiber design is limited to a step-index profile, resulting in low data rates.

### 1.8.3 Fibers for Long-Wavelength Operation

Fibers also can be fabricated that operate well beyond wavelengths of current commercial interest. They are fabricated from glass compositions whose bonding characteristics and band structures allow use far into the infrared region, where losses can theoretically be as low as 0.001 dB/km. Glass compositions and manufacturing techniques capable of achieving this ultralow loss at a usable operating window—at the $CO_2$ laser wavelength of 10.6 μm, for example—are being researched.

Long-wavelength fibers currently are fabricated from three material systems: heavy-metal fluorides, chalcogenide glasses, and crystalline materials. Heavy-metal fluoride compositions center around stable glass-forming regions of zirconium fluoride. Losses from these glass-forming systems are currently in the 5- to 10-dB/km range at wavelengths of 2 to 2.2 μm. As these fibers are directly melted in a double-crucible forming apparatus, contamination of the fibers is orders of magnitude greater than for commercially produced optical fibers that are formed by chemical vapor deposition and that maintains chemical purity. Thus, the major loss mechanism of these fibers is attributed to metal impurities.

Some fluoride fibers have been produced using beryllium fluoride ($BeF_2$) in a chemical vapor delivery process. In this case, $OH^-$ absorption from fabrication techniques currently limits the low loss achievable in the infrared spectrum.

Fibers formed from chalcogenide glasses, such as arsenic/sulfur (As/S), and crystalline materials, such as silver bromide (AgBr) core, silver chloride (AgCl) cladding, suffer from those impurities already discussed, as well as from structural impurities. In chalcogenide glass systems (e.g., As/S), electronic defects limit low-loss transmission. Physical defects such as grain boundaries also impart high loss by scattering light.

### 1.8.4 Optically Active Fibers

Fibers that successfully transmit in the infrared have been produced by doping commercial compositions with trace amounts of rare earth compounds,[34] for example, erbium, neodymium, and europium. Dissolved in the glass matrix, these trace dopants undergo an electronic transition when excited optically, producing coherent light at the wavelength of the electronic decay. Fibers doped in this manner can be fabricated as lasers that easily couple light into optical fibers because they have similar physical dimensions. Such optically active components are being developed for eventual application as in-line amplifiers for communication systems. A rare-earth-doped fiber would be pumped at one wavelength and would amplify at another wavelength, which would be used for signal transmission, for example, at 1535 nm.

## ACKNOWLEDGMENTS

I would like to express my appreciation to my colleagues whose discussions and assistance contributed to the understandable presentation of this complex subject. In particular, I would like to express my appreciation to Mr. Les Button for his effective presentation of propagation theory and to Dr. Scott Glaesemann for his concise treatment of the theory of strength and reliability in fibers. I would also like to acknowledge those on the Corning Corporate Review Staff for their many suggestions which made this work more focused and technically accurate. The following figures were redrawn from figures originally drawn by Corning Inc.: Figs. 1.6, 1.7, 1.10, 1.13–1.16, 1.18, 1.23, 1.25, 1.26, and 1.28–1.35.

## REFERENCES

1. N. S. Kapany, *Fiber Optics, Principles and Applications,* Academic Press, New York, 1967, p. 29.
2. M. Born and E. Wolf, *Principles of Optics, Electromagnetic Theory of Propagation, Interference and Diffraction of Light,* Pergamon Press, New York, 1975, pp. 1–2.
3. J. Gowar, *Optical Communication Systems,* Prentice-Hall International, New York, 1984, p. 117.
4. J. Gowar, *Optical Communication Systems,* Prentice-Hall International, New York, 1984, p. 133.
5. D. Gloge and E. A. Marcatili, "Multimode Theory of Graded-Core Fibers," *Bell System Technical Journal,* vol. 52, no. 9, 1973, pp. 1563–1578.
6. R. O. Olshansky, "Propagation in Glass Optical Waveguides," *Reviews of Modern Physics,* vol. 51, no. 2, 1979.
7. S. Suzuki et al., "Characteristics of Graded Index Fiber by VAD Method," *Proceedings of the Fifth European Conference on Optical Communication,* Amsterdam, 1979.
8. D. A. Nolan, R. M. Hawk, and D. B. Keck, "Multimode Concatenation Modal Group Analysis," *IEEE Journal of Lightwave Technology,* vol. LT-5, no. 12, 1987.
9. R. Olshansky and S. M. Oaks, "Differential Mode Delay Measurement," *Fourth European Conference on Optical Communications,* Genoa, 1978.
10. J. C. Lapp, V. Bhagavatula, and A. J. Morrow, "Segmented-Core Single-Mode Fiber Optimized for Bending Performance," paper WQ16, *Optical Fiber Communications Conference,* New Orleans, 1988.
11. R. G. Sommer, R. D. DeLuca, and G. E. Burke, "New Glass System for Low Loss Optical Waveguides," *Electronic Letters,* vol. 12, no. 16, 1976, p. 408.
12. K. Yoshida, S. Sentsui, and T. Kuroha, "Low Loss Fibre Prepared Under High Deposition Rate by Modified CVD Techniques," *Electronic Letters,* vol. 13, no. 20, 1977, p. 608.
13. G. Mie, "Beitrage zur Optik Truber Medien, Speziell Koloidaler Metallosungen," *Ann. Phys.*, vol. 4, no. 25, 1908, p. 377.
14. F. Urbach, "The Long Wavelength Edge of Photographic Sensitivity and of the Electronic Absorption of Solids," *Physical Review,* vol. 92, 1953, p. 1324.

15. J. D. Rush et al., "Hydrogen Related Degradation in Optical Fibers—System Implications and Practical Solutions," *British Telecommunications Journal,* vol. 2, no. 4, 1984, pp. 57–61.

16. J. A. Wall, H. Posen, and R. Jaeger, "Radiation Hardening of Optical Fibers Using Multidopants Sb/P/Ce," *Advances in Ceramics,* vol. 2 (Physics of Fiber Optics) 393-7, 57-1, 1981.

17. E. J. Friebele, R. E. Jaeger, G. H. Sigel, Jr., and M. E. Gingerich, "Effect of Ionizing Radiation on the Optical Attenuation in Polymer-Clad Silica Fiber Optic Waveguides," *Applied Physics Letters,* vol. 32, 1978.

18. E. J. Friebele et al., "Radiation-Resistant Low –OH Content Silica Core Fibers," *IEEE Journal of Lightwave Technology,* vol. LT-1, no. 3, 1983, p. 462.

19. E. J. Friebele, R. J. Ginther, and G. H. Sigel, Jr., "Radiation Protection of Fiber Optics Materials: Effects of Oxidation and Reduction," *Applied Physics Letters,* vol. 24, no. 9, 1974.

20. E. J. Friebele and M. E. Gingerich, "Photobleaching Effects in Optical Fiber Waveguides," *Applied Optics,* vol. 20, no. 19, 1981, p. 3448.

21. E. A. J. Marcatili, "Bends in Optical Dielectric Guides," *Bell System Technical Journal,* vol. 48, no. 7, 1969, p. 2103.

22. A. W. Snyder, I. White, and D. S. Mitchell, "Radiation From Bent Optical Waveguides," *Electronics Letters,* vol. 11, 1975, p. 332.

23. D. B. Keck, "Optical Fiber Waveguides," Chap. 1 in M. Barnoski (ed.), *Fundamentals of Optical Fiber Communications,* 2d ed., Academic Press, New York, 1981, p. 78.

24. R. B. Dyott and J. R. Stern, "Group Delay in Glass-Fiber Waveguides," *Electronics Letters,* vol. 7, 1971, p. 82.

25. C. K. Kao, "Optical Fibre and Cables," Chap. 5 in M. J. Howes and D. V. Morgan (eds.), *Optical Fibre Communications,* Wiley, New York, 1980, p. 210.

26. A. A. Griffith, "The Phenomena of Rupture and Flow in Solids," *Roy. Soc. Phil. Trans.*, vol. 221, 1921, p. 163.

27. B. R. Lawn and T. R. Wilshaw, *Fracture of Brittle Materials,* Cambridge University Press, London, 1975.

28. W. Weibull, "A Statistical Distribution Function of Wide Applicability," *Journal of Applied Mechanics,* vol. 18, no. 9, 1951, p. 293–297.

29. B. Bergman, "On the Estimation of the Weibull Modulus," *Journal of Material Science,* vol. 3, no. 8, 1984, pp. 689–692.

30. T. A. Michalske and S. W. Freiman, "A Molecular Interpretation of Stress Corrosion in Silica," *Nature,* vol. 295, no. 5849, 1982, pp. 511–512.

31. A. G. Evans, "Analysis of Strength Degradation After Sustained Loading," *Journal of the American Ceramic Society,* vol. 57, no. 9, 1974, pp. 410–411.

32. S. T. Gulati et al., "Fiber Optics Reliability: Benign and Adverse Environments," *Proceedings of the Society of Instrumentation Engineers,* vol. 842, no. 22, 1987.

33. K. E. Lu, M. T. Lee, D. R. Powers, and G. S. Glaesemann, "Hermetically Coated Optical Fibers," paper PD1-1, *Optical Fiber Communications Conference Technical Digest,* New Orleans, 1988.

34. I. Kaminow, "Polarization in Optical Fibers," *IEEE Journal of Quantum Electronics,* vol. 17, no. 1, 1981.

# CHAPTER 2
# FIBER-OPTIC CABLES

M. M. Ramsay
*STC Technology, Ltd.*

## 2.1 INTRODUCTION AND BACKGROUND

Optical fibers are inherently brittle and without protection become fragile, but have to be made into cables that will withstand installation and the often hostile environments in which they must operate reliably. Working environments that have been met range from the almost constant 2 to 4°C of the seabed (for deep-sea submarine cables) to the rapid temperature changes (from −55 to +155°C) of some avionic applications; from normal pressures up to 70 MN/m$^2$; and from ambient air to hot, corrosive liquids. The number of fibers in a cable may vary from one, as in rack wiring, to several thousand, as in distribution cables now being proposed. The cable may be all-dielectric, containing only fibers, or may incorporate numerous metallic conductors or other, more complex elements.

Such diversity of applications can be accommodated only with a wide variety of cable designs. The state of the art is such that some straightforward applications can be handled by several designs. However, all optical-fiber cables can be judged by their success in meeting two basic criteria: (1) they must not significantly degrade the transmission properties of the fiber, and (2) they must maintain the integrity of the fiber during manufacture, installation, and service.

The invention of the cladded waveguide in 1954 by Van Heel[1] was the key that brought dielectric waveguides from the realm of scientific curiosity to practical exploitation. A *cladding*—one that is sufficiently thick and has a slightly lower refractive index than the core—ensures that any evanescent fields associated with the guided energy will decay to negligible levels at the surface of the composite waveguide, so that the waveguide can be supported without incurring prohibitive losses. However, despite the enormous step forward that this invention represented, the problems of cabling such waveguides—or optical fibers—were still formidable.

Van Heel proposed a plastic-clad, plastic-cored fiber. Such fibers, although still in use today, were soon to be replaced by multicomponent glass fibers for all but the very shortest ranges. More recently, it became evident that silica offered the potential for lower losses, and attention switched to this material. At first, plastic-clad silica was considered for large-core, step-index fiber, but currently various configurations of silica and doped silica are favored. Although present-day focus is on these fibers, the inherent principles can be applied to other types, with the necessary adjustments to detail.

### 2.1.1 Cabling Objectives

The cabling engineer has two main tasks:

- To minimize optical attenuation increments associated with the manufacture and use of cables
- To maintain the physical integrity of the fiber during the cabling process and installation and in service

*Attenuation Increments.* To understand the problem of cabling-induced attenuation increments, it is necessary to examine the propagation characteristics of dielectric waveguides and how these may be modified by the cabling process.

First, any dielectric waveguide will radiate when its axis is not straight. For a particular fiber waveguide, the amount of radiation is critically dependent on the fiber's radius of curvature and can rapidly increase from totally negligible to quite prohibitive, with a comparatively small reduction in fiber bend radius. Although this rapid increase will occur at some radius for every fiber waveguide, for certain waveguides it will occur at a much greater radius than for others. Historically, emphasis shifted from step-index fibers (the first were made from two multicomponent glasses or plastic-clad silica) to graded-index fibers of multicomponent glasses or silica with various dopants, and, finally, to single-mode fibers, almost invariably made from some configuration of silica and doped silica. It is simpler, however, to consider loss mechanisms, first for single-mode fibers, and then expand these mechanisms to encompass multimode fibers.

A qualitative picture of bend-induced radiation can be obtained for a low-order-mode waveguide by considering the field distribution associated with a curved fiber waveguide (Fig. 2.1*a*). At first approximation, the field distribution is described by a curve similar to the refractive index profile within the core, and has an exponential decay within the cladding. Since the associated wave front is guided, it must remain perpendicular to the fiber axis, and so there must be some radius ($R + X_R$ in Fig. 2.1*a*) at which the phase velocity of the guided radiation reaches the velocity of light in that medium. Beyond this point, the wave front cannot remain perpendicular to the fiber axis, and guiding cannot occur. Therefore, the evanescent energy represented by the field at greater distances from the center of curvature will radiate. This simple picture is complicated, in reality, by a shift of the field in relation to the fiber axis if the bend radius is steady, and by oscillation of the field in relation to the axis when transient conditions pertain. It is also oversimplified, in that it implies loss to a radiating continuum, whereas fiber waveguides radiate into discrete modes. Nonetheless, the success of mode field diameter as a measure of the susceptibility of single-mode fibers to microbending loss does substantiate it to a certain extent.[2,3]

Similarly, any energy in the evanescent field can be lost to absorption if this field penetrates into the cladding as far as the transition from deposited material to the original silica carrying tube, when used. If, at this transition point, there is also an increase in the real part of the refractive index, further loss can occur from quantum tunneling (Fig. 2.1*b*).

For some fiber waveguides, a change in refractive index is introduced in the cladding comparatively close to the core. Such a stratagem is used by the waveguide designer, for instance, to make significant reductions in waveguide dispersion, but is important to the cable engineer because it modifies mode cutoff wavelengths. As mode cutoff is approached, a greater proportion of evanescent energy is carried in the cladding until the mode is just short of cutoff, when it

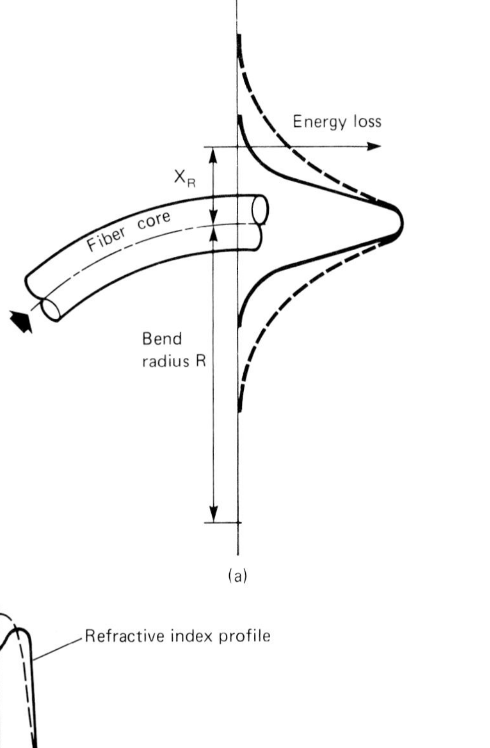


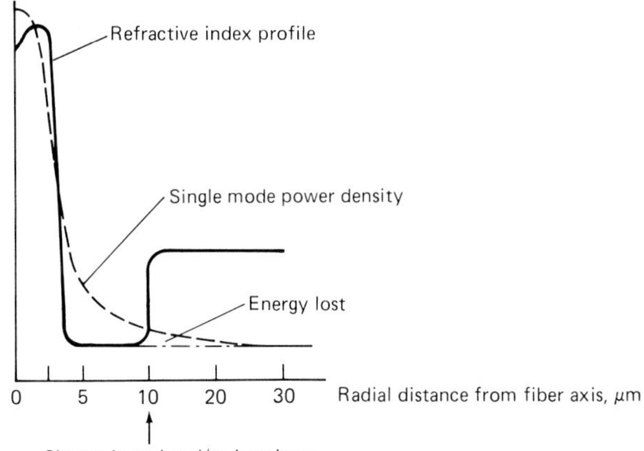


FIGURE 2.1 (*a*) Field distributions in curved optical fiber waveguide for a bound mode (solid line) and a mode near cutoff (dashed line). (*Reproduced with permission of STC Technology Ltd.*) (*b*) Energy loss due to quantum tunneling.

becomes, in effect, a plane wave traveling in the cladding and hence very susceptible to minor perturbations of the fiber axis. It is quite possible—indeed likely—that a fiber with stepped cladding index (W guide) will have a cutoff point for all modes and no dominant mode, a characteristic of normal single-mode fibers.[4]

For multimode fibers, there are always some modes close to cutoff. Clearly, more energy radiates from these modes than from the more tightly bound, lower-order modes. The close connection between cutoff and radiation has been elegantly demonstrated by Midwinter and Reeve in an experiment using small-core, step-index fiber with a carefully set radius of curvature.[5] The waveguide geometry chosen was such that tens—rather than hundreds or thousands—of modes were carried. A wavelength scan of the scattered power spectrum of this fiber showed clearly delineated peaks that could be assigned unambiguously to the expected cutoff wavelength of the individual modes.

Like the bound and radiating modes, a third class of mode—the so-called *leaky* modes—can be represented by highly skewed rays, such that (1) the critical angle is exceeded in either the circumferential or meridional plane (but not both) or (2) a quantum-tunneling loss of radiation is apparent. For small-core fibers, the radiation loss is rapid and is not relevant in cabling applications. For larger-core fibers, however, these modes may radiate for distances of a kilometer or more. These modes can be important because of their influence on measurement results in typical cabling experiments (of under a kilometer). In telecommunications applications, where multikilometer lengths are typical, their effect is quite negligible.

If energy lost from leaky modes and modes near cutoff represented the total loss of energy, these modes would not be of great concern when cabling multimode fibers. However, minor perturbations of the fiber geometry or of its axis can cause energy to be coupled between adjacent modes. Such perturbations are usually random—as opposed to the regular perturbations introduced in cable design and manufacture, which are too low-level and long in mechanical wavelength to be significant—and can be regarded as a complex mechanical disturbance composed of many frequencies, each representing a sinusoidal displacement. This situation was analyzed by Gloge,[6] who showed that the coupling is strongest when the perturbing wavelength coincides with the beat length between adjacent modes.

In early experiments, this situation led to unexpectedly good results (low radiation losses and significantly decreased dispersion), which led to unjustified optimism. This arose because, with step-index and quasi-step-index fibers (or even graded-index fibers with a profile exponent greater than optimum), the required coupling energy rises with increasing mode order. Hence, it is possible that the mechanical perturbations in these experiments supplied only sufficient energy to couple low-order modes, which, by averaging propagation times between them, significantly reduced the dispersion without coupling energy to the radiating modes (and thereby incurred an increase in attenuation). Needless to say, results relying on such a delicate balance—at a time when the technology was such that neither mechanism involved was under close control—were seldom repeatable.

However, if a graded-index profile is used to bring all modes as close as possible to the same group velocity, the coupling energy also tends to be a single value. Since perturbation theory involving statistically random energy coupling and negligible change in propagation constants has been used in the analysis,[7] it tends to break down for the ideal case of identical propagation constants, but the trend is clear. For such an idealized fiber, the beat length $\Lambda$ is given by

$$\Lambda = \pi d n_1 (n_1^2 - n_2^2)^{-1/2} \tag{2.1}$$

where $d$ is the core diameter and $n_1$ and $n_2$ are core and cladding indices, respectively. Typically, $\Lambda$ is about 0.8 mm, which corresponds to probable conditions that would arise from microbends introduced by processes such as plastic extrusion. (See Chap. 1 for discussions on microbend and macrobend losses.)

***Fiber Physical Integrity.*** Although silica has a high Young's modulus and quite exceptional elasticity, the small cross section of a normal fiber means that it contributes very little to the cable's strength, except in certain, very specialized military cables. Usually the load and the Young's modulus of the strength member determine the strain or elongation of the cable, and hence that of the fiber. What matters most is the maximum strain that the fiber will withstand. The answer to the simple question of how much strain the fiber will withstand is, unfortunately, very complex.

Glass fiber, unlike steel or copper wire or plastic cable elements, does not have a well-defined or controllable breaking strain because failure is initiated in flaws that are usually at the surface, but that may also be within the fiber. These flaws cause local intensification of any applied stresses. A stress intensification factor $K_I$ can be defined,[8] which reaches a critical value $K_{I_c}$ such that, when fracture occurs,

$$\sigma_f = \frac{K_{I_c}}{Y\sqrt{a}} \tag{2.2}$$

where $\sigma_f$ = applied stress at fracture
$a$ = crack depth
$Y$ = geometric factor

Usually these flaws are extremely small. For a typical telecommunication fiber, a flaw 1-μm deep will cause failure at about 1 percent strain. Flaws may arise from a number of sources during fabrication, for instance: imperfections in the substrate tube, when used, or scratches introduced by handling, discontinuities during deposition, stresses introduced during collapse or pulling, impurities implanted from the furnace during pulling, and foreign matter trapped during coating. Tests on a number of fiber lengths will give a range of breaking strains depending on the largest flaw in each length tested. Results must then be treated statistically and are usually presented in the form of an ordered plot showing the cumulative failure probability as a function of either stress or strain, as in Fig. 2.2. These results are for 10-m lengths of fiber. Had longer lengths of fiber been used, lower strengths would have been recorded because of the likelihood of finding larger flaws in each length. This can be seen qualitatively by noting that if the samples had been tested as one single fiber of 1470-m length, it would have failed at 2 percent strain, rather than the 5.5 percent median of the 10-m lengths. However, even a length of 1.47 km is orders of magnitude less than the total length of fiber to be installed in a sizable telecommunication system, and the samples in Fig. 2.2 contained none of the larger flaws that occur infrequently but are nonetheless significant for systems containing hundreds or thousands of kilometers of optical fiber.

The implication of Eq. (2.2) is that a flaw in a fiber will cause an almost immediate fracture at, or above, some critical strain. Below this critical strain, although there will be no immediate fracture, the flaw can grow until it reaches a

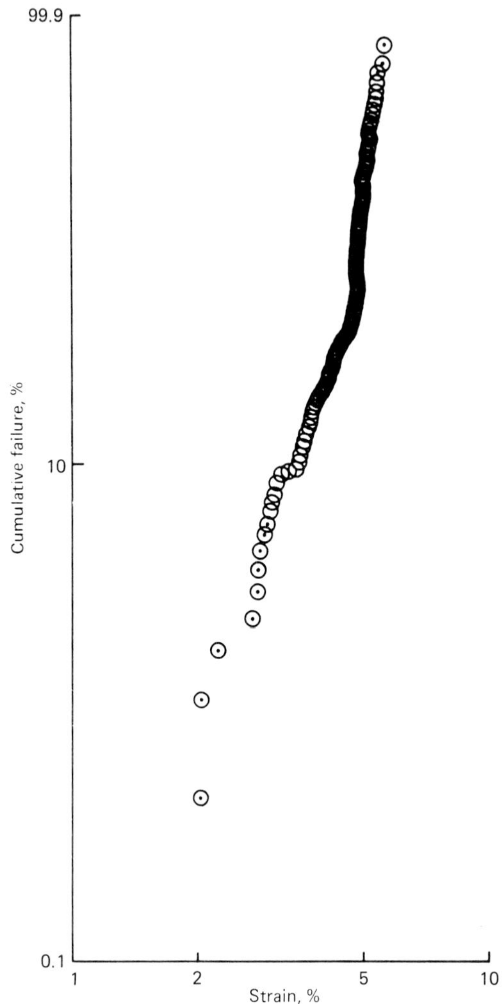


**FIGURE 2.2** Weibull plot of strain at fiber break (147 samples; 10-m gauge length tests). (*Reproduced with permission of STC Technology Ltd.*)

size at which it causes the fiber to break. This phenomenon has been studied both theoretically and empirically. The proposal most commonly used[9]—although not universally accepted—is for a growth law of the form

$$v = AK_I^n \tag{2.3}$$

where $v$ is the velocity of crack growth and $A$ and $n$ are constants. This can be combined with equations of the form of Eq. (2.2) and integrated and manipulated to give

$$t_f = BS_i^{n-2}\sigma^{-n} \tag{2.4}$$

where $t_f$ = time to failure
$S_i$ = inert strength (i.e., stress required to produce instant failure or failure in absence of any crack growth)
$B$ = constant related to both $A$ and $n$
$\sigma$ = applied stress

The time-to-failure is thus inversely proportional to the $n$th power of the applied stress or strain. Values of $n$ usually lie in the range of 14 to 25. The lowest value corresponds to 100 percent relative humidity, and the higher values correspond to dry laboratory conditions. With special precautions, much higher values of $n$ can be obtained.

The time-to-failure is therefore extremely sensitive to changes in both $\sigma$ and $n$. As a result, it is important to obtain accurate values for $n$ and to demonstrate that this power law can be extrapolated to a cable service life of 20 years. Both are difficult; because of the large statistical spreads, many kilometers of fiber must be tested to determine $n$ accurately for any given set of conditions, and recent evidence has shown that there are more complex, time-related changes.[10] Further evidence of the effects of more complex changes in coated fibers subjected to high humidity can be seen in Fig. 2.3, where—contrary to expectation—the smallest flaws (found in the strongest sections of the fiber) actually seem to be strengthened after exposure to high humidity at negligible stress. This is due to blunting of the crack tip by corrosion of Si-O to SiOH. (See Chap. 1, Sec. 1.73, for further discussion of crack growth.)

Because of these uncertainties, the fracture mechanics described must still be used with extreme care. However, there is a way of avoiding the use of fiber con-

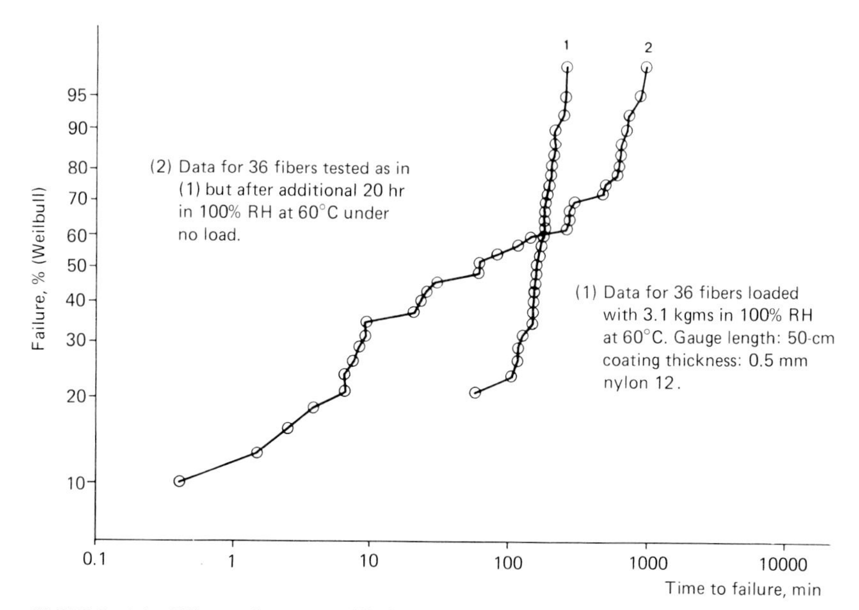


**FIGURE 2.3** Effect of water diffusion through fiber coating on static fatigue. *(Reproduced with permission of STC Technology Ltd.)*

taining the largest flaws, and that is to subject all fiber to a critical load or strain before use and to use only lengths which survive this test, known as a *proof test.* Thus major flaws can be removed. And, because the time of the proof test is short (typically around 1 s) and the conditions are under control, fracture mechanics can be used to examine what is happening during the proof test.

Analysis of cumulative probability of failure plots, such as that in Fig. 2.2, indicates that flaws can be split into two families. In the high-strength region, there is a family with a narrow distribution probably associated with the nature of the surface of the pulled silica fiber. The lower skirt of this distribution runs into the family of larger flaws. Here the number of flaws of any size is a continuous function of the applied strain for instantaneous fracture. For simplicity, let us assume that the number of flaws is directly proportional to the fracture strain. There is some circumstantial evidence that this may be close to the truth, but assuming any other continuous function does not alter the following argument, since there will be only small errors in the final probabilities. With this assumption, the plot in Fig. 2.4 can be used with the vertical scale taken either as the inert strength (fracture strain to cause an instantaneous break) or as the number of flaws. This figure shows diagrammatically the relationship between inert strength and proof-test level for our standard test: 0.6 percent strain applied for 0.6 s. For simplicity, straight lines have been drawn between the calculated points in the diagram, to trace the growth of any flaw. In reality, these lines should have an exponent of 25, since it has been assumed that $n = 25$, a likely value for a laboratory test. The diagram also shows what happens if the test is applied a second time. Now, if the number of breaks $N$ that have occurred up to the end of the proof test is known, this is equivalent to a 1.205 percent strain. The number $N_2$, which will occur on the second proof test, is equivalent to $(1.241 - 1.205)N = 3.6$ percent $N$. Let us say, for the sake of argument, that 28 breaks have occurred in the manufacturing and testing of 100 km of fiber. If this fiber is retested, only one break would be expected (i.e., $N_2 = 1$).

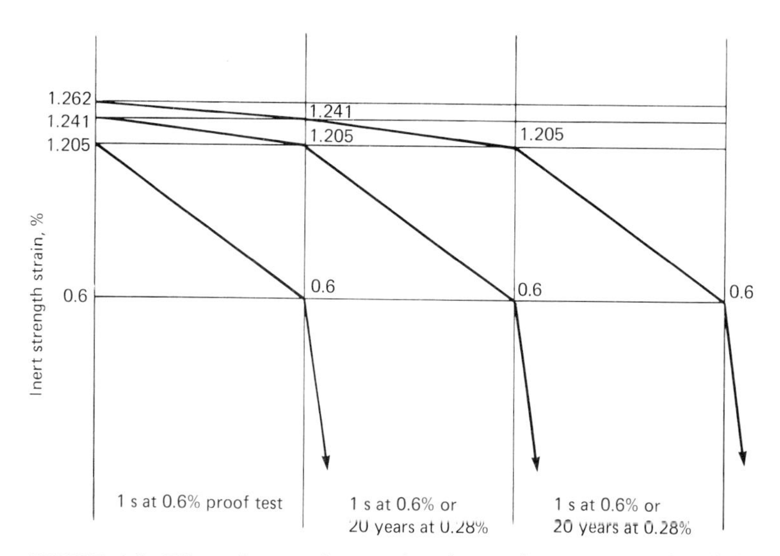


**FIGURE 2.4** Effect of successive proof testing on inert strength of optical fiber. (*Reproduced with permission of STC Technology Ltd.*)

Figure 2.5 shows a rather more informative plot of a similar analysis. An idealized—but slightly more sophisticated—distribution of flaws is assumed, and probabilities of failure are calculated for different proof test strains. As in Fig. 2.4, an inert strength of 0.6 percent strain is shown after a 0.6 percent proof test "guaranteeing" fiber strength. However, the strength at even a low probability of failure is much higher—1.4 percent strain for 0.1 percent failure probability for the length of fiber represented in Fig. 2.5. This illustrates the very large margins that can be gained if a low probability of failure is acceptable, rather than an absolute guarantee of survival.

So far in this discussion it has been assumed that the applied proof stress or strain is removed instantaneously. This will never happen. Allowance must therefore be made for a possible increase in crack size when the stress for strain is reduced. If this reduction is rapid, it introduces only a very small—but not infinitesimal—probability that subsequent fiber failure may occur at stress or strain levels below those expected from the proof test.

In summary, if a proof-strain level is specified, then a guaranteed minimum breaking strain is set. If the duration of the proof test is also specified and the

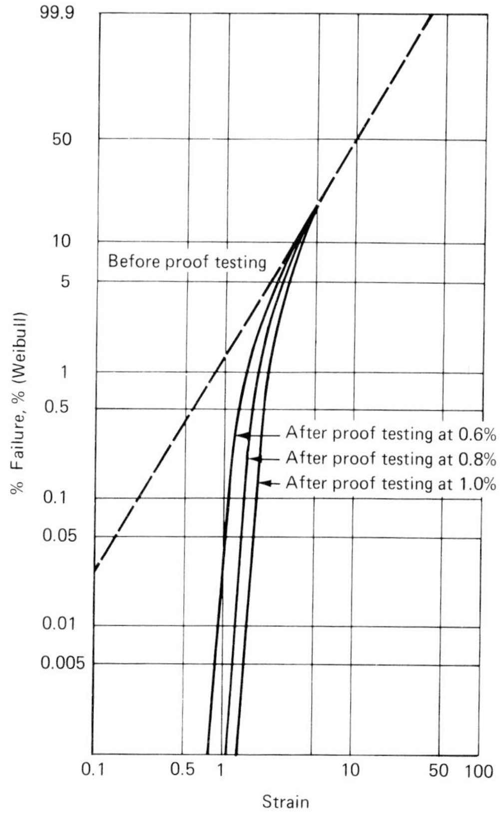


FIGURE 2.5 Inert strength before and after proof testing. *(Reproduced with permission of STC Technology Ltd.)*

number of breaks found during manufacture and test is known, then the probability of failure can be calculated for a similar strain and time.

During service, however, much lower strains are likely to be encountered for much longer times. Here the new complexities in static fatigue, which are still being discovered, complicate the picture. Nonetheless, a combination of theory, experience, and safety margins (to cover uncertainties) can be used to facilitate the design, manufacture, installation, and use of fiber-optic cables. From Eq. (2.4), it can be shown that stress and time-to-failure are related:

$$\frac{t_1}{t_2} = \left(\frac{\sigma_2}{\sigma_1}\right)^n \tag{2.5}$$

where the subscripts refer to two corresponding sets of time and stress. In this way, differing times and stresses can be referred to the proof test conditions:

$$t_p = \sum t_x \left(\frac{\sigma_n}{\sigma_p}\right)^n \tag{2.6}$$

where $t_p$ and $\sigma_p$ refer to the equivalent time at proof test conditions and the proof stress, respectively. By building up a complete scenario of the fiber life, and referring each stage to standard proof test conditions, a probability of failure can be calculated from results such as those depicted in Fig. 2.5.

For many industrial and military applications, high stresses may be introduced by sharp cable bends. Often it is quite impractical to raise the proof level to the point where survival at such bends can be guaranteed. Fortunately, in most installations, the fiber surface area subjected to tensile stress at such bends is very small, and reference to Fig. 2.2 shows that, by far, the major portion of the fiber will be sufficiently strong. The Appendix shows that the survival probability at such bends can be derived from tensile data by using a logarithmic transform of the Weibull ordinate of cumulative probability of failure to relate one stress to another. Such calculations can be used to determine a minimum bend radius that will ensure a chosen probability of survival.

### 2.1.2 Cable Design Principles

With this background, we are now in a position to summarize the design principles that must be applied when optical fibers are to be incorporated in a cable. First, for any particular fiber waveguide, there is a minimum radius of curvature that must be exceeded under all conditions of use. In general, it is comparatively easy to ensure that the designed radii of curvature will always considerably exceed this minimum; however, Sec. 2.3.1 presents a particular sample design with complex geometry in which the minimum radius of curvature is not intuitively obvious. However, it is the inadvertent microbends introduced by the cabling process that are of paramount importance. The steps by which these can be minimized will form a major topic in the following sections.

It is also essential, if a long service life is to be achieved, that the limitations imposed by stress corrosion (static fatigue) be understood. Although the time-to-failure can be calculated, as outlined, it must be borne in mind that this time is inversely proportional to the $n$th power of the applied stress or strain and that values of $n$ can lie anywhere in the range of 14 to 25 for normal conditions. The time-to-failure is therefore very sensitive to changes in either $n$ or $\sigma$. It is important both to know $n$ and $\sigma$ accurately and to demonstrate the validity of the power law for cable service life periods. Both are difficult, and there appear to be more complex, time-related changes that are not accounted for by the simple inverse-power theory.[11] It is therefore prudent to apply this theory with adequate engineering safety margins and to err on the side of caution. Having stated this, one must admit that the number of fiber breaks reported, to date, during the installation of cable in the telecommunications market is far lower than the simple theory predicts.

## 2.2 FIBER BUFFERS

If fibers are to be used successfully in current cable designs, it is essential that their pristine strength be preserved. This requires the application of a protective coating (buffer) as soon as practicable and within a clean environment, in such a way that the fiber surface is not scratched or marked. The coating must be applied before the fiber touches any other surface, and it must protect the fiber from abrasion during subsequent processing. Because the coating application is an intimate part of the fiber-drawing process, constraints imposed by fiber drawing play a very large part in the choice of coating (or coatings). However, the coating also plays an essential part in the design of the cable as a whole.

In order to meet present pulling and cabling needs, the coating(s) should provide:

1. Fast application and rapid cure
2. Uniform application and concentricity
3. A contribution to microbending protection
4. Fiber strength preservation
5. Long-term durability in a wide range of environments

The first requirement is entirely one of fiber production economics. The second is for complete protection and preservation of the fiber strength and to provide the symmetrical coating needed as part of the microbend protection design. The third requirement is essential to minimize any inequality in the forces acting on a fiber due to contraction of other cable elements during manufacture or at low temperature. (At temperatures below −20°C, many candidate materials show marked changes in modulus and can become quite brittle. This significantly alters their contribution to the mechanical filtering essential to overcome microbending.) The fourth, strength preservation, is not a simple matter and depends on the combination of longitudinal tensile stress applied to the fiber and the amount of corrosive element (OH) present at the fiber surface. (Hermetic coatings to protect the fiber are often considered, but generally play little part in other aspects of cable performance.) The fifth requirement is intended to assure broader product application.

### 2.2.1 Polydimethyl Siloxane

Possible materials for on-line coating have been reviewed by Lawson.[11] Thermally cured polydimethyl silicones (e.g., Sylgard) were the most successful of the early coatings. In comparison with uncoated fibers, they offer a marked increase in usable fiber strength, they have a very small temperature dependence (their moduli remain largely unchanged from −60 to +85°C), and they remain largely usable at high temperatures. Their durability is good, and their low modulus properties provide a useful element in any mechanical filter system. Their major deficiencies—limited application rates, long curing time, and short pot life—are all deficiencies with regard to fiber pulling, rather than cable making. For a long time, they remained the coating by which others were judged. Ultraviolet-cured silicones based on polydimethyl siloxane avoid many of these deficiencies and still find extensive use.

### 2.2.2 Silicone Oils

Silicone oils, applied in thin films of less than 5-μm thickness, provide useful lubricity for subsequent cabling processes, but have poor durability and do not contribute to microbending protection. Various lacquers can be applied at fast rates in fairly thin films, but the resulting fibers do not show high strength and, again, these films contribute little to microbending protection.

### 2.2.3 Extrudates

A variety of fluoropolymer and polyester extrudates have been investigated as primary coating buffers. These are difficult to apply on-line with fiber pulling—at a stage before the fiber has passed over any capstans or pulleys—and have therefore not given high fiber strength. They have, however, proved very useful as secondary coatings and, as such, will be discussed later.

### 2.2.4 Acrylates

The use of acrylate resin coatings was pioneered by workers in the Bell System.[12] These are now finding widespread acceptance because they offer considerable advantages in fiber drawing speed and economics. They also offer more versatile buffering techniques to the cable manufacturer.

A major advantage springs from the versatility of the chemistry that can be used to cure acrylates.[13] Curing is based on the free-radical polymerization of unsaturated vinyl (double bonds), using combinations of acrylate monomers and higher molecular weight oligomers. In this way, it is possible to formulate durable coatings with a wide range of moduli. For convenience in fiber pulling, it is usual to employ materials tailored for uv curing.

A well-favored system employs a low-modulus inner coating with a high-modulus outer coating, giving a package that provides a tough, smooth exterior coating capable of withstanding subsequent handling and abrasion, with an excellent contribution to protection against microbends.[14] Some fiber manufacturers favor a single, hard acrylate coating as, in general, fibers with this protection give a higher yield during proof test. Use of such fibers means that protection against microbending must be provided elsewhere in the cable structure.

The versatility of acrylate coatings offers the cable manufacturer several advantages. On the one hand, they offer excellent adhesion to the fiber surface—giving high yields at high proof-strain levels—with the possibility of removal by organic solvents. On the other hand, they can be formulated to give minimal bonding to the fiber to facilitate rapid and safe mechanical stripping. They offer low-modulus coatings for excellent microbend protection or high-modulus coatings for excellent protection in loose-tube and other cable constructions. They offer a very low free surface energy (to give a low coefficient of friction with other cable elements) or good adhesion characteristics, when overlaid with extrusion coatings such as nylon, polyester elastomers (Hytrel) and fluorinated polymers (Halar, Tefzel, etc.).

## 2.3 ENCAPSULATION

Usually—but not invariably—some form of encapsulation is used to complete the packaging of the optical fiber. An exception is the double, on-line coating (described in Sec. 2.2.4), which can be used with strength member and sheath to complete the cable. But such applications are specialized and are applicable only where the two on-line coatings have been designed to form a mechanical filter, following the principles laid down by Gloge and described in Sec. 2.3.3.

In designing the encapsulation, the cable engineer should have in mind the two main tasks:

1. To complete the mechanical filter, giving the fiber protection from microbending induced during subsequent cabling operations by other cable elements (thus minimizing incremental increases in attenuation)
2. To combat stress corrosion by introducing some degree of strain relief

### 2.3.1 Loose-Tube Technique

Perhaps the most obvious technique for protecting the primary-coated fiber is to incorporate the fiber loosely in a tube, as in Fig. 2.6, relying on an air gap to isolate the fiber from any external forces. Early experiments showed catastrophic attenuation increments when the temperature was lowered, however. Attention was given to matching the thermal coefficients of fiber and tube by draw-down techniques with polypropylene or terephthalate (Arnite), so that very low coefficients could be obtained over a limited temperature range. This technique has been made to work well under conditions in which the cable is not subjected to cyclic loading, handling, vibration, or temperature fluctuations outside the tube's stable range.[15]

Unfortunately, the inside surface of the tube is never smooth, and if, as a result of the cable's environment, pressurized contact arises between the fiber and the inside surface of the tube, losses due to microbending can occur. It is quite probable, in view of the material and its mode of extrusion, that the mechanical spectrum of the tube's inside surface will contain substantial amplitudes at and around the critical beat length for mode coupling. Obviously, if long vertical runs of these cables are installed, the fiber will move and attenuation increments will result.

If, during coating, it is arranged that the fiber is wound at a large radius of curvature, a certain amount of strain relief is obtained in the final cable.[16] This arises because, at zero cable extension, the fiber can be made to lie on a helix of greater radius than that of the tube axis. But with increasing cable strain, it

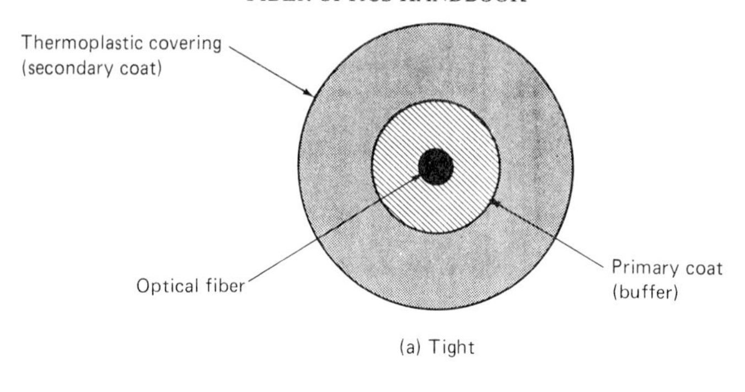


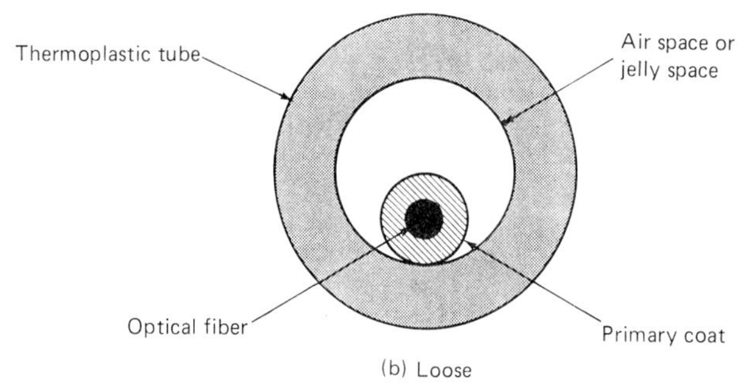


**FIGURE 2.6** Comparison of fiber encapsulation by (*a*) tight-buffer and (*b*) loose-tube techniques (cross section). (*Reproduced with permission of STC Technology Ltd.*)

moves to a helix of smaller radius. The disposition of the fiber and the relationship of fiber strain and cable strain is illustrated in Fig. 2.7. The maximum cable strain $\Delta L/L$ that can be incurred without fiber strain is given by

$$\frac{\Delta L}{L} = \frac{(S^2 + \pi^2 d_f^2)^{1/2} - (S^2 + \pi^2 d_i^2)^{1/2}}{(S^2 + 4\pi^2 R^2)^{1/2}} \tag{2.7}$$

where $L$ = cable length
$d_f$ = maximum pitch-circle diameter for fiber axis
$d_i$ = minimum pitch-circle diameter for fiber axis
$R$ = tube-axis pitch-circle radius
$S$ = pitch length

Typically this amounts to about 0.2 percent for cables where $R$ is about 5 mm, but can be as high as 0.4 percent.

Greater strain relief can be obtained by using much larger loose tubes. These are difficult to lay helically at a small radius of curvature, but, with a large enough tube-to-fiber diameter ratio, the fiber can be laid helically within the tube by feeding into the tube a length of fiber that is longer than the axial length of the tube. This is known as *overfeeding*. If the tube is laid along the cable axis, the relationship between the tube-to fiber diameter ratio, helical angle of lay, radius of curvature, strain relief, etc., can be determined from the geometry. The additional length of the helical path in relation to the axial length is given by

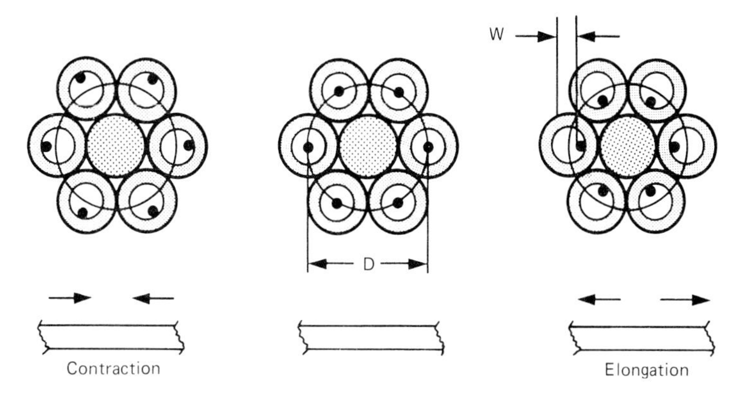


(a) Fiber location in loose tube

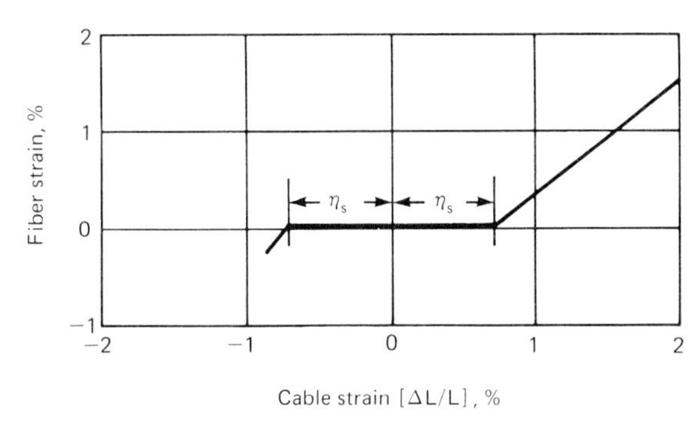


(b) Strain-free window

**FIGURE 2.7** Fiber strain as a function of cable strain for a six-fiber stranded cable. (*a*) Illustration of fiber location with elongation and contraction of the cable; (*b*) plot of fiber strain as a function of cable strain. $\eta_S$ is the allowed elongation/contraction window as determined by clearance $W$ and pitch-circle diameter $D$. (*Siecor Corp.*)

$$\frac{\Delta L}{L} = \frac{(2\pi R)^2}{2S^2} \tag{2.8}$$

This may be expressed as a percentage (the usual way of expressing strain relief):

$$\frac{\Delta L}{L} = \frac{50(2\pi R)^2}{S^2} \quad \% \tag{2.9}$$

In the same circumstances, one can also calculate the radius of curvature of the fiber $r$:

$$r = R\left[1 + \frac{S^2}{(4\pi R)^2}\right] \tag{2.10}$$

Alternatively, one can calculate $r$ in terms of the lay angle $\theta$ rather than lay length:

$$r = \frac{R}{\sin^2 \theta} \tag{2.11}$$

By substituting for $S$ from Eq. (2.10), one can plot the fiber radius of curvature for the amount of strain relief in a given internal tube diameter, bearing in mind that the internal tube diameter must equal the sum of the helical pitch diameter and the fiber-coating diameter.

Experience has shown that there may be difficulties if the designer is too ambitious in trying to achieve large strain relief, but—properly used—the technique can give good results. It is particularly useful with high-fiber-count cables, as discussed in Sec. 2.8.

One way of increasing fiber strain relief is to employ a cable in which fibers are laid helically within tubes that are also laid helically. This construction provides the potential of obtaining strain relief as indicated in Eqs. (2.7) and (2.9). The fiber in such a cable now lies as a helix on the surface of a toroid having a helical axis (as in Fig. 2.8). The geometry pertaining to such a fiber is not immediately obvious. However, the required parameters, such as minimum radius of curvature, can be obtained by numerical analysis from the cartesian coordinate system for such a helix.

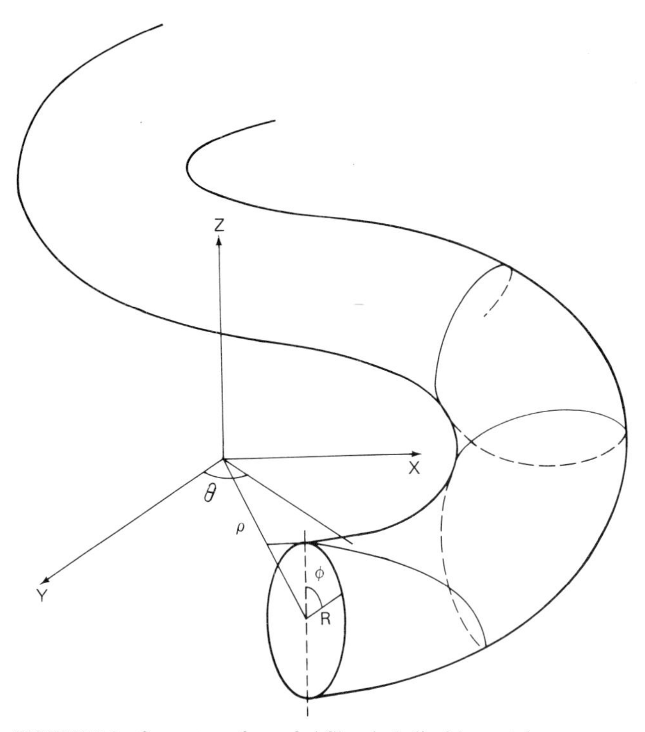


**FIGURE 2.8** Geometry of overfed fiber in helical loose tube.

This geometry can be described in much the same manner used to describe steel ropes and armoring in the conventional cable industry:

$$x = (\rho + R \cos \psi) \frac{\sin k\psi}{\rho} \tag{2.12a}$$

$$y = \rho\left(1 - \cos \frac{k\psi}{\rho}\right) + R\left(1 - \cos \psi \cos \frac{k\psi}{\rho}\right) \tag{2.12b}$$

and

$$z = R \sin \psi \tag{2.12c}$$

where $k = l/2\pi$
$l$ = lay length of fiber helix
$\psi$ = angular cylindrical coordinate of fiber in tube
$\rho$ = radius of curvature of tube
$R$ = radius at which fiber axis lies within tube

The angular variation along the length of the toroid is given by

$$\theta = \frac{k\psi}{\rho} \tag{2.12d}$$

### 2.3.2 Gel-Filled, Loose-Tube Technique

Some of the disadvantages of loose-tube constructions, such as difficulties in water blocking and long-term fiber movement, can be overcome by introducing a gel between the fiber and the tube.[17] However, if most of the advantages of the loose-tube technique are not to be lost, the fiber must still be able to move within the tube. In this respect, a filling compound with a low viscosity would be preferable. But such a material would not be mechanically stable in a cable structure. On the other hand, a high-viscosity material would impair the fiber's ability to move rapidly and, thus, to respond quickly to changes in cable stress. Such changes in cable stress can occur suddenly during installation. It is also desirable that the filling compound remain stable at temperatures that may be met in subsequent manufacturing processes, as well as through the temperature range of the cable's working environment.

Traditionally, these requirements were met somewhat inadequately by using petroleum jelly, but new ranges of filling compounds have now been developed. Some of these are based on petroleum jelly, modified to improve its characteristics. Others are based on polybutene. In general, such compounds must be raised to temperatures in excess of 100°C, if the loose tubes containing optical fibers are to be filled. This can generate problems, such as restrictions in the choice of other coating materials and the possibility of voids forming because of thermal contraction while the cable cools down to ambient temperature.

To alleviate these problems, a number of thixotropic gel compounds have been developed.[18] Typically, these compounds consist of a synthetic oil base with mineral fillers, gel stabilizers, antioxidants, and other additives. The gels developed for fiber-optic applications have been formulated to give stability over temperature ranges as great as −55 to +160°C. They also have very fine granular structure and excellent homogeneity to ensure that the filling compounds them-

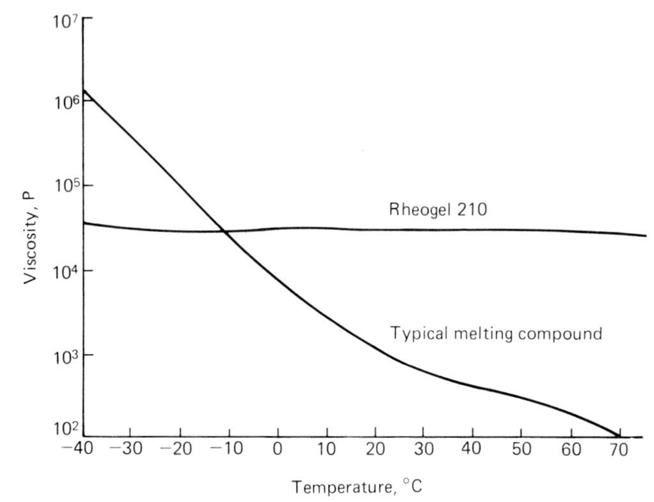


**FIGURE 2.9** Viscosity vs. temperature for Rheogel 210 (comparative). *(Reproduced with permission of STC Technology Ltd.)*

selves do not introduce microbending loss. The viscosity variation with temperature is shown for one such compound in Fig. 2.9. The considerable reduction in viscosity for this compound, when subjected to pumping shear, is shown in Fig. 2.10. Note not only the reduction in viscosity with shear rate in Fig. 2.10 but also the considerable reduction from the viscosity in Fig. 2.9, which was determined at a negligible shear rate.

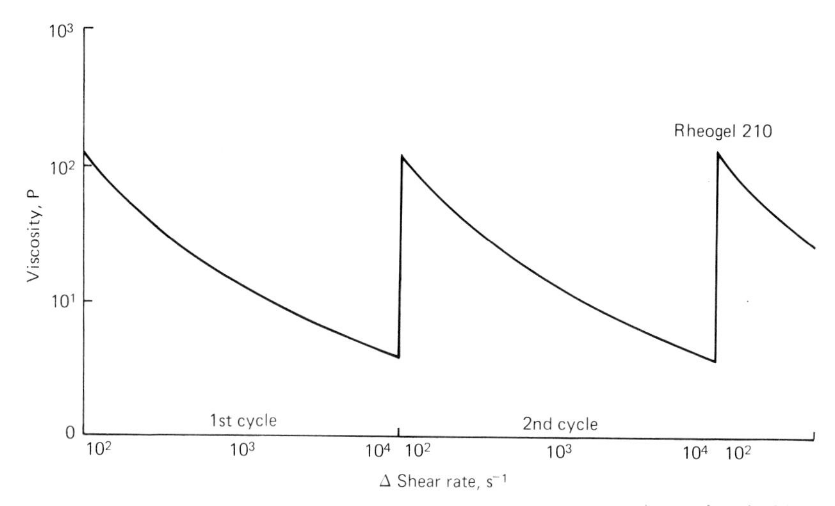


**FIGURE 2.10** Effect of pumping shear on viscosity of Rheogel 210. *(Reproduced with permission of STC Technology Ltd.)*

Because of its low viscosity at high shear forces, it is comparatively easy to inject this gel during conventional extrusion of the loose containing tube. When injection is complete and the shear force has been removed, the gel returns to its high-viscosity state. Similarly, during cable elongation or compression, movements of the fiber can occur rapidly, when the threshold has been overcome. But the gel is not disturbed by the fiber movement, and again returns to its high-viscosity state when the stress is removed.

With such compounds, most of the disadvantages of (unfilled) loose-tube construction can be overcome. Cables have been used, for example, with vertical runs of several kilometers without the slightest sign of the filling shifting.

### 2.3.3 Tight-Buffer (Multilayer-Coating) Technique

Conceptually, multilayer coatings, since they have an inner low-modulus coating surrounded by an outer higher-modulus coating (Fig. 2-6*b*) can be considered a logical progression from the gel-filled, loose tube. Geometrically, they look similar. Historically, they came into favor long before loose tubes. Early mechanical analysis showed the considerable benefits that could be obtained with multilayer coatings.[19] Correctly designed, such coatings offer effective mechanical isolation, with negligible incremental attenuation. This analysis was applied to optical-fiber packaging in cables[20] and gives a complete picture of the mechanical effects of microbends and the complex relationships between applied stress on a microscale and perturbations of the waveguide axis, which lead to attenuation increments as outlined in Sec. 2.1.

When there is continuous contact between a fiber and its surrounding buffer and encapsulation, the following relationship (based on Gloge's analysis,[19] which itself is based on the theory of thin elastic beams) applies:

$$\frac{2H}{D}\frac{d^4x}{dz^4} + 2x = \nu_1 - \nu_2 - \nu_3 + \nu_4 \tag{2.13}$$

where $H$ = modulus of rigidity
$D$ = effective modulus of rigidity of outer surfaces
$\nu_1, \ldots, \nu_4$ = statistics of surface variations of fiber, its coatings, and neighboring surfaces
$z$ = distance along fiber axis
$x$ = displacement of axis

Gloge plotted the incremental loss for a 120-μm multimode fiber, with an 80-μm core as a function of encapsulation radius for a hard coating (modulus = 100 kg/m$^2$), soft coating (modulus = 1 kg/m$^2$), hard coating outside a soft coating, and soft coating outside a hard coating. Figure 2.11, from Gloge's paper, shows the advantages of two-layer coatings.

The fiber chosen by Gloge had a geometry that made it susceptible to microbend loss. However, it is possible to scale the loss to correspond to other graded-index, multimode fibers, following an analysis derived by Olshansky.[21] Olshansky used a simpler model, based on the optical effect of a single perturbation rather than a statistical variation of the surface. For this simpler model, he was able to derive a relationship for the excess loss for a standard, graded-index, multimode fiber. The excess loss $\gamma$ is given by

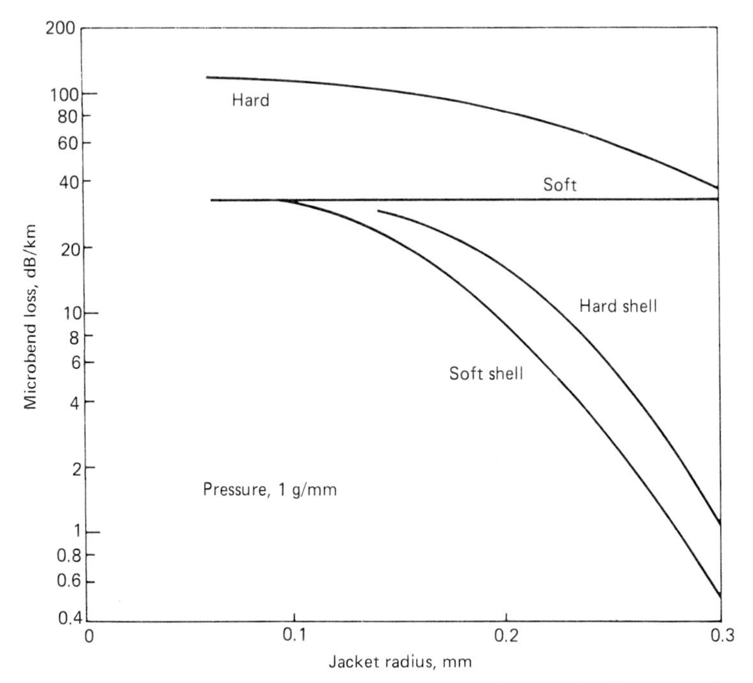


**FIGURE 2.11** Microbend loss vs. outside coating diameter for 80/120-μm fiber. *(After Gloge.[19])*

$$\gamma = Kd^4D^{-6}\left(\frac{2n_1^2}{(n_1^2 - n_2^2)}\right)^3 \qquad \text{dB/km} \tag{2.14}$$

where $K$ = normalizing constant
$D$ = fiber diameter
$d$ = core diameter
$n_1$ = core refractive index
$n_2$ = cladding refractive index

Practical tests of this relationship show very good correlation with experimentally determined attenuation increments. An exception is that $K$ reduces dramatically for cables where attenuation is extremely low (which is consistent with Gardner's analysis[20]).

For practical reasons, a hard coating outside a soft coating has been preferred. This provides a package with good abrasion resistance, one that is not easily damaged in transit or subsequent cabling operations. An early configuration that was much favored used a buffer coating of polymethyl siloxane (Sylgard), between 50 and 100 μm in thickness, surrounded by an extruded polymer coating to an outside diameter between 0.5 and 1.0 mm. Materials used for the outer coating have included polyethylene, polyethylene terephthalate (Arnite), polyester elastomers (Hytrel), nylon 11, nylon 12, and fluorinated polymers (Halar, Tefzel).

Exactly the same analysis can be applied to the double on-line coatings described in the previous section and for the same practical reason: a hard layer around a soft, inner layer is the preferred configuration.

If any of these packaged fibers is stranded helically, as in loose-tube cables, some strain relief will arise from movement of the fiber within the low-modulus inner coating. Unlike thixotropic gels, however, the modulus is always nonzero, so the strain relief is much more limited. The permissible fiber movement is usually more limited, as well. This is not the sum total of obtainable strain relief, however. It is possible, during application of the secondary coating, to apply longitudinal compression to the fiber, a strain offset. This situation and its controlling mechanisms have been thoroughly investigated for nylon 12, as comprehensively reported by Barnes et al.[22] Typically, compression of around 0.25 percent is obtainable, which is within the strain relief range of loose-tube packaging. Similar compressions, ranging from 0.1 to 0.3 percent, are obtainable with most of the other materials listed.

The reasons why some fiber manufacturers prefer to use a single fairly high modulus epoxy acrylate coating were outlined in Sec. 2.2.4. Referring again to Fig. 2.11, it can be seen why such a coating works well in a thixotropic-gel-filled loose tube: the combination of acrylate and gel forms the pair labeled *soft shell* by Gloge. However, if such a fiber is coated with a comparatively high modulus coating, such as nylon 12, it gets only the protection of a single thick, hard layer, which is very much less than that which a double coat provides. If, alternatively, a soft buffer is applied over the hard inner coating, and then a polymer is extruded over the two as a final encapsulation, much better microbend protection can be obtained. This is because the inner two layers provide the protection of a soft shell, and the outer two provide the protection of a hard shell. If the middle buffer layer is sufficiently thick to accommodate the statistics of all the coating surfaces, the two mechanical filters are decoupled, and the two filtering processes become additive.

### 2.3.4 Multifiber Coatings

Sometimes it is necessary to build a cable with very high fiber density, such as 100 fibers within a cable outside diameter (OD) of less than 20 mm. To handle this case, it is possible to encapsulate several fibers at a time within a common secondary coating. It is comparatively straightforward to insert, say, 12 on-line-coated fibers within a loose tube having an inside diameter (ID) of less than 2 mm. With gel filling, this technique has produced very satisfactory, high-fiber-count cables. Two other configurations are also favored: one based on a ribbon design and the other based on fibers helically wrapped around a strength member. In the ribbon construction, 5, 10, or 12 primary-coated fibers are laid alongside each other within a single ribbon coating. This technique has been extensively developed at AT&T (Bell Laboratories and Western Electric),[23] which has reported improvements over a period of some years.[24] Figure 2.12 shows a typical ribbon construction. Incremental losses are reported at 0.0 ± 0.1 dB/km during ribboning, and 0.1 ± 0.1 dB/km during subsequent cabling operations, with a fully filled cable. Many other versions with such coatings have been reported subsequently.

The second configuration has received less attention but offers similar potential. This compact unit construction, illustrated in Fig. 2.13, was first published in

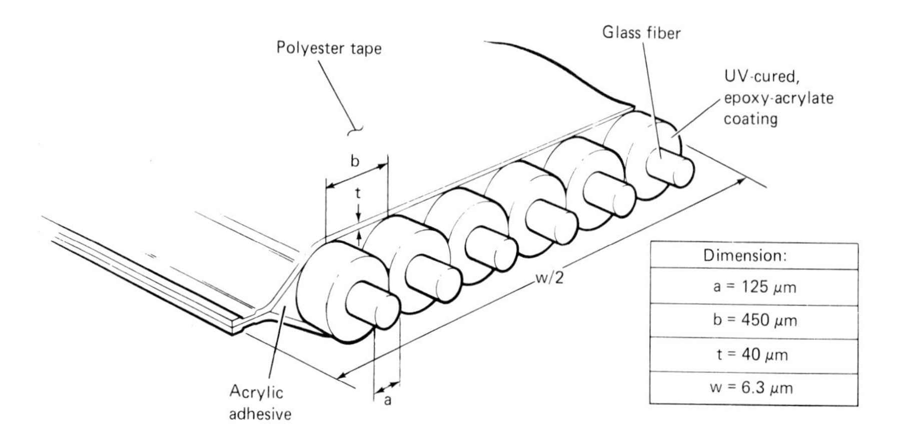


**FIGURE 2.12** Cross section of a 12-fiber, adhesive-sandwich ribbon coating. *(Reproduced with permission of STC Technology Ltd.)*

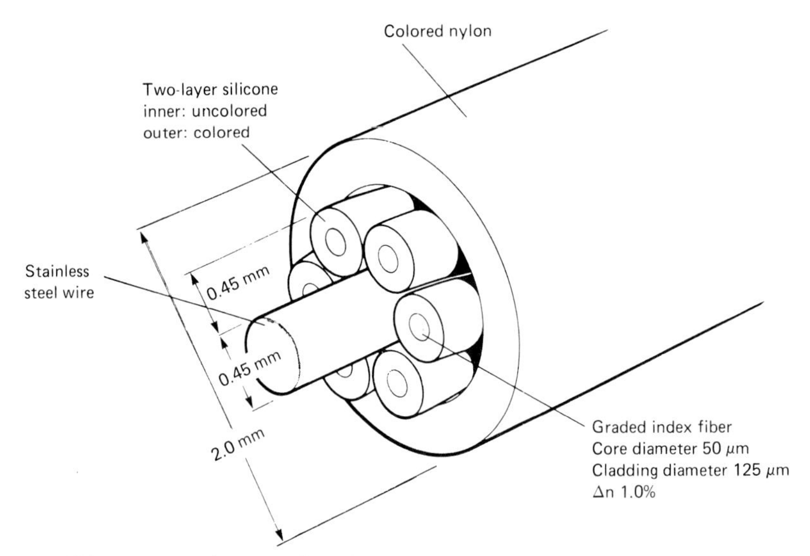


**FIGURE 2.13** Cross section of a six-fiber high-density cylindrical coating. *(Reproduced with permission of STC Technology Ltd.)*

1979,[25] with further progress being reported soon thereafter.[26] With this configuration, cabling losses of 0.1 dB/km can be achieved with over 200 fibers in a 21-mm-OD cable.

## 2.4 STRENGTH MEMBERS

Silica is characterized by a high Young's modulus and a very large elastic range. However, it is brittle, and the considerations outlined in Sec. 2.1.2 indicate why this operation must be confined to comparatively small strains, if long life is to be assured. Moreover, the total cross section of fibers in a cable is generally small, so their contribution to overall cable strength is low, despite their high modulus. It is therefore usual to incorporate high-modulus elements as strength members.

In conventional cables with copper conductors, the conductors act, in many cases, as the strength members and may be strained to several percent during extreme loading. In this case, all of the cable elements accommodate the strain by plastic deformation. Because optical fibers are subject to static fatigue, as discussed in Sec. 2.1.1, it is essential that all cable elements be used within their elastic range to avoid any permanent elongation, which would drastically shorten fiber life. Therefore, the strength member should have the following:

1. High Young's modulus
2. Greater strain at yield than the maximum designed cable strain
3. Low weight per unit length
4. Flexibility

Other properties may also be relevant, including stability of properties over a wide temperature range, high compressive modulus, and low or zero conductivity.

At least five types of material have been considered in the construction of strength members—all characterized by a high elastic modulus:

1. Steel wires
2. Plastic monofilaments
3. Textile fibers
4. Glass fibers
5. Carbon fibers

Of these, very little has been done with carbon fibers, since their cost in long lengths is high, even in relation to the most exotic fiber waveguides. And, although aligned plastic monofilaments were extensively used in the early days, difficulties with wide-temperature-range performance have limited their application. Three main classes of strength member therefore remain for consideration, as follows.

### 2.4.1 Steel Wires

Both carbon steel and stainless steel have been widely employed for many years in the cable industry, both as strength members and for armoring. They are readily available and offer high ultimate strength at low cost. The various grades available range in tensile strength from 0.5 to 3.0 GN/m$^2$, with an inverse range in strain at break. The Young's modulus is very similar for all carbon steels and only slightly less for stainless steel (20 GN/m$^2$). The choice, therefore, becomes one of ensuring that the elastic range is sufficient to meet all eventualities during cable manufacture and life.

Because both tensile and compressive moduli are high, single steel wires are very stiff. Therefore, with a central strength member, it is usual to employ stranded constructions in all but the very smallest cables. It is necessary to ensure that the strand design does not introduce a reduction of modulus at low strain or unacceptable torsional effects under load. Fortunately, strands with the desired properties are readily available. A notable exception is cables in which strength members are incorporated into the outer sheath—a design derived from armoring practice in conventional cables. A combination of helical lay of strength members and the use of multiple thin wires gives an acceptable flexibility for most telecommunications applications.

Steel does have two potential disadvantages as a strength member for optical cables:

1. It cannot be used in applications where the dielectric nature of the waveguide will be fully exploited to ensure a nonconducting cable.
2. Its high specific gravity substantially increases cable weight.

### 2.4.2 Textile Fibers

The conventional cable industry has made use of yarns and rovings of many filamentary fibers of different materials. Typically, these filaments are about 10 μm in diameter and are laid up in twisted or parallel configurations. The filament molecules can be highly oriented during manufacture to give high longitudinal elastic modulus and strength. Both polyamides (nylons) and polyethylene terephthalate (Terylene, Dacron, etc.) have been used extensively for cable reinforcement. Because of their filamentary nature, they are useful as cable fillers and cushioning elements. Despite their comparatively high modulus (about 100 $GN/m^2$), however, they are seldom used as strength members because they exhibit significant creep under load.

A notable exception is Kevlar, an aromatic polyester developed by the DuPont Company. This is available in two forms, but it is the high modulus form, Kevlar 49, that is of most interest to the optical-cable engineer. Basically, Kevlar 49 is produced as a 12-μm-diameter filament with a specific gravity around 1.45, a tensile strength greater than 2.7 $GN/m^2$, and a tensile modulus greater than 120 $GN/m^2$. These filaments show a creep rate under load that is little more than that of steel wire. Kevlar 49 is available either as untreated yarns or as cords, in which the filaments are embedded in a carrying epoxy or other potting compound or size.

It would be unrealistic to expect a packing density greater than 80 percent for yarns in a normal cable construction. Similar or lower densities are achieved with cords. However, with a modulus nearly as great as that of steel and a specific gravity about one-fifth that of steel, Kevlar is a very attractive strength member. Its properties are remarkably uniform over a wide temperature range and it shows only a small, negative coefficient of expansion.

Kevlar's principal disadvantage arises from its very low compressive modulus, which is some 3 orders of magnitude less than its tensile modulus and—because of the small filament diameter—is totally negligible when it is used in yarn form. Thus Kevlar can provide excellent cable flexibility, but at the expense of providing no protection against fiber buckling. Fiber buckling can occur at low temperatures and is discussed in Sec. 2.7.1.

### 2.4.3 Glass-Reinforced Plastics

Many different types of glass-based strength members are available. The fractional quantity of glass within each structure varies considerably, as does the binding medium and overall construction method. The binding medium, which is generally a polyester or epoxy resin, determines the usable temperature range.

The modulus of a glass-reinforced plastic strength member is largely determined by the type of glass used for reinforcement and the glass fill factor. S-glass, with a tensile modulus of 72 GN/m$^2$, and E-glass, with a modulus of 76 GN/m$^2$, are the most common types of glass used. S-glass members have the lowest coefficient of thermal expansion, but most glasses are below the coefficient for steel. In comparison with textile fibers, all glass fibers show extremely low creep. In addition, the compressive modulus is usually comparable to the tensile modulus, giving good protection against buckling at low temperatures.

Because a multiplicity of fibers are used, the elastic range is usually between 2 and 3 percent, showing excellent linearity up to the breaking point. In use, care must be exercised so that cable bending radii do not fall to the point where the glass strength members break. A good cable design will ensure that this is not likely to happen.

### 2.4.4 Strength Member Summary

The main features of the strength members described above are summarized in Table 2.1.

**TABLE 2.1** Summary of Strength Member Features

| Strength member | Young's modulus, GN/m$^2$ | Elastic range, % | Specific gravity | Coefficient of thermal expansion, per °C | Temperature range, °C |
|---|---|---|---|---|---|
| Steel | 210 | 2.0–2.0 | 7.8 | $1 \times 10^{-5}$ | $<600$ |
| Kevlar 49 yarn | 120 | 2.2–2.8 | 1.45 | $-2 \times 10^{-6}$ | $<250$ |
| Kevlar 49 cord | 60 | 2.3–2.5 | 1.1 | $-1 \times 10^{-6}$ | $<180$ |
| Glass-reinforced plastic | 45–60 | 3 | 2.0 | $5 \times 10^{-7}$ to $1 \times 10^{-5}$ | $<100$ |

## 2.5 CHARACTERISTICS OF MATERIALS FOR SHEATHS AND OTHER COMPONENTS

The materials used for sheathing, cushioning layers, fillers, wraps, etc., for optical-fiber cables are usually chosen on exactly the same grounds as those used for conventional cables. In certain respects, however, the criteria may be even more stringent for optical-fiber cables. In the telecommunications industry, it is usual for the operating company to specify the sheath material in detail, even, in some instances, specifying the processing procedures.

Usually the material supplier will determine the chemical and mechanical properties of such materials by measurements on molded plaques. However, this does not give an adequate picture, particularly in regard to environmental stress cracking. In addition to the raw material properties of the sheath, it is necessary to specify:

*Processing conditions:* Extrusion, cooling, and winding

*Storage conditions:* Temperature variations, exposure to uv and chemicals, etc.

*Installation conditions:* Lubricants, ground pollution, etc.

Bonicel and his colleagues[27] have investigated a variety of the commonly used materials, including low-density polyethylene, EVA copolymers, high-density polyethylene, and linear low-density polyethylene. They conclude that, in the sheath, these materials are anisotropic and have a structure different from that of molded plaques. Regardless of which test method was used, the crack resistance in the cable sheath was found to be inferior to that measured on plaques. Although the range of materials tested was limited, there is little doubt that similar considerations apply to most of the commonly used materials. It is therefore essential that, wherever possible, sheath materials be evaluated in sheath form.

One of the dominant criteria, when a sheath material is being chosen for a conventional cable, is price. Because the sheath forms a significant portion of the cable cross section, its price and ease of extrusion are of paramount importance. This usually restricts the choice. With optical-fiber cables, on the other hand, the cable can be smaller and the fiber cost becomes a larger portion of the overall cost, so the choice of sheath material can be much wider. This is particularly true when very small cables are being considered. In addition to polyvinyl chloride (PVC) and the materials listed above, which are commonly used, polyurethane, polyester elastomers (Hytrel), nylon, neoprene, and a variety of fluorinated polymers can also be considered.

### 2.5.1 Influence of Additives

The caveat sounded above, as to the importance of not just material properties but also of extrusion conditions, applies even more strongly when materials are not just being used in their natural state, but with additives to give color, uv stabilization, fire retardance, etc. These additives influence the mechanical properties to some extent. Figure 2.14 illustrates the difference in mechanical behavior that can occur with Tefzel with a comparatively minor change of color additive. As will be seen in Sec. 2.71, even such relatively minor changes can significantly alter cable performance when cables with a low longitudinal compressive modulus are cycled over a wide temperature range.

### 2.5.2 Flammability

Flammability considerations for optical-fiber cables are similar to those for conventional cables, except that, in general, the smaller sizes of optical-fiber cables open the field to more expensive materials. Conventionally, the high standard of fire retardance in the wire and cable industry has been achieved by the use of halogenated polymers and additives.[28] More recently, a trend has developed toward the development and use of nonhalogenated materials, especially for indoor, industrial, and some military applications.

During the early stages of a fire, a cable will experience heating. Behavior during this stage will be moderated by the specific heat of the sheathing compound. Inorganic hydrates can be used as fillers to increase the specific heat and improve

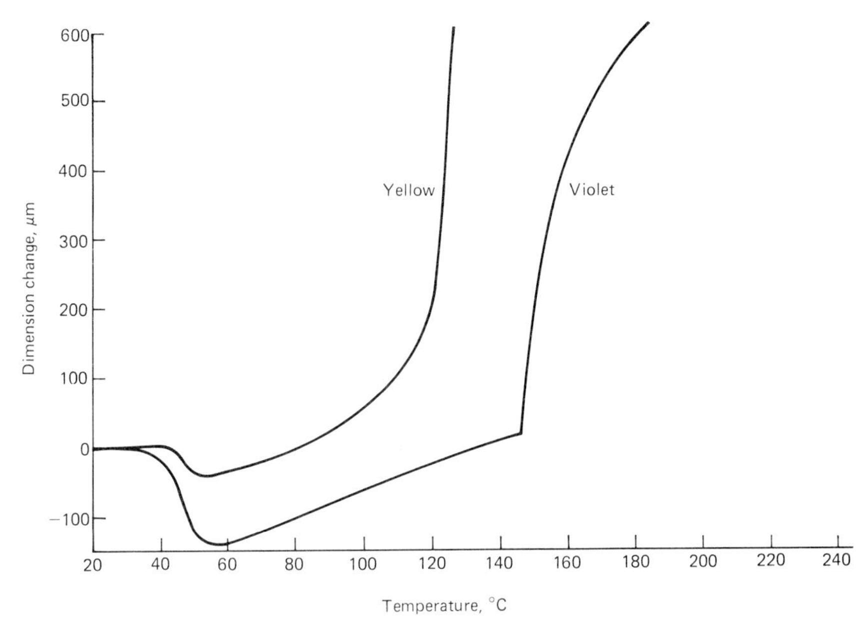


FIGURE 2.14 Dimensional change of Tefzel as a function of temperature for two color additives. (*Reproduced with permission of STC Technology Ltd.*)

initial flame resistance. Alumina trihydrate is one such filler, but high loadings are required to give effective flame retardance, and the effects on material properties such as modulus may be significant. Later in the fire, polymer decomposition commences. This usually takes place progressively over a temperature range. Fluorine forms an extremely strong bond with carbon, so fluorinated polymers are flame-resistant in the same sense that they are difficult to ignite. It is not unusual to find limiting oxygen index near 100 in such materials. Chlorine and bromine are effective, to a lesser extent, in the same way. Phosphorus compounds and other additives can be effective. Some decompose to leave a char residue which prevents the access of oxygen to the combustible material and provides a thermally insulating layer.

At a still later stage, combustion and propagation will occur. Here the presence of heavy halogens can inhibit oxygen access, but they may facilitate the formation of carbon, which would account for the high level of smoke associated with burning PVC.

### 2.5.3 Toxicity and Aggressivity

While halogenated polymers and flame-retardant additives have proved reasonably effective in achieving flame retardance, their behavior in a fire environment can lead to devastating effects. The aggressivity and toxicity of the comparatively small quantities of the products of combustion have lead to extensive damage. To combat these effects, specifications now aim not only at a high level of resistance to ignition and flame propagation but also at low smoke evolution, low acid and toxic gas evolution, and low heat of combustion. The current philosophy is that cables are rarely the source of a fire, but they can add considerably to the damage when they are involved in a fire, so that care needs to be taken to minimize these effects.

There has been considerable progress in reducing smoke evolution while maintaining the critical properties of halogenated polymers and additives.[29] Nonetheless, there is now considerable emphasis on the use of halogen-free flame-retardant polymers and compounds. Three such compounds that have been investigated as wire insulation could be considered for sheathing small optical-fiber cables.[30] They are:

*Noryl:* Modified polyphenylene oxides compounded by General Electric Plastics.

*PEEK:* Polyether-ether ketone, an ICI high-performance thermoplastic polymer, which, unfortunately, requires processing temperatures in the 270 to 400°C range, limiting its applicability. PEEK is otherwise very attractive for its high strength and outstanding chemical fluid resistance, with low acid and smoke evolution.

*ULtem:* An amorphous polyetherimide from General Electric Plastics, with lower extrusion temperatures than PEEK but otherwise similar properties, except for a lower solvent resistance.

For applications where a cheaper material is essential, it is possible, by choice of ethylene copolymer and the inorganic hydrate filler described in Sec. 2.5.2, to significantly improve upon the fire-related characteristics of PVC, without any danger of acid gas evolution and with very low smoke generation.

## 2.6 SPECIFIC PROTECTION TECHNIQUES

### 2.6.1 Water Blocking

Traditionally, petroleum jelly has been widely used by the cable industry as a water-blocking filler. It is an inexpensive compound and there is extensive experience to demonstrate its compatibility with many other cable materials. It presents a major problem, however, in that it has a continuously changing viscosity over its usable temperature range (approximately −10 to +80°C). This affects ease of application and presents problems in subsequent sheathing operations.

Silicone greases can provide an alternative, since they can offer uniform viscosity over temperature ranges as great as −50 to +200°C. The same discussions presented in Sec. 2.3.2 (which lead to a preference for thixotropic gels as a filling for loose tubes) apply equally to cable structures. Thixotropic gels, therefore, are now frequently used to provide water blocking in cables.

It is necessary, though, to distinguish between conditions where the head of water may only be a few meters and the situation in submarine applications where pressures may be as high as 7 MPa. Under the latter conditions, the usable thixotropic gels flow rapidly, and recourse has to be made to materials such as Hyvis 2000, a high-viscosity polyisobutylene. In common with most polybutene fillers, it can be inserted only at elevated temperature, such as +120°C, with a consequent contraction on cooling. This raises one of the recurrent problems with many of the conventional polybutene-based fillers: any voids introduced by contraction not only lead to incomplete water blocking but also present inhomogeneous uniaxial stresses to the fibers that can lead to attenuation increments.

A further water-blocking technique is available for small optical-fiber cables. If the fibers are first packaged in plastics with a high working temperature range—silicone resin or fluorinated plastics, for instance—they and a strength member can be embedded by extrusion in a solid plastic rod. With many materials—but particularly nylon 12—it is possible to arrange the extrusion conditions so that sufficient contraction occurs to form an effective seal.[22] Under suitable conditions, this technique and the Hyvis 2000 blocking technique can also ensure gas blocking.

### 2.6.2 Rodent Protection

Thin plastic-coated metal tapes have been used for shielding and armoring purposes for a long time in the telecommunications cable industry. Initially, a polyethylene-and-aluminum laminate tape was used as a barrier to moisture ingress. Later it was found that use of an adhesive copolymer of ethylene and acrylic acid resulted in a superior bonded jacket because it provided a strong chemical bond between the copolymer coating and the metal.[31] In the mid-1970s, the search for an economical rodent-resistant sheath eventually led to the wide use of copolymer-coated corrugated steel as an effective solution.

The most usual attack is from small rodents whose chewing kinematics are such that cables under 15-mm OD are particularly vulnerable. Most optical-fiber cables fall in this size range. Cogelia and coworkers[32] found that a thickness of at least 150 μm of carbon steel, or 75 μm of stainless steel, is required to protect these small cables. It has been found that a tape of corrugated, coated metal can be applied to the cable core by longitudinally forming the tape around the core with an overlapping seam. The steel is bonded to the polyethylene sheath during sheath extrusion and the copolymer coating then seals the seam. Sealing the seam considerably strengthens the steel protection and reduces the possibility of rodent penetration to the cable core. Most of these rodents, however, are creatures of habit, and will return to damaged cable to continue chewing. Since it is likely that the steel will eventually be exposed, and may then be weakened by corrosion, it may be preferable to use a stainless-steel tape. Workers at Western Electric developed a suitable adhesive coating system for combining stainless steel with a polyethylene sheath.[33] However, because of hydrogen evolution and the effects of lightning, it has been found preferable to use a combination of copper and stainless-steel tapes.[34]

### 2.6.3 Hydrogen Control

The effects of hydrogen on silica-based optical fibers are due either to the interstitial absorption of hydrogen or to the permanent formation of Si-OH bonds in the glass. The interstitial effects are reversible and will be reduced by any subsequent out-diffusion of the hydrogen, whereas bond formation is permanent in almost all instances. It is possible to reduce the permanent effects to negligible levels by good fiber design, particularly by elimination of any phosphate dopant or contaminant. (See Chap. 1.)

In certain adverse environments, where nuclear radiation levels are significant, the combination of hydrogen and radiation will produce permanent absorption.[35] Usually, though, it is only the molecular absorption that need be considered. Molecular absorption produces a broad loss increase centered at 1.24 μm. This loss can rise to 6 to 8 dB/km, if the silica fiber is saturated in 1 atm of hydrogen at room temperature. The attenuation increase at 1.3 μm is about 0.2 dB/km, but can rise considerably further at higher temperatures.

There are three main potential sources of hydrogen generation in optical-fiber cables:

1. Polymer degradation
2. Metallic outgassing
3. Galvanic corrosion

Most commercially available polymers can emit hydrogen, often with little or no thermal stimulation. Because these polymers are not simple repeated molecular structures, but contain antioxidants, stabilizers, etc., analysis is difficult. However, determination of hydrogen evolution rates has been carried out on a practical basis.[34,36,37]

The amount of hydrogen retained in metals at a variety of trapping sites is very dependent on the original manufacturing conditions and subsequent working. Most hydrogen is tightly bound and will be released only at elevated temperatures. Galvanic action becomes important whenever an electrolyte, such as ionized water, is present. Hydrogen will be produced in this way by any cable structure containing dissimilar metals.

For most underground telecommunication cable systems, where temperature excursions are low, the effects of hydrogen are likely to be negligible, but it is prudent to ensure that, where dissimilar metals may meet ionized water, the electric potential between them be as small as possible.[34] In addition, all curable polymers must be fully cured. In more severe applications, such as submarine cables[38] and other high-pressure environments where hydrogen partial pressures may be much higher, or in cables for use at elevated temperatures, or in situations where they may be subject to nuclear radiation, hydrogen control can assume far greater importance.

## 2.7 ENVIRONMENTAL EFFECTS

### 2.7.1 Temperature

To operate over a specified temperature range, optical-fiber cables have to meet all the requirements pertinent to conventional cables. The variation of plastic properties with temperature must be considered, for example. In addition, with optical fibers, consideration has to be given to changes in length. Since the coefficient of expansion of most plastics is some 2 orders of magnitude greater than that of silica, and since other changes can occur in plastics, affecting their physical dimensions, it is necessary to examine dimensional changes in plastics closely. The coefficient of expansion of steel- or glass-reinforced plastics is about an order of magnitude less than that of most plastics. If the overall cable modulus is high and is dominated by a material of low coefficient of expansion, the cable is more likely to meet a wider temperature range specification.

Upper temperature limits are likely to be set by the behavior of the plastic components. While all of the conventional materials can be used up to 40°C, most materials can be used up to 70°C. Whereas Hytrel—with additives—can be used above 100°C, fluorinated polymers can be used up to 155°C. Above this temperature, materials like polyether-ether ketone can be used, but extreme care is required in order to apply them over the on-line fiber coatings. With proper design, it is possible to assure that the fibers are not taken outside their strain relief window because of cable expansion (see Sec. 2.3).

For good low-temperature operation, plastics must be chosen with physical properties that do not change so radically that the mechanical buffering provided by the encapsulation is lost. For most cables, however, there will be some low temperature at which a large attenuation increment becomes apparent. This onset of increased attenuation is often ascribed to low-temperature microbending loss, but this does not explain a loss that suddenly appears at some critical temperature, rises steeply with small, further reductions in temperature, or can even increase while the cable is held at a steady, low temperature.

A mechanism likely to cause such behavior was first demonstrated by Reeve.[39] He showed that a fiber held in a loose tube takes up a helical configuration when it is subject to longitudinal compression. Since most plastics have a temperature coefficient of expansion at least 2 orders of magnitude greater than that of silica, compression is to be expected with reduced temperature. At regular intervals the fiber helix will flip from a left-hand to a right-hand orientation, so the overall torsion in the fiber remains near zero. Similar behavior has also been observed with tight-buffered fibers, particularly in the early days, when buffering was not very well controlled. This behavior is to be expected when one considers that the fiber is a long thin strut, which will exhibit a classic Euler instability when subjected to longitudinal compression.

Lenahan[40] was able to show that if the fiber is treated as a beam in an elastic medium, a force and moment balance yields the differential equation

$$E_f I \frac{d^4 y}{dz^4} + F \frac{d^2 y}{dz^2} + ky = 0 \tag{2.15}$$

where $E_f$ = fiber modulus
$I$ = moment of inertia of fiber
$F$ = compressive force
$k$ = spring constant of fiber

By calculating the compressive force at a given temperature from the material properties of the fiber and coating, the temperature at which the minimum buckling force arises can be determined:

$$F_{\min} = 2(E_f I k)^{1/2} \tag{2.16}$$

The calculated results rank the performance of different coating designs reliably, but also indicate that, at temperatures at which there is usually an onset of incremental loss, the thermally induced fiber strain falls short of the minimum buckling strain. Additional critical parameters are the consistency of the mechanical properties of the coatings and their concentricity, as well as fiber uniformity.

Once the collapse has been initiated, the collapsed portion of the fiber assumes a helical configuration whose dimensions are determined by the fiber and tube dimensions in one case and by the fiber's and coating's mechanical parameters in the other. The helical configuration grows from the point of initiation as further compression is applied, changing from left-hand to right-hand as torsion builds up in the fiber. But, because the compression is relieved on adjacent portions of the fiber, they tend to remain straight. Reeve was able to demonstrate, by using a multimode fiber with a fully filled aperture, that the greatest energy loss was radiation, which occurred at the points where the helix changed sign, and that this loss was associated with the higher-order modes.

When a configuration of this type is involved, there are two main considerations: (1) the mechanisms that will determine the onset of instability and (2) the mechanisms that will cause loss, once a helical configuration is established. Since it is likely that the plastics in a cable will have the greatest coefficients of expansion, the compressive strain they impose on the fibers will be determined by the compressive modulus of the overall cable and the temperature coefficients of the material dominating this compressive modulus. Here a strength member, such as an S-glass-reinforced plastic or steel, has a considerable advantage over Kevlar 49, whose compressive modulus is extremely low.

In the loose-tube case, the maximum tolerable compression will be determined by the fiber and tube dimensions, the fiber uniformity, and the degree of overfeed that has been incorporated to achieve some strain relief. In the case of tight buffering, however, instability can be postponed by increased support from the coatings. From this point of view alone, the ideal coatings would give a maximum radial support, i.e., high effective radial modulus, with high radial compression and minimum longitudinal compression. These requirements, however, are not consistent with achieving low microbending loss and some strain relief in the cable, so a compromise must be reached. Once the reversing helical configuration has been established, loss can occur through various mechanisms. Reeve highlighted the loss that occurs as the helix switches from left-hand to right-hand and vice versa. He also found some loss associated with the helices. Since these losses were predominantly from higher-order modes, the loss in normal transmission would be governed by the rate of transfer of energy to the high-order modes, which would follow the same pattern as microbending loss. Also, in both the loose-tube and tight-buffered case, microbending loss can increase as the helix presses the fiber more firmly against its adjacent surfaces.

In summary, therefore, in the case of tight-buffered fibers, the temperature below which attenuation increments will grow is set by the fiber uniformity, the concentricity and homogeneity of all coatings as well as the consistency of the coatings, and the radial compression—and hence support exerted by the coatings on the fiber. Meanwhile, the rate at which the increments increase at temperatures below this critical point are governed by mechanisms that are very similar to those governing microbending.

### 2.7.2 Longitudinal Stress

It is clear from the preceding section that longitudinal compression presents a major problem to optical fibers. Fortunately, it is not a common situation for installed optical-fiber cables and can be met by using adequate strength members. Uniaxial transverse compression is usually countered by conventional armoring techniques or can be met with suitable fully filled designs (see Sec. 2.6.1). However, longitudinal tensile stress is a common occurrence, and, because any fiber strain adversely affects fiber life, it must be dealt with adequately by all successful designs. Section 2.4 shows how strength members can be used to increase overall cable modulus and limit the cable strains imposed by particular cable loads.

Other aspects of cable design can help reduce stress. In Sec. 2.3.1, the use of fibers in helically laid loose tubes was shown to produce a window in which fibers remained strain-free, even though cable strain could be as high as 0.5 percent. The same techniques can be used in cable design with loose-tube, gel-filled, or tight-buffered fibers. The encapsulated fibers are laid in the cable with a helical or S-Z configuration (i.e., a helix with regular changes of the direction of rotation) in such a way that they are able to move radially or straighten if the cable is subjected to strain. In this way, significant cable strains can be accommodated without imparting anything but the most trivial strains to the fibers. Often, encapsulated fibers are allowed to move circumferentially as well as radially, but this is not necessary to obtain strain relief. Depending upon the constraints of cable design, either Eq. (2.7) or—in the simplest case—Eq. (2.8) can be used to calculate the available strain relief, if the cable restraints are substituted for the encapsulant restraints in Eq. (2.7).

### 2.7.3 Pressure

The loose-tube and gel-filled techniques for fiber encapsulation described in Secs. 2.3.1 and 2.3.2 provide some protection against the effects of hydrostatic pressure. However, at some critical pressure, which may be well below 100 bars, the tube can collapse and catastrophic attenuation increments result. Exactly the same considerations apply to complete cables, where similar protection techniques are used. The critical pressure at which collapse will occur can be calculated from[41]

$$P_{cr} = \frac{2E}{1 - \mu^2}\left(\frac{t}{d}\right)^3 \tag{2.17}$$

where $E$ = modulus of tube
$\mu$ = Poisson's ratio for tube
$d$ = tube diameter
$t$ = tube thickness

Increasing the wall thickness of the tube provides a rapid increase in critical pressure as long as only elastic buckling needs to be considered. There comes a point, however, when the maximum circumferential stress on the inner wall of the tube exceeds the yield point of the material and inelastic buckling occurs. This reduces the critical pressure given by Eq. (2.17). An approximate value for this reduced pressure can be obtained by substituting the tangential modulus of the tube material:

$$P_{cr} = \frac{2E_T}{1 - \mu^2}\left(\frac{t}{d}\right)^3 \tag{2.18}$$

This equation implies that, when the material yields and the tangential modulus becomes zero, the critical pressure also falls to zero.

With a suitable choice of materials and dimensions, tubes can thus, in principle, be manufactured to meet any pressure a fiber-optic cable is likely to meet (see Sec. 2.8.2). Some materials that are mechanically suitable, however, are not practical.

The group velocity of propagation in optical fibers, fortunately, is largely independent of hydrostatic pressure. And if, as in the majority of today's systems, optical phase information is not used, free flooding (i.e., pressure balancing) of the cable may ensure satisfactory operation at high pressures.

The fiber surface itself, however, should not be directly exposed to water because of the danger of fiber strength reduction. The tight buffering described in Sec. 2.3.3 will satisfactorily transfer hydrostatic pressure to the fiber surface (i.e., the tight-buffering system described can reduce any uniaxial stress). Figure 2.15 shows the effect of hydrostatic pressure on such a packaged fiber up to pressures corresponding to the deep ocean.

When dissimilar metals are present in the cable structure or when watertight layers must be used, free flooding is not permissible. Instead, the cable must be filled with a thixotropic gel or a solid.

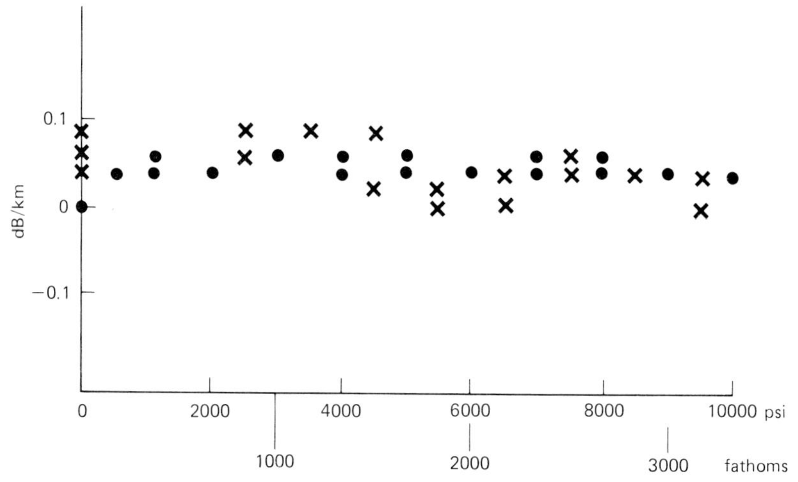


**FIGURE 2.15** Incremental attenuation vs. hydrostatic pressure for a tight-buffered, 50/125-μm graded-index fiber. (*Reproduced with permission of STC Technology Ltd.*)

## 2.8 CABLE DESIGN EXAMPLES

A number of cable design examples are given to illustrate the points that have been raised. The examples are chosen primarily for illustration and not because they are necessarily thought to have particular merit.

### 2.8.1 General-Purpose Cables

An early design of cable is shown in Fig. 2.16. In the example shown, tight-buffered fibers lie helically around a nylon-coated steel strength member; tapes and an aluminum-plastic water barrier surround the fibers, and the cable is complete with a polyethylene sheath. In this particular cable, the sheath is pressure-extruded, a fact that—combined with tight buffering—gives an impression that

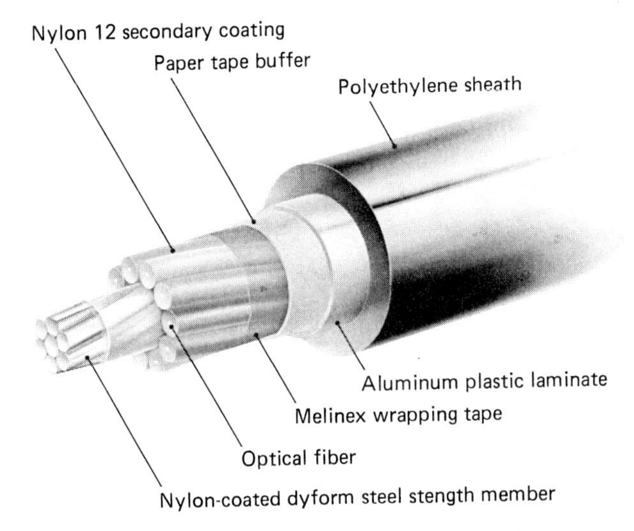


**FIGURE 2.16** Optical-fiber cable for duct installation. (*Reproduced with permission of STC Technology Ltd.*)

the cable is a tight one. However, as in most successful examples, the radial pressure is low. This is because, on extrusion, the sheath solidifies first at its outer surface. As the solidification front proceeds inward, the contraction associated with the phase change pulls the inner surface away from the cable core. Therefore, attenuation increments can be kept to very low levels, as shown in Fig. 2.17.[42] Note in this figure that the change in cutoff is due to small movements of the cable. The change in the –OH peak at 1390 nm occurred because the –OH

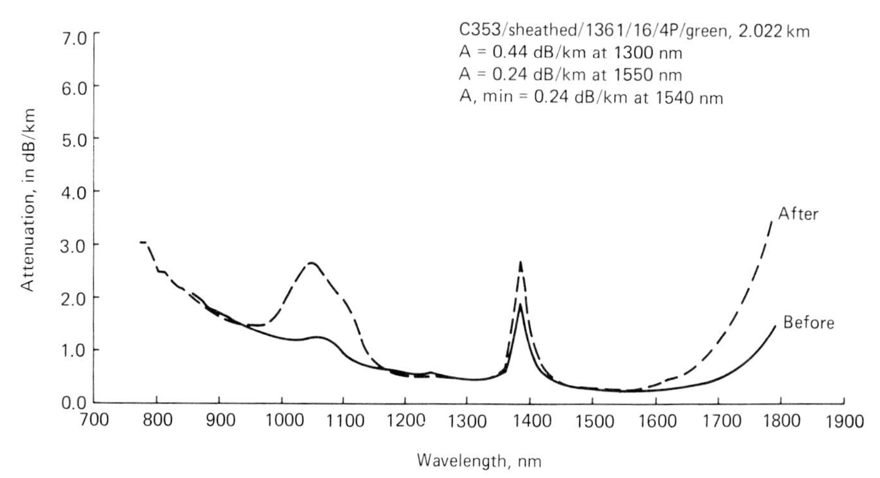


**FIGURE 2.17** Spectral attenuation of single-mode fiber before and after cabling. (*Reproduced with permission of STC Technology Ltd.*)

content of the fiber was not uniform and not all of the fiber was cabled. The change near 1700 nm represents macrobend losses due to the cabling process.

The same general design has also been used with fibers packaged in loose tubes and in gel-filled loose tubes. It has also been used with a variety of dielectric strength members and without the aluminum-plastic laminate for situations where an all-dielectric cable is required.

### 2.8.2 Submarine Cables

The requirements for a deep-water submarine optical cable were identified by Worthington in 1980.[43] He chose a design with a fiber elongation of 0.2 percent at the maximum anticipated tensile loads of 100 kN, with a pressure-resistant aluminum tube that showed no deformation at pressures up to 100 GN/m$^2$. A development of this design[38] (for transoceanic telephonic communications) is shown in Fig. 2.18. Here, to reduce the problems of hydrogen generation discussed in Sec. 2.6.3, a composite copper tube is used, rather than an aluminum tube. Two layers of steel strength members are used to achieve torque balance and a greater working load. Polyethylene insulation is used to isolate the power supply to the repeaters from the seawater. For the cable ends, which are laid in shallower water, a further armoring of polypropylene servings and steel armor wires is added.

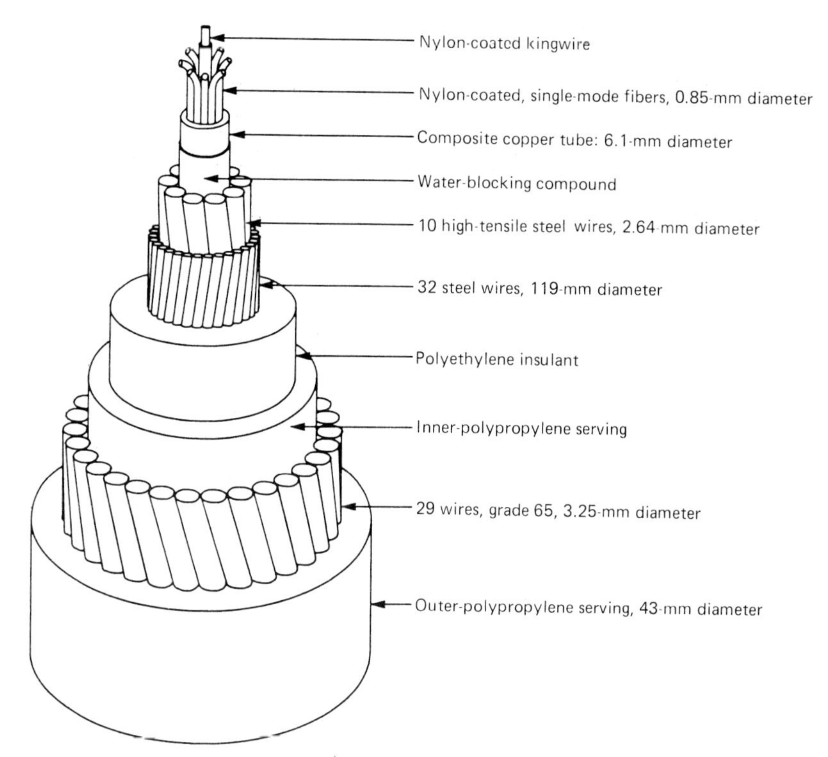


FIGURE 2.18 Deep-water submarine cable incorporating optical fibers. (*Reproduced with permission of STC Technology Ltd.*)

### 2.8.3 Industrial Cables

Many industrial applications can be met with general-purpose cables, particularly those versions that use tight buffering. This gives adequate protection from the high uniaxial pressures and impacts that are found in many industrial environments. Also, the use of dielectric constructions gives immunity from electromagnetic interference and ground loop problems. Special cables have also been developed. One, illustrated in Fig. 2.19, has tight-buffered units extruded into an easily separated tape sheath, so that individual units can very simply be split off to serve different destinations.

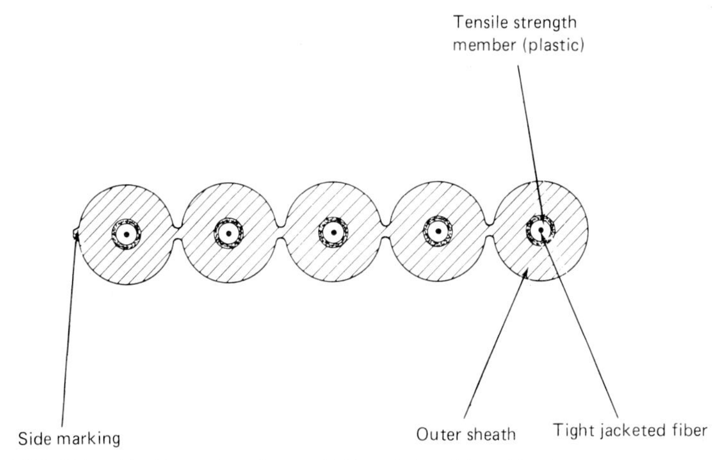


**FIGURE 2.19** Example of a multifiber industrial cable (cross section). (*Reproduced with permission of STC Technology Ltd.*)

### 2.8.4 Aerial Cables

Aerial installation can be achieved by lashing a general-purpose cable to a messenger wire already installed on suitable poles. Such techniques allow most of the smaller telecommunications cables to be used in aerial installations. A slightly more specialized, similar cable can be made by adding a separate, parallel steel strength member, the two being held together in a single extrusion having a dumbbell cross section. Such designs, although simple in concept, can show marked deficiencies when wind velocity and icing conditions are taken into consideration. These conditions are best met with a smooth cable of circular cross section, but if long lifetimes are to be achieved, there must be adequate strain relief for the fibers, as quite large, long-term loadings will be present.

A method of achieving extra strain relief was described in Sec. 2.3.1. However, such cables are not very practical because of the difficulty of constraining large-enough tubes to lie helically around a strength member. Also, the cable will be large in relation to the strength member, resulting in increased strains for wind and ice loading that could negate the advantages of extra strain relief. A practical, self-supporting design that overcomes this disadvantage is shown in Fig. 2.20.

This design employs a slotted S-glass-reinforced polymer rod as a lightweight strength member. The loose tube containing the helically laid fibers is laid as a

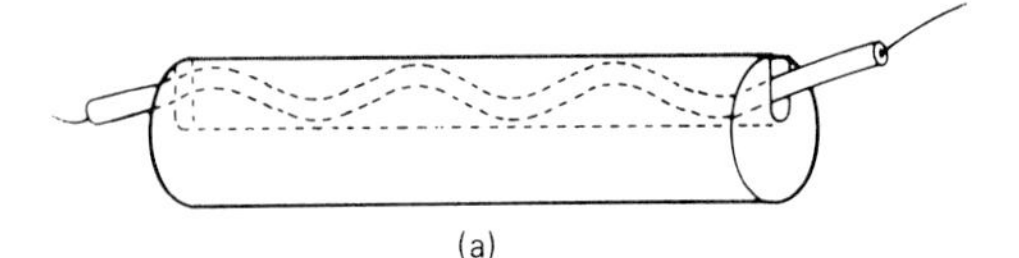


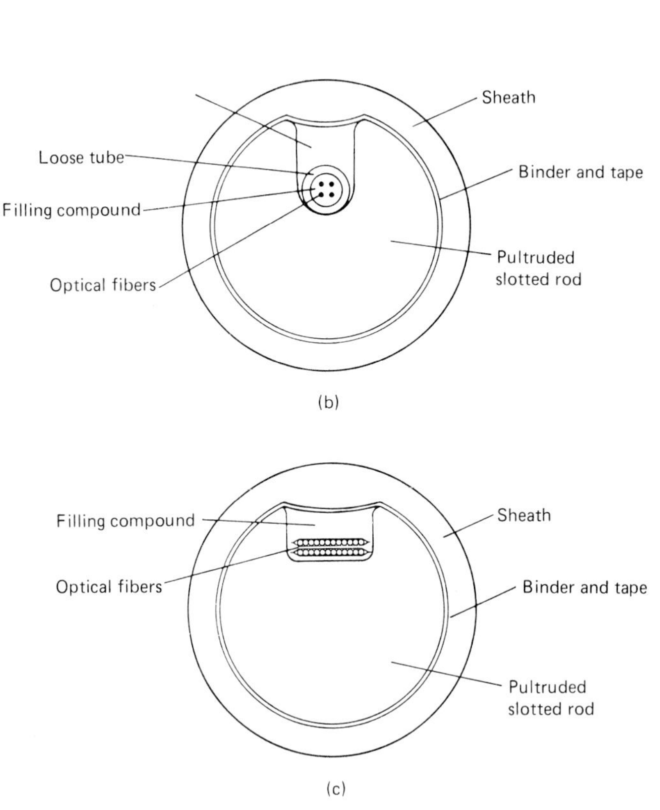


**FIGURE 2.20** Examples of aerial cable construction techniques. (*a*) Helically laid fiber in sinusoidally laid loose tube (in slotted rod); (*b*) end view of sheathed rod containing minibundle of fibers; (*c*) end view of sheathed rod containing fiber ribbons. (*Reproduced with permission of STC Technology Ltd.*)

sinusoid within the slot (as in Fig. 2.20*a*). This can be regarded as a two-dimensional projection of the primary helix discussed in Sec. 2.3.1 and illustrated in Fig. 2.8. With the double strain relief, the cable can withstand a load of 22.5 kN, which is sufficient for a 1-km self-supporting span in tropical climates, without imposing any longitudinal fiber strain. The cable gives adequate margins for wind and ice loading on 0.5-km spans, allowing this all-dielectric cable to be used on the same towers as existing power lines.

For high-fiber-count cables, the design can be used with the fibers laid in tapes, as in Fig. 2.20*c*, but at the expense of some strain relief.

### 2.8.5 Buried Cables

In early applications, general-purpose cables were buried or laid under conditions in which they were susceptible to rodent attack. It quickly became apparent that rodents favor optical cables as much as they do conventional cables. The protection mechanisms discussed in Sec. 2.6.2 should therefore be employed.

One of the first optical cables to employ such protection was described by Hope and coworkers in 1981.[44] This cable also illustrates one of the presently favored techniques for obtaining more cost-effective high-fiber-count cables for telecommunications applications. In this technique, shown in Fig. 2.21 with the rodent-protection sheath, multiple fibers are laid, either singly, in ribbons, or in bundles, within slots in the cable core. These slots are formed in the core either in a helical or S-Z configuration. If the fibers are overfed, they can be made to lie at a radius greater than that of the bottom of the slot. Strain relief is then obtained in a manner exactly analogous to that detailed for loose-tube fibers in Sec. 2.3.1.

### 2.8.6 Telecommunications Cables

In many ways, the cables employed in the telecommunications field operate in a most favorable environment. Temperature excursions are small, disturbance after laying is rare, and vibration is usually absent. On the other hand, cost-effective designs are essential. Early installations used cables of the type described in Sec. 2.8.1 or cables in which fibers were organized within an armored tube. Figure 2.22 shows a multiunit design, based on the early concept of tight-buffered fibers helically laid around a strength member, and built up into a 320-fiber cable. Described in 1980 by Schmidt and Zwick,[45] the shear bulk of such a

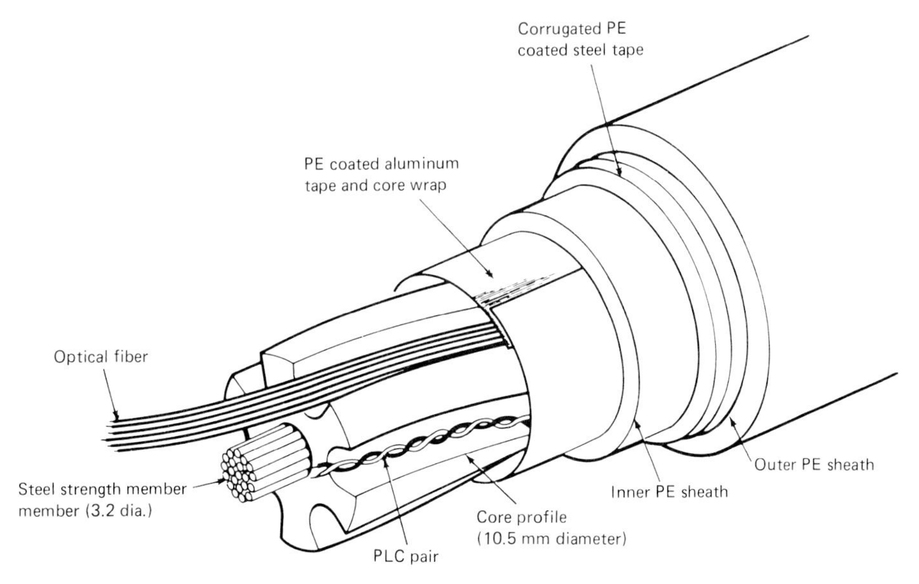


FIGURE 2.21 Slotted-core cable with rodent protection. (*Reproduced with permission of STC Technology Ltd.*)

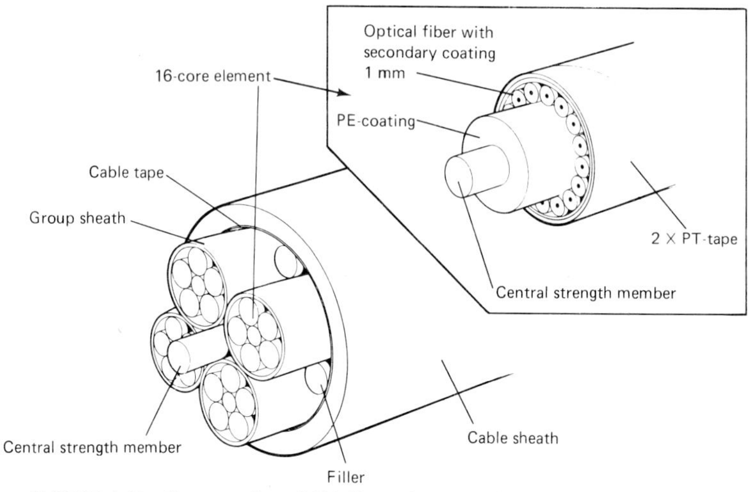


**FIGURE 2.22** Cross section of 320-fiber telecommunications cable. *(Reproduced with permission of STC Technology Ltd.)*

cable inevitably leads to high cable costs. This cable perhaps represents the final development of such bulky cables.

The cable shown in Fig. 2.21—with or without rodent protection—offers the possibility of a much higher packing density and hence a more cost-effective design. Another cost-effective design, but of different concept, is shown in Fig. 2.23. This cable, described by Patel and Gartside in 1985,[46] contains minibundles of optical fibers, each identified with a color-coded binder. In this technique, pioneered by Bark and Lawrence,[47] the bundles are laid with overfeed in a single, large-diameter tube, which is filled with a soft gel. The tube is protected with a

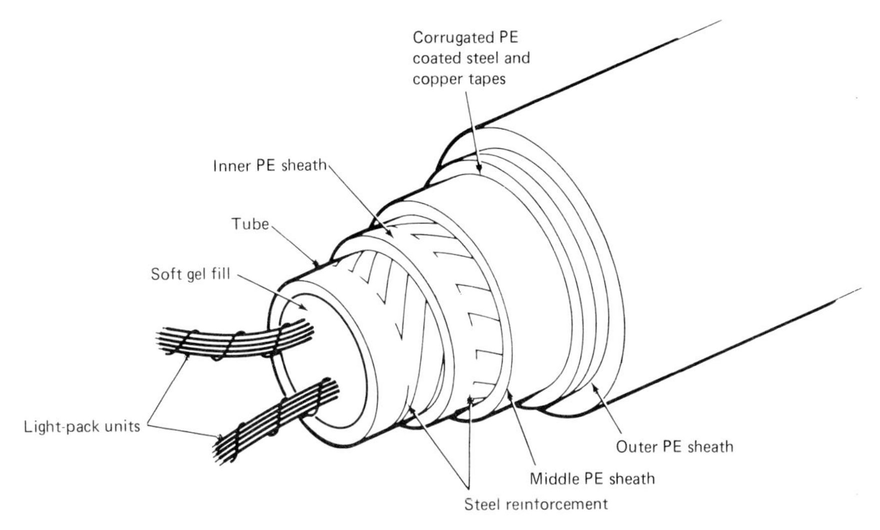


**FIGURE 2.23** Lightpack cable with rodent protection. *(Reproduced with permission of STC Technology Ltd.)*

reinforced polyethylene sheath, with the cross-ply reinforcing steel elements forming the cable strength member. The cable may be completed with corrugated stainless steel and copper tape bonded to a second polyethylene sheath to provide lightning and rodent protection.

In comparison with earlier designs, from which this cable has been derived,[48] the use of fibers in bundles—rather than fibers arranged in ribbons (as described in Sec. 2.3.4)—provides the possibility of greater strain relief. An additional advantage is that the attenuation increments can be as low as those illustrated in Fig. 2.15.

## 2.8.7 Military Communications Cables

Cables for use in tactical military situations are required to withstand the very rigorous conditions of military transport, storage, and use. These conditions can include:

1. Handling, laying, and recovering at short intervals in extreme climatic conditions
2. Immersion in water and use in high humidity
3. Exposure to extreme cold
4. High-impact, compressive, abrasive, twisting, and snatching forces
5. The effects of nuclear, chemical, and biological attacks

A cable compatible with such environmental conditions was described in 1982.[49] Its cross section is shown in Fig. 2.24. The cable used two tight-buffered, nylon-coated fibers and two nylon monofilament fillers in a "quad" construction

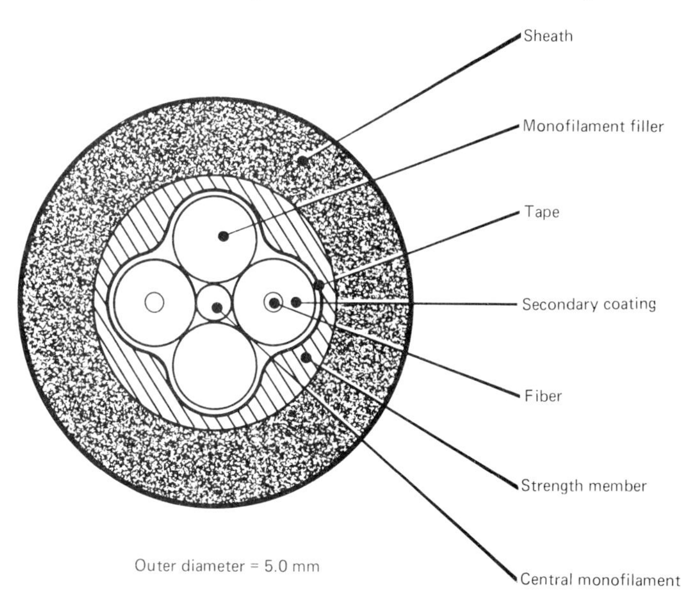


**FIGURE 2.24** Optical-fiber cable for military field communications with a "quad" construction. *(Reproduced with permission of STC Technology Ltd.)*

(four fibers were also used in other applications) surrounded by 20 strands of 1580-denier aromatic polyamide (Kevlar 49) fiber and a flame-retardant Hytrel sheath. The cable performed well in military rough-handling trials. Further developments have been reported.[50]

### 2.8.8 Special-Purpose Cables

A wide variety of special-purpose cables have appeared. Many are special applications of cable designs already described above. For instance, Brataas[51] described a cable in which an optical cable similar to those described in Sec. 2.8.1 is incorporated within a large, submerged, three-phase power cable, in place of one of the fillers. Similarly, other cables have been incorporated with the ground wire or other conductors in overhead power lines, using techniques like those described in Sec. 2.8.2.

However, there is a family[52] of single-fiber-cable designs (for a range of military applications) that does not fall into any of the preceding categories. A typical cross section of such a cable is shown in Fig. 2.25. The cables in this family range in diameter from 0.55 to 3.5 mm. The smallest of these cables has been tested in simulated high-speed deployment conditions at payout speeds approaching mach 1. The largest, which includes triple coating of the packaged fiber, gel filling of the Kevlar strand, and a Hytrel sheath, shows very low attenuation increments, even at wavelengths longer than 1.55 μm (Fig. 2.26). It also shows virtually no attenuation increments with hydrostatic pressure up to 69 MN/m$^2$.

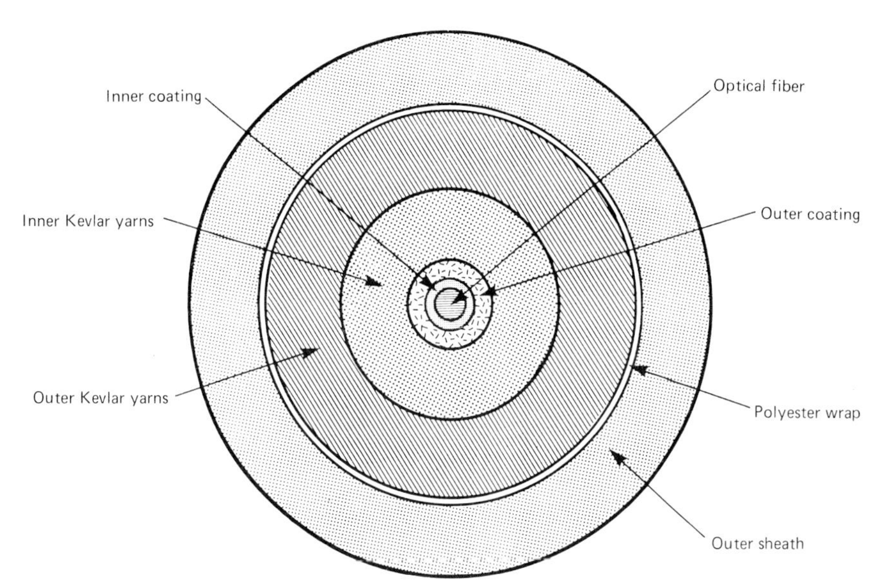


FIGURE 2.25 Special-purpose single-fiber cable (cross section). *(Reproduced with permission of STC Technology Ltd.)*

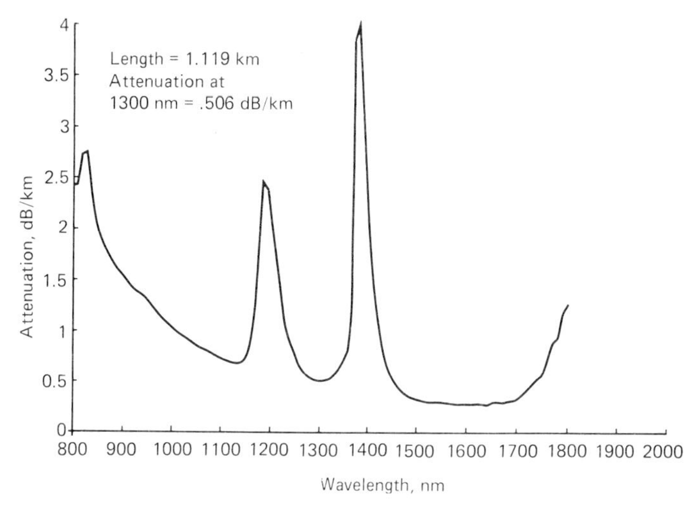


**FIGURE 2.26** Spectral attenuation of single-fiber cable (military application). (*Reproduced with permission of STC Technology Ltd.*)

### 2.8.9 Summary

The choice of cable type will be influenced by the application and environment in which the cable will be used. For the telecommunications market, which is comparatively benign, the choice is dominated by cost. Three generic designs have come to the fore in this market. Loose-tube cables (Sec. 2.3.1) have developed into modern designs such as that shown in Fig. 2.27. Slotted-core or open-channel cables, such as that described in Sec. 2.8.5, continue to be featured largely in the market, whereas the Lightpack described by Patel and Gartside[46] is now a powerful contender.

It will be noted that in all of these designs, the coated fiber is loose in the cable structure. The same trend is discernible in telecommunications submarine cables.

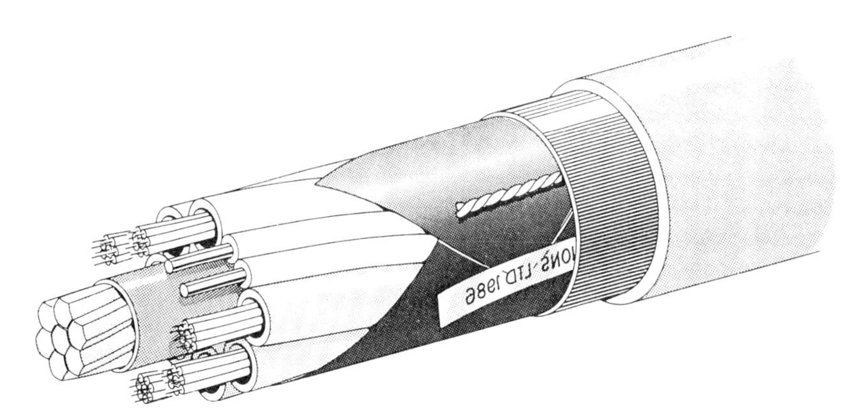

**FIGURE 2.27** Loose-tube cable. (*Reproduced with permission of STC Technology Ltd.*)

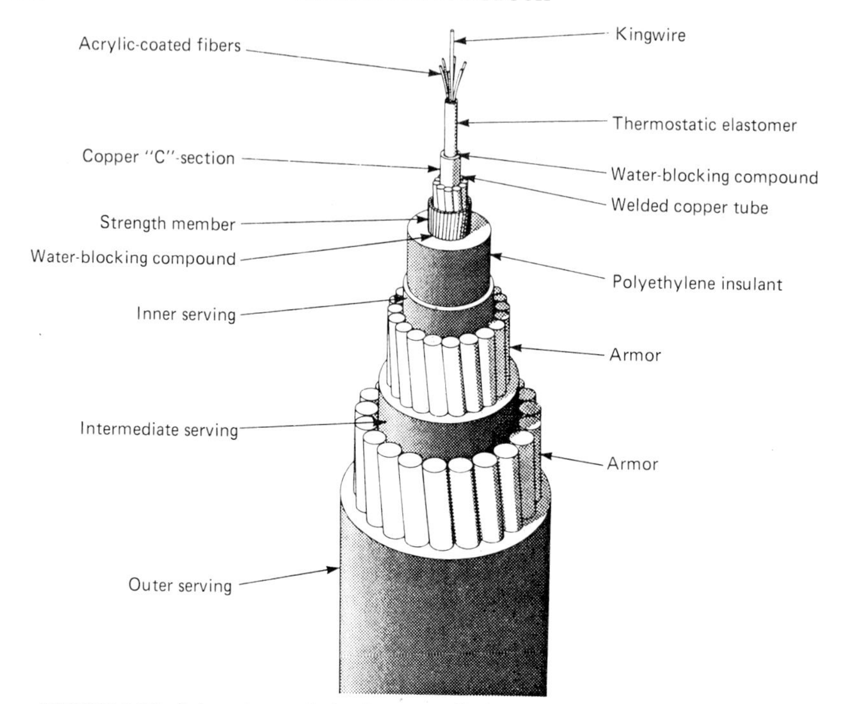


**FIGURE 2.28** Submarine optical telecommunications cable. (*Reproduced with permission of STC Technology Ltd.*)

Figure 2.28 shows the reduced material content of a more recent cable design, a trend that is still continuing.

In many ways, the military market is much more demanding and the applications much more varied. Cables may have to be nonmetallic and may have to operate over wide temperature ranges, such as −65 to +200°C for some avionic applications. For missile guidance, the overall cable diameter may be as little as 0.5 mm, or a coated fiber can be used. At the other extreme, dynamic-strain cables for use in naval towing applications can be just as large as the submarine cables in Sec. 2.8.2. For military-field communication applications, the cables must be extremely tough, flexible, nonmetallic, and fire-retardant. Cables may have to meet pressures as great as those experienced by submarine cables but at the same time be light, flexible, and nonmetallic. Usually such requirements can be met only with tight-buffered coatings, and this type of coating dominates most military applications. See Secs. 2.8.7 and 2.8.8.

The industrial market falls between these two. Conditions are seldom as severe as they can be in the military environment and seldom as benign as those in a duct purposely built for the telecommunications market. The designs chosen can range from versions (usually smaller) of telecommunication cables to the general-purpose design in Sec. 2.8.1 to the special industrial cable in Sec. 2.8.3 to some of the military cables in Sec. 2.8.7. Sometimes, for special applications, the requirements for a few data and communication links can be met by incorporating a fiber-optic element in an existing cable design. Figure 2.29 shows a high-voltage

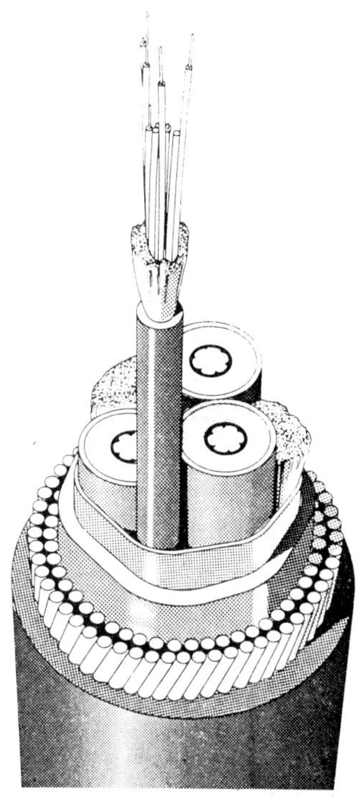

FIGURE 2.29 Offshore three-phase power cable with optical insert. (*Reproduced with permission of STC Technology Ltd.*)

(11-kV) three-phase power cable for underwater operation in which a fiber-optic element has been incorporated.

It is perhaps a sign of a maturing market that for some uses, certain designs are becoming dominant. However, as outlined in the introduction, Sec. 2.1, the diversity of applications can be met only by a wide variety of cable designs.

## APPENDIX: RELATIONSHIP OF FAILURE DISTRIBUTIONS IN BENDING AND TENSILE STRAIN*

Suppose, for a fiber of length $L$, that the probability of failure in tension at a strain of $e$ or less is $\Phi(e)$ and that the corresponding probability of survival is $\Psi(e) = 1 - \Phi(e)$. Suppose also that the failure distribution is given by a Weibull distribution of slope $m$ and scale parameter $e_0$ so that $\ln \ln (1/\Psi) = m \ln (e/e_0)$. Therefore

*By I. F. Scanlan, STC Technology, Ltd.

$$\ln \frac{1}{\Psi} = \left(\frac{e}{e_0}\right)^m \tag{2.19}$$

and

$$\ln \Psi = -\left(\frac{e}{e_0}\right)^m \tag{2.20}$$

Now, suppose that the surface of the fiber is divided into parallel strips, each $d\theta$ wide and $L$ long, as in Fig. 2.30. There will be $2\pi/d\theta$ strips around the entire

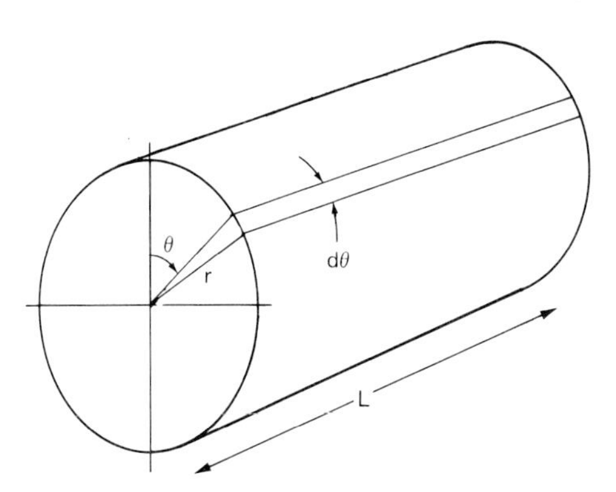


**FIGURE 2.30** Perspective view of fiber with strip $r\,d\theta$ wide and $L$ long.

circumference. Also suppose that the fiber is in uniform tension $e$ and that the probability of survival of each strip is $\psi(e)$. The probability that the fiber as a whole survives, $\Psi(e)$, is the probability that the strips all survive simultaneously, which is the product of their individual probabilities:

$$\Psi(e) = \psi(e)^{2\pi/d\theta} \qquad \text{and} \qquad \psi(e) = \Psi(e)^{d\theta/2\pi} \tag{2.21}$$

Now consider a fiber bent to a radius $R$ as shown in Fig. 2.31. $BB'$ is in the plane of bending and $AA'$ is the neutral plane, containing the neutral axis. The bending strain $e$ at $B$ is $r/R$, so that the strain in a strip at angle $\theta$ is $e \cos \theta$.

As before, the probability of survival of the fiber will be the probability that all the strips survive simultaneously. But now the strips are at different strains, which are expressed:

$$\Psi\ (\text{bending}) = \prod_0^{\pi} \psi^2(e \cos \theta) = \prod_0^{\pi} \Psi^{d\theta/\pi}(e \cos \theta) \tag{2.22}$$

Taking the logarithm of each side gives

$$\ln \Psi\ (\text{bending}) = \frac{1}{\pi}\int_0^{\pi} \ln \Psi\ (e \cos \theta)\, d\theta \tag{2.23}$$

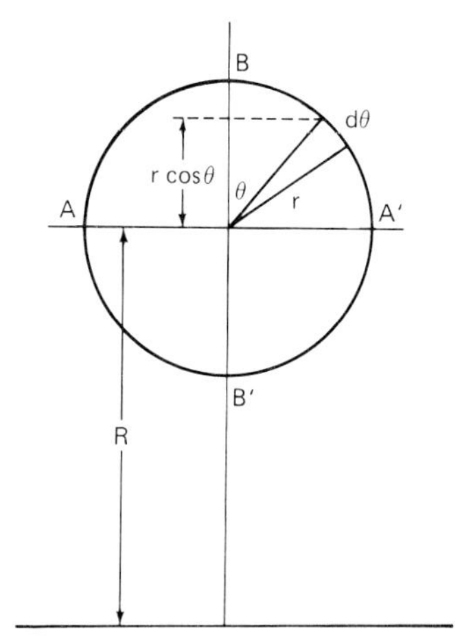


**FIGURE 2.31** Cross section of fiber bent to radius $R$.

Hence it can be shown that

$$\ln \frac{1}{\Psi} \text{(bending)} = \frac{1}{\pi}\left(\frac{e}{e_0}\right)^m \int_0^{\pi/2} \cos^m \theta \, d\theta \qquad (2.24)$$

because the integral

$$\int_{\pi/2}^{\pi} = 0$$

since that part of the fiber surface is in compression and, therefore, $\ln \Psi = 0$.

The integral in Eq. (2.24) is a multiplying constant, which is a function of the original Weibull slope $m$ (tension), but is independent of the applied maximum bending strain $e$. Therefore,

$$\ln \ln \frac{1}{\Psi} \text{(bending)} = m \ln \frac{e}{e_0} + \ln \frac{K(m)}{\pi} \qquad (2.25)$$

In Eq. (2.25), the left-hand side is the Weibull ordinate and the right-hand side is the original Weibull plot moved up by $\ln [K(m)/\pi]$. Since $K(m)$ is less than 1, the curve, in fact, is moved downward.

> Therefore, for the specific case where the failure distribution in tension is a Weibull function, the expected failure distribution in bending is also a Weibull function—but displaced by $\ln (K(m)/\pi)$.[53]

Although the two distributions may not, in general, be exact Weibull functions, usually a sufficiently accurate approximation can be made to a Weibull function. Occasionally, when this cannot be done, the bending failure distribution can be found only by integrating Eq. (2.25) piecemeal.

This analysis looks at the end result, comparing stresses in bending with tensile stress. A more complex analysis, considering not only the flaw distribution in relation to the bending stress but also the differential rates of flaw growth, was performed by Fox.[54]

## ACKNOWLEDGMENT

The discussion of strength members in Sec. 2.4 was adapted from CSELT, *Optical Fibre Communication* (McGraw-Hill, 1981).

## REFERENCES

1. A. C. S. Van Heel, "A New Method of Transporting Optical Images without Aberrations," *Nature,* vol. 73, p. 39, Jan. 2, 1954.
2. F. Alard, P. Lamouler, D. Moutonnet, and P. Sansonetti, "The Mode Spot Size: A Universal Parameter for Single Mode Fibre Properties," Comm. AIV-2, *Proceedings of the Eighth European Conference on Optical Communications,* pp. 89–92, Cannes, 1982.
3. A. D. Pearson, P. D. Lazay, W. S. Reed, and M. J. Saunders, "Bandwidth Optimization of Depressed Index Single Mode Fibre by Means of a Parametric Study," Comm. AIV-3, *Proceedings of the Eighth European Conference on Optical Communications,* pp. 93–97, Cannes, 1982.
4. M. M. Ramsay, G. A. Hockham, and K. C. Kao, "Propagation in Optical Fibre Waveguides," *Electrical Communication,* vol. 50, no. 3, pp. 162–269, 1975.
5. J. E. Midwinter and M. H. Reeve, "A Technique for the Study of Mode Cutoffs in Multimode Optical Fibres," *Opto-Electronics,* vol. 6, no. 5, pp. 411–416, Sept. 1974.
6. D. Gloge, "Impulse Response of Clad Optical Multimode Fibers," *Bell System Technical Journal,* vol. 52, no. 6, pp. 801–816, July 1973.
7. J. E. Midwinter, *Optical Fibers for Transmission,* Wiley, New York, 1979.
8. G. R. Irwin, "Fracture," in S. Flugge (ed.), *Encyclopedia of Physics,* vol. 6, pp. 251–259, Springer-Verlag, Berlin, 1958.
9. A. G. Evans, "Slow Crack Growth in Brittle Materials under Dynamic Loading Conditions," *International Journal of Fracture,* vol. 10, no. 2, pp. 251–259, June 1974.
10. J. D. Helfinstine and F. Quan, "Optical Fiber Strength/Fatigue Experiments," *Optics and Laser Technology,* vol. 14, no. 3, pp. 133–166, June 1982.
11. K. R. Lawson, "Contributions and Effects of Coatings on Optical Fibers," Photon '83, *Proceedings of the International Conference on Optical Fibers and Their Application,* Paris, May 1983.
12. H. Schonhorn, C. R. Kurkjian, R. E. Jeagar, H. N. Vizirani, N. V. Albarino, and F. V. DiMarcello, "Epoxy-Acrylate-Coated Fixed Silica Fibers with Tensile Strengths > 500 ksi," *Applied Physics Letters,* vol. 29, no. 11, pp. 712–714, 1976.
13. G. Pasternuk, "Fundamental Aspects of UV and Electron Beam Curing," *Journal of Radiation Curing,* vol 9, no. 3, pp. 12–19, July 1982.
14. L. G. Amos, D. K. Cardon, and R. A. Miller, "Improvements in Optical Fibre Coatings," *Proceedings of Plastics in Telecommunication III,* pp. 13-1–13-8, London, Sept. 1982.
15. C. A. Jackson, M. H. Reeve, and A. G. Dunn, "Oriented Polymer for Optical Fibre Packaging," Comm. VI-2, *Proceedings of the Second European Conference on Optical Communications,* pp. 175–176, Paris, 1976.

16. P. R. Bark, U. Oestreich, and G. H. Zeidler, "Stress-Strain Behaviour in Optical Fibre Cables," *Proceedings of the 28th International Wire and Cable Symposium,* pp. 385–390, Cherry Hill, N.J., Nov. 1979.
17. O. R. Bressner, A. J. H. Seenan, and S. H. K. in't Veld, "Water Blocking in Optical Cables," *Proceedings of the 29th International Wire and Cable Symposium,* pp. 290–298, Cherry Hill, N.J., Nov. 1980.
18. J. R. Bury and D. A. Joiner, "Versatile High Performance Filling Compound for Telecoms Cable Applications," *Proceedings of the 34th International Wire and Cable Symposium,* pp. 38–43, Cherry Hill, N.J., Nov. 1985.
19. D. Gloge, "Optical-Fiber Packaging and Its Influence on Fiber Straightness and Loss," *Bell System Technical Journal,* vol. 54, no. 2, pp. 245–262, Feb. 1975.
20. W. B. Gardner, "Microbending Loss in Optical Fibers," *Bell System Technical Journal,* vol. 54, no. 2, pp. 457–465, Feb. 1975.
21. R. Olshansky, "Distortion Loss in Cabled Optical Fibers," *Applied Optics,* vol. 14, no. 1, pp. 20–21, Jan. 1975.
22. S. R. Barnes, P. G. Hale, J. N. Russell, and S. V. Wolfe, "Processing and Characterization of Tight Nylon Secondary Coatings for Optical Fibres," *Proceedings of Plastics in Telecommunications III,* pp. 15-1–15-12, London, 1982.
23. M. J. Saunders and W. L. Parham, "Adhesive Sandwich Optical Fiber Ribbons," *Bell System Technical Journal,* vol. 56, no. 6, pp. 1013–1014, 1977.
24. M. R. Santana et al., "Transmission Loss Performance of Three Optical Coatings in a Ribbon Structure," paper WCC4, *Digest of the Fifth Topical Meeting on Optical Fiber Communication,* Apr. 1982.
25. K. Inada et al., "High Density Low-Loss Fibre Unit and Cabling," Comm. 7.5, *Proceedings of the Fifth European Conference on Optical Communications,* Amsterdam, Sept. 1979.
26. K. Ishihara et al., "Low Loss High-Density Optical Cable with Nylon Extruded Fiber Units," *Electronics Letters,* vol. 17, no. 2, pp. 69–71, Jan. 1981.
27. J. P. Bonicel, D. Bosc, and J. J. de Ballet, "Investigation of Cable Sheath Materials," *Proceedings of Plastics in Telecommunications III,* pp. 12-1–12-20, London, Sept. 1982.
28. C. J. Hildo, "Flammability Handbook for Plastics," Technical Publishing Co., 1981.
29. J. J. Duffy and C. S. Icardo, "New Reduced Smoke Flame Retardant Wire and Cable Formulations Containing Halogenated Additives," *Proceedings of the 33d International Wire and Cable Symposium,* pp. 14–18, Reno, Nov. 1984.
30. J. R. Bury and B. A. Cranfield, "Development of Flame Retardant Low Aggressivity Cables," *Proceedings 32d International Wire and Cable Symposium,* pp. 193–199, Cherry Hill, N.J., Nov., 1982.
31. K. E. Bow and P. V. Bakhru, "Bonded Sheath with Co-Polymer Coated Steel for Optical Fiber Cable," *Proceedings of Plastics in Telecommunications III,* pp. 11–10, London, Sept. 1982.
32. N. J. Cogelia, G. K. La Voie, and T. F. Glan, "Rodent Biting Pressure and Chewing Action and Their Effects on Wire and Cable Sheath, " *Proceedings of the 25th International Wire and Cable Symposium,* pp. 117–124, Cherry Hill, N.J., Nov. 1976.
33. W. C. L. Weinraub, D. D. Davis, and M. D. Kinard, "A Rodent and Lighting Protective Sheath for Fiber Optic Cables," *Proceedings of the 32d International Wire and Cable Symposium,* pp. 243–249, Cherry Hill, N.J., Nov. 1983.
34. E. W. Mires, D. L. Philen, W. D. Reents, and D. A. Meade, "Hydrogen Susceptibility Studies Pertaining to Optical Fiber Cables," postdeadline paper W13-1-4, *Digest of the Seventh Topical Meeting on Optical Fiber Communications,* New Orleans, Jan. 1984.
35. A. Robinson and S. Day, "Combined Effects of $H_2$ and Ionizing Radiation," *Proceedings of the IEE Colloquium on Optical Fiber Environmental Effects,* London, Oct. 1985.

36. W. E. Dennis, D. A. Sierawski, and D. N. Ingebrigtson, "Hydrogen Evolving Tendencies of Cable Fillers and Optical Fiber Coatings," *Proceedings of the 33d International Wire and Cable Symposium,* pp. 401–417, Reno, Nov. 1984.

37. S. R. Barnes, S. P. Riley, and S. V. Wolfe, "Hydrogen Evolution from Silicone Resins for Primary Coating Applications," *IEE Proceedings,* vol. 132, no. 3, pp. 169–171, June 1985.

38. S. R. Barnes, N. J. Pitt, and S. Hornung, "Prediction of Optical Cable Losses due to Hydrogen," *Proceedings of the 34th International Wire and Cable Symposium,* pp. 102–106, Cherry Hill, N.J., Nov. 1985.

39. M. H. Reeve, "Investigation of Optical Fiber Buckling in Loose-Tube Packaging," *Electronics Letters,* vol. 14, no. 3, pp. 47–48, Feb. 1978.

40. T. A. Lenahan, "Thermal Buckling of Dual-Coated Fiber," *AT&T Technical Journal,* vol. 64, no. 7, pp. 1565–1584, 1985.

41. F. B. Seely and J. O. Smith, *Advanced Mechanics of Materials,* Wiley, New York, 1952.

42. P. G. Hale et al., "A Refined Tight Design Duct Cable for Low Loss Monomode Fiber," *Proceedings of the 31st International Wire and Cable Symposium,* pp. 371–376, Cherry Hill, N.J., 1982.

43. P. Worthington, "Application of Optical Fibre Systems in Underwater Services," *Proceedings of the IEE Conf. on Submarine Telecommunication Systems,* London, Feb. 1980.

44. T. S. Hope, R. J. Williams, and K. Abe, "Developments in Slotted Core Optical Fiber Cables," *Proceedings of the 30th International Wire and Cable Symposium,* pp. 220–227, Cherry Hill, N.J., Nov. 1981.

45. W. Schmidt and U. Zwick, "Fabrication and Test of Optical Fibre Cables for Military and PTT Applications," *Proceedings of the 29th International Wire and Cable Symposium,* pp. 306–311, Cherry Hill, N.J., Nov. 1980.

46. P. D. Patel and C. H. Gartside III, "Compact Lightguide Cable Design," *Proceedings of the 34th International Wire and Cable Symposium,* pp. 21–27, Cherry Hill, N.J., Nov. 1985.

47. P. R. Bark and D. O. Lawrence, "Low and High Fiber Count Single Mode Cables, Design and Performance," *Technical Digest of the Optical Fiber Conference,* paper TV12, New Orleans, 1983.

48. P. F. Sagen and M. R. Santana, "Design and Performance of a Cross-Ply Lightguide Cable Sheath," *Proceedings of the 28th International Wire and Cable Symposium,* pp. 391–395, Cherry Hill, N.J., Nov. 1979.

49. M. M. Ramsay and M. A. Bedgood, "Rugged Fiber-Optic Cables for Military Use," *IERE Conference Proceedings,* vol. 53, Fibre Optics, pp. 81–90, London, 1982.

50. M. A. Bedgood et al., "The Development and Testing of a Tactical Military Fiber Optic Cable," *Proceedings of the 35th International Wire and Cable Symposium,* Reno, Nov. 1986.

51. T. Brataas, "Standard Telephone og Kabelfabrik Laboratory," *Electrical Communication,* vol. 55, no. 4, pp. 382–385, 1980.

52. M. M. Ramsay, P. G. Hale, and R. Sutehall, "Single Fibre Cables for Military Applications," *Proceedings of the 34th International Wire and Cable Symposium,* pp. 9–15, Cherry Hill, N.J., Nov. 1985.

53. I. F. Scanlan, J. G. Titchmarsh, J. G. Lamb, and W. J. Duncan, "Optical Fibre under Strain: Establishing the Survival Probability in Service from Tensile Strength Data," *Proceedings of the IEE Colloquium on Implementation and Reliability of Optical Fibre Links,* London, June 1984.

54. M. Fox, "Theoretical Fatigue Life of Helically Stranded Optical Fibres," *Optical and Quantum Electronics,* vol. 15, pp. 253–260, 1983.

# CHAPTER 3

# FIBER-OPTIC SPLICES, CONNECTORS, AND COUPLERS

**Pierangelo Morra and Emilio Vezzoni**
*Centro Studi e Laboratori Telecomunicazioni S.p.A., Torino, Italy*

## 3.1 COUPLING LOSSES

### 3.1.1 Definition of Coupling Losses

An ideal joint between two identical optical fibers should completely restore the optical continuity of the guiding medium, so as not to reduce system performance. Actual joints, however, always suffer from imperfections, resulting in a reduction of the power transmitted across the joint.

If $P_0$ and $P_1$ are, respectively, the core-guided power before and after the joint (Fig. 3.1), the coupling (or transmission or insertion) efficiency $\eta$ of the joint is expressed by

$$\eta = \frac{P_1}{P_0} \tag{3.1}$$

and the fraction of power $\Lambda$ which is lost is

$$\Lambda = \frac{P_0 - P_1}{P_0} = 1 - \eta \tag{3.2}$$

The corresponding coupling (or transmission) loss $L$, expressed in decibels, is usually defined as

$$L = -10 \log \eta \tag{3.3}$$

and is the main parameter that determines the quality of fiber joints. To avoid uncertainties, the conditions under which $P_1$ and $P_0$ are considered or measured must be carefully defined, particularly for multimode fibers (see Sec. 3.1.3).

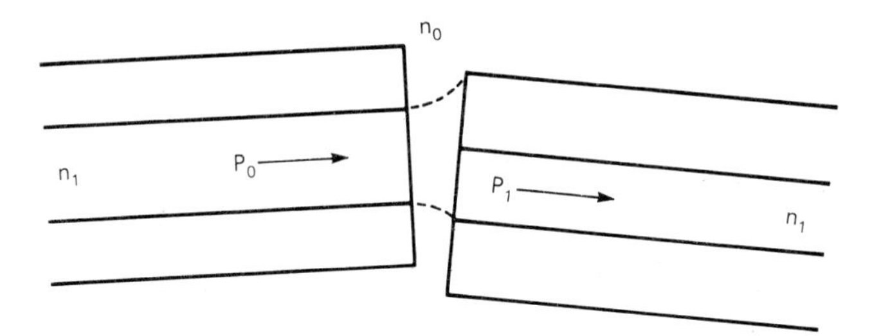


**FIGURE 3.1** Optical power transmission at an optical fiber joint.

### 3.1.2 Causes of Coupling Losses

There are several causes for reduced coupling efficiency between two optical fibers. These include reflections at glass-air interfaces, alignment errors, poor fiber-end quality, and mismatches between the parameters of the fibers being coupled.

Reflection losses, also known as Fresnel losses, are caused by the abrupt change of refractive index that occurs at the end of each fiber in a joint, assuming that the fiber ends are separated by a gap (usually air). If $n_1$ and $n_0$ are the refractive indexes of the fiber core and gap medium, respectively, of a step index fiber, the ratio $R$ of reflected-to-incident power at each fiber end, for rays at normal incidence, is given by

$$R = \left(\frac{n_1 - n_0}{n_1 + n_0}\right)^2 \tag{3.4}$$

Because of the limited acceptance angle of optical fibers, Eq. (3.4) can be used as an acceptable approximation, even if the rays are not exactly at normal incidence. Taking into account the reflections at the double fiber-air-fiber interface, the power transmission efficiency $\eta_F$ is

$$\eta_F = (1 - R)^2 = \left[\frac{4n_1n_0}{(n_1 + n_0)^2}\right]^2 \tag{3.5}$$

and corresponding reflection (or Fresnel) losses $L_F$, expressed in decibels, are

$$L_F = -10 \log \eta_F \tag{3.6}$$

Since $n_0$ is 1.00 in air, and since the refractive index of the fiber core is about 1.45 to 1.50, typical reflection losses $L_F$ at a joint range from 0.30 to 0.35 dB. If the gap is filled with an index-matching fluid or gel whose refractive index closely matches that of the core (i.e., $n_0 \approx n_1$), Fresnel losses can disappear.

Alignment errors and poor fiber-end quality are produced directly by defects in the jointing technique, rather than by inherent fiber characteristics. Therefore, the coupling losses they introduce are sometimes referred to as *extrinsic losses.*

Alignment errors are caused by mechanical imperfections in the jointing technique and are illustrated in Fig. 3.2*a*. They include:

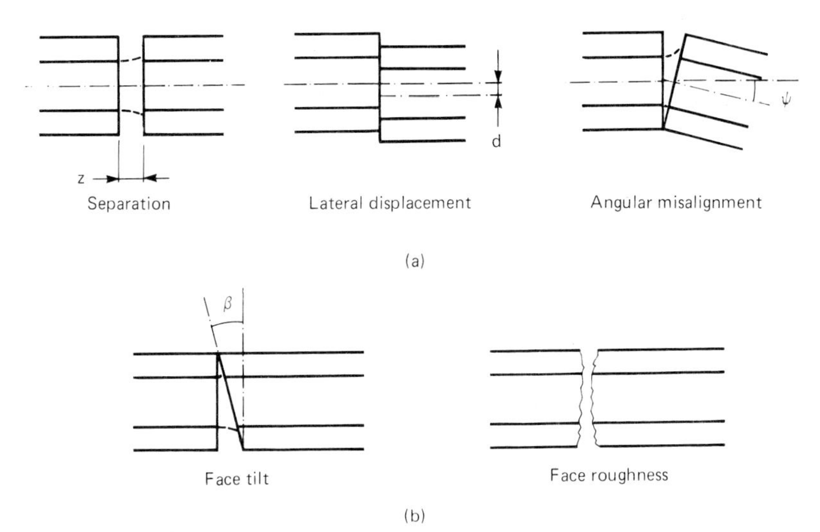


FIGURE 3.2 (*a*) Joint misalignments; (*b*) examples of improper end preparation that cause extrinsic coupling losses.

- Separation between the fiber ends
- Lateral displacement (relative) of the core axes
- Angular misalignment of the fiber axes

Furthermore, extrinsic losses can be the result of the poor quality of fiber-end faces, which, ideally, should be flat, smooth, and perfectly perpendicular to the fiber axis. Improper preparation techniques can result in nonperpendicular and uneven end faces (Fig. 3.2*b*).

In addition to extrinsic losses, there is another family of losses related to mismatches between fiber parameters (Fig. 3.3) such as:

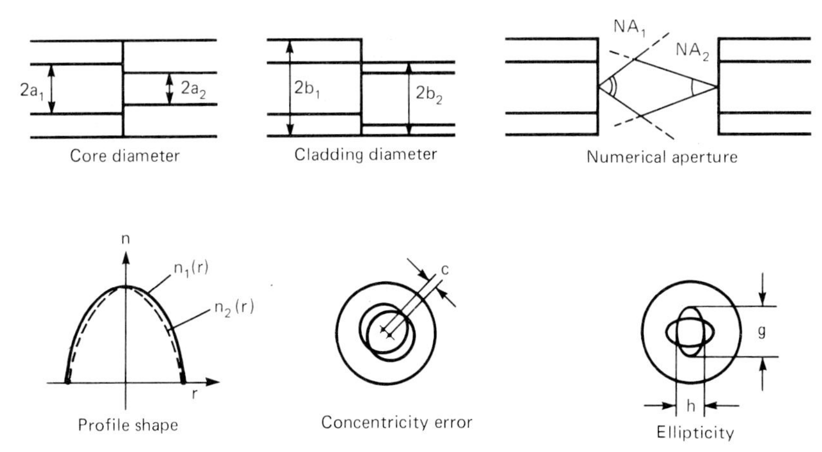


FIGURE 3.3 Mismatches between optical and geometrical fiber parameters that cause intrinsic coupling losses.

- Core diameter
- Cladding diameter
- Index difference between core and cladding (i.e., mismatch in numerical aperture)
- Profile shape

or mismatches related to geometrical imperfections, such as:

- Concentricity error between core and cladding
- Core ellipticity

All of these mismatches are associated with the tolerances in the fiber fabrication process. Since these losses are produced as a result of inherent fiber characteristics and are independent of the jointing technology, they are commonly known as *intrinsic losses.* It is worth noting that, for multimode fibers, core diameter mismatch results in coupling loss only when the launching fiber core is larger than the receiving fiber core (i.e., $a_1 > a_2$). Should this condition be reversed (i.e., $a_1 < a_2$), no loss would occur (from this mechanism), since the total power emitted by the first fiber would be intercepted by the second one. Similar considerations hold as well for numerical aperture mismatch: the coupling loss occurs only when $NA_1 > NA_2$. This is not true for single-mode fibers, as will be clarified in Sec. 3.1.5.

### 3.1.3 Coupling Losses and Mode Power Distribution in Multimode Fibers

For joints between multimode fibers, the power distribution among different propagating modes must be taken into account, because it can be responsible for:

- Variation of coupling efficiency
- Variation of the attenuation in the fiber following the joint

The mode power distribution (MPD) for a given core section depends on both the radiation pattern of the source at the launching end and on the degree of mode coupling (i.e., the transfer of power from one mode to another that occurs along the fiber). Mode coupling is produced by any perturbation or microscopic irregularity that makes the fiber depart from ideal circular waveguiding conditions, and is stronger for low-quality fibers. While propagating along the fiber, the MPD is subject to evolution, until it reaches an equilibrium condition (or steady state) that remains unchanged from that point on, unless significant further perturbations occur. The minimum fiber length required to produce the equilibrium condition depends on the fiber quality (i.e., the presence of microscopic perturbations), and can be as much as several kilometers for high-quality fibers.

In general, the power transmission efficiency across a fiber joint in the presence of misalignments or mismatches depends on the spatial distribution of power in the core; hence, MPD strongly influences coupling losses. Furthermore, a misaligned joint that perturbs the power distribution among the modes can change the attenuation in the fiber following the joint because the fiber attenuation is also a function of MPD. For example, the steady state that has been established in a long fiber can be perturbed by misalignments or intrinsic mismatches that may be present in a connection. If, for instance, higher-order modes (which have inherently higher losses) are excited, the attenuation of the fiber following the joint is increased with respect to the value that would be obtained under unperturbed equilibrium conditions. This attenuation increase can be regarded as a supplementary loss produced by the joint, and therefore part of the total coupling loss. After the radiation has traveled a sufficient distance in the fiber to reach modal equilibrium again, no additional attenuation is produced by this mechanism (unless other perturbations occur).

Two extreme MPDs are usually considered: the *uniform* MPD (corresponding to the uniform excitation of all the modes of the fiber) and the *equilibrium* MPD (corresponding to the steady-state distribution previously discussed). Analytical expressions for such MPDs under representative conditions follow.

Consider a generic multimode fiber having a circular core cross section of radius $a$ and having a refractive index profile following the power law:

$$n(r) = n_1\left[1 - 2\Delta\left(\frac{r}{a}\right)^{\alpha}\right]^{1/2} \qquad \text{for } r \leq a \tag{3.7a}$$

$$n(r) = n_1(1 - 2\Delta)^{1/2} \qquad \text{for } r > a \tag{3.7b}$$

where $r$ = radial coordinate
$n_1$ = refractive index at center of the core
$n_2$ = refractive index of cladding
$\Delta$ = relative index difference $[\approx (n_1 - n_2)/n_1]$
$\alpha$ = refractive-index profile-shape parameter

For step-index fibers, $\alpha$ is $\infty$, and for graded-index parabolic profile fibers, $\alpha$ is 2.0.

Modes propagating along the fiber core can be identified and grouped by means of a mode parameter $\chi$, which ranges between 0 and 1, and, within the limits of ray optics, is defined[1] by

$$\chi = \left[\left(\frac{r}{a}\right)^{\alpha} + \left(\frac{\sin\theta}{\sin\theta_c}\right)^2\right]^{(\alpha+2)/2\alpha} \tag{3.8a}$$

where $\theta$ = propagation angle and $\theta_c$ = critical value of $\theta$ inside the fiber.

The group of modes defined by a specified $\chi$ value can be considered as a group of rays that, while propagating, change angle $\theta$ and position $r$ according to Eq. (3.8*a*). For step-index and parabolic graded-index fibers, Eq. (3.8*a*) reduces, respectively, to

$$\chi = \frac{\sin\theta}{\sin\theta_c} \qquad \text{step} \tag{3.8b}$$

and

$$\chi = \left(\frac{r}{a}\right)^2 + \left(\frac{\sin\theta}{\sin\theta_c}\right)^2 \qquad \text{parabolic} \tag{3.8c}$$

The power $P$ that is carried by each mode individually (i.e., MPD) can be expressed as a function $P(\chi)$ of the mode parameter $\chi$. The total power $P_t$ carried by the fiber is

$$P_t = \int_0^1 P(\chi)N(\chi)\, d\chi \tag{3.9}$$

where $N(\chi)$ is the number of individual modes identified by the same mode parameter $\chi$.

In the case of uniform MPD, each mode carries the same amount of power. In this case, regardless of index profile,

$$P(\chi) = \text{constant} \tag{3.10}$$

Analytical expressions for equilibrium MPD are[2]

$$P(\chi) = A_s J_0(2.405\sqrt{\chi}) \tag{3.11a}$$

and

$$P(\chi) = A_p \frac{J_1(3.832\sqrt{\chi})}{\sqrt{\chi}} \tag{3.11b}$$

$J_0$ and $J_1$ are the Bessel functions of 0th and 1st order, respectively, while $A_s$ and $A_p$ are proper constants. The former expression applies better to step-index fibers and the latter applies preferably to parabolic graded-index fibers.[3]

### 3.1.4 Coupling Losses in Multimode Fibers

Simple analytical expressions relating misalignments or mismatches to coupling losses are needed to define corresponding tolerances for efficient coupling devices. A general approach to the problem, based on ray optics,[3,4] is difficult to handle, however. Therefore, only simple—yet reasonable—approximations[3–8] are, in the following, provided for the limited purpose of clarifying the influence of certain important parameters on coupling efficiency. These simplified expressions are listed in Table 3.1. Each expression represents the value of $\Lambda$ (i.e., the fraction of input power that is lost because of the joint) as a function of each type of misalignment or mismatch. These are suitably normalized with respect to fiber parameters as follows:

$$s = \frac{z}{a}\frac{\text{NA}}{n_0} = \text{normalized longitudinal separation} \tag{3.12}$$

$$u = \frac{d}{a} = \text{normalized lateral displacement} \tag{3.13}$$

$$t = \frac{n_0 \sin \psi}{\text{NA}} = \text{normalized angular misalignment} \tag{3.14}$$

$$p = \frac{a_2}{a_1} = \text{core radius ratio } (\leq 1) \tag{3.15}$$

$$q = \frac{\text{NA}_2}{\text{NA}_1} = \text{numerical aperture ratio } (\leq 1) \tag{3.16}$$

**TABLE 3.1** Fraction of Power $\Lambda$ Lost Due to Coupling Misalignments and Mismatches (Small Values)

*For various MPD and measurement conditions, in multimode optical fibers having step or parabolic refractive index profile. Coupling loss* $L = -10 \log (1 - \Lambda)$.

| Type of misalignment or mismatch | I Uniform MPD | | II Equilibrium MPD, loss at the joint | | III Equilibirum MPD, loss far from the joint | |
|---|---|---|---|---|---|---|
| | Step | Graded | Step | Graded | Step | Graded |
| Separation: | | | | | | |
| $s = \frac{z}{a} \frac{NA}{n_0}$ | $0.42s$ | $0.5s$ | | $0.6s^2$ | | $0.4s^2$ |
| Lateral displacement: | | | | | | |
| $u = \frac{d}{a}$ | $0.64u$ | $0.85u$ | $0.64u$ | $0.92u^2$ | $0.64u$ | $1.84u^2$ |
| Angular misalignment: | | | | | | |
| $t = n_0 \frac{\sin \psi}{NA}$ | $0.64t$ | $0.85t$ | $0.72t^2$ | $0.92t^2$ | $1.44t^2$ | $1.84t^2$ |
| Radius mismatch: | | | | | | |
| $p = \frac{a_2}{a_1} \quad (\leq 1)$ | $1 - p^2$ | $1 - p^2$ | $1 - p^2$ | $2.45(1 - p)^2$ | $1 - p^2$ | $1 - p$ |
| Numerical aperture mismatch: | | | | | | |
| $q = \frac{NA_2}{NA_1} \quad (\leq 1)$ | $1 - q^2$ | $1 - q^2$ | $1.44(1 - q)^2$ | $2.45(1 - q)^2$ | | $1 - q$ |
| Ellipticity: | | | | | | |
| $e = 1 - \frac{h}{g} \quad (h \leq g)$ | $0.64e$ | $0.64e$ | $0.64e$ | $0.6e$ | $0.64e$ | $0.4e$ |

$$e = 1 - \frac{h}{g} = \text{ellipticity} \tag{3.17}$$

where $z$, $d$, and $\psi$ = absolute misalignments as defined in Fig. 3.2*a*
$a$ = core radius ($a_1$ and $a_2$, if mismatched, as defined in Fig. 3.3)
NA = fiber numerical aperture ($NA_1$ and $NA_2$, if mismatched, as defined in Fig. 3.3)
$n_0$ = refractive index of the medium between the fiber ends
$h$ and $g$ = minor and major axes, respectively, of the elliptical fiber core ($h = g$ when the core is perfectly circular)

When ellipticity is considered, it is assumed that each fiber has the same values of $h$ and $g$ and that the worst-case mismatch condition applies, that is, a relative rotation of 90° between the major axes of the two cores.

Core and numerical aperture mismatches are often expressed as relative values:

$$e_a = \frac{a_1 - a_2}{a_1} = 1 - p \tag{3.18}$$

$$e_{NA} = \frac{NA_1 - NA_2}{NA_1} = 1 - q \tag{3.19}$$

When numerical aperture mismatch is considered, the two fibers are assumed to follow the same profile law [Eq. (3.7)] and are assumed to have the same $\alpha$ value, but with different relative index differences, $\Delta_1$ and $\Delta_2$.

Three significant conditions are considered in Table 3.1, referring to the MPD at the joint and to the position, after the joint, where coupling losses are measured:

I—uniform MPD, with loss measured just after the joint (first and second columns)

II—equilibrium MPD, with loss measured just after the joint (third and fourth columns)

III—equilibrium MPD, with loss measured far enough from the joint that equilibrium conditions have been restored (fifth and sixth columns)

Although actual joints usually operate under intermediate conditions, these extreme conditions are nonetheless useful to delimit the expected losses when coupling devices are designed. In principle, the first configuration can be approached when two very short fibers are coupled and the source is not highly directive [e.g., a surface light-emitting diode (LED)]. The second configuration is approximated by a joint following a long fiber, just before the detector. The third configuration is representative of a joint between two long fibers.

Both step-index and parabolic graded-index profiles are taken into account, for mode power distributions as in Eqs. (3.10) and (3.11). In general, the relationships in Table 3.1 are subject to some limitations and simplifications:

- Fibers are assumed to be weakly guiding ($\Delta \ll 1$).
- Misalignments and mismatches are assumed to be small.
- The effect of leaky modes is neglected.

The second assumption is particularly reasonable, since, in the case of low-loss connectors, small misalignments and mismatches are design goals.

The coupling loss $L$ is directly obtained from the loss value $\Lambda$ determined via Table 3.1 and Eqs. (3.2) and (3.3):

$$L = -10 \log (1 - \Lambda) \quad \text{dB} \tag{3.20}$$

For the third modal configuration (fifth and sixth columns), $\Lambda$ and, therefore, $L$ already include the effect of a long fiber following the joint. Fresnel losses $L_F$, determined from Eq. (3.6), should be added to $L$ to establish the total loss $L_T$. Some examples and appropriate comments are, in the following, provided.

***Separation.*** A typical parabolic graded-index optical fiber for telecommunications is considered, whose parameters are:

- $a = 25$ μm
- NA $= 0.20$
- $n_1 = 1.46$

For uniform MPD and a 10-μm separation, coupling loss $L$, with and without index matching, can be calculated by means of Table 3.1. Fiber parameters and the subsequently determined coupling loss are:

- For $z = 10$ μm and $n_0 = 1$: $s = 0.08$, $\Lambda = 0.50$ s $= 4\%$, and $L = 0.18$ dB
- For $z = 10$ μm and $n_0 = 1.46$: $s = 0.05$, $\Lambda = 0.50$ s $= 2.7\%$, and $L = 0.12$ dB

The presence of index matching, which reduces the output angle from the launching fiber, decreases the loss $L$. Fresnel losses $L_F$ of 0.31 dB should be added to $L$ in the first case to obtain a total loss $L_T$ of 0.49 dB. Should the equilibrium MPD and a long fiber after the joint be assumed, the above figures would be reduced by an order of magnitude.

Since even demountable connectors can easily bring the fibers almost into contact (i.e., within a few micrometers), separation is usually not a critical parameter, unless interference effects occur (see Sec. 3.3.3).

***Lateral Displacement.*** Figure 3.4 shows the coupling losses $L(u)$ due to lateral displacement $u$, calculated according to the relationships in Table 3.1, for a parabolic graded-index profile and for the three specified MPDs and measurement conditions (indicated as I, II, and III).

For the same fiber parameters as in the previous example, and allowing for equilibrium MPD (as in a long fiber following a joint), the loss induced by a 5-μm lateral displacement is calculated on the basis of Table 3.1:

- For $d = 5$ μm: $u = 0.2$, $\Lambda = 1.84u^2 = 7.4\%$ and $L = 0.33$ dB

The presence of index matching does not affect the loss $L$ (although it eliminates Fresnel losses).

It appears that, in order to keep connnector loss below 0.3 dB (excluding Fresnel losses), lateral displacements smaller than 5 μm must be maintained. This tolerance requires accurate machining of the parts for a demountable connector. If the more critical, uniform MPD were to be considered, this tolerance would be further tightened.

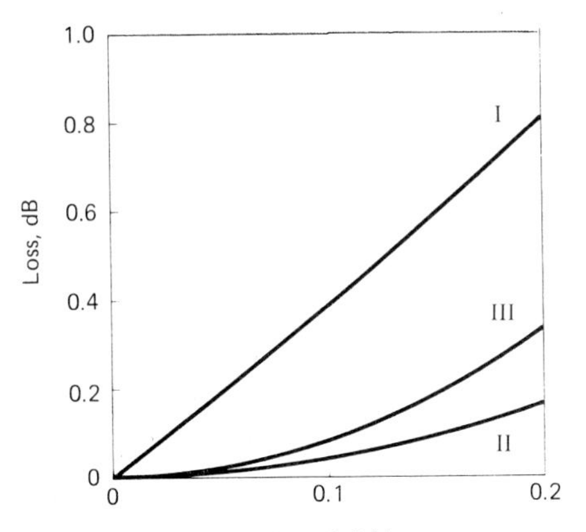


**FIGURE 3.4** Coupling loss for parabolic graded-index optical fibers as a function of normalized lateral displacement or normalized angular displacement, for the three measurement conditions (I, II, and III) listed in Table 3.1.

*Angular Misalignment.* The same curves in Fig. 3.4 represent the coupling losses $L(t)$ that are produced by angular misalignment (given the assumption of a parabolic, graded-index profile) for the three specified MPDs and measurement conditions, as the numerical coefficients are the same for both $L(u)$ and $L(t)$.

With the same fiber parameters as in the previous examples and assuming equilibrium MPD and a long fiber after the joint, the loss induced by a 1° angular misalignment can be calculated from Table 3.1:

- For $\psi = 1°$ and $n_0 = 1$: $t = 0.09$, $\Lambda = 1.84t^2 = 1.4\%$, and $L = 0.06$ dB
- For $\psi = 1°$ and $n_0 = 1.46$: $t = 0.13$, $\Lambda = 1.84t^2 = 3.0\%$, and $L = 0.13$ dB

The presence of index matching reduces the emission angle from the fiber and increases the sensitivity of the joint to angular misalignments.

Since, in the case of demountable connectors, misalignment $\psi$ can be kept within 1° without great effort, the corresponding loss is relatively low. Uniform MPD would result in higher losses. Realistically, pure angular misalignment loss is impossible (if fiber-end faces are perpendicular to the corresponding axes), since the fiber ends cannot be tilted and still be in close-enough proximity (across the entire fiber end) to ensure capture of all radiation. Therefore, losses due to angular misalignment are always concomitant with some amount of end separation (see Fig. 3.2*a*) that increases total loss.

*Core Radius Mismatch.* Figure 3.5 shows the coupling losses $L(p)$ for parabolic graded-index fibers resulting from core radius mismatch. The curves are calculated according to the relationships in Table 3.1 and represent the three specified MPDs and measurement conditions (indicated as I, II, and III).

As an example two parabolic graded-index optical fibers are considered, with $a_1 = 25$ μm and $a_2 = 23.75$ μm (i.e., $e_a = 5\%$). Assuming equilibrium MPD and a long following fiber, the corresponding loss can be calculated from Table 3.1:

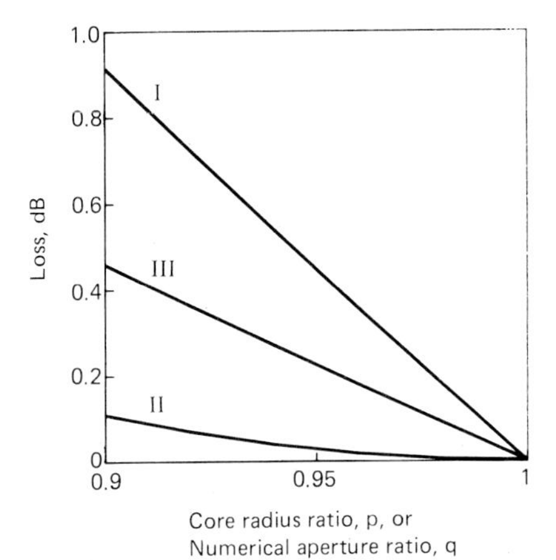


**FIGURE 3.5** Coupling loss for parabolic graded-index optical fibers resulting from core radius mismatch or numerical aperture mismatch, for the three MPD and measurement conditions listed in Table 3.1.

- For $p = 0.95$: $\Lambda = 1 - p = 5\%$ and $L = 0.22$ dB

Again, higher loss would occur if uniform MPD were considered.

*Numerical Aperture Mismatch.* The same curves in Fig. 3.5 represent the coupling losses $L(q)$ due to numerical aperture mismatch, since the numerical coefficients are the same for both $L(p)$ and $L(q)$. Again, a parabolic graded-index profile is assumed, and the curves represent the three specified MPD and measurement conditions in Table 3.1.

As an example, two graded-index optical fibers are considered, with $NA_1 = 0.20$ and $NA_2 = 0.19$ (i.e., $e_{NA} = 5\%$). Assuming equilibrium MPD and a long fiber after the joint, loss can be calculated by means of Table 3.1:

- For $q = 0.95$: $\Lambda = 1 - q = 5\%$ and $L = 0.22$ dB

As expected, uniform MPD would result in higher loss.

*Ellipticity.* As an example of the effect of ellipticity, assume that two parabolic graded-index optical fibers have the same degree of ellipticity (but with major axes at 90°). Referring to Fig. 3.3, minor and major axes are $h = 48.6$ μm and $g = 51$ μm, for an ellipticity $e = 5\%$. Given uniform MPD, loss is calculated in accordance with Table 3.1:

- For $e = 0.05$: $\Lambda = 0.64e = 3.2\%$ and $L = 0.14$ dB

*Reference Surface Mismatch.* A mismatch between the diameters $2b_1$ and $2b_2$ of the outer fiber reference surfaces (usually the cladding, but sometimes the primary coating), as illustrated in Fig. 3.3, can be important. For instance, when a V-groove structure is used for fiber alignment, this type of mismatch results in a lateral offset between the centers of the cores (the cores are assumed to be centered within their respective reference surfaces). If $2\beta$ is the angle of the V groove, the amount of lateral displacement $d$ between centers of the cores is given by $d = (b_1 - b_2)/\sin\beta$. The relationships in Table 3.1 referring to lateral displacement $d$ can be applied. A narrow groove will aggravate the effect of reference surface mismatch.

***Core–Reference-Surface Concentricity Error.*** Once the amount of concentricity error between core and reference surface (cladding or, sometimes, primary coating) of the two fibers is known, the coupling loss, corresponding to the worst case relative displacement $c$ (Fig. 3.3) of the two core axes, can be calculated.

***Refractive Index Profile Mismatch.*** Differences between the shapes of the index profiles, i.e., differences between the exponent $\alpha$ in Eq. (3.7) or resulting from profiles described by laws that are inconsistent with the power law given in Eq. (3.7), can introduce coupling losses. Simple, general expressions, however, are not available in this case.

### 3.1.5 Coupling Losses in Single-Mode Fibers

Single-mode fibers are more sensitive to alignment errors than multimode fibers because of the smaller size of the waveguide core.

An exact evaluation of coupling losses for arbitrary index profiles can be performed by numerical techniques.[9] When small coupling errors are considered (which is realistic, since high-performance single-mode systems are not compatible with high coupling losses), accurate analytical expressions can be obtained,[10] but these must be adjusted for the specific index profile in question.

Nonetheless, straightforward relationships can be obtained[11] when the field distribution within the fiber is approximated by the gaussian function:

$$E(r) = E_0 e^{-r^2/w^2} \tag{3.21}$$

where $r$ = radial coordinate and $w$ = appropriate half-width parameter [corresponding to the $1/e$ value of the relative field amplitude, $E(r)/E_0$]. The best match between gaussian and exact field distributions, for step-index profiles, is obtained when $w$ satisfies the relationship

$$\frac{w}{a} = 0.65 + \frac{1.619}{V^{3/2}} + \frac{2.879}{V^6} \tag{3.22}$$

where $a$ = core radius and $V$ = normalized frequency [$V = (2\pi a/\lambda)(n_1^2 - n_2^2)^{1/2}$].

Different refractive index profiles require different optimum $w/a$ values. The mode field radius $w$ can be obtained experimentally, for instance, by fitting a gaussian curve to the measured near-field intensity pattern for a given fiber. The gaussian approximation is good enough in the frequency range of practical interest to allow a meaningful evaluation of the influence of misalignments and mismatches on single-mode fiber coupling efficiency. Care should be taken, however, when considering more complex refractive index profiles (e.g., profiles having a parabolic shape with a central dip or profiles for multiple-core fibers).

Within the limits of the gaussian approximation, the coupling efficiency that results after allowing for a combination of:

Separation $z$
Lateral displacement $d$
Angular misalignment $\psi$
Mismatch between mode field radii $w_1$ and $w_2$

is expressed by[12]

$$\eta = \left(\frac{4D}{B}\right)e^{-AC/B} \tag{3.23}$$

where
$$A = (kw_1)^2/2 \qquad B = G^2 + (D + 1)^2$$
$$C = (D + 1)F^2 + 2DFG \sin\psi + D(G^2 + D + 1)\sin^2\psi$$
$$D = (w_2/w_1)^2 \qquad F = 2d/kw_1^2$$
$$G = 2z/kw_1^2 \qquad k = 2\pi n_0/\lambda$$

As usual, $n_0$ is the refractive index of the medium between the fiber ends and $\lambda$ is the wavelength. Coupling loss in decibels is given, as usual, by $L = -10 \log \eta$.

Simpler expressions can be obtained,[11] consistent with Eq. (3.23), when only one misalignment or mismatch is present at one time, or for a simple combination of misalignments. These expressions follow.

***Separation.*** For separation only and no mode field radius mismatch (i.e., $w_1 = w_2 = w$):

$$\eta = \frac{1}{Z^2 + 1} \tag{3.24}$$

where $Z = z\lambda/2\pi n_0 w^2$.

For example, assume $\lambda$ is 1.3 μm, $n_0$ is 1, and $w$ is 5 μm. If the separation $z$ is 10 μm, the coupling loss will be $L = -10 \log 0.99 = 0.03$ dB. [*Note:* This value does not include Fresnel losses, which are significant. Also, interference effects (see Sec. 3.3.3) are neglected here.]

***Lateral Displacement.*** For lateral displacement only and no mode field radius mismatch (i.e., $w_1 = w_2 = w$):

$$\eta = e^{-U^2} \qquad U = \frac{d}{w} \tag{3.25}$$

For example, consider the same assumptions as in the previous case, except that the separation is zero. If the lateral displacement is 1 μm, the coupling loss will be $L = -10 \log 0.96 = 0.17$ dB (excluding Fresnel losses).

***Angular Misalignment.*** For angular misalignment only and no mode field radius mismatch (i.e., $w_1 = w_2 = w$):

$$\eta = e^{-T^2} \qquad T = \frac{\sin\psi}{\lambda/n_0\pi w} \tag{3.26}$$

For example, keep the same general assumptions as in the first case, and let the separation and the displacement be zero. If the angular misalignment is 1°, the coupling loss would be $L = -10 \log 0.96 = 0.19$ dB (excluding Fresnel losses).

*Combination of Lateral Displacement and Angular Misalignment.* For a combination of lateral displacement and angular misalignment and no mode field radius mismatch (i.e., $w_1 = w_2 = w$):

$$\eta = e^{-(U^2 + T^2)} \tag{3.27}$$

For example, keep the same general assumptions as in the first case, and let the separation be zero. If the angular misalignment $\psi$ is 1° and the displacement $d$ is 1 μm, the coupling loss would be $L = -10 \log 0.92 = 0.37$ dB (excluding Fresnel losses).

*Mode Field Radius Mismatch.* For mode field radius mismatch only:

$$\eta = \frac{4}{[(w_2/w_1) + (w_1/w_2)]^2} \tag{3.28}$$

It is worth noting that single-mode fibers suffer from radial mismatch for both $w_1 > w_2$ and $w_1 < w_2$. This is specifically different from the case with multimode fibers; with single-mode joints, loss occurs even if the launching fiber has a smaller mode field radius than the receiving fiber.

As an example, keep the same general assumptions as in the first case. If $w_1$ is 4.5 μm and $w_2$ is 5.5 μm, the coupling loss due to core radius mismatch will be $L = -10 \log 0.96 = 0.17$ dB (excluding Fresnel losses).

Should the mismatch be reversed (i.e., $w_1$ is 5.5 μm and $w_2$ is 4.5 μm), the same loss would occur.

Figure 3.6*a* and *b* illustrates coupling loss as a function of misalignment parameters and is obtained from Eqs. (3.24) to (3.26). Figure 3.6*c* illustrates coupling loss as a function of mode field radius ratio $w_1/w_2$ and is obtained from Eq. (3.28). Note that Fresnel losses are not considered.

## 3.2 SPLICES

Fiber splicing is needed to permanently join optical cable sections in those cases when the system span is longer than available factory cable lengths or when installable cable sections are restricted by other, practical limitations. Because fiber splices must establish transmission continuity along the optical link, obtaining and maintaining the lowest possible splicing loss is usually a primary objective. This is particularly true for links operating at 1.3 μm or 1.55 μm over long distances.

The splicing procedure must be as free of alignment errors as possible. Figure 3.7 illustrates the effect of lateral displacement, the most troublesome of the various types of misalignment, on splicing loss. The fibers represented, 50-μm (core size) graded-index multimode and 10-μm (core size) single-mode, are widely used. As shown, it is necessary to keep lateral displacement below about 3 μm and 0.75 μm, respectively, in order to achieve splicing losses of less than 0.1 dB.

Likewise, good matching of the optical and geometrical properties of the fi-

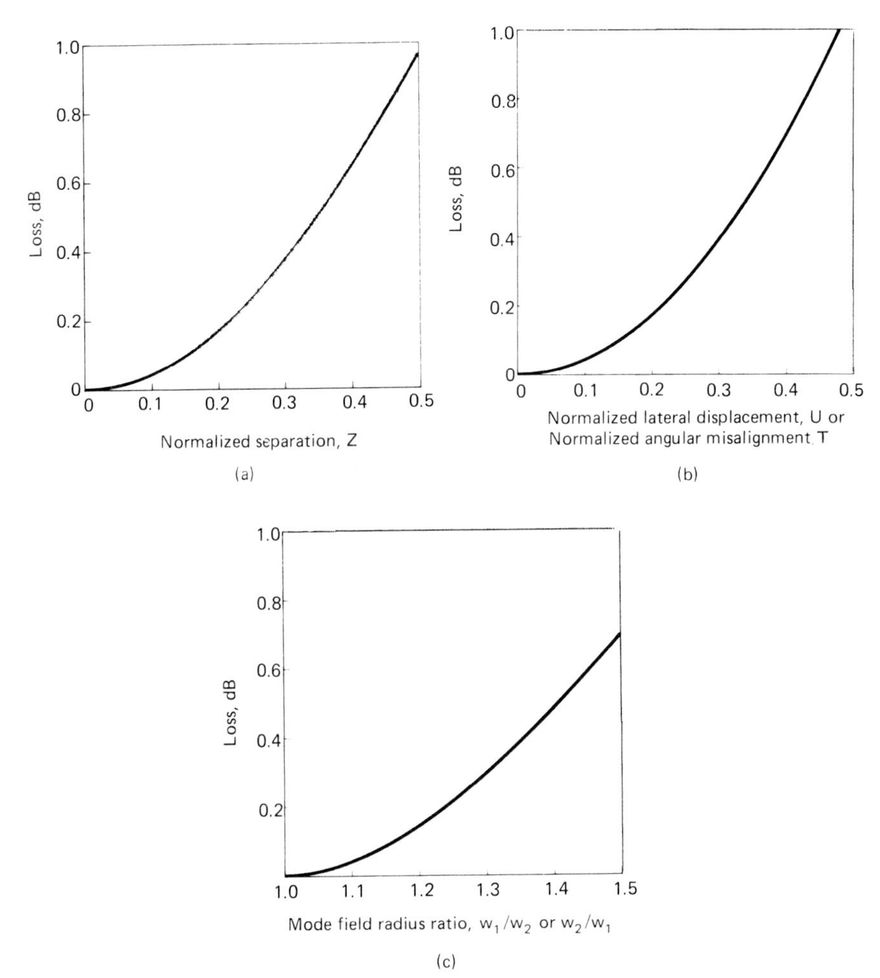


**FIGURE 3.6** Coupling loss for single-mode fibers as a function of (*a*) normalized separation $Z$; (*b*) normalized lateral displacement $U$ or normalized angular misalignment $T$; (*c*) mode field radius mismatch, expressed by the ratio $w_1/w_2$ or $w_2/w_1$.

bers is necessary in order to avoid the introduction of additional, intrinsic losses. Good matching can be achieved only by maintaining stringent fiber tolerances during the production process.

Furthermore, low-loss splices call for high-quality preparation of the fiber ends, which must be smooth, flat, and perpendicular to the fiber axis. A simple and widely used method to attain these criteria consists of scoring the fiber with a diamond blade, then bending the fiber at an appropriate radius of curvature and applying tension (as in Fig. 3.8). This generates a clean transverse break such that the surfaces of the resultant fiber faces are of optical grade. Various fiber-cleaving tools based on this technique have been developed to ensure consistent, acceptable results.

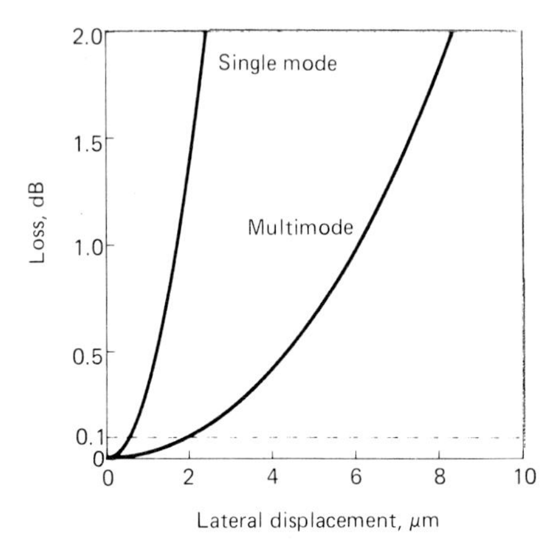


**FIGURE 3.7** Splicing loss as a function of lateral displacement for typical single-mode (10-μm core size) and multimode (50-μm core size, graded-index) fibers.

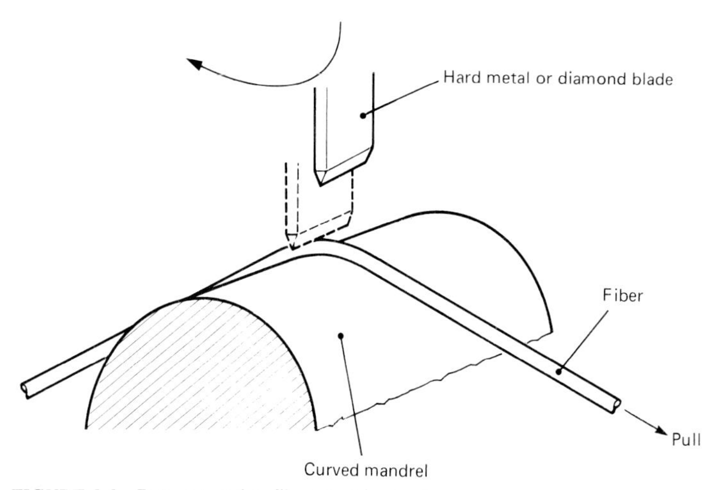


**FIGURE 3.8** Representative fiber-cleaving technique.

Current splicing techniques can be divided into two general categories: mechanical splicing and fusion splicing (i.e., welding).

### 3.2.1 Mechanical Splices

Mechanical splices make use of suitable fixtures to obtain good alignment and to permanently hold the fiber ends in alignment. Many devices have been developed to serve these goals. The most popular devices can be divided into two broad categories: capillary splices and groove splices.

A glass or ceramic capillary, with an inner diameter only slightly larger than the outer diameter of the fiber, provides for a very simple splicing device, as shown in Fig. 3.9. The fiber ends are gently inserted into the capillary and a

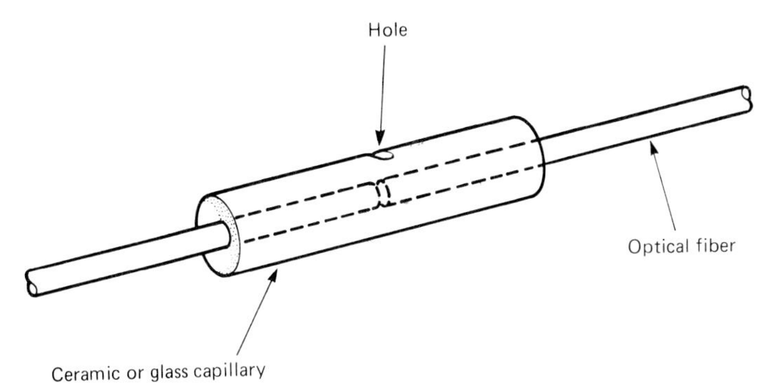


**FIGURE 3.9** Representative capillary splicing device.

transparent glue (usually epoxy adhesive) is then injected through a small, transverse hole. The transparent glue ensures both efficient bonding and good index matching, thus substantially eliminating Fresnel losses while securing the stability of the splice. The critical feature of this splicing technique is the inside diameter of the capillary, as too large a clearance between the fiber and the capillary bore results in significant misalignment, and too tight a clearance makes the fiber insertion difficult.

A similar technique, one relying on a different alignment principle, adopts a loose capillary with a square cross section.[13] The two fibers are inserted into the square capillary and bent in the same direction, and the fiber ends are thus forced against one of the four inner corners of the capillary, as shown in Fig. 3.10.

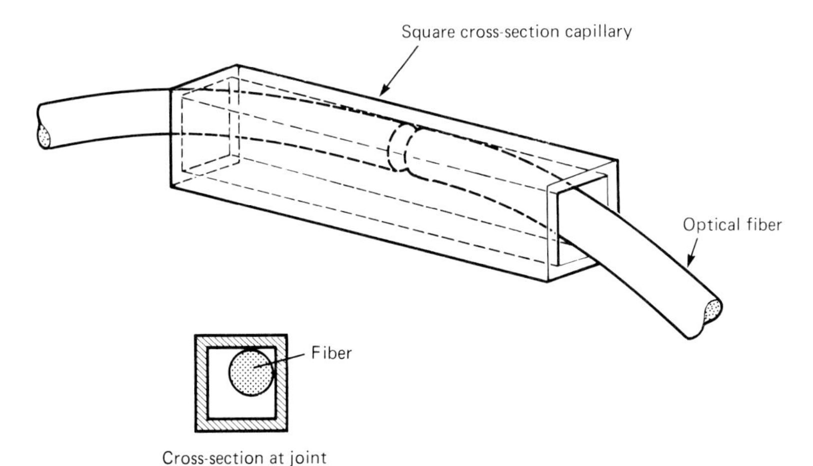


**FIGURE 3.10** Square-section, loose capillary splicing device. *(Adapted from Ref. 13 and reproduced with permission from the AT&T Technical Journal, copyright 1977, AT&T.)*

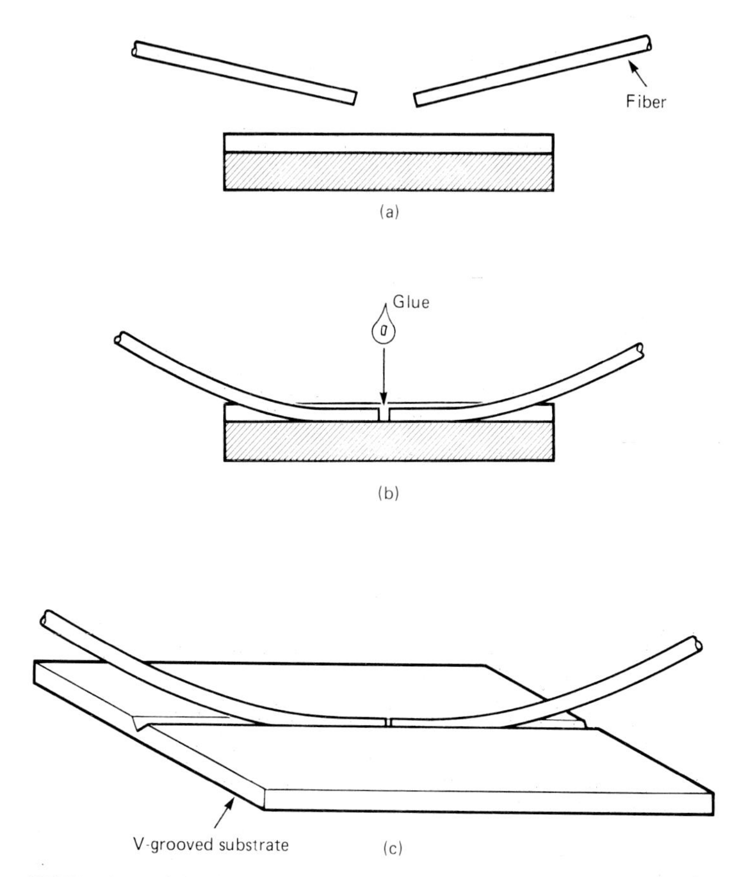


FIGURE 3.11 Steps in the V-groove splice technique: (*a*) Placement of the fibers; (*b*) immobilization of the fibers; (*c*) the completed splice.

Other commonly used splices rely on an open groove to carry out fiber alignment. Basically, the reference cylindrical surfaces of the fiber (e.g., cladding or primary coating) are kept in position by the two plane surfaces of a V groove, as shown in Fig. 3.11. Bonding and index matching are usually ensured by a transparent epoxy adhesive.

A suitable groove can be obtained by setting two precision pins close to each other.[14] If an appropriate pin diameter has been selected, the fibers are nested by the cusp, but project over the pins (as in Fig. 3.12). A flat spring applies pressure on the fibers so that the fibers seat in the groove, and an adhesive is provided for both fiber bonding and index matching.

Three pins[15] can be used as well to grip the fiber, as in Fig. 3.13. The pins are held in position and tightened by means of an elastic band or shrinkable tube.

It is noted that, when a V-groove structure is used for fiber splicing, a difference in the outer diameter of the two fibers results in a lateral displacement of the cores.

The V-groove principle can be applied to realize an elastomeric splice,[16] as in

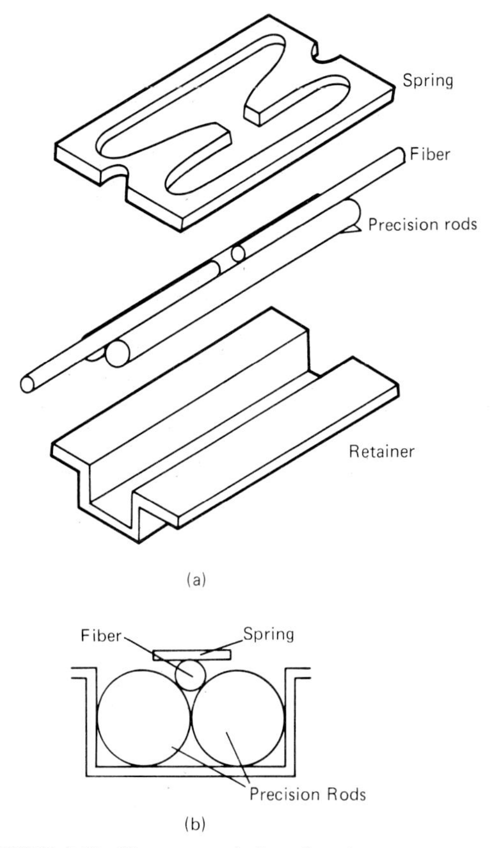


**FIGURE 3.12** V-groove technique based on two rods (Springroove). (*a*) Exploded view, showing spring, fibers on rods, and retainer; (*b*) cross section.

Fig. 3.14, in which the fiber is pressed between a grooved surface and a flat one. The depth of the 60° groove is such that the fiber is not completely buried. Once the two parts are inserted into a sleeve, the upper part, pressing against the lower part, holds the fibers tightly in the triangular cavity that is formed. A resilient spring action is produced by the elastomeric constitution of the material forming the upper and lower halves of the splice.

Most of these techniques, properly carried out with multimode fibers, result in splice losses on the order of 0.1 dB or less. Some of them can be applied to single-mode fibers as well.

All of these techniques make use of passive alignment, relying on precision reference surfaces (i.e., grooves or cylindrical holes). If very low splicing losses are required, particularly with single-mode fibers, active alignment can be utilized. In this case, each bare fiber end is inserted and bonded within a precision capillary, whose face is then optically polished. After preparation, the two capillaries are butted and the cores are precisely aligned by means of a micropositioning jig. The core alignment procedure can be performed by injecting

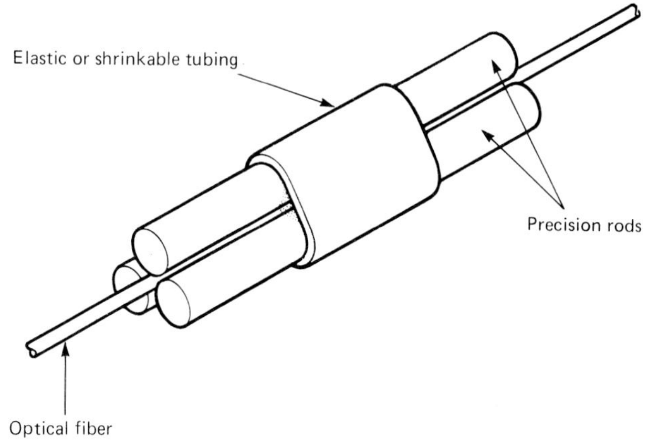


**FIGURE 3.13** A V-groove fiber splice using three rods.

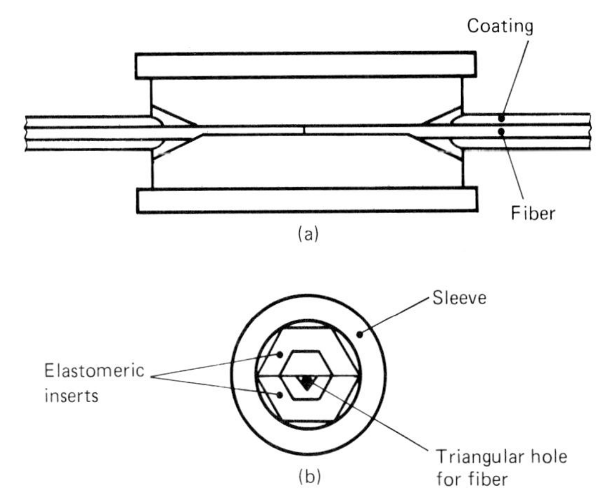


**FIGURE 3.14** Elastomeric splice. (*a*) Longitudinal section, with fibers in place; (*b*) cross section. (*Adapted from Ref. 16.*)

optical power into the input fiber and detecting the transferred radiation following the joint, using methods similar to those adopted for precision fusion splicing (see Sec. 3.2.2). Once the transferred power has been maximized (by adjusting the relative position of the launching and receiving assemblies), the two capillaries are fixed in position with an ultraviolet-curable resin. Once the resin has cured, an external mechanical reinforcement is applied. This approach inherently avoids the effects of core concentricity errors. Alternatively, a precision sleeve can be used to passively align the two capillaries. In this case, a partial compensation of core-cladding nonconcentricity can be obtained by reciprocal axial rotation of the capillaries.

This technique can be considered a valid alternative to fiber fusion for very low loss, single-mode fiber splicing. Comparative advantages include the lack of need for electric discharge and the absence of any core distortion (since no glass fusion takes place). On the other hand, a somewhat longer procedure is required. By using the active technique, splicing losses as low as 0.02 dB have been achieved for single-mode fibers.

### 3.2.2 Fusion Splices

An efficient way to permanently join optical fibers in the field relies on fiber fusion. By this technique, the two fiber ends are welded together with the possibility of recreating almost perfect continuity of the dielectric waveguide.

Arc discharge is widely used as a heating source, although a microflame or $CO_2$ laser is also used, with good results. The basic setup for fusion splicing by arc discharge is shown in Fig. 3.15. The splicing process goes through several

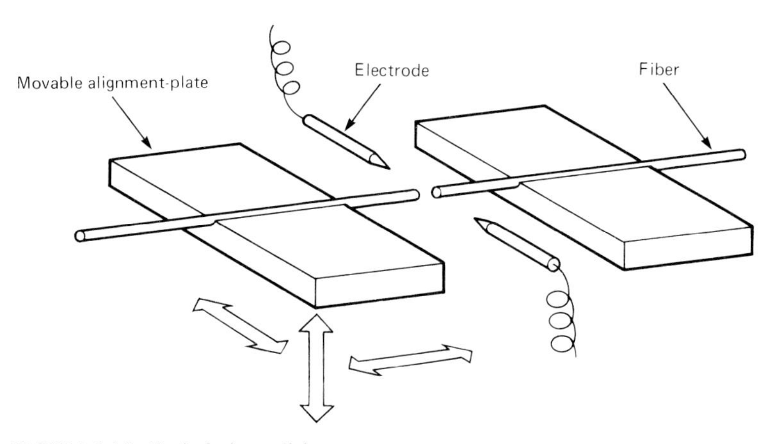


**FIGURE 3.15** Basic fusion-splicing apparatus.

steps, as in Fig. 3.16. First, the fiber ends are cleaved carefully and then the fibers are aligned by means of a precision jig. The initial alignment of the fibers is usually established with the aid of an inspection microscope.

The next step is to apply a short discharge while the fibers are separated by a short gap. This step eliminates possible surface defects, perhaps due to imperfect cleaving, by "fire polishing" the fiber ends. Removal of defects reduces the possibility of core distortion or bubble formation during the next step.

After accurate alignment of the fibers (described in the following paragraph), the final step involves butting the fibers together with the appropriate pressure while the main discharge is applied, thus producing the desired fusion splice.

A peculiar effect that occurs during the fusion process is the self-alignment of the fiber ends, as shown in Fig. 3.17. During fusion, when the glass becomes soft, surface tension tends to align the outer surface of the fibers. Although this alignment mechanism can compensate for small transverse displacements, it usually brings about a small core distortion (Fig. 3.17*c*). The effects of this distortion are usually negligible for multimode fibers, but can be detrimental for single-mode fibers.

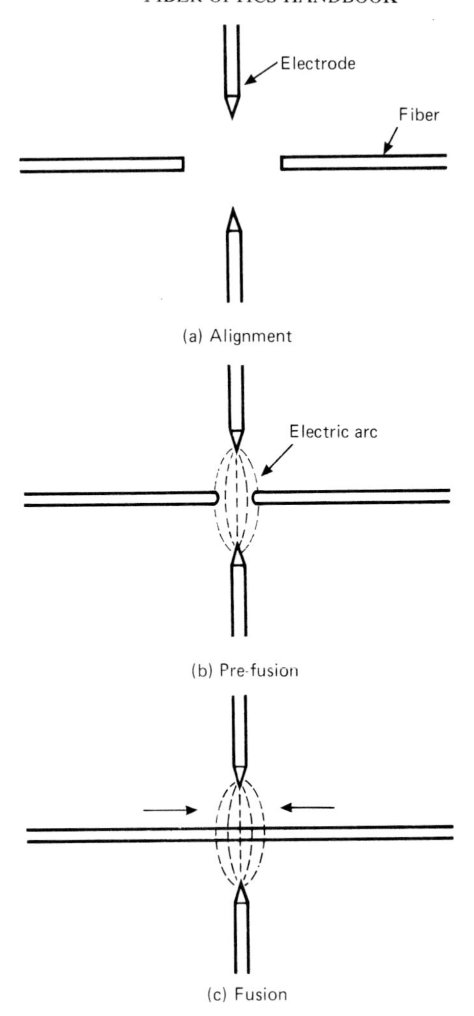


**FIGURE 3.16** Fusion-splicing steps.

The prolonged softening of an extended zone of the fiber ends, if some initial displacement is present, can result in significant core distortion in single-mode fibers (as shown in Fig. 3.18*a*) because of surface tension effects. The consequence is increased splice loss. Further, if some core-to-cladding concentricity error is present and the two cores are actively aligned, the self-alignment effect during fusion can distort the cores, as shown in Fig. 3.18*b*. To avoid such impairments, the so-called *quick-and-narrow fusion* technique[17] can be used. This technique consists of reducing the arc discharge duration and limiting the extension of the heated zone by decreasing the gap between the electrodes. Proper optimization of parameters prevents core distortion (Fig. 3.18*c*), as intended, but, because of the narrower fusion zone, the mechanical strength of the splice is somewhat lower.

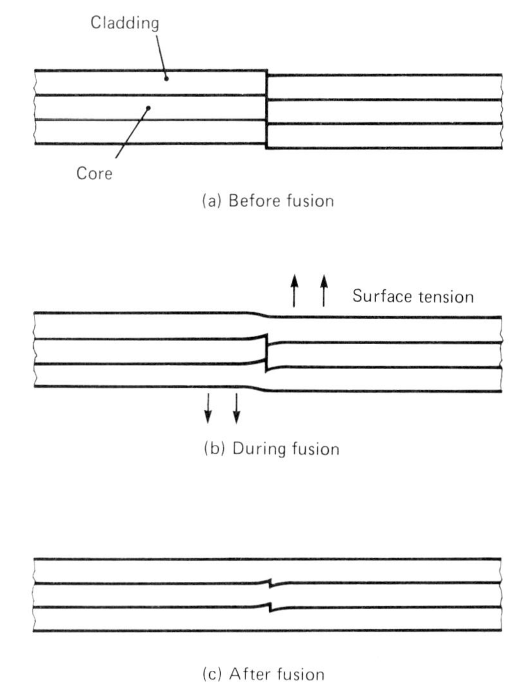


**FIGURE 3.17** Stages in the fusion-splice process, illustrating self-alignment.

The variation of coupling efficiency during the execution of a fusion splice is shown in Fig. 3.19. At $t = T_0$, as the two fibers are butted, a sharp increase in efficiency occurs. This is a result of physical contact of the fiber ends, which eliminates Fresnel losses. From $T_0$ to $T_1$, the efficiency decreases, since the applied pressure results in some fiber deformation. During the arc discharge, the softening of the glass gradually eliminates this deformation and the efficiency increases, up to a maximum, which is indicated at $T_2$. If the discharge continues beyond this point, a reduction in coupling efficiency occurs. This is a result of surface tension effects and the diffusion of core material (with the consequent perturbation of refractive index profile).

For multimode fibers, initial alignment is often ensured by a precision fiber holder, provided that the cladding surface provides an adequate reference. Slight misalignments are compensated by the self-alignment effect already mentioned. Poor-quality splices can be identified after fusion by visual inspection. The presence of bubbles, neck deformations, distortions, and imperfections can be seen through the built-in microscope that is provided in commercially available fusion splicers.

In the case of single-mode fibers, a simple alignment procedure, based on the assumption that the fiber cladding is a reference surface, usually is not reliable. Since some degree of core-to-cladding concentricity error is unavoidable, a core-to-core alignment is necessary. This can be performed in either a passive or active manner.

Passive alignment requires some means to identify the core position inside the

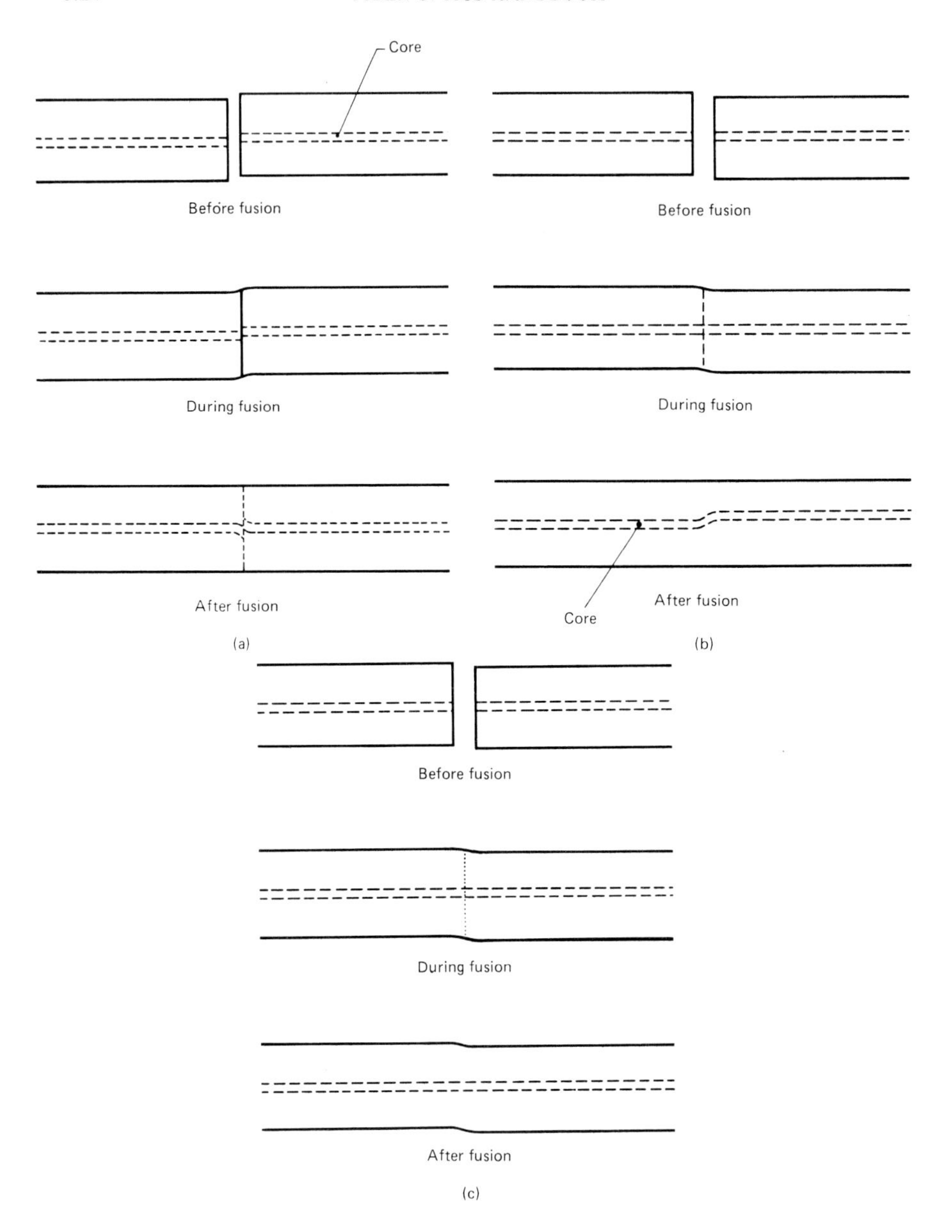


**FIGURE 3.18** (*a*) Stages in fusion splicing of single-mode fibers, illustrating significant core distortion as a result of prolonged fusion. (*b*) Core misalignment due to core-cladding nonconcentricity. (*c*) Stages in the *quick-and-narrow* fusion technique. (*Adapted from Ref. 17.*)

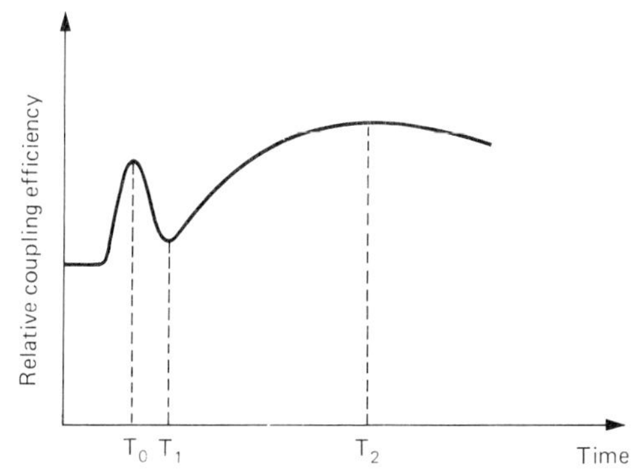


**FIGURE 3.19** Coupling efficiency as a function of time during fusion splicing.

fiber. This can be done, for instance, by exploiting ultraviolet (uv)-fluorescence in the core or by analyzing the fiber image collected through an optical microscope and a TV camera.[18] These methods introduce some degree of complexity to the splicing machine, but, on the other hand, do not require supportive operations outside the splicing apparatus.

Active alignment requires optical power to be injected into the core before the joint and to be extracted and measured after the joint, in order to optimize power transfer from one fiber to the other. The simplest way to carry out this procedure is to launch optical power into one (free) end of the cable and to receive it at the opposite (free) end. This requires a transmitter and receiver, each at some distance from the operator (in a typical installation), and a service link which allows the operator to read the receiver output level while optimizing the core alignment. An obvious disadvantage of this technique is the requirement for apparatus at three separate locations.

To avoid at least the remote receiver, local detection may be used in the vicinity of the joint. Either lost or transmitted power can be detected and used as an indication to optimize transmitted power.

In the former case, part of the power that is transferred from the core of the launching fiber to the cladding of the receiving fiber (because of misalignments) is detected. Its level is at a minimum when the fibers are perfectly aligned and virtually all power is transmitted. A higher alignment sensitivity can be achieved by monitoring the minimum of the lost power (Fig. 3.20), rather than the maximum of the transmitted power.[19]

The concept for a device designed to collect the light scattered into the cladding[20] is represented in Fig. 3.21. In this technique, optical power is collected immediately after the joint from a short stretch of fiber that is placed within a glass groove. The glass groove conducts the collected optical power to the detector.

In a different technique, local information on transmitted power is obtained by simply bending the fiber after the joint; this causes power transfer from guided to radiant modes (i.e., causes some of the guided power to flow outside the core region, and it thus becomes available for local detection).

The principle behind a local power extractor[21] is represented in Fig. 3.22. The primary buffered fiber is reverse-bent over two rods that have a radius of curvature of a few millimeters. A detector is placed close to the second bend, as illus-

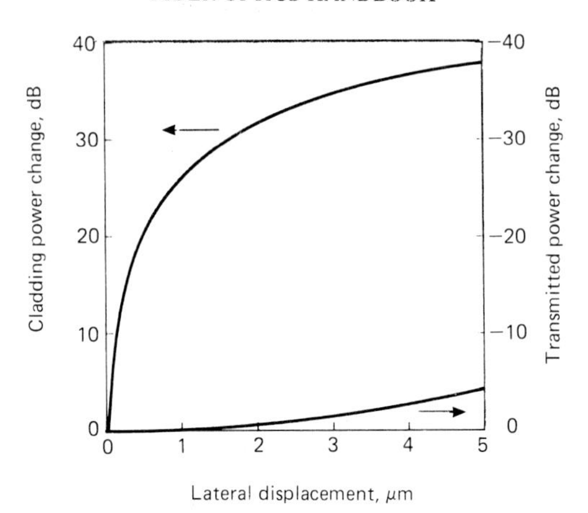


**FIGURE 3.20** Change of transferred-into-cladding and transmitted optical power at a splice, as a function of lateral displacement.

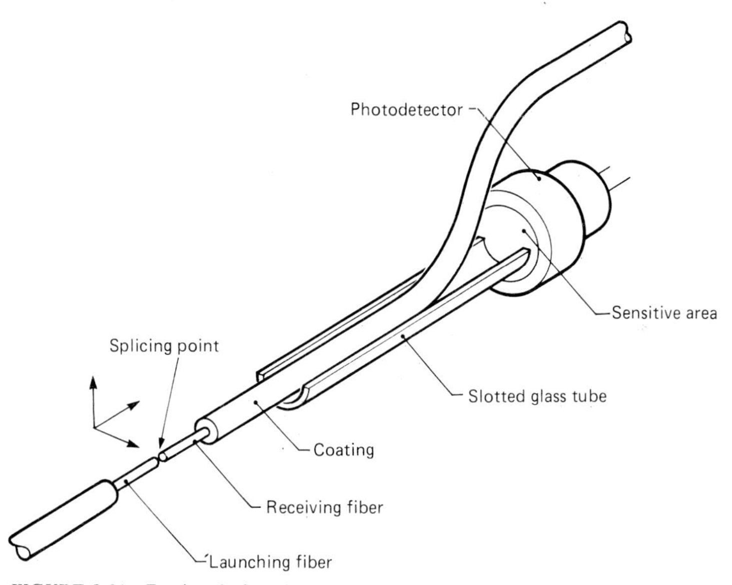


**FIGURE 3.21** Device designed to collect optical power scattered into the fiber cladding after a splice in preparation. (*Adapted from Ref. 20.*)

trated, and the extracted power (which is proportional to the power transmitted by the core) is measured. In this technique, the optimum fiber alignment corresponds to the maximum extracted power. This technique can also be used in a reciprocal way for local injection of optical power. It is therefore possible to achieve completely local support of an active core-to-core alignment process. Both local detection and injection preferably should be performed through the primary fiber coating, provided that it is transparent at the operating wavelength.

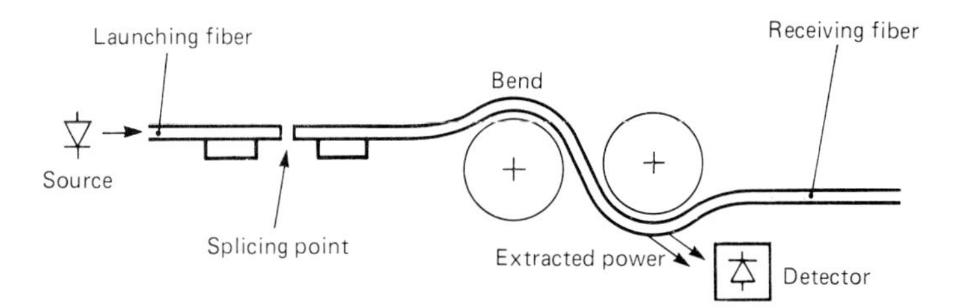


**FIGURE 3.22** Schematic for a local power extractor. (*Adapted From Ref. 21.*)

Removal of the fiber's protective layers, particularly the coating in direct contact with the fiber itself, should be avoided.

Fusion splicing is usually carried out in the field by means of battery-powered portable splicing machines. Sometimes the whole or part of the sequence of operations is performed automatically. Optimum conditions for low-loss splicing depend on the specific kind of fiber that must be spliced. Key parameters include glass composition (e.g., softening point), time and intensity for both prefusion and fusion arc discharge, and pressure stroke on the fibers after contact. Optimum fusion conditions are usually determined experimentally.

Besides low coupling losses, the mechanical integrity of a fiber in the region of a fusion splice must be ensured. With respect to the bare fiber, the mechanical strength of a fusion-spliced fiber is considerably reduced, even to less than 40 percent of the original fiber's strength. This reduction is mainly caused by damage and stresses, possibly produced during end preparation (particularly coating removal and cutting), and by thermal shock during fusion. Particular care should be taken to avoid contamination of the fiber ends from the local environment during handling and to avoid stress on the fiber caused by the positioning fixtures.

The mechanical properties of fused fibers depend on splicing parameters as well (e.g., fusion time and temperature), which also influence optical properties. Nonetheless, very low losses and high mechanical strength are difficult to achieve simultaneously. The *quick fusion* technique, for instance, avoids core distortion effects (thus improving coupling efficiency), but, at the same time, reduces the extent of the fused region, resulting in a somewhat lower mechanical strength. From the mechanical point of view, the quality of splices obtained by flame fusion is generally considered better than that obtained by arc fusion.

Usually, a tradeoff is reached, depending on the particular application. In submarine links, for instance, where reliability is the dominant concern, high-strength splices are required, even if coupling efficiency is not fully optimized.

After the fusion has been completed, suitable reinforcement is needed to restore adequate mechanical strength and to protect the splice from the effects of handling and other environmental hazards. Two techniques are currently popular:

- The U-shaped reinforcement (Fig. 3.23*a*), in which the splice and some length of coated fiber are laid in a channeled rod that is backfilled with an adhesive resin
- The thermosetting reinforcement (Fig. 3.23*b*), in which the splice is overlaid with a thermosetting tube, which in turn is encapsulated (along with a steel strength member) within a shrinkable tube[22]

In the latter technique, the inner, thermosetting tube is molded around the splice as the outer tube is shrunk by heating action.

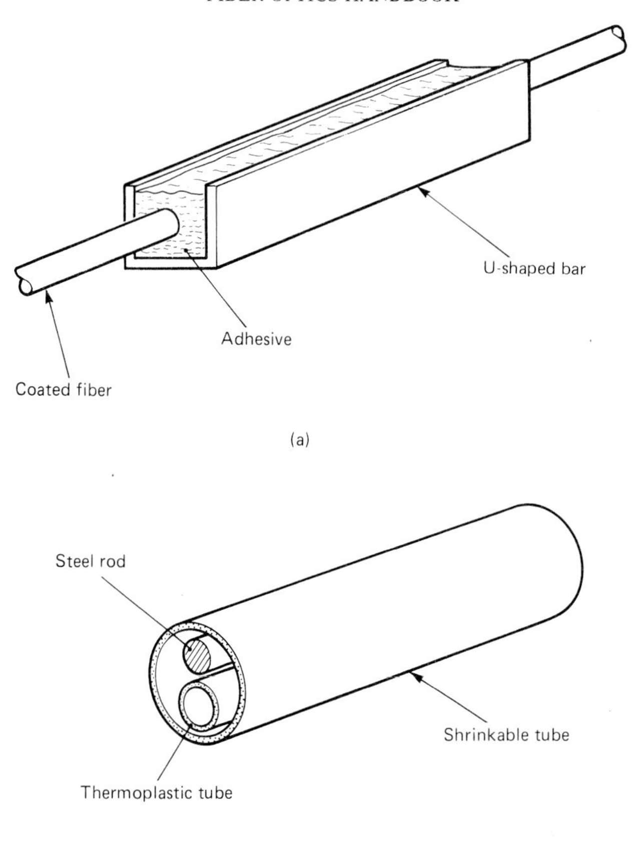


**FIGURE 3.23** Typical splice reinforcements. (*a*) Adhesive-filled channel; (*b*) thermosetting tube (stiffened). (*Adapted from Ref. 22.*)

### 3.2.3 Mass Splicing

Splicing cables with a small number of fibers can often be carried out by individual splicing of single fibers. On the other hand, for applications involving high-fiber-count cables (such as those in the telephone subscriber network, where fiber counts can be on the order of 100), mass splicing techniques are required to reduce installation costs.

Since high-count cables often adopt multiribbon structures (see Sec. 2.3.4 in the previous chapter), corresponding splicing techniques are necessary to maintain the appropriate fiber array distribution. In principle, a linear array of fibers can be aligned with precise multigroove substrates. Each bare fiber, previously prepared, is seated in a groove and is butted to its corresponding opposite. Adhesive is used for bonding and index matching. A cover plate holds the fibers in position, ensuring mechanical stability (Fig. 3.24).

Actual splices for multiribbon cables rely on silicon chips, preferentially etched to attain precision grooves. Many double-sided microgroove chips can be stacked, as illustrated in Fig. 3.25*a*. Mass splicing of 12 × 12 multiribbon cables

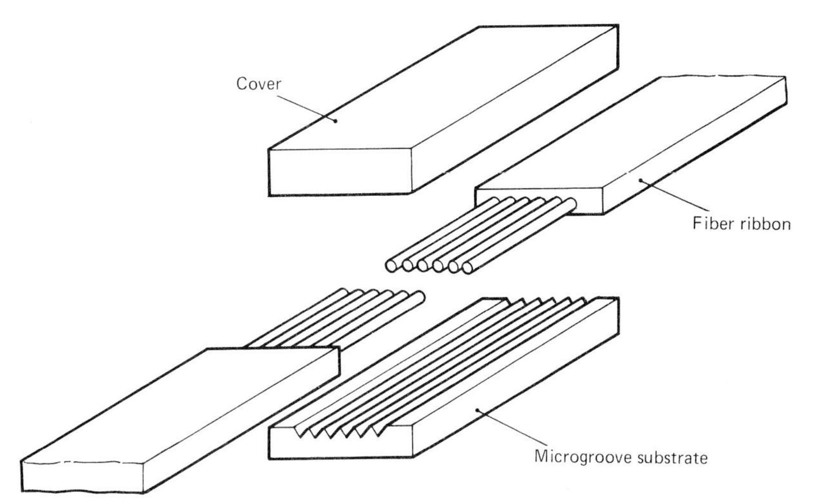


**FIGURE 3.24** Representative multifiber ribbon cable splice using the microgroove technique.

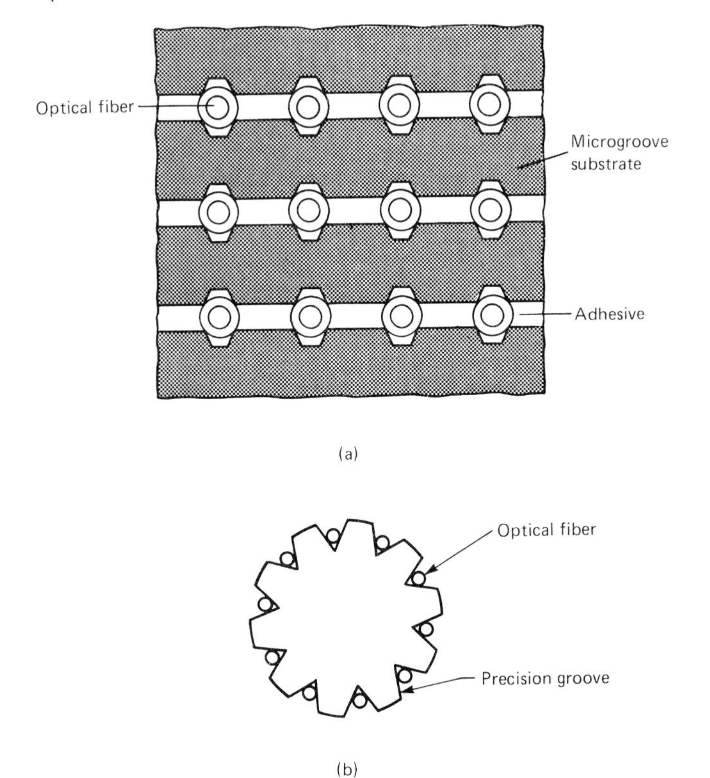


**FIGURE 3.25** Cross sections of mass splicing devices. (*a*) Array microgroove splice using silicon chip substrate (*adapted from Ref. 23, reproduced with permission from the AT&T Technical Journal, copyright 1978, AT&T*); (*b*) precision-grooved cylindrical substrate (*adapted from Ref. 24, copyright © 1983 IEEE*).

(144 fibers total) with coupling losses lower than 0.3 dB for multimode fibers is possible.[23] Alternatively, suitably grooved plastic substrates can be obtained by precision molding.

Multiple splicing techniques can be adapted to circular cable structures[24] as well. In one technique, precision alignment grooves are created along the external surface of a cylindrical core, as illustrated in Fig. 3.25*b*. Fibers are seated and bonded into the grooves, and then each cable end is sawed to obtain a mass quasi-polishing of the end surfaces. After this preparation, the two cable ends are mated, aligned, and assembled—with index-matching material—within proper precision shells. To improve alignment accuracy, the two precision-grooved cores are obtained by cutting one grooved cylindrical substrate in half. The resultant splice halves actually constitute a matched-pair connector.

Mass fusion splicing of ribbon cables is possible, as well, by placing a fiber array, rather than a single fiber, between electrodes. In order to attain uniform, low coupling losses, a careful control of temperature distribution over all fiber ends is necessary.

## 3.3 CONNECTORS

### 3.3.1 Role and General Characteristics of Connectors

A connector for optical fibers is a jointing device that ensures efficient coupling between two fiber ends or two groups of fiber ends, permitting easy manual mating and demating, whenever necessary.

Permanent splices are typically found along a transmission line, whereas demountable connectors are more likely to be located at a distribution frame and at the transmitter and receiver (Fig. 3.26). This is to allow easy reconfiguration of the link or to permit easy servicing of the terminal equipment.

The key parameters defining the quality of a fiber connector for a given transmission system include the following:

- Insertion loss
- Ease of field assembly
- Environmental stability
- Loss repeatability
- Reliability

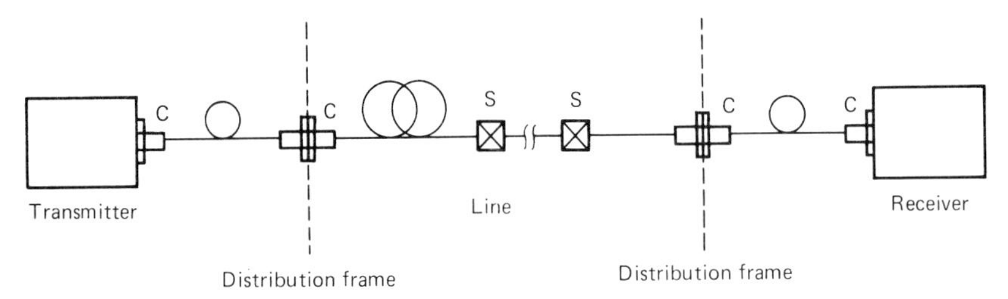


FIGURE 3.26 Schematic of a representative communication link, showing the location of splices (*S*) and connectors (*C*).

- System perturbations (optical feedback, modal noise)
- Cost

Since it is usually impossible to optimize these parameters simultaneously, the design or the choice of a connector is typically the result of a tradeoff among conflicting needs.

To achieve low coupling losses, the connector must accurately align the cores of the two fiber ends with good repeatability. The degree of mechanical precision that is required to keep losses below a specified value depends on the fiber characteristics, according to the relationships described in Secs. 3.1.4 and 3.1.5. Transverse displacement and angular misalignment are critical parameters, whereas end separation is less significant, as, with ease, the fiber ends can be brought virtually into physical contact.

As an example, the core-positioning accuracy that is required to keep the insertion loss of a connector below 0.3 dB is evaluated. A typical graded-index, 50-μm-core, 0.20-NA (numerical aperture), multimode optical fiber is considered first; equilibrium mode power distribution is assumed at the joint, which is followed by a long fiber. From the corresponding formulas of Table 3.1, core alignment tolerances that meet the stated requirement can be derived. For instance, lateral displacement (alone) should be kept below about 5 μm, and angular misalignment (alone) should be kept below about 2°. Since both misalignments are usually found, even tighter tolerances than those just cited should be maintained.

The assumption of a given modal configuration in multimode fibers to establish tolerances is somewhat arbitrary. Under real operating conditions, equilibrium mode power distribution is approached only after a long fiber section. Just after the transmitter, the actual distribution is less clearly defined and depends on many factors, such as source emission pattern, source-to-fiber coupling technique, and pigtail characteristics. Transmitter-end connectors can show higher insertion loss than the receiver-end connectors, but the actual loss value is subject to mode distribution uncertainty.

To make the characterization of connectors meaningful when multimode fibers are used, conditions of modal excitation must be specified. Equilibrium mode distribution is often adopted to test connectors because of the relative ease in reproducing experimental conditions. However, measured losses under such test conditions can be optimistic in comparison with losses incurred during actual use.

In addition to the loss due to limited mechanical precision, other factors, including intrinsic loss and Fresnel loss can contribute to the total loss of a connector.

Intrinsic losses, due to mismatched and uneven fiber characteristics (see Fig. 3.3 and Table 3.1), depend on production tolerances. Fresnel losses (about 0.3 dB) can be reduced by the introduction of suitable index-matching fluids or gels that, unfortunately, can collect dirt and dust and, therefore, are not used routinely.

Because of their smaller core sizes, low-loss connectors for single-mode fibers require more stringent tolerances. Within the limits of gaussian approximation, the evaluation of alignment tolerances for single-mode fibers is unequivocal. From Eq. (3.27), one can draw constant-loss curves in the plane of angular misalignment and lateral displacement, assuming perfect matching between the mode field radii. Figure 3.27 shows such curves for single-mode fibers operating in the 0.85-μm and 1.3-μm wavelength regions. The corresponding typical $1/e$ mode field radii $w$ are about 3 and 5 μm, respectively, for currently used fibers.

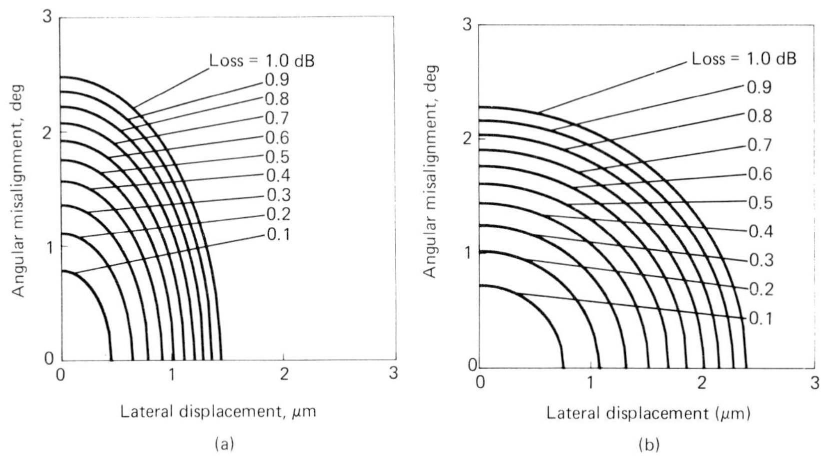


**FIGURE 3.27** Constant-loss curves for angular misalignment and lateral displacement between two single-mode fibers, calculated from Eq. (3.27). (*a*) 0.85-μm-wavelength fiber (mode field radius $w$ = 3 μm) and (*b*) 1.3-μm-wavelength fiber (mode field radius $w$ = 5 μm).

Taking only core positioning errors into account and assuming an angular misalignment within 0.5° (a reasonable value in many actual conditions), if the target loss is 0.3 dB, the lateral displacement must be kept below 0.7 μm and 1.3 μm for wavelengths of 0.85 μm and 1.3 μm, respectively. That is, submicron tolerances must be achieved.

Most of the currently used connectors for both multimode and single-mode fibers adopt the general plug-adapter-plug configuration shown in Fig. 3.28; plug-

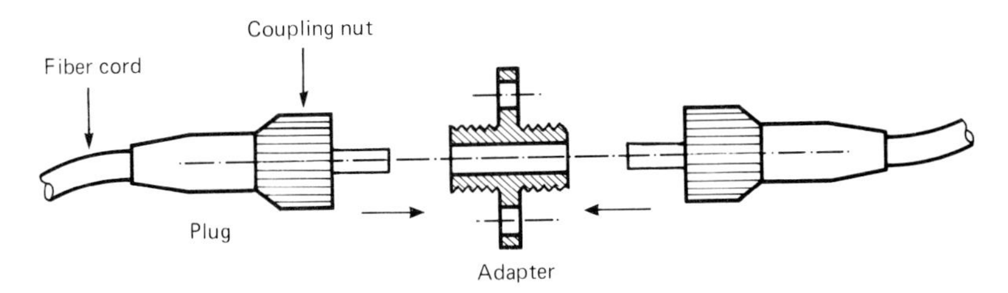


**FIGURE 3.28** General plug-adapter-plug connector configuration.

receptacle connectors are available, as well. The mechanical connection between the plugs and the adapter is made with threaded nuts or bayonet locks. Some fiber connectors mimic the mechanical structure of widely accepted standard electrical connectors (e.g., BNC, SMA).

### 3.3.2 Connector Technology

The main feature distinguishing connectors from one another is the particular technique that is used to align fiber cores. Different approaches are possible, such as those listed below:

- Direct alignment of the bare fiber ends
- Indirect alignment of the fibers
- Expanded-beam coupling

Direct alignment of the two bare fiber ends, such as that attained by means of a precision V groove, directly uses the cladding surface of the fiber as a precision reference surface for alignment, as shown in Fig. 3.29*a*. This is a simple and potentially low-cost technique. The intrinsic disadvantages, on the other hand, are the vulnerability of the bare fibers (which do require some kind of protection for handling) and the possible formation of debris after repeated insertions. In addition, V-groove alignment suffers from any mismatch of the cladding diameters.

Indirect alignment of the fibers is accomplished by accurately positioning each fiber within a precision plug, usually in the shape of a cylinder or truncated cone. These plugs are, in turn, aligned by a precision adapter, as in Fig. 3.29*b*. This approach has the advantage of great robustness. On the other hand, a highly accurate positioning of the fiber core with respect to the reference outer surface is required, and this can be expensive.

Expanded-beam coupling is accomplished by means of two lenses, which form an image of the input fiber core on the output fiber core, as shown in Fig. 3.29*c*. Every plug incorporates one lens, and the output of each plug is in the form of a

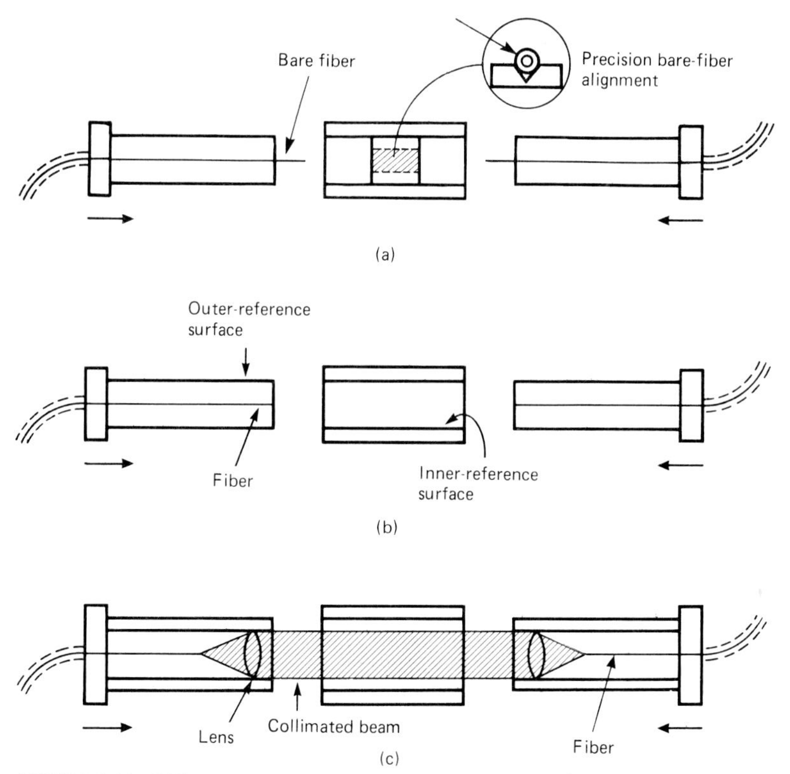


FIGURE 3.29 Different alignment techniques in connectors. (*a*) Direct alignment of bare fibers; (*b*) indirect alignment by reference surfaces; (*c*) beam expansion.

collimated beam. Because the lens system works on a 1:1 image transfer basis, the required lateral positioning accuracy for the fiber inside the plug is the same as for a fiber-to-fiber butt coupling. But, because of the larger optical beam section, the lateral displacement between assembled plugs inside the adapter is much less critical and the effect of dust particles is reduced. These connectors, therefore, are particularly suitable for operation in severe (i.e., industrial and military) environments. On the other hand, the tolerance on angular misalignment becomes correspondingly more stringent, the presence of lenses can add losses (due to reflections and aberrations), and the cost can increase for expanded beam coupling.

Most of the commercial connectors for telecommunications make use of the reference surface technique (Fig. 3.29*b*). The inherent problem of accurately positioning the core axis with respect to the outer precision reference surface can be handled in different ways. Figure 3.30 shows the following three main approaches to core centering within a cylindrical plug:

- Using a precision hole in the ferrule
- Micropositioning the core with respect to the ferrule axis
- Machining the ferrule to reestablish its coaxiality with the core

The precision hole technique, illustrated in Fig. 3.30*a*, requires the creation of a hole, by some accurate mechanical means, that precisely fits the cladding outer surface and is exactly concentric with the outer surface of the plug. Concentric-

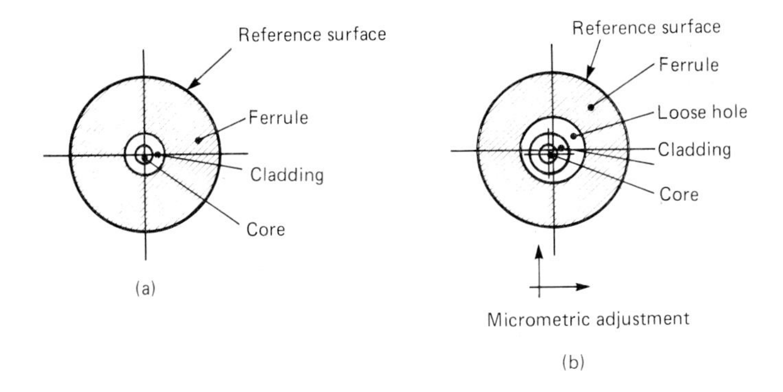


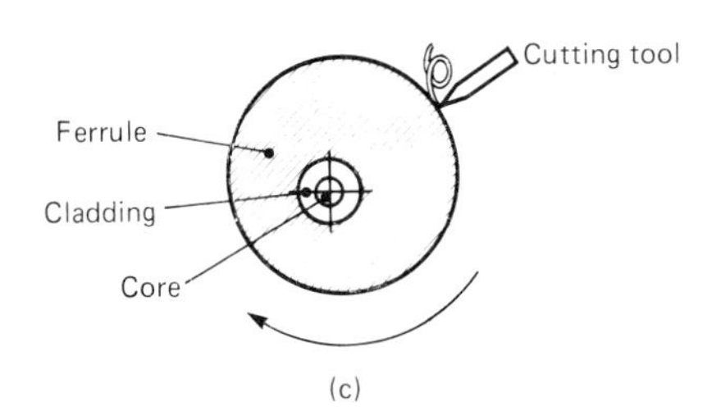


**FIGURE 3.30** Methods of centering fiber core with respect to an outer reference surface. (*a*) Precision hole; (*b*) adjustment of core position; (*c*) machining of plug.

ities within a few micrometers or even a fraction of a micrometer must be achieved, depending upon the application. Provided that the fiber has satisfactory core-to-cladding concentricity, this connector is inherently field-installable.

Another method requires the micropositioning of the fiber core within a relatively loose hole in the ferrule, as represented in Fig. 3.30*b*. The required core concentricity (relative to the ferrule) can be achieved by sending light into the fiber, inspecting the core position by optical means (e.g., by a microscope, TV camera, and monitor), and moving the fiber until the core and a proper target are centered. The fiber is then glued in place. High accuracy can be achieved independently of intrinsically poor core-to-cladding concentricity of the fiber. On the other hand, the fiber alignment process is time-consuming and requires a complex apparatus. Pigtails terminated with this type of connector in the factory then can be joined to fibers in the field.

A third method is represented in Fig. 3.30*c*. In this method, the fiber is inserted into the plug, without particular regard to centering with respect to the existing outer plug surface. After the fiber is fixed, the outer surface of the plug is accurately machined to create a reference surface that is perfectly concentric with the fiber core. A small lathe or grinding machine can be used for this purpose. An optical system is needed to visualize the fiber core or the central hole position with the necessary resolution. Sometimes, only a precisely centered hole is generated by a similar machining technique, so that the connector can then be assembled in the field. This procedure results in high alignment accuracy but, in comparison to other methods, requires more time and more expensive tools.

Different materials can be used to fabricate the precision parts of connectors: metal, ceramic, and plastic.

Metals commonly used in metal connectors are nickel-plated brass, stainless steel, tungsten carbide, and special alloys. An inner ferrule, made of a hard material (e.g., ceramic) and containing a precision, centered hole, is often inserted concentrically with the external surface of the plug.

Connectors can also be made of plastic components (sometimes containing other reinforcing materials) and can be produced by molding techniques using high-precision dies. They are suitable for low-cost, high-volume production. Good mechanical precision is attainable by means of highly accurate molding techniques.

In addition to positioning accuracy, a low-loss connector requires properly finished fiber-end faces. The faces must be perpendicular to the optical axis, and they must be flat and smooth. To achieve this goal, the fiber-end face is usually polished after it has been fixed in the plug. Abrasive surfaces of different grades are used, typically ranging from (optically) coarse silicon carbide to fine, 0.3-μm alumina. Polishing can be accomplished manually or by machine. Proper tool movement is necessary to obtain a relatively random abrasive action (i.e., an optically smooth surface) and to produce a flat surface, as well.

Although fiber cleavage is an intrinsically faster and hence cheaper fiber-finishing technique, it is often not compatible with assembly procedures and it does not always provide consistently acceptable surface quality. Therefore, polishing is usually preferred.

### 3.3.3 Interference Effects in Connectors

Because of the mirror-like quality of the fiber ends, the small air gap at the center of the connector behaves like a plane-mirror, Fabry-Perot resonator, and, as a

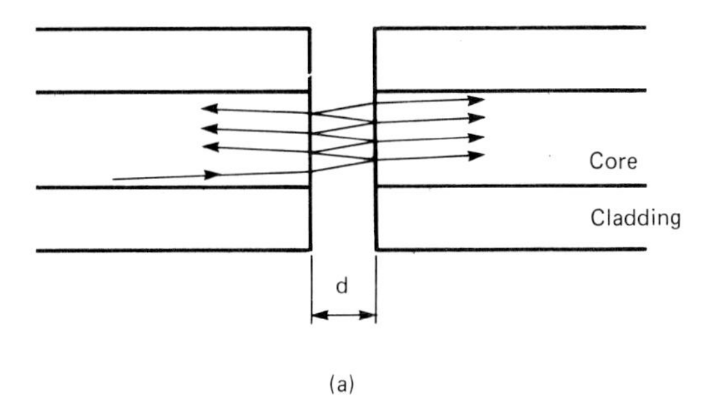


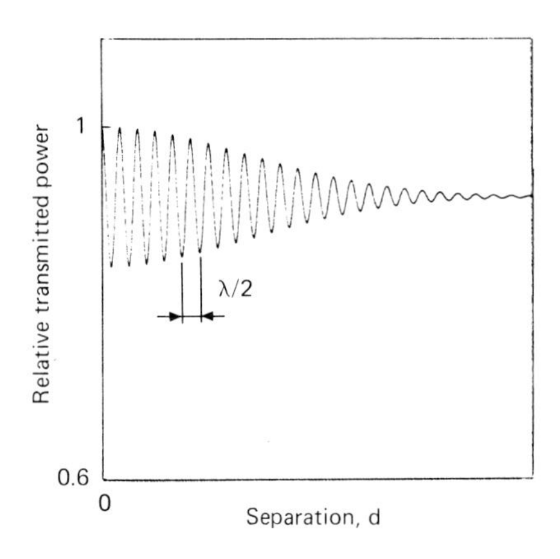


**FIGURE 3.31** Interference effects in connectors. (*a*) Multiple beam generation through Fresnel reflection; (*b*) transmitted power as a function of fiber separation, according to Eq. (3.29).

result, multiple-beam interference takes place. Such an interference effect produces periodic fluctuations of transmitted power when the end face separation is changed or the source wavelength shifts.

This condition for a multimode fiber is illustrated in Fig. 3.31. In the multimode case, assuming uniform mode power distribution and a gaussian source spectrum, the power transmission coefficient through the resonator can be expressed by[25]

$$T = 1 - 2R + 2RAB \cos\left[\pi\left(\frac{2n_0 d}{\lambda}\right)(1 + C)\right] \tag{3.29}$$

where $A = \dfrac{\sin\,[(C - 1)(2n_0 d)/\lambda]}{(C - 1)(2n_0 d)/\lambda}$

$$B = \exp\left\{-\pi\left[\left(\frac{\Delta\lambda}{\lambda}\right)(1 + C)\left(\frac{n_0 d}{\lambda}\right)\right]^2\right\}$$

$$C = \left[1 - \left(\frac{\text{NA}}{n_0}\right)^2\right]^{1/2}$$

R = preflection coefficient of each fiber end (about 4 percent in air)
$n_0$ = refractive index of the gap material
$d$ = gap width
$\lambda$ = center wavelength of the source spectrum
$\Delta\lambda$ = source spectral width
NA = numerical aperture of the fiber

The coefficient $C$ takes into account the different propagation angles of the rays exiting the fiber, which slightly reduce the oscillation period with respect to $\lambda/2$. $A$ and $B$ are simplified envelope functions resulting from the simplifying assumptions of uniform mode excitation and gaussian source spectrum, respectively.

The first term of Eq. (3.29) assumes a perfect connection. The second term takes Fresnel losses into account, whereas the third term introduces fluctuations as a function of $d$ and $\lambda$. Since $A$ and $B$ are $\leq 1$, $T$ can fluctuate between 1 and $1 - 4R \approx 84\%$ in air (Fig. 3.31*b*). Corresponding reflection losses fluctuate between 0 and about 0.7 dB (i.e., around the familiar 0.35-dB Fresnel loss).

The envelopes $A$ and $B$ tend to reduce the amplitude of such fluctuations when the gap width $d$ is increased. For instance, the short coherence length (corresponding to large $\Delta\lambda$ in $B$) of an LED limits the relevance of interference effects to gaps shorter than 10 to 15 μm. Unfortunately, this is the particular region of interest for low-loss butt connectors.

In actual connectors, this oscillatory effect is superimposed on the average joint efficiency, which decreases if the separation $d$ is increased. Imperfections in fiber-end parallelism or flatness reduce the theoretical modulation depth given by Eq. (3.29).

A similar behavior is common in single-mode fibers as well, but without the limitation set by the presence of many modes.

Interference effects, which are modified by even very small separation changes (on the order of a fraction of a wavelength), can add a significant degree of uncertainty to the measurement of connector losses. If the two core faces were placed in contact, thus eliminating the air gap ($d = 0$), Fresnel losses and the loss uncertainty would be eliminated.

### 3.3.4 Connector Contributions to System Noise

To the overall system noise, connectors can contribute the two following types of noise:

- Modal noise
- Reflection noise

When coherent light is injected into a multimode fiber, a multiple spot pattern, the so-called speckle pattern, can be observed in the fiber core. This pattern is produced by interference effects among the propagating modes, which after traveling along the fiber, present different phases.

If the speckle pattern changes, the coupling efficiency varies as well, since the coupling efficiency of a misaligned connector depends upon the power distribution in the core (Fig. 3.32). Therefore, any change in source or propagation con-

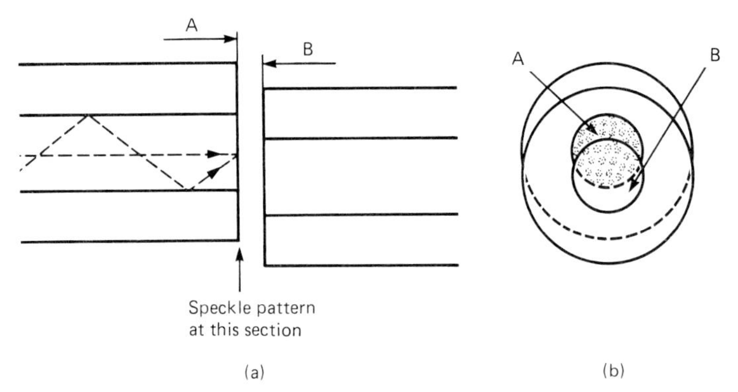


**FIGURE 3.32** Modal noise generation in a misaligned joint. (*a*) Side view; (*b*) cross section, showing the speckle pattern on the fiber core.

ditions altering the phases of the different modes results in a speckle pattern variation and thus a change in coupling efficiency, which is perceived as intensity noise.[26] Possible fluctuations of more than $\pm 0.5$ dB can be produced by a misaligned connector showing an average loss of 1.0 dB.

For the above-mentioned reasons, thermal drifts (which shift the source wavelength) or stress and mechanical vibrations (which affect the propagation inside the fiber) can generate slow fluctuations of the signal (i.e., low-frequency noise components). On the other hand, high-frequency noise components can result from source phase noise, which the misaligned connector transforms into intensity noise. Also, direct intensity modulation of a laser diode produces a modulation of the emitted wavelength, which is transformed into modulation of the coupling efficiency at the connector, thus introducing nonlinear distortion.[27] These noise and intermodulation contributions reach levels that, although not critical for digital transmission, can be intolerable for high-quality analog transmission.

Modal noise can be avoided by using low-coherence sources, such as LEDs or broad-spectrum, multilongitudinal-mode laser diodes (preferably gain-guided). Indeed, when the temporal delay among the different fiber modes increases above the coherence time of the source, the speckle pattern reduces and tends to disappear because of the lack of coherence among the modes. From this point of view, connectors near the transmitter are more critical than those farther along the link.

Although seemingly contradictory, modal noise can be produced in single-mode fibers, as well, whenever a significant amount of the $LP_{11}$ mode is excited (in addition to the fundamental $LP_{01}$ mode) because of the presence of some fiber section having a cutoff wavelength $\lambda_C$ higher than the operating wavelength $\lambda_0$. This condition is shown in Fig. 3.33, where fibers $F_1$ and $F_3$ operate above the cutoff, while $F_2$ is below the cutoff. Connector $C_A$ excites the higher-order $LP_{11}$ mode in the central fiber, and connector $C_B$ affects spatial filtering, potentially giving rise to modal noise. Since the $LP_{11}$ mode suffers high transmission losses, modal noise is more likely to be seen in short jumper cables.

Reflection noise is produced in both multimode and single-mode fibers by the feedback power that is reflected back from a connector into the laser diode (see Sec. 5.4.3). Laser emission intensity, wavelength, and frequency response can be perturbed, up to levels affecting the performance of analog or high bit-rate digital systems. Furthermore, uncontrolled feedback, producing wavelength fluctua-

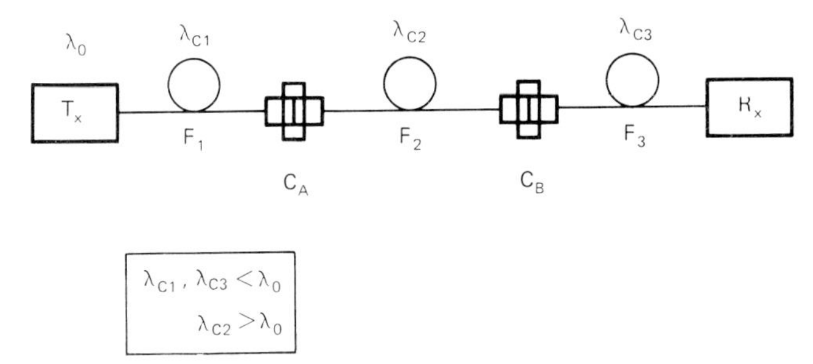


FIGURE 3.33 Condition giving rise to modal noise in a single-mode fiber link.

tions, can enhance modal noise. To avoid feedback, index-matching fluids or antireflection coatings can be used at the connectors. The introduction of a tilt in the end faces of the fibers or a connector structure allowing the two fiber ends to come into contact are more practical and efficient solutions that significantly reduce undesired reflections (see Sec. 3.3.5, "Antireflection Connectors").

### 3.3.5 Types of Connectors

Some interesting structures that have been developed in order to achieve efficient optical-fiber connection, are, in the following discussion, reviewed. Sketches are provided of some basic concepts, many of which have been employed in widely used commercial connectors. Some of the principles originally developed for multimode fiber connectors have been subsequently adapted to achieve the higher precision required for single-mode fiber connectors.

Unless otherwise specified, most of the connector types reviewed are intended for telecommunications use. They generally have losses lower than 1.0 dB, usually averaging about 0.5 dB if applied to graded-index multimode fibers (50-μm core size, 0.20 numerical aperture) without index matching. If Fresnel reflection can be prevented, these figures can be reduced to values below 0.3 dB. Similar values can often be obtained by the improved designs that are used for single-mode fibers.

*Ferrule Connectors.* A simple, generic connector design is shown in Fig. 3.34*a*. The fiber is inserted into a cylindrical plug (or ferrule) having a precision axial hole and is fixed by adhesive, usually epoxy resin. The two ferrules are aligned by a precision sleeve and are pushed toward each other by axial springs. Often, a mechanical stop or spacer is provided to prevent the fiber ends from coming into contact with each other. The connector is locked by threaded or bayonet-locking shells. To allow both accurate fiber positioning and easy fiber insertion and bonding, actual designs use a tapered, nozzle-type ferrule or a watch jewel on the ferrule tip, as in Fig. 3.34*b* and *c*. Simple connectors of this type usually have insertion losses higher than 1.0 dB if applied to standard, multimode fibers.

*Ferrule Connectors with Actively Aligned Core.* Accurate placement of the fiber core within the connector ferrule can be accomplished by active alignment techniques. One technique[28] uses micromanipulation, as illustrated in Fig. 3.35*a*. In this technique, the fiber is bonded in a glass capillary tube, which is then bonded

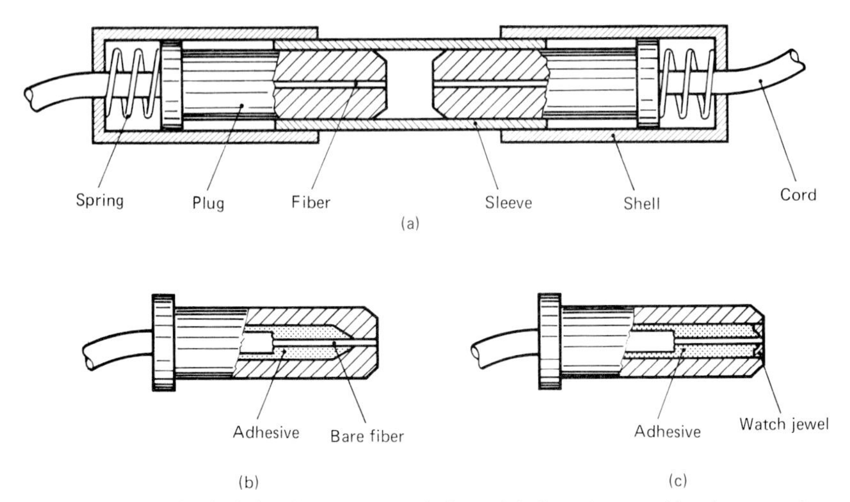


**FIGURE 3.34** Typical ferrule connector designs. (*a*) Generic assembly, incorporating sleeve alignment technique; (*b*) detail of nozzle ferrule; (*c*) detail of watch-jewel ferrule.

within a metal tube. This prevents fiber flexure and allows the fiber to be handled more easily. After this preparation, the fiber assembly is inserted into the loose hole of the main ferrule and is centered by means of micromanipulators. Proper centering is determined by means of an observation system. Once centered, the fiber is bonded in place.

In an alternative technique, shown in Fig. 3.35*b,* the fiber is bonded inside the main ferrule, without particular regard for centering. The plug is then machined by means of a small, precision lathe, while an optical monitoring system assures the concentricity of the external surface of the plug (i.e., the reference surface for the plug) with the core. A refinement of this technique can be applied to single-mode fiber connectors, permitting submicron accuracy in core alignment.[29]

Compliant alignment sleeves[30], such as those shown in Fig. 3.36, can compensate for slight differences between the absolute values of the plug diameters, providing greater assurance that the two plugs are located on the same axis.

***Triple-Ball Connector.*** The precision needed to center the fiber inside the plug can be achieved by localizing the fiber in the clearance that is created by placing three accurately ground tungsten carbide balls in a bushing,[31] as shown in Fig. 3.37. If the two plugs are engaged at a relative rotation of 60°, the corresponding sets of balls nest into each other, thereby locating their clearance spaces (and thus the fiber ends) on a common axis. To assure angular alignment, a double set of balls can be used inside each plug.

*Ceramic Capillary Connector.* Accurate fiber positioning can be achieved by capturing a ceramic capillary within the tip of the plug ferrule,[32] as in Fig. 3.38*a*. By precision machining , good concentricity can be obtained between the outer plug surface and the capillary hole. If the clearance between the fiber and the hole is made very small by selecting the proper hole size (i.e., within 1.0 μm) for a given fiber, the average concentricity error between core and outer surface can be kept below 2.0 μm. Furthermore, since the capillary hole can be as long as several

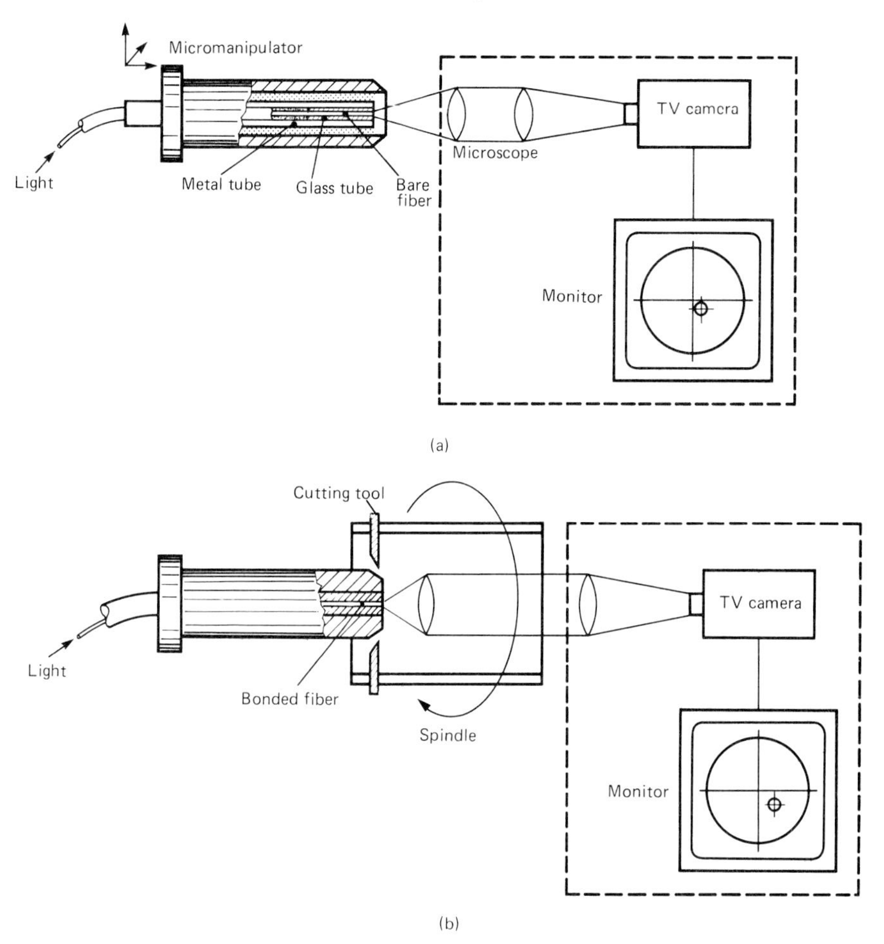


**FIGURE 3.35** Active core alignment techniques. (*a*) Micromanipulation of core within ferrule (*adapted from Ref. 28*); (*b*) micromachining of ferrule after fiber is bonded in place (*adapted from Ref. 29*).

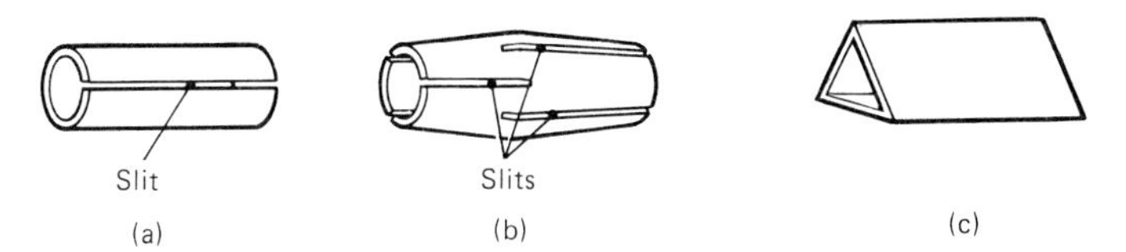


**FIGURE 3.36** Compliant alignment sleeves: (*a*) slit cylinder (*adapted from Ref. 28*); (*b*) modified slit cylinder (*adapted from Ref. 30*); (*c*) triangular section (*adapted from Ref. 29*).

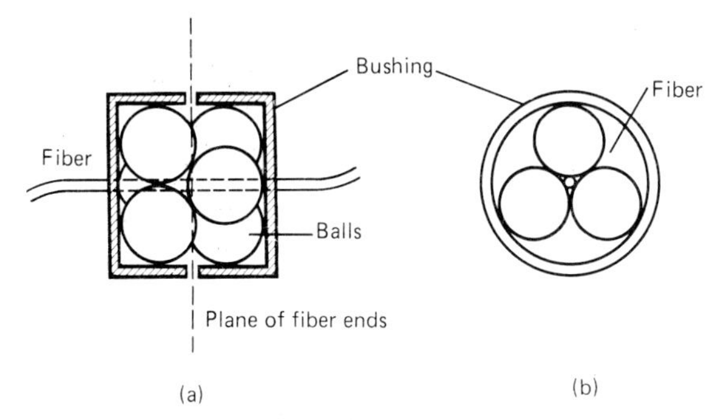


**FIGURE 3.37** Triple-ball connector. (*a*) Side view of mated connector; (*b*) cross section of connector half. (*Adapted from Ref. 31.*)

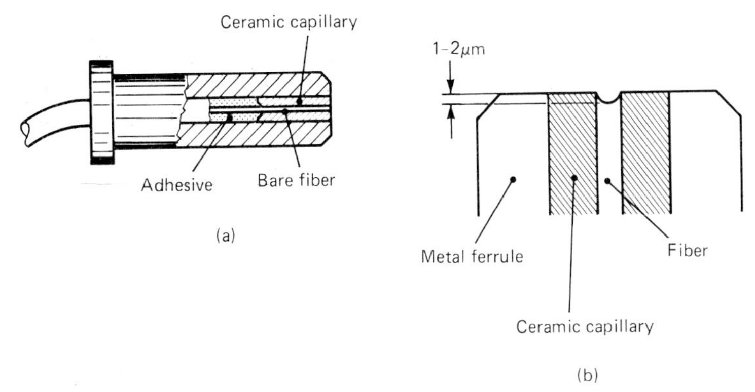


**FIGURE 3.38** Ceramic capillary connector. (*a*) Side view (section) (*adapted from Ref. 32*); (*b*) detail of plug end, showing face concavity (*adapted from Ref. 33; copyright © 1982 IEEE*).

millimeters, angular misalignment between fiber and plug ferrule axes can be kept to less than 0.5°.

These connectors can be assembled directly in the field, with no need of costly and time-consuming fiber adjustment or ferrule machining. The ceramic material that is used for precision capillaries has inherently good thermal, mechanical, and chemical characteristics. This assures high performance stability under inclement environmental conditions.

The plug end is finished by ordinary grinding and polishing techniques. If proper buff polishing is applied, the fiber is slightly undercut (see Fig. 3.38*b*), because of the difference in material hardness between the fiber and the ceramic capillary. The small face concavity that results actually protects the fiber tip from scratching during handling and reduces interference effects, thus increasing transmission stability vs. temperature and wavelength changes.[33] The critical steps in the assembly process include selecting the precise capillary hole and bonding the fiber. In order to avoid overstressing and breaking the fiber, a low-viscosity epoxy resin should be used.

This technique has been extended to single-mode fibers, with enhancements in mechanical accuracy yielding values for average concentricity and parallelism between outer surface of the plug and the capillary hole of about 0.4 μm and 0.1°, respectively.[34]

The main advantage of this approach is that it allows assembling connectors in the field, with an easy and fast procedure, while achieving low coupling losses. When single-mode fibers are connected, intrinsic high core-to-cladding concentricity is required to take advantage of the strict mechanical tolerances of the connector.

*Deformable Insert Connector.* The deformable insert technique,[35] as shown in Fig. 3.39, is an alternative to the use of epoxy resin for fiber bonding. The accurately machined, hard metal ferrule is terminated with an insert made of a relatively soft metal alloy and having a precision concentric hole. After the fiber is inserted into the hole, a special tool compresses and deforms the insert and then tightly grips the fiber. As usual, a precision sleeve ensures proper alignment of the plugs. This technique can be applied to both multimode and single-mode optical fibers. In the latter case, asymmetrical compression can be used to correct slight core eccentricities.

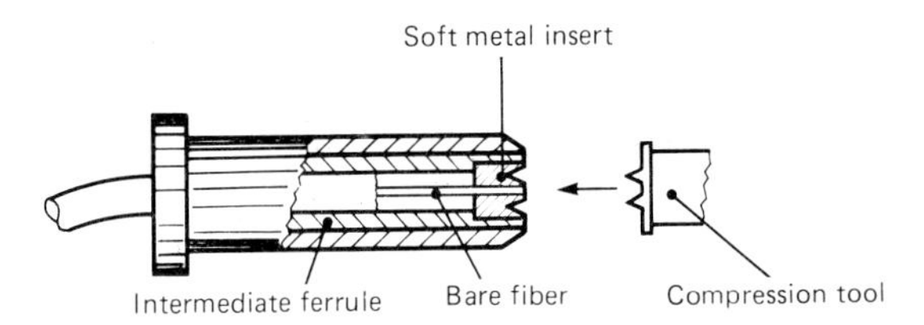


**FIGURE 3.39** Deformable insert connector. *(Adapted from Ref. 35, copyright © 1986 Hewlett-Packard Company; reproduced with permission.)*

*Plastic Molded Ferrule Connector.* When low cost is an objective, plastic connectors for field assembly can be mass-produced by precision transfer molding.[36] The die used to fabricate plug ferrules incorporating a precision microhole is electroformed on a high-quality master, with a transfer precision that can approach 0.5 μm. Alignment sleeves are prepared by a similar technique. Resin, filled with glass fibers or silica powder, is used as a molding material. The connectors can be assembled under field conditions; the fiber is inserted into the precision hole in the plug and bonded by epoxy, and the plug end is then polished. Extremely accurate control of molding conditions can lead to average eccentricities lower than 1.0 μm and outer diameter variations within ± 1.0 μm. This technology could also be suitable for the production of low-loss, single-mode fiber connectors.

***Biconical Connector.*** The geometrical features of the biconical design are shown in Fig. 3.40. The terminus of the plug has the shape of a truncated cone, and the sleeve is a corresponding double conical cavity.[37] The conical design of the plug requires accurate concentric alignment between the fiber core and the conical surface, but does not require precision control of the outer diameter of the plug. Fine adjustment of the plug length is required, however, and this is accomplished by a final grinding and polishing of the connector-end face. An advantage of the

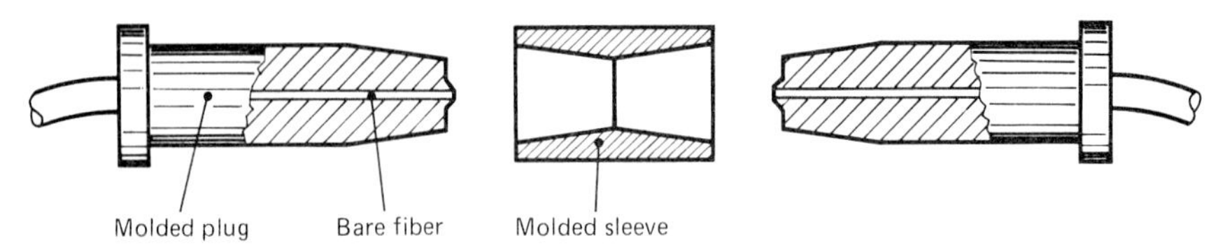


FIGURE 3.40 Biconical connector. (*Adapted from Ref. 37, copyright 1978 AT&T, reproduced with permission from the AT&T Technical Journal.*)

conical shape is that it ensures that the plug and sleeve come into contact only when the plug is almost completely seated into the conical sleeve. This reduces the abrasive wear that is induced with repeated insertions.

Material and production processes are important features of this connector. Both plugs and sleeves are produced by means of a precision transfer-molding process using thermosetting plastic material (silica-filled epoxy resin). This is important for low-cost mass production.

To produce factory-assembled pigtail connectors, the plastic conical plug can be molded directly on the fiber; alternatively, it can be molded around a wire mandrel that creates a precision hole about 1.0 μm wider than the fiber to be terminated. The latter approach yields a field-installable plug.

Both single-mode and multimode fibers can be connected by this technique. For single-mode fibers, the necessary concentricity is achieved by precisely grinding the conical surface of the plug. Residual eccentricity and angular misalignment between the inner hole and the conical surface can be reduced to values as low as 0.4 μm and 0.1°, respectively.

If the plugs are long enough to touch each other when seated into the sleeve, Fresnel losses disappear, with average losses of less than 0.3 dB being possible, even for field-installed, single-mode connectors. To achieve these results requires careful control of the final end-face polishing. To prevent crushing the fibers that are in contact, a calibrated axial spring is provided which ensures a constant and proper seating force. Figure 3.41 shows an actual biconical connector.

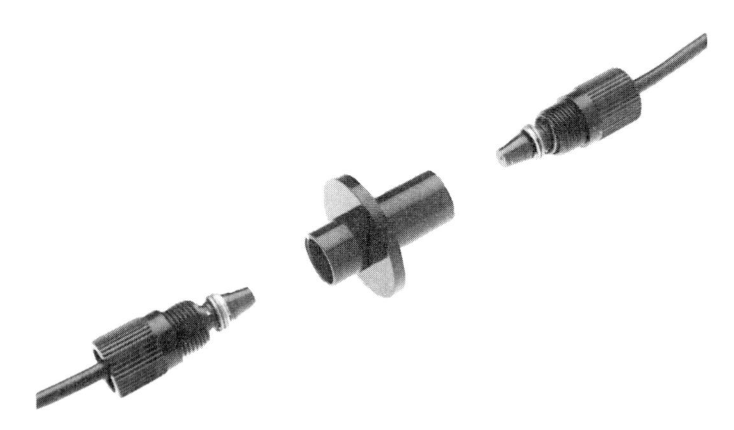

FIGURE 3.41 Photograph of a biconical connector. (*Courtesy of SIRTI S.p.A., Milano; connector produced under license from AT&T.*)

*V-Shaped Plug Connector.* The reference surfaces of a connector need not be cylindrical or conical. V-shaped plugs and corresponding V-grooved sleeves, as shown in Fig. 3.42, can be used for precise alignment.[38] To achieve good cou-

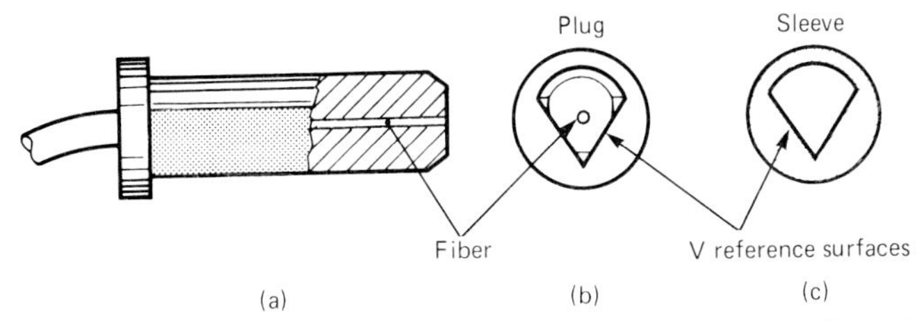


**FIGURE 3.42** V-shaped plug connector. (*a*) Side view of plug; (*b*) end view of plug; (*c*) cross section of corresponding alignment sleeve. (*Adapted from Ref. 38.*)

pling efficiency, the fiber must be accurately positioned with respect to the plane surfaces forming the V. A specific tool is required for the fiber-positioning procedure. Proper transverse spring action is required to ensure the contact of the plug-and-sleeve reference V surfaces.

*Ball and Socket Connector.* To avoid clearances between plug and sleeve, like those that can occur in cylindrical, conical, or prismatic coupling geometries, plugs can be aligned by the ball-and-socket approach,[39] shown in Fig. 3.43. The

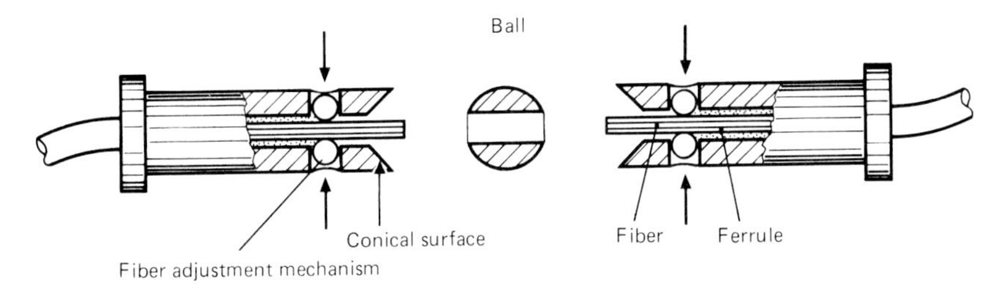


**FIGURE 3.43** Ball-and-socket connector. (*Adapted from Ref. 39.*)

basic concept relies on a concave, conical plug-end nesting on a central alignment ball. The fiber, protected by a ferrule, protrudes from the vertex of the concave cone, slipping into a hole that runs through the ball. To achieve accurate core-to-plug concentricity, the assembly procedure requires adjustment of the core position within the plug. This is accomplished by means of active alignment that employs trimming screws and a special microscope tool. It is worth noting that, with this design, it is the contact between conical and spherical surfaces that exerts the necessary alignment action.

***Four-Rod Connector.*** Efficient coupling between two optical fibers can be attained by direct use of the cladding surface as the reference surface for alignment.[40] The alignment guide is made of four glass rods, with flared ends to make the fiber insertion easier, as shown in Fig. 3.44*a*. The bent guide forces the two fiber ends to be mated in the cusp formed by two adjacent rods, as in Fig.

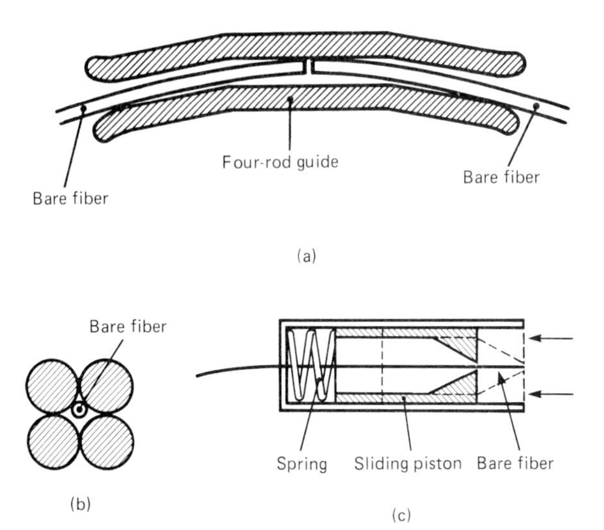


**FIGURE 3.44** Four-rod connector. (*a*) Side view or alignment guide; (*b*) cross section of alignment guide; (*c*) schematic of plug with sliding piston protecting the bare fiber. (*Adapted from Ref. 40.*)

3.44*b*. Since this structure implies the use of a bare, fragile fiber, a protective piston covers the fiber end when the plug is not engaged. When the plug is seated in the receptacle, the piston is pushed back, as in Fig. 3.44*c*, thus permitting the insertion of the fiber into the alignment guide.

The nature of this design is such that the fiber ends should preferably be cleaved. A built-in fluid reservoir can be provided to feed the alignment guide with index-matching fluid, thereby preventing reflection losses. Both single and multimode fibers can be mated by this technique. Since neither adhesive bonding nor polishing is needed, fast connector assembly is possible in the field.

***Expanded-Beam Connectors.*** The general characteristics of expanded-beam connectors, which make use of a pair of lenses, were discussed in Sec. 3.3.2. Spherical aberration of the lens and reflections from the additional optical surfaces can easily increase connector losses to values exceeding 1.0 dB. Nonetheless, by carefully specifying lens shape, lens material, and antireflection (AR) coatings, this figure can be significantly reduced.

Figure 3.45 shows some kinds of lenses that are used or can be used inside lensed connectors. Ball and rod lenses (Fig. 3.45*a* and *b*) have about the same imaging quality as graded-index lenses (Fig. 3.45*c*) if high-refractive-index (greater than 1.7) materials are used, since this choice reduces spherical aberration. Proper lens design can theoretically lead to losses as low as 0.2 dB for typical multimode fibers.[41] Experimental results come rather close to this value. The use of AR-coated sapphire balls ($n = 1.75$) can result in 0.5-dB coupling loss, even with single-mode fibers.[42] Similar results (0.6 dB) can be predicted[43] for mass-produced, high-quality aspheric glass lenses (Fig. 3.45*d*).

Other approaches make use of fluid-filled plastic cavities (Fig. 3.45*e* and *f*) to obtain or enhance the lens effect. Finally, fibers can be coupled by means of glass

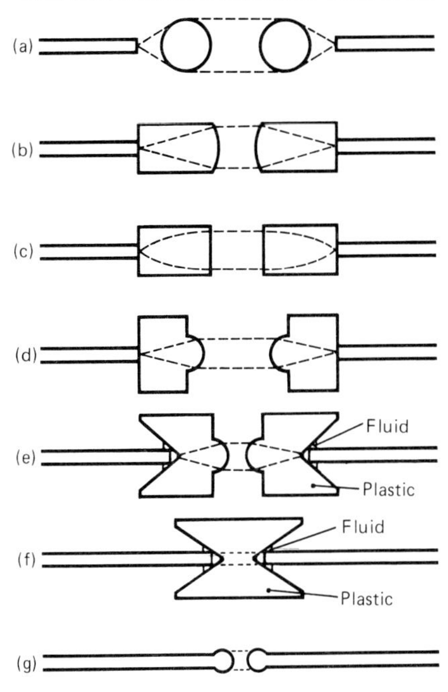


**FIGURE 3.45** Lensed connector techniques. (*a, b*) Ball and rod lenses; (*c*) graded-index lens; (d) aspheric glass lens; (*e, f*) fluid-filled plastic lenses; (*g*) fused lens.

beads (Fig. 3.45*g*) that are formed directly on the fiber ends by arc discharge or similar techniques, such as those that are used for fusion splicing.

*Antireflection Connectors.* In order to avoid additional source noise problems (see Sec. 3.3.4), feedback from connectors must be prevented or substantially reduced. This can be accomplished in one of several ways:

- By reducing reflections through the use of antireflection coating of the fiber ends or by the use of index-matching fluids, gels, or pads
- By bringing the two fiber ends into contact so as to remove the air gap
- By tilting the fiber ends at an appropriate angle so as to frustrate feedback by producing angular misalignment of the reflected beam

Coating of the fiber ends is usually not practical, and index-matching fluids can easily attract dirt and dust.

Physical contact of the fiber ends is easier to achieve if the contacting surfaces have limited area. The ends of ceramic capillary cylindrical plugs can be polished in such a way as to obtain a slightly convex spherical surface,[44] as in Fig. 3.46*a*. Return losses of at least 28.0 dB can be achieved, meaning that the gap can be reduced to less than approximately two one-hundredths of a wavelength. (By comparison, a conventional connector has a worst-case return loss of 8.0 dB, due to interference effects.) Correspondingly, average insertion losses are reduced from 0.5 to less than 0.2 dB for single-mode fibers, and from about 0.4 to 0.1 dB

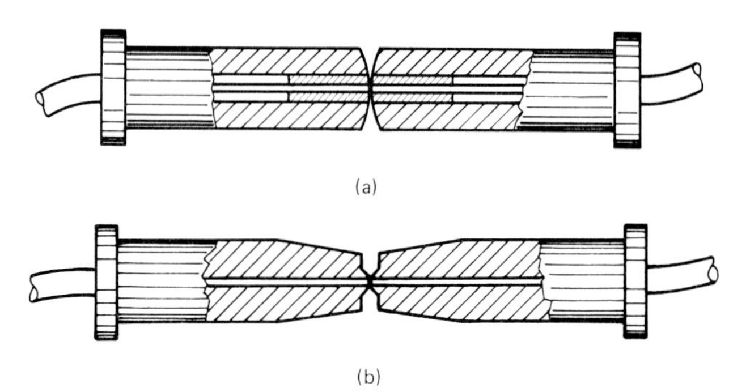


FIGURE 3.46 Plugs allowing physical contact of the fiber ends. (*a*) Convex-polished ceramic capillary plugs (*adapted from Ref. 44*); (*b*) biconical plugs (*adapted from Ref. 45*).

for multimode fibers. The variation in loss for repeated connections and disconnections becomes negligible, because of the elimination of interference effects. Physical contact can be easily achieved with biconical plugs,[45] as well, by virtue of the slight central protrusion in the region of the fiber end, as shown in Fig. 3.46*b*.

The third approach requires that the fiber-end face be tilted, as in Fig. 3.47, in order to misalign the reflected beam by an angle large enough to introduce the desired feedback suppression. If the plug is polished at an angle of 8° with respect

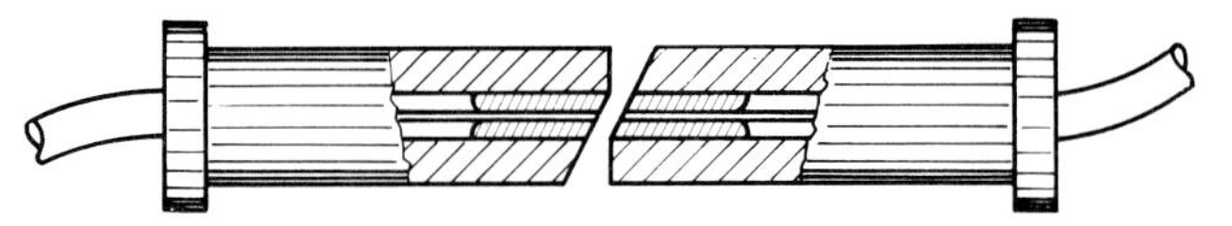

FIGURE 3.47 Plugs utilizing tilted faces to suppress feedback. (*Adapted from Ref. 46.*)

to the ferrule axis, a return loss of 40 dB is achieved, if typical 50-μm core size, multimode, graded-index fibers are used.[46] A 40-dB return loss is considered high enough to be compatible with high-performance analog transmission systems. In comparison to a corresponding nontilted connector, the connector loss is only 0.1 dB higher, provided that the plugs are not axially rotated from the optimum mating orientation.

For single-mode fibers, a refined approach[47] requires tilting the fiber axis by an angle $\theta_1$ with respect to the mechanical axis of the connector and tilting the plug-end face by an angle $\theta_2$ with respect to the normal plane, as shown in Fig. 3.48*a*. Tilt angles $\theta_1$ and $\theta_2$ are chosen such that the beam exiting a fiber is parallel to the connector axis and enters the other fiber without suffering any loss due to angular misalignment. The reflected beams, on the other hand, are strongly misaligned with respect to the fiber axis and therefore suffer high return losses, independent of the relative rotational position of the two plugs, as illus-

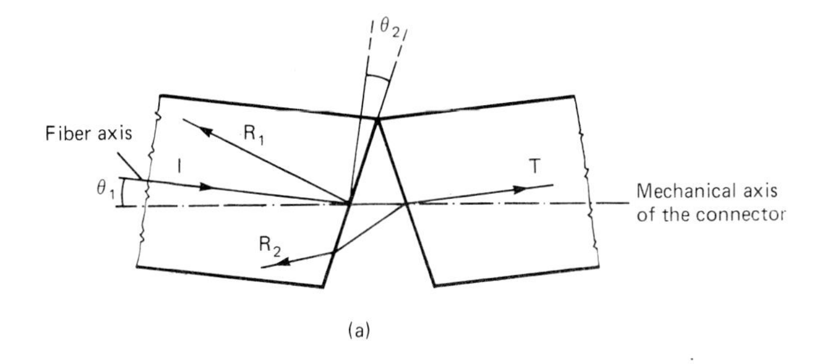


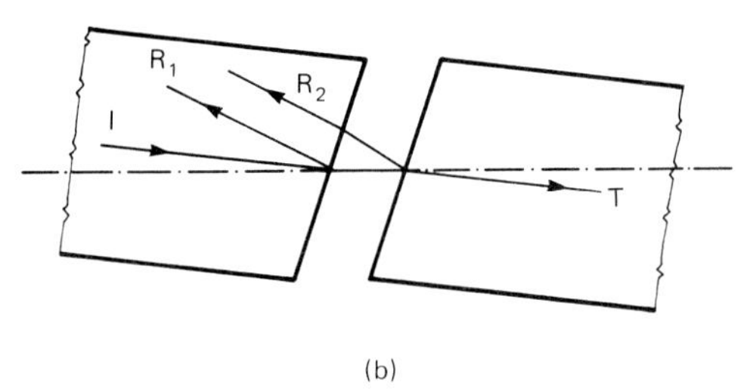


**FIGURE 3.48** Principle of the double-tilt (fiber axis and fiber face) approach to increase the return loss of the connector. $I$ is the incident beam; $T$ is the transmitted beam; and $R_1$ and $R_2$ are the reflected beams. This configuration is effective for any rotation of the plugs around their mechanical axis: two extreme conditions (*a*) and (*b*) are shown here. (*Adapted from Ref. 47.*)

trated in Fig. 3.48*b*. Applying this concept to biconical connectors results in return losses greater than 32 dB, with an average value of 38 dB, if the angle between the reflected beam and the fiber axis is about 8°.

The average insertion loss of 0.7 dB is caused by the slight separation between the fiber ends associated with the tilt and is therefore inherent to this approach.

***Multiple Connectors.*** To reduce assembly time, a number of fibers from a multifiber cable can be grouped and simultaneously joined by means of a multifiber connector. Two techniques that are unique to multiple fiber use are described below.

High-precision-grooved silicon chips, as in Fig. 3.49*a* and *b*, have been used[48] to precisely position the fiber arrays of a multiribbon cable. With this approach, 12 fibers are seated on one chip, with the possibility of stacking many 12-fiber arrays. Spring clips and metal-backed chips with complementary grooves are used to accommodate and secure this multiple-fiber structure. Even single-mode fibers can be mass-connected with low losses by this technique.

Alternatively, plastic molded plug ferrules, as in Fig. 3.49*c*, can be manufactured to accommodate a linear array of 5 or 10 fibers.[49] Two large-diameter guide pins are used to align the plugs. Single-mode fibers can be connected as well, with low losses.

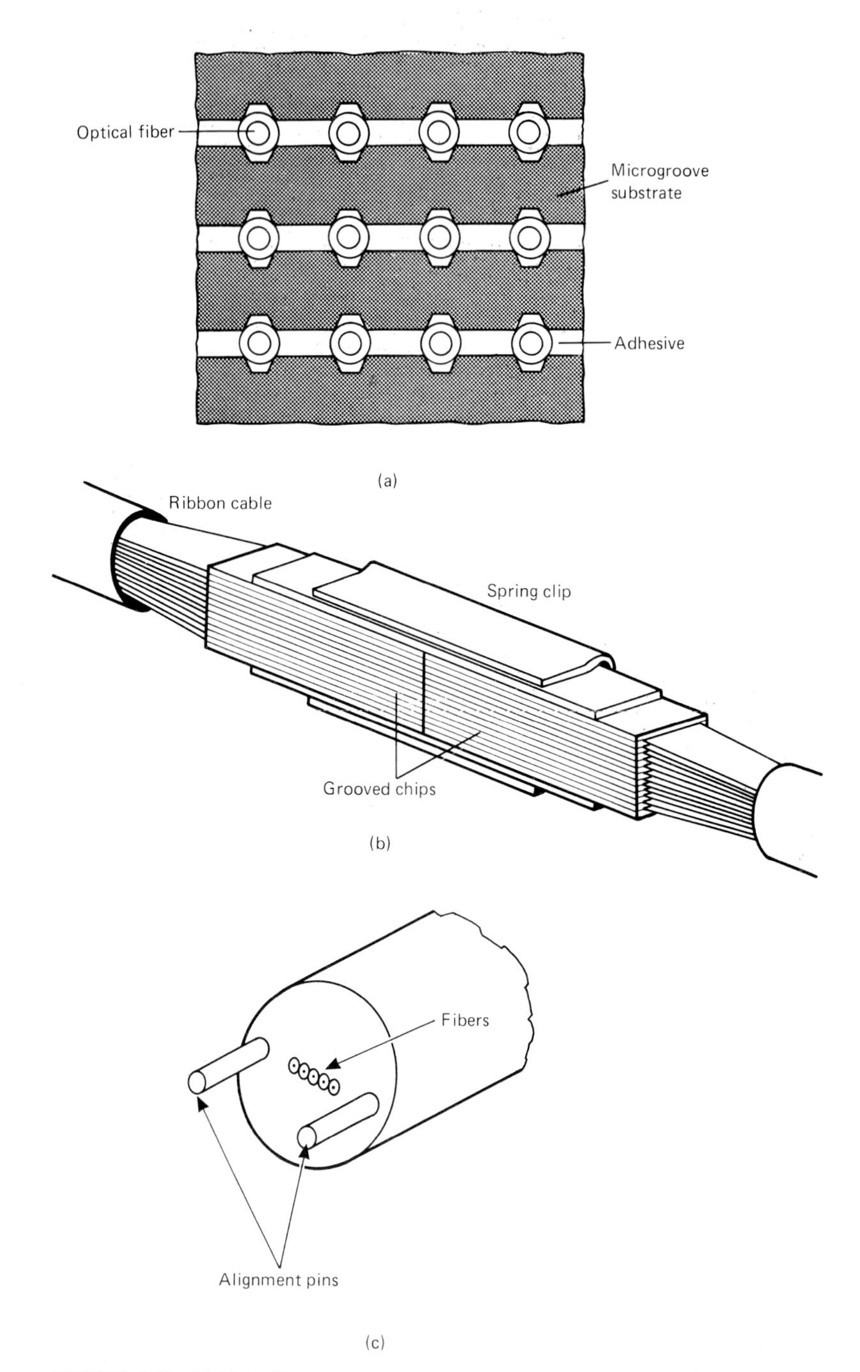


**FIGURE 3.49** Multiple fiber connectors. (*a*) Cross section of grooved chip connector (*adapted from Ref. 23 and reproduced with permission from the AT&T Technical Journal, copyright 1978, AT&T*); (*b*) grooved chip assembly (*adapted from Ref. 48 and reproduced with permission from the AT&T Technical Journal, copyright 1975, AT&T*); (*c*) molded plastic multifiber connector (*adapted from Ref. 49*).

## 3.4 OPTICAL-FIBER DIRECTIONAL COUPLERS

### 3.4.1 General Considerations

Besides simple point-to-point links, optical-fiber communications often makes use of more complex topologies (e.g., ring, star, and bus), mainly in local-area networks. For these applications, special functions are needed, such as distribution of the same signal to several subscribers, insertion of many signals into one line, or exploitation of the same fiber for a bidirectional link. These and other operations can be carried out directly on the optical signal, avoiding any optical to electrical conversion, by making use of passive optical devices that are known as optical directional couplers.

A schematic representation of one of these devices, having $N$ input ports and $M$ output ports, is shown in Fig. 3.50. Optical couplers are usually provided with

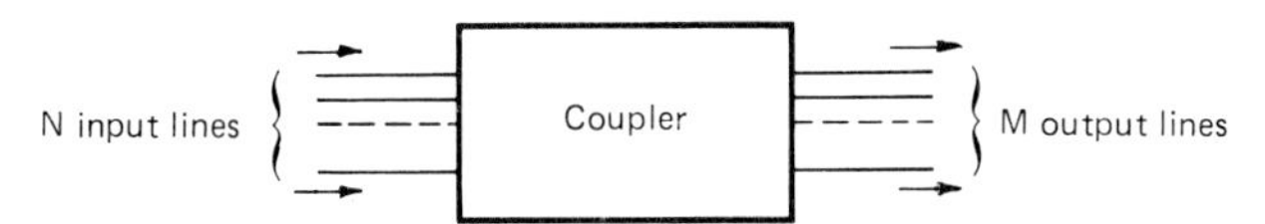


**FIGURE 3.50** General representation of an $N \times M$ optical coupler.

connectors to allow an easy connection to optical-fiber lines and to active or passive devices.

According to their characteristics, different names are used for optical couplers. A *Y coupler* (or optical splitter) is a device that divides the optical power carried by one (input) fiber, sharing it between two output fibers (Fig. 3.51). When one of the two output ports splits off only a small amount of the total power, the device is sometimes called a *T coupler,* or optical tap, and can be used as a monitor of the power level that is carried by the main line. When these devices are used to combine the signals carried by two fibers into one, they are sometimes referred to as *optical combiners* (Fig. 3.52). Both functions, combining and splitting, are carried out by the so-called *X coupler* (or $2 \times 2$ directional coupler) that is shown in Fig. 3.53.

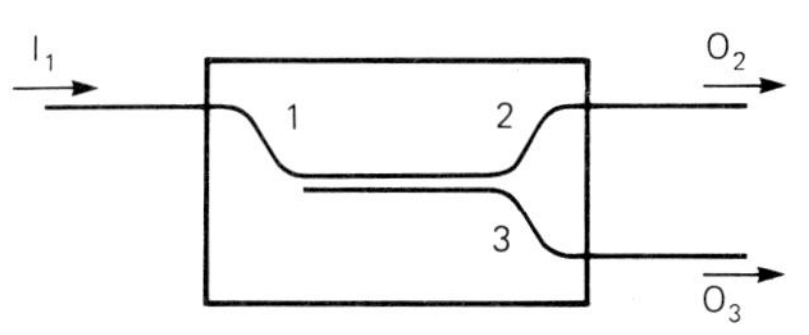


**FIGURE 3.51** Schematic representation of a Y coupler (or optical splitter).

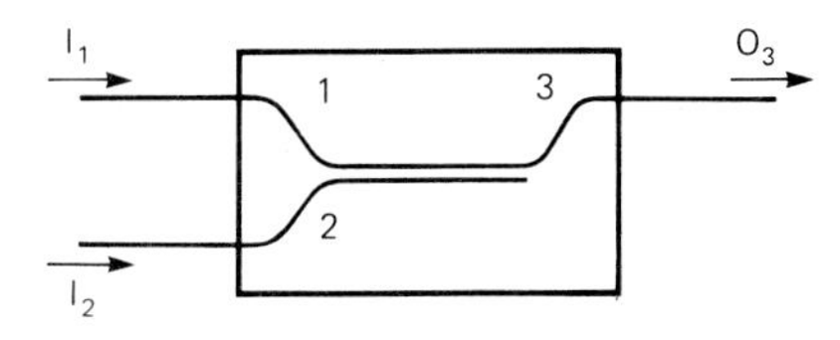


**FIGURE 3.52** Schematic representation of an optical combiner.

Multiport devices having input (or output) ports in excess of 2 are usually called *star couplers* (Fig. 3.54), while the term *tree coupler* sometimes applies to $1 \times M$ (or $M \times 1$) couplers, when $M > 2$ (Fig. 3.55).

Star couplers and tree couplers are mainly used in local-area networks and usually are supposed to produce uniform division of the input power at the output ports.

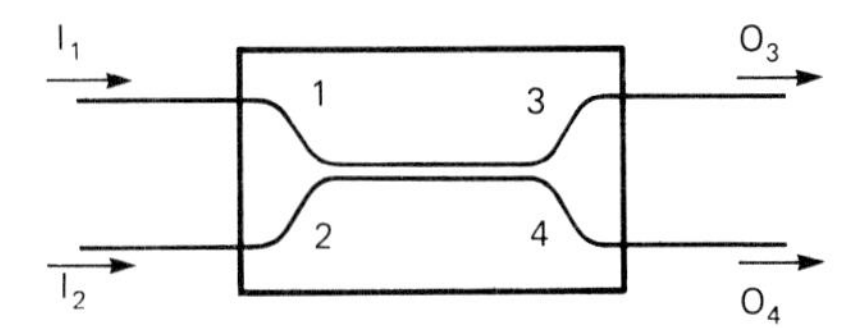


**FIGURE 3.53** Schematic representation of an optical X coupler (or 2 × 2 directional coupler).

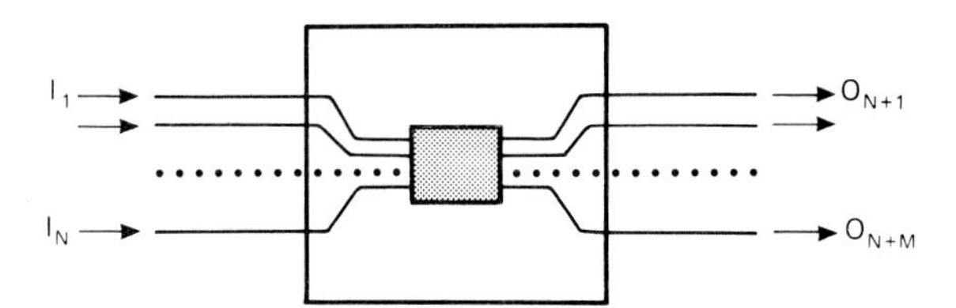


**FIGURE 3.54** Schematic representation of an $N \times M$ star coupler.

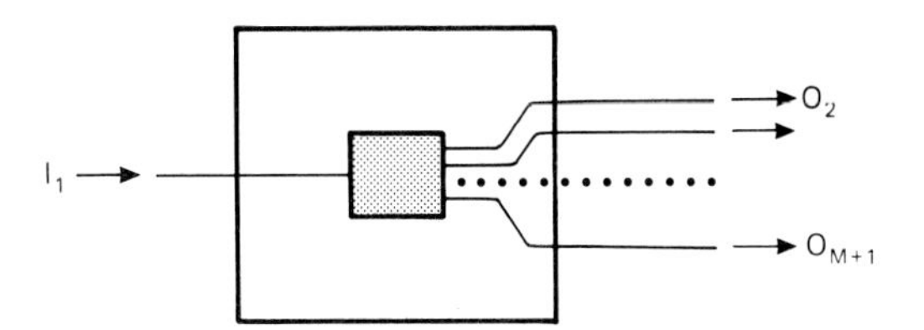


**FIGURE 3.55** Schematic representation of a $1 \times M$ tree coupler.

Optical couplers are directive, in the sense that no power transfer should take place, at least ideally, from one input port to any other input port. Often, they are also symmetrical, in that the same fraction of power is transmitted through the coupler if the optical path from an input port to an output port is reversed (i.e., from the former output port to the former input port).

The fundamental parameters describing optical coupler characteristics are presented in the following. With reference to the schematic representation in Fig. 3.56, if $I_i$ is the input power entering the generic $i$th input port and $O_j$ is the output power exiting the generic $j$th port, $\Sigma_i I_i$ and $\Sigma_j O_j$ are the total powers, respectively, entering and exiting the coupler. The difference power $\Delta P$

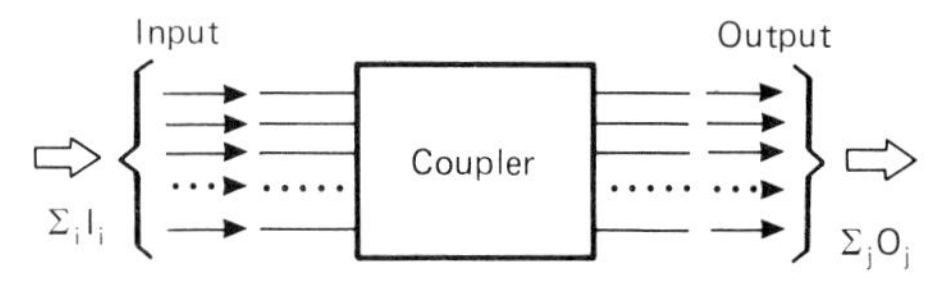


**FIGURE 3.56** Schematic representation of the power flux entering and exiting the optical coupler.

($\Delta P = \Sigma_i I_i - \Sigma_j O_j$) is the amount of power that is lost along the optical paths from the input to the output ports because of reflections, scattering, and various imperfections. The transmission efficiency $\eta$ of the coupler is defined as

$$\eta = \frac{\Sigma_j O_j}{\Sigma_i I_i} \tag{3.30}$$

while the fraction of power lost is

$$\Lambda = \frac{\Delta P}{\Sigma_i I_i} = 1 - \eta \tag{3.31}$$

The excess loss of the device, expressed in decibels, is

$$L = -10 \log \eta \tag{3.32}$$

The output power at the $j$th output port $O_j$ can be expressed by

$$O_j = \Sigma_i A_{ji} I_i \tag{3.33}$$

where the coupling coefficient between the $i$th input port and the $j$th output port, $A_{ji}$, represents the fraction of the power entering the $i$th input port which exits the $j$th output port. $A_{ji}$ actually takes into account both the branching effect of the coupler and the waste of power inside the coupler caused by reflections, scattering, and various imperfections. The insertion loss in decibels along the path from the $i$th to the $j$th port is then given by

$$L_{ji} = -10 \log A_{ji} \tag{3.34}$$

The directivity of an optical coupler is measured by the isolation between its input ports. When power is launched into the $i$th port, the isolation between the $i$th and the $k$th input port (Fig. 3.57) is usually defined, in decibels, as

$$D_{ki} = -10 \log \frac{R_k}{I_i} \tag{3.35}$$

where $R_k$ is the amount of undesired return power at the $k$th port, when all other ports are index-matched (if possible) to avoid reflections. In actual operating conditions, the presence of unmatched reflecting surfaces at some ports can significantly reduce isolation.

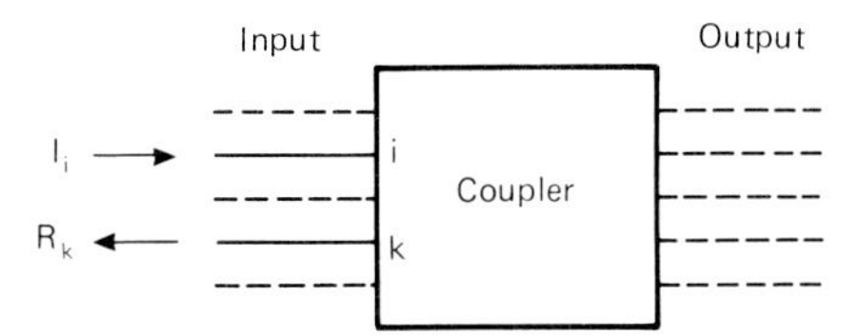


**FIGURE 3.57** Schematic representation of the undesired coupling between two input ports.

The simple splitter of Fig. 3.51 is analyzed in the following. The power $I_1$ that is launched into input port 1 is divided into $O_2$ and $O_3$ at the corresponding output ports 2 and 3. The splitting ratio $S$ is defined as

$$S = \frac{O_2}{O_3} \tag{3.36}$$

and is often expressed in the form of a proportion, e.g., 1:1, 3:1, 10:1. Insertion losses along the paths 1–2 and 1–3 are given, respectively, by

$$L_{21} = -10 \log \frac{O_2}{I_1} = -10 \log A_{21} \tag{3.37}$$

$$L_{31} = -10 \log \frac{O_3}{I_1} = -10 \log A_{31} \tag{3.38}$$

The same device can be used in the reverse way, as a combiner. In this case, if the device is symmetrical, $L_{12} = L_{21}$ and $L_{13} = L_{31}$.

Star couplers, which are supposed to have the same coupling coefficient between each pair of input and output ports, are characterized by an average insertion loss and by the deviation from this average value, which can be expected at each individual port. For example, a 4 × 4 star coupler could have an average insertion loss of 6 dB, assuming (ideally) no excess loss and a maximum deviation of 5 percent from this value at each port, because of nonuniform coupling among channels.

Optical couplers can be produced according to different principles and, hence, according to different technologies: micro-optics or coupled-fiber, single-mode or multimode, and symmetrical or asymmetrical. In the following, an overview is presented of some of these devices and of their properties.

### 3.4.2 Micro-Optical Couplers

Splitting or combining of optical signals can be carried out by operating on collimated beams, at the output of the optical fibers, as is shown in Fig. 3.58 for a generic X-coupler configuration. Light is collimated and focused by lenses, while the splitting or combining effect relies on a partially reflecting mirror whose reflectivity $R$ has the desired value.

Actual couplers make extensive use of miniaturized optical devices such as microlenses (see Sec. 3.6) and especially graded-refractive-index (GRIN) rods, which allow for simple fabrication of coupling devices. The basic structure of a micro-optical X coupler,[50] corresponding to the schematic configuration of Fig. 3.58, is shown in Fig. 3.59. It makes use of two 0.25-pitch GRIN rod lenses, which are used in an offset configuration for collimation and focusing. The splitting effect is provided by a partially reflecting coating that is deposited on the intermediate GRIN rod facet.

Since the reflection coefficient is, with good approximation, the same for all the input rays, these micro-optical couplers are relatively independent of the mode power distribution in the fiber at the input port. Therefore these devices are particularly attractive for multimode fibers. For multimode fibers, typical excess losses of micro-optical couplers are on the order of 0.5 to 1.0 dB.

Single-mode couplers make preferential use of a different technology (refer to

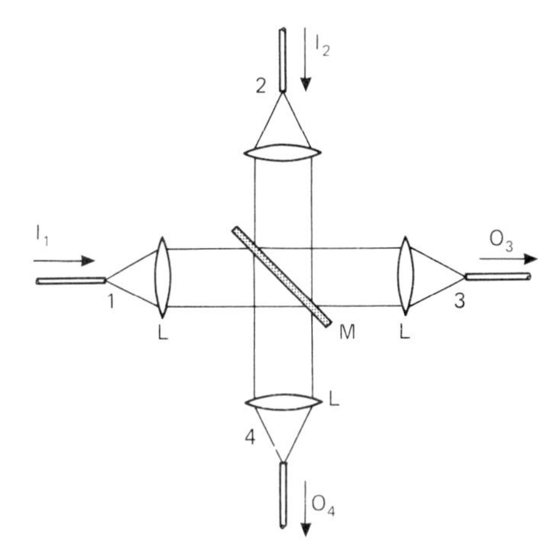


**FIGURE 3.58** Basic collimated beam configuration of an X coupler. $L$ indicates lenses and $M$ is a partially reflecting mirror.

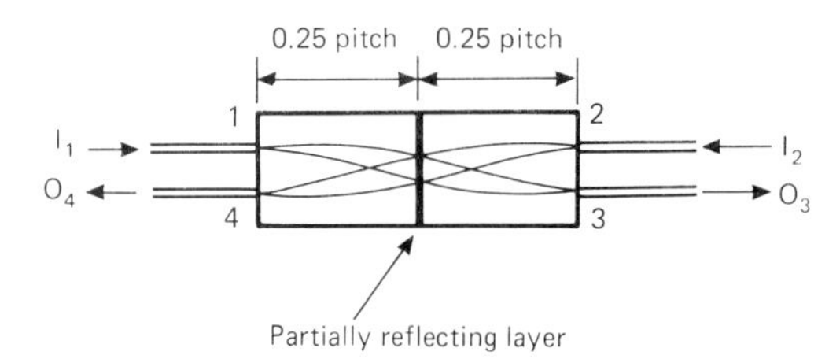


**FIGURE 3.59** Schematic section of a micro-optical X coupler using GRIN rod lenses. *(Adapted from Ref. 50.)*

coupled fibers in Sec. 3.4.4), since, in this case, stable and accurate positioning of the micro-optical elements would become much more critical. Furthermore, the detrimental effects of lens aberrations would be significant.

Star couplers, as well, can be fabricated by micro-optical techniques, in accordance with the schematic representation of Fig. 3.60. The input fibers are coupled to an *optical mixer* (which is a guiding structure that conveys the input power, distributes it across its section as uniformly as possible, and shares it among the output fibers). Such a device has an inherent excess loss that is produced by the fractional covering of the core areas with respect to the output surface of the optical mixer. The optical mixer is usually shaped as a planar waveguide of transparent, high-refractive-index material.[51]

### 3.4.3 Coupled-Fiber Devices (Multimode Fibers)

A simple and effective X coupler can be fabricated by tapering, twisting, and fusing together, for some length, two multimode fibers. The process results in the so-called *fused biconical taper coupler* (FBT). The working principle of such a coupler can be explained with reference to the longitudinal section in the coupling

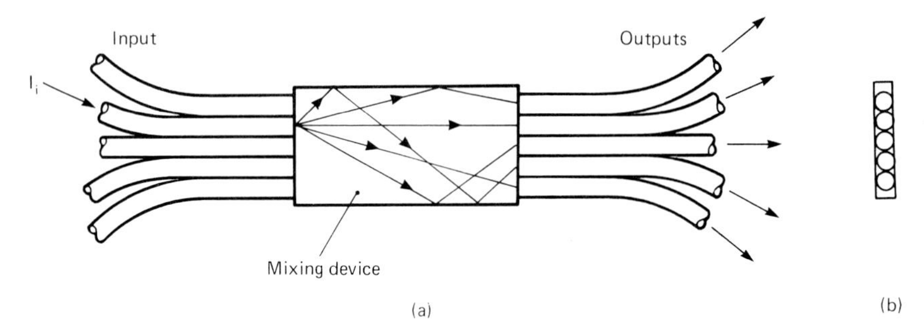


**FIGURE 3.60** Micro-optical star coupler. (*a*) Schematic longitudinal section; (*b*) transverse section in the plane where fibers are butt-coupled to the mixing region. (*Adapted from Ref. 51.*)

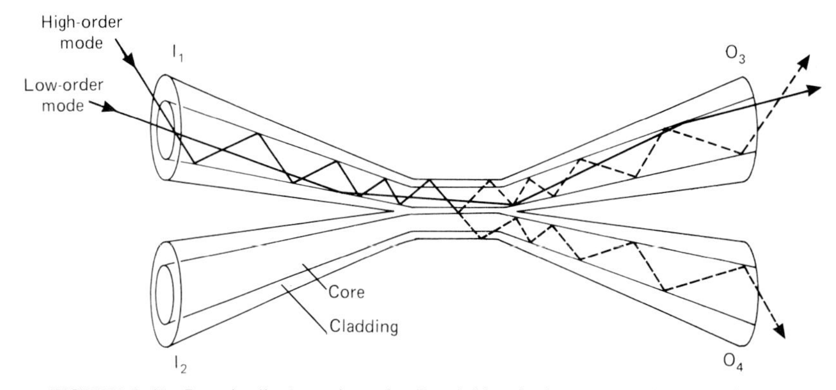


**FIGURE 3.61** Longitudinal section of a fused biconical taper coupler, showing the ray paths along the coupling region.

region (Fig. 3.61). Step-index fibers and meridional rays are assumed, only for the sake of simplicity.

The optical power $I_1$ that is launched into port 1 propagates, distributing itself among the guided modes that are compatible with the guiding structure. When the guided modes reach the tapered region, they see a reduced effective numerical aperture. Therefore, some of the higher-order modes are no longer guided by the core and diffuse into the cladding with a spatial distribution that, in the fused common region, is nearly uniform. After this region, the optical power in the cladding is split by the bifurcation into two equal parts, each of which finds a reversed taper structure and, because of the corresponding increase in the effective numerical aperture, is progressively re-collected and guided by the cores. By this mechanism, part of the input power is extracted from the core of the first fiber, divided into two parts, and guided again, finally reaching output ports 3 and 4. Lower-order modes, on the other hand, are never extracted from the core of the input fiber and are transferred directly to output port 3.

Some important features of FBT couplers can be clarified by reference to the basic properties of the tapered fiber, as shown in Fig. 3.62. For this example, the simplifying assumptions of step-index and meridional rays have been made.

A ray propagating along the taper (characterized by a taper angle $\beta$) will un-

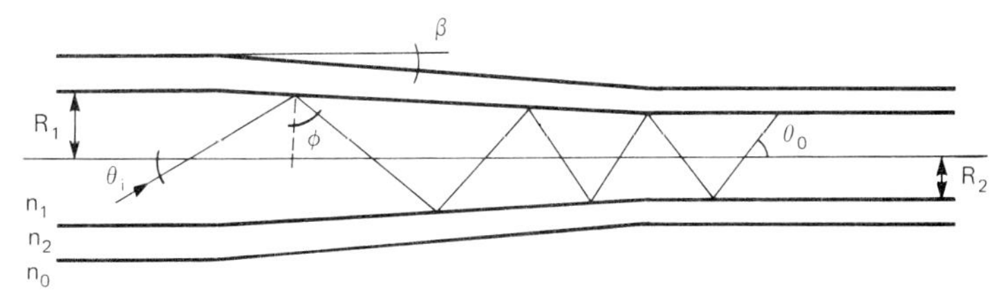


FIGURE 3.62 Section of a tapered fiber. (*Adapted from Ref. 52.*)

dergo an increase in its propagation angle (i.e., the angle formed with the fiber axis). Starting at an initial value $\theta_i$, the propagation angle will increase by $2\beta$ at each reflection at the core-cladding interface. As long as the incidence angle $\phi$ is larger than the critical angle (for total internal reflection), the ray is still guided and emerges from the taper region, forming an angle $\theta_o$ with the fiber axis. If the taper is smooth and the number of reflections is high, $\theta_o$ is given by

$$\sin \theta_o = \frac{R_1}{R_2} \sin \theta_i \tag{3.39}$$

where $R_1$ and $R_2$ are, respectively, the core radii before and after the taper. Therefore, the propagation angle of the ray at the output of the taper depends on the tapering ratio $R_1/R_2$. When the incidence angle $\phi$ becomes smaller than its critical value, the ray is no longer guided by the core, yet it can still be confined within the cladding, provided that

$$\sin \theta_o \le \sin \theta_{oM} = \frac{(n_1^2 - n_0^2)^{1/2}}{n_1} \tag{3.40}$$

where $n_0$ and $n_1$ are the refractive indices of the outer medium and of the core, respectively, and $\theta_{oM}$ is the maximum propagation angle within which rays are still confined. At the input of the coupler, the maximum launching angle $\theta_{iM}$ is limited by the numerical aperture of the fiber, according to

$$n_1 \sin \theta_{iM} = \text{NA} = (n_1^2 - n_2^2)^{1/2} \tag{3.41}$$

To avoid radiation outside the cladding, by any accepted ray, $\theta_o(\theta_{iM})$ must be limited:

$$\theta_o(\theta_{iM}) \le \theta_{oM} \tag{3.42}$$

Therefore, from Eqs. (3.39) to (3.41),

$$\frac{R_1}{R_2} \le \left(\frac{n_1^2 - n_0^2}{n_1^2 - n_2^2}\right)^{1/2} \tag{3.43}$$

which sets an upper limit to the taper ratio, $R_1/R_2$. When this condition is not met, light exits the coupler, giving rise to losses. The rays that are always guided by the launching fiber—and hence do not contribute to coupling—are simply defined by

$$n_1 \sin \theta_o \le (n_1^2 - n_2^2)^{1/2} \tag{3.44}$$

which, by means of Eq. (3.39), becomes

$$n_1 \sin \theta_i \le \left(\frac{R_2}{R_1}\right)(n_1^2 - n_2^2)^{1/2} \tag{3.45}$$

Therefore, the numerical aperture for the tapered structure, $\mathrm{NA}_T$, can be defined:

$$\mathrm{NA}_T = n_1 \sin \theta_{iT} = \left(\frac{R_2}{R_1}\right)(n_1^2 - n_2^2)^{1/2} = \left(\frac{R_2}{R_1}\right)\mathrm{NA} \tag{3.46}$$

For the tapered fiber in Fig. 3.62, the numerical aperture is reduced by a factor $R_2/R_1$ ($< 1$), with respect to its original value, NA. Therefore, rays can be classified on the basis of their input propagation angle, $\theta_i$:

- Rays with $\theta_i \le \theta_{iT}$ are transmitted directly from port 1 to port 3.
- Rays with $\theta_{iT} \le \theta_i \le \theta_{iM}$ are shared between ports 3 and 4.

It is clear that, in fused biconical taper couplers, only rays propagating at high angles (corresponding to high-order modes) are coupled to the adjacent fiber, while the low-order rays are kept confined within the launching fiber.

Assuming a lambertian source (i.e., uniform excitation of the guided rays), coupling coefficients between the ports, as defined in Sec. 3.4.1, can be easily calculated:

$$A_{31} = \frac{1}{2}\left[1 + \left(\frac{R_2}{R_1}\right)^2\right] \tag{3.47}$$

$$A_{41} = \frac{1}{2}\left[1 - \left(\frac{R_2}{R_1}\right)^2\right] \tag{3.48}$$

$A_{31}$ and $A_{41}$ represent the fractions of input power that exit ports 3 and 4, respectively. The behavior of these coefficients is represented in Fig. 3.63; the 1:1 partition ratio between the output branches would occur only for $R_2/R_1 = 0$, which conflicts with Eq. (3.43) and, furthermore, would be unrealistic. If Eq. (3.43) is taken into account, $A_{41}$ is bounded:

$$A_{41} \le \frac{1}{2}\left(1 - \frac{\mathrm{NA}^2}{n_1^2 - 1}\right) \tag{3.49}$$

For a step-index fiber with $n_1 = 1.47$ and $\mathrm{NA} = 0.22$, the minimum allowed value of $R_2/R_1$ is 0.2, corresponding to $A_{41} = 0.48$.

Typical excess losses of fused biconical taper couplers are in the range of 0.5 to 1.0 dB, for graded-index, 50-μm core, 125-μm cladding fibers.

Multiport, $N \times N$ fused biconical taper star couplers can be fabricated by a

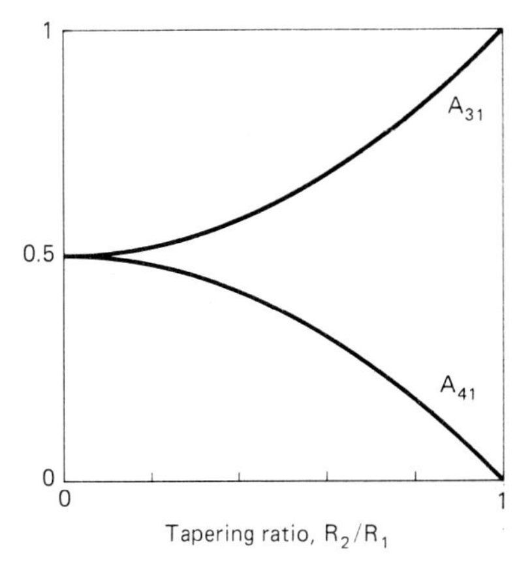


FIGURE 3.63 Coupling coefficients $A_{31}$ and $A_{41}$ vs. tapering ratio $R_2/R_1$.

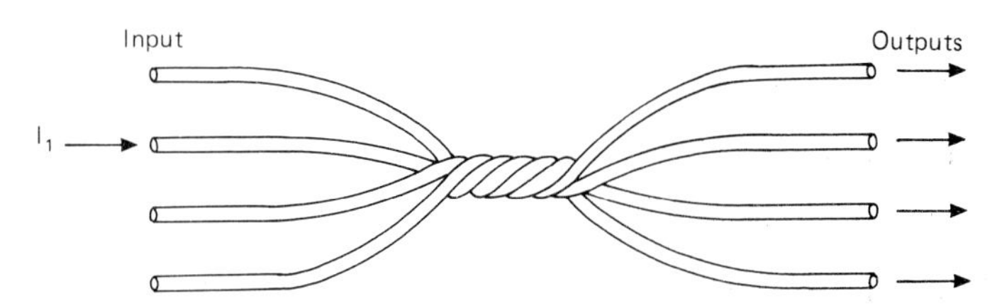


FIGURE 3.64 Schematic structure of a fused biconical taper star coupler. (*Adapted from Ref. 53.*)

similar technique and operate according to the principles described. Figure 3.64 shows the schematic of a transmissive star coupler.[53]

### 3.4.4 Coupled-Fiber Devices (Single-Mode Fibers)

Efficient couplers can be fabricated by direct coupling of single-mode fibers, as well. Several theoretical approaches have been implemented to analyze the behavior of two parallel and adjacent single-mode waveguides. For instance, the coupled-waveguide structure represented in Fig. 3.65 has been analyzed[54] under the hypothesis of weak coupling, assuming that the resulting field distribution, associated with the modes of the composite waveguide, can be expressed by a linear superposition of the fields of the single, uncoupled waveguides.

In the particular case in which the propagation constants of the fundamental modes of the single, uncoupled waveguides are equal, i.e., $\beta_1 = \beta_2 = \beta$, the general solution of the coupled-mode equations consists of the overlap of two new modes having propagation constants $\beta_+$ and $\beta_-$, given by

$$\beta_+ = \beta + \Delta\beta$$
$$\beta_- = \beta - \Delta\beta \tag{3.50}$$

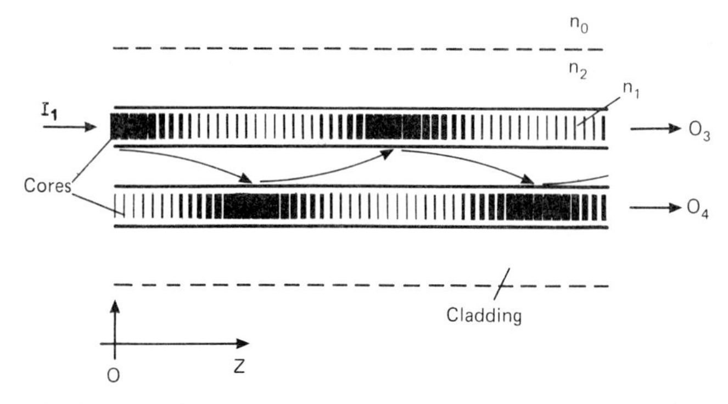


**FIGURE 3.65** Schematic structure and power transfer mechanism in a generic single-mode fiber coupler.

where $\Delta\beta$ represents the coupling coefficient between the two waveguides. That is, in a first-order approximation, two normal modes, whose propagation constants are given by Eq. (3.50), can be supported by the composite waveguide. We can note that full transfer of optical power from one waveguide to the other is possible only if $\beta_1 = \beta_2$. That is, substantial exchange of power occurs only between modes with the same propagation constant. If power is assumed to be launched through port 1, and where $z = 0$ defines the beginning of the interaction zone, the corresponding optical power levels along $z$ in the two output branches $O_3$ and $O_4$ are subsequently given by

$$O_3(z) = I_1 \cos^2(\Delta\beta z)$$
$$O_4(z) = I_1 \sin^2(\Delta\beta z) \tag{3.51}$$

where $I_i$ is the input power entering port 1 and $\Delta\beta$ is the phase delay appearing in Eq. (3.50). Equation (3.51) shows clearly that optical power is periodically transferred along $z$ from one waveguide to the other. $O_3(z)$ and $O_4(z)$ are represented in Fig. 3.66. The particular value of $z$, $L_c$, corresponding to a complete power transfer (e.g., from waveguide 1 to waveguide 2), is given by

$$L_c = \frac{\pi}{2\,\Delta\beta} \tag{3.52}$$

and is usually called the *coupling length*.

Single-mode, optical-fiber couplers are usually manufactured according to one of two methods. The first method consists of polishing the fibers to remove the cladding to a predetermined extent, until the cores can be brought into close proximity. This results in the coupled-waveguide structure that is illustrated in Fig. 3.67.

In the second method, two single-mode fibers are fused and tapered, sometimes with chemical etching (as already described in Sec. 3.4.3 for the case of multimode fused biconical taper couplers). Along the interaction region, the coupling process is favored by both the short distance between the cores and the broadening of the mode field diameter of the guided mode (see Fig. 3.68) that is produced by the core size reduction [Eq. (3.22)]. Excess losses of these fused

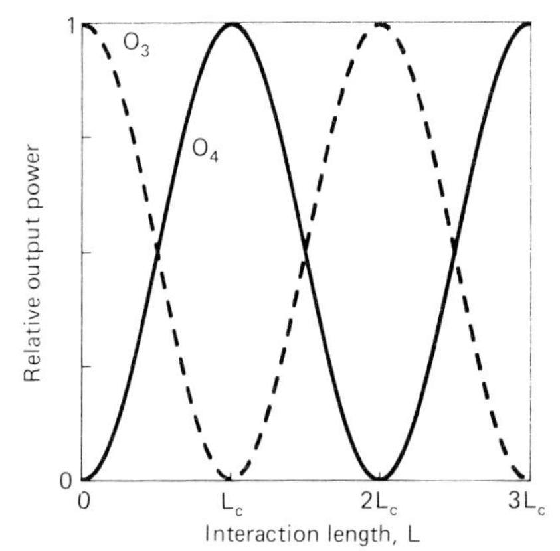


**FIGURE 3.66** Transmitted and coupled power $O_3$ and $O_4$ (respectively) vs. interaction length in the single-mode coupler of Fig. 3.65.

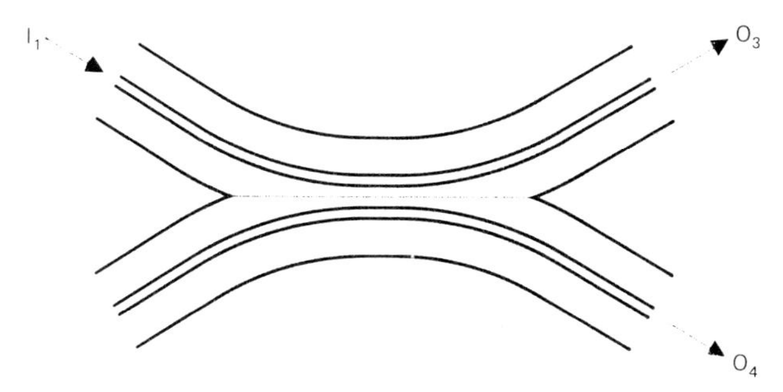


**FIGURE 3.67** Schematic structure of a polished, single-mode fiber coupler.

biconical taper single-mode couplers are very low, typically on the order of 0.1 dB or less.

In both structures, the coupling coefficient $\Delta\beta$ is actually $z$-dependent. In both cases, however, relationships of the same type as Eq. (3.51) can be obtained.

A simplified model[56] can be adopted to describe the behavior of a single-mode, fused biconical taper coupler, relying on the assumption that, in the fused central region, the two cores have become so diminutive as to no longer produce actual guiding. On the other hand, in this region, strong waveguiding results from the index difference existing between the cladding material ($n_2$) and the outer medium ($n_0$), such that they therefore become the new core and cladding, respectively. This central coupling region is assumed to have a rectangular $b \times 2b$ cross section.

Under the hypothesis that only the two fundamental modes of the waveguide can propagate in the coupling region, the power branching between the two output ports, 3 and 4, is still described by Eq. (3.51), where the coupling coefficient is expressed by

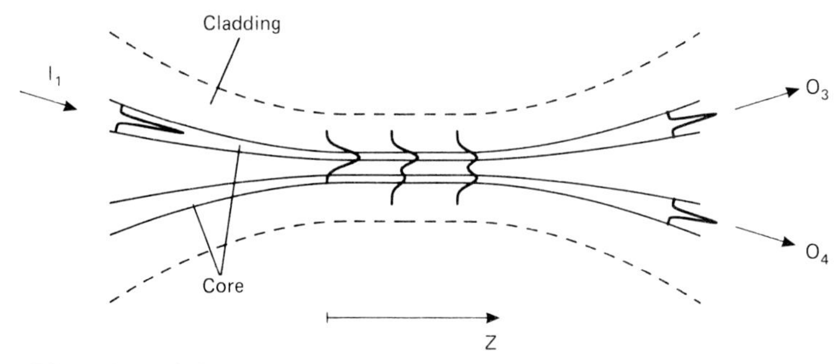


**FIGURE 3.68** Schematic structure of a single-mode fused biconical taper coupler, showing the change of the mode field diameter along the tapered region. *(Adapted from Ref. 55.)*

$$\Delta\beta = \frac{3\pi\lambda}{32 n_2 b^2} \frac{1}{(1 + 1/V)^2} \tag{3.53}$$

where $V = (2\pi b/\lambda)(n_2^2 - n_0^2)^{1/2}$

Equation (3.53) clearly displays the dependence of $\Delta\beta$ on geometrical and optical characteristics of the structure.

### 3.4.5 Asymmetrical Couplers (Multimode Fibers)

In a symmetrical X coupler, it is impossible to combine into one output all of the power coming from two input ports. This can be understood with the help of Fig. 3.69.

When the coupler is used as a splitter (Fig. 3.69*a*), the output powers are $O_3 = A_{31}I_1$ and $O_4 = A_{41}I_1$, while, of course, $A_{31} + A_{41} \leq 1$. In the ideal case, $A_{31} + A_{41} = 1$ and $O_3 + O_4 = I_1$ (i.e., no excess loss occurs in the ideal splitting process). If, for example, $A_{31} = A_{41} = 0.5$, then $O_3 = O_4 = I_1/2$.

If the same device is used in the reverse way (Fig. 3.69*b*), to combine $I_3$ and $I_4$ into port 1, the output of the device can be calculated on the basis of symmetry (i.e., $A_{13} = A_{31}$ and $A_{14} = A_{41}$). One can then write $O_1 = A_{31}I_3 + A_{41}I_4$. If, for example, $A_{31} = A_{41} = 0.5$, then $O_1 = I_3/2 + I_4/2$. Therefore, some of the total input power $(I_3 + I_4)$ is not available at the output of port 1. Looking only at the effect of power combination from ports 3 and 4 into port 1, this appears as a loss;

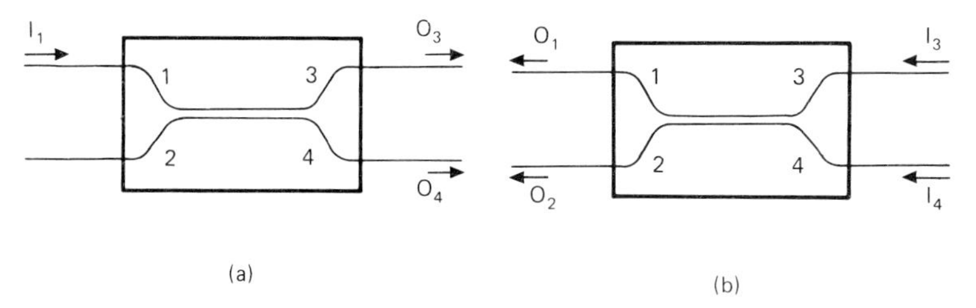


**FIGURE 3.69** Schematic of an optical coupler used (*a*) in the forward direction, as a splitter, and (*b*) in the reverse direction, as a combiner.

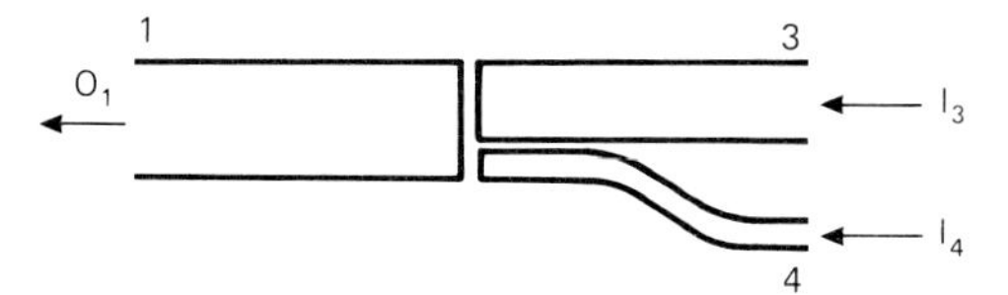


**FIGURE 3.70** Schematic representation of an asymmetrical combiner, making use of multimode fibers with different sizes. The cladding thickness is assumed to be negligible in this illustration. Ports are numbered in the same order as in Fig. 3.69.

actually, the power lacking at port 1 has simply been diverted because of the branching effect, and exits via port 2.

This condition is usually met in fused biconical taper Y and X couplers that are symmetrical. If the assumption of symmetry is removed (i.e., $A_{13} \neq A_{31}$ and $A_{14} \neq A_{41}$), it is possible, when using multimode fibers, to eliminate the apparent loss mentioned. A coupler can then be realized where, ideally, $A_{13} = 1$ and $A_{14} = 1$. In this case, $O_1 = A_{13}I_3 + A_{14}I_4 = I_3 + I_4$. This is impossible in typical fused biconical taper couplers in which the input and output ports are made with fibers having the same size, since this results in a symmetrical device.

Asymmetrical couplers (splitters and combiners) can be made from multimode fibers or planar waveguides having different core sizes, as represented in principle in Fig. 3.70. In this case, symmetry no longer holds: $A_{13} = 1$, while $A_{31} < 1$. Similarly, $A_{14} = 1$, while $A_{41} < 1$. Signal insertion, in principle, can be lossless in such a structure.

## 3.5 DEVICES FOR WAVELENGTH-DIVISION MULTIPLEXING (WDM)

### 3.5.1 Introduction to Wavelength-Division Multiplexing

Wavelength-division multiplexing (WDM) is a possible means to increase transmission capacity of optical fibers. The basic idea in this multiplexing technique consists of transmitting different signals along the same fiber, where the signals are emitted from separate sources having different wavelengths. At the receiver end, the signals are discriminated from each other on the basis of wavelength.

The advantages of WDM techniques include the ability to support bidirectional information flow on a single fiber and the ability to upgrade the capacity of existing fiber systems (without laying additional fibers.)

Basic configurations for both unidirectional and bidirectional WDM transmission systems are illustrated in Fig. 3.71, which shows the use of wavelength-division multiplexers and demultiplexers. These devices are multiport couplers with $N \times 1$ and $1 \times N$ terminals, respectively, whose coupling coefficients $A_{ij}$ (see Sec. 3.4.1) between input and output ports depend on wavelength. This dependence is exploited at the transmitter end to reduce the insertion loss for the input channels and at the receiver end to discriminate the different wavelengths.

Actually, the use of a wavelength-selective coupler as a multiplexer is not strictly necessary; a simple, $N \times 1$ (nonwavelength-selective) coupler could be used as a multiplexer. In the case of symmetrical devices, this would involve

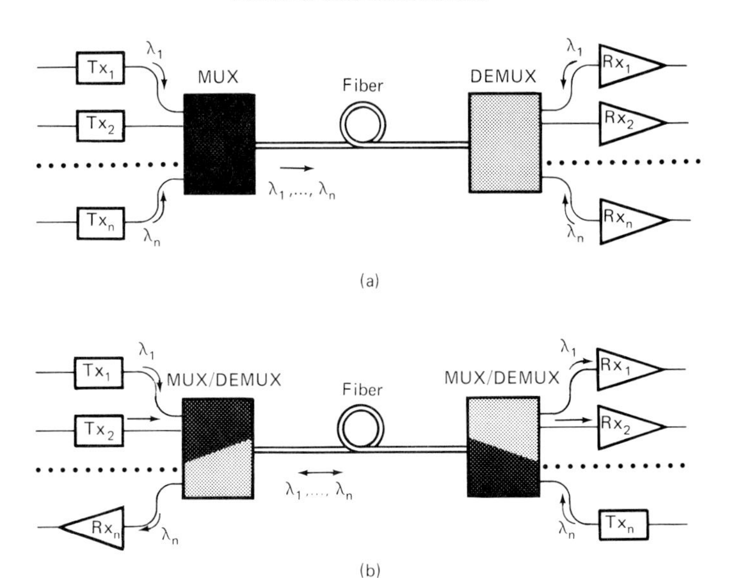


**FIGURE 3.71** Transmission systems making use of wavelength-division multiplexing in (*a*) unidirectional and (*b*) bidirectional configurations.

higher insertion losses, since no more than $1/N$ of the power at each input port would be coupled to the line. Wavelength selection, however, must necessarily be present at the receiver end, to distinguish the different channels on the basis of their wavelength.

From Fig. 3.72, which shows general configurations for the wavelength-division multiplexer, wavelength-division demultiplexer, and hybrid wavelength-division multiplexer-demultiplexer, some useful parameters can be defined.

In a wavelength-division multiplexer, the insertion loss $L(\lambda_i)$ for a channel operating at a given wavelength $\lambda_i$ is defined as the attenuation that the signal at wavelength $\lambda_i$ undergoes, from the input port nominally specified for $\lambda_i$, to the output port. With reference to Fig. 3.72*a*, the insertion loss $L(\lambda_i)$, is defined:

$$L(\lambda_i) = -10 \log \frac{O(\lambda_i)}{I_i(\lambda_i)} \tag{3.54}$$

where $I_i(\lambda_i)$ is the signal, at the wavelength $\lambda_i$, entering the $i$th input port (i.e., the port that nominally must accept $\lambda_i$) and $O(\lambda_i)$ is the signal, at the same wavelength $\lambda_i$, exiting the common output port (i.e., the port that connects the multiplexer to the line).

In a wavelength-division demultiplexer, the insertion loss $L(\lambda_i)$ for a channel operating at a given wavelength $\lambda_i$ (i.e., the attenuation that the signal at that wavelength $\lambda_i$ undergoes from the input port to the output port nominally specified for $\lambda_i$) is defined with reference to Fig. 3.72*b* as

$$L(\lambda_i) = -10 \log \frac{O_i(\lambda_i)}{I(\lambda_i)} \tag{3.55}$$

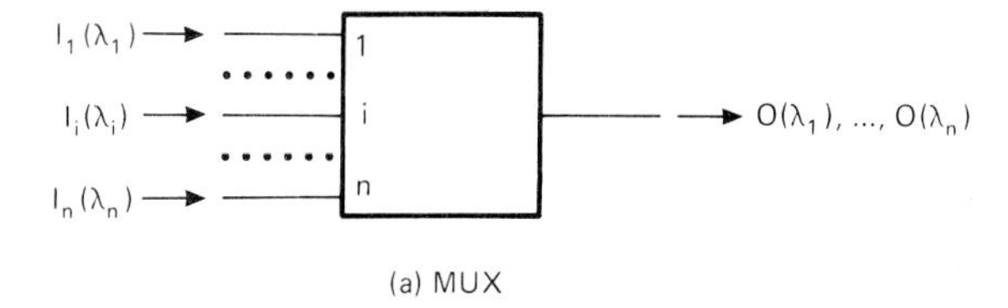


(a) MUX

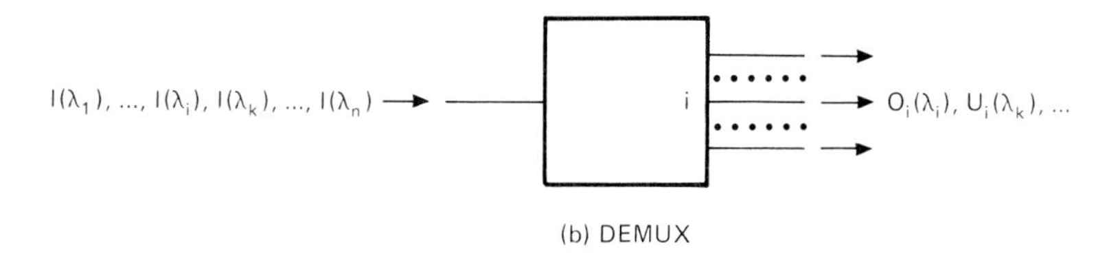


(b) DEMUX

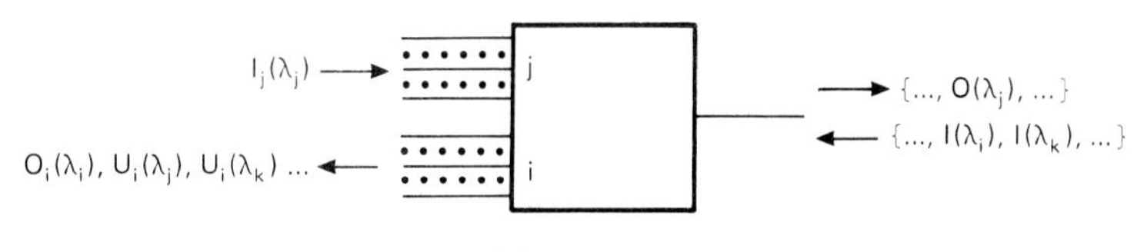


(c) MUX/DEMUX

FIGURE 3.72 General configurations of the wavelength-division (*a*) multiplexer, (*b*) demultiplexer, and (*c*) hybrid multiplexing-demultiplexing device.

where $O_i(\lambda_i)$ is the signal, at the wavelength $\lambda_i$, exiting the $i$th output port (i.e., the port that, nominally, must transmit only $\lambda_i$) and $I(\lambda_i)$ is the signal, at the same wavelength $\lambda_i$, entering the common input port (i.e., the port that connects the line to the demultiplexer).

In an ideal demultiplexer, at the output port that transmits the wavelength $\lambda_i$, there should be no power leakage from the channels operating at wavelengths different from $\lambda_i$. Actual devices suffer, anyway, from a certain degree of crosstalk among the channels, which should be kept as small as possible. In this sense, the quality of a demultiplexer can be expressed by the attenuation of the crosstalk that exists from the channel operating at the wavelength $\lambda_k$ to the channel that should transmit only the wavelength $\lambda_i$. This parameter, $D_i(\lambda_k)$, is defined by

$$D_i(\lambda_k) = -10 \log \frac{U_i(\lambda_k)}{I(\lambda_k)} \qquad k \neq i \tag{3.56}$$

where $U_i(\lambda_k)$ is the amount of undesired power at the wavelength $\lambda_k$ that leaksinto the $i$th channel and exits the corresponding $i$th port, nominally intended to operate only at the wavelength $\lambda_i$.

In the hybrid device of Fig. 3.72*c*, Eqs. (3.54) and (3.55) apply to multiplexed and demultiplexed channels, respectively. A definition of the crosstalk attenuation similar to Eq. (3.56), that applies only to the demultiplexed channels holds as well; yet, in the case of Fig. 3.72*c,* two different kinds of crosstalk at the *i*th port, interfering with the nominal output signal $O_i(\lambda_i)$, must be considered:

1. Crosstalk produced by other multiplexed channels that are received through the line, such as $U_i(\lambda_k)$, which is produced by $I(\lambda_k)$
2. Crosstalk produced by the channels that enter the device through the input ports and are multiplexed inside the same device, such as $U_i(\lambda_j)$, which is produced by $I_j(\lambda_j)$

These contributions are sometimes referred to as far-end and near-end crosstalk, respectively.

Since these devices are usually based on a reversible structure (at least in principle), the following discussion will consider only wavelength-division demultiplexers in detail, on the assumption that the corresponding multiplexing function can be obtained with the same device by simply exchanging input and output directions. In the bidirectional configuration, the same device simultaneously accomplishes both multiplexing and demultiplexing functions.

Wavelength-division demultiplexers (and multiplexers) can be classified in two broad families, according to their fabrication technology:

Micro-optical devices

Coupled-fiber devices

In the former case, different wavelengths are spatially discriminated from one beam (or combined into one beam) by making use of micro-optical assemblies (i.e., including miniaturized lenses and other optical components), operating according to principles that will be described in the following. These demultiplexing devices, which are typically suitable for multimode fibers, usually rely on either interference filters or optical components producing angular dispersion, as schematically shown in Fig. 3.73*a* and *b*. The ideal interference filter shown in Fig. 3.73*a* is an optical component that transmits only one wavelength and reflects all the others. To demultiplex more than two channels, then, two or more cascaded filters are necessary. The angular dispersion device (Fig. 3.73*b*), on the other hand, produces the simultaneous spatial separation of all the different wavelengths that are present in the input beam.

The second family of demultiplexers (coupled-fiber devices) relies upon the coupling that takes place between the fields propagating along two adjacent fiber cores, according to the principle already described in Sec. 3.4.4 for fiber cou-

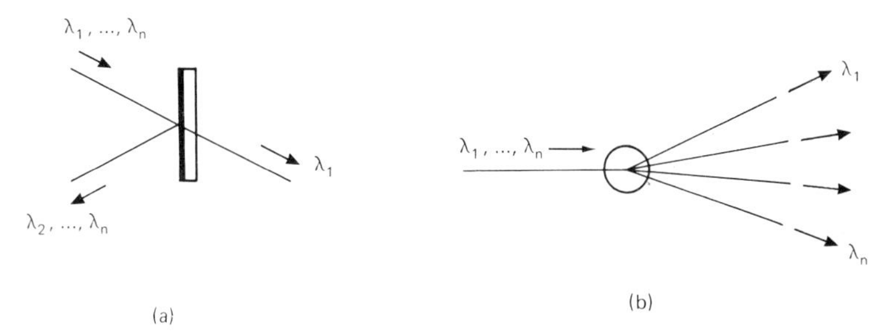


**FIGURE 3.73** Operating principles of wavelength-division demultiplexers making use of (*a*) interference filtering and (*b*) angular dispersion.

plers. Since this kind of coupling inherently depends on wavelength, demultiplexing can be accomplished. Only single-mode devices are produced using this technique.

### 3.5.2 Micro-Optical Devices for WDM Based on Interference Filters

Interference filters that are used inside multiplexers and demultiplexers for WDM applications typically consist of thin-film, multilayered structures, based on alternating high-and-low refractive index layers. While propagating through these structures, light undergoes multiple reflections, giving rise to interference effects that result in constructive or destructive interference depending on wavelength. Therefore, devices can be built to produce a high transmittance in a given wavelength range and a high reflectance (i.e., nearly zero transmittance) outside this region. By proper design of the multilayer parameters, interference filters with a broad variety of spectral characteristics can be obtained.

According to spectral characteristics, filters can be classified in two large families:

- Sharp-cut filters, i.e., those having short-pass- or long-pass-type spectral transmittance (as shown in Fig. 3.74*a* and *b*), characterized by the cutoff wavelength, $\lambda_c$

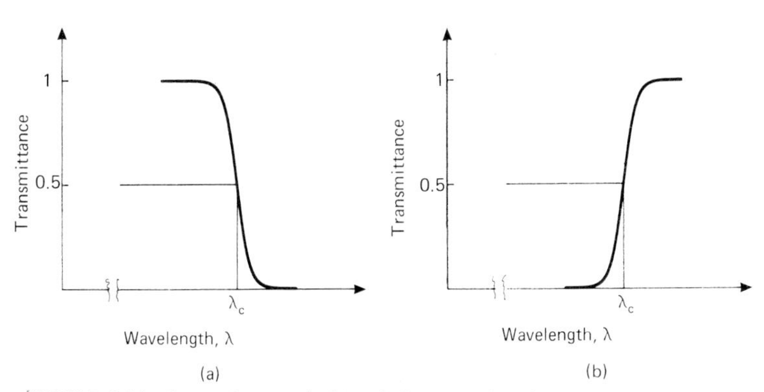


FIGURE 3.74 Spectral transmission of sharp-cut interference filters. (*a*) Low-pass and (*b*) high-pass.

- Bandpass filters, i.e., those having bandpass-type spectral transmittance (as shown in Fig. 3.75), characterized by the center wavelength $\lambda_0$ and the full-width, half-maximum (FWHM) optical bandwidth $\Delta\lambda$

The former class of filters is typically employed in two-channel devices, to multiplex (or demultiplex) two significantly different wavelengths, such as 850 and 1300 nm or 1300 and 1550 nm. Because of the large useful spectral bandwidth, they can be effectively used with broad-spectrum sources (i.e., LEDs).

The latter class of filters, on the other hand, is preferably used in WDM devices operating with more than two channels and is suitable for narrow-spectrum sources (i.e., laser diodes).

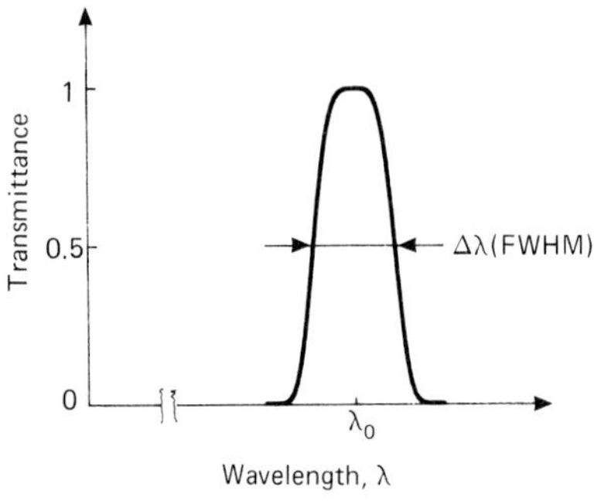


FIGURE 3.75 Spectral transmission of a bandpass interference filter.

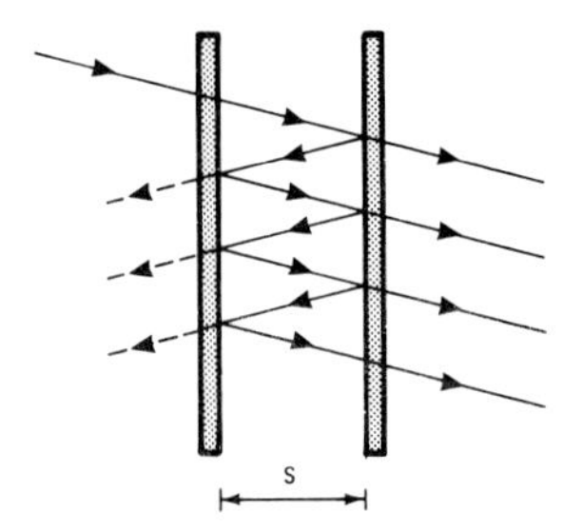


FIGURE 3.76 Structure of a Fabry-Perot resonator.

Most interference filters operate according to the principle of the Fabry-Perot resonator, basically consisting of a transparent, nonabsorbing dielectric medium enclosed between two reflectors (Fig. 3.76). When a beam impinges on such a device, multipath interference takes place: if the round trip $2s$ between two consecutive reflections on the same mirror equals a given wavelength $\lambda_0$ of the incident beam (or an integer multiple of it), constructive multiple interference takes place and transmission reaches a maximum for that wavelength. Reflectivity has a complementary behavior. Therefore, different wavelengths are transmitted or reflected by the resonator according to their values with respect to the resonance wavelengths.

As shown in Fig. 3.77, typical Fabry-Perot resonators have a transmission vs. wavelength characteristic significantly different from zero only in a restricted wavelength region around $\lambda_0$ (and around its submultiples), while, outside that

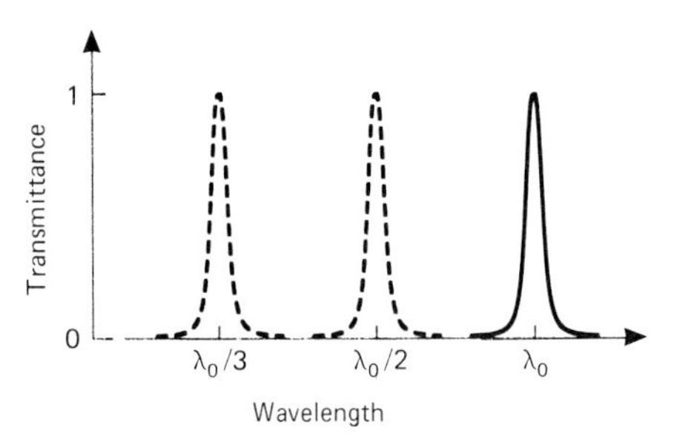


FIGURE 3.77 Transmission vs. wavelength response of a Fabry-Perot resonator.

region, any radiation is almost completely reflected. If necessary, transmission windows at submultiples of $\lambda_0$ can be eliminated (for instance, by means of additional glass filters).

To reduce losses and achieve high reflectivities (which results in a narrow-bandwidth spectral response), reflectors of practical Fabry-Perot resonators usually incorporate dielectric multilayers, as in Fig. 3.78, that are obtained by deposition of thin films. The thin-film layers, with a thickness of one-quarter the desired maximum transmission wavelength, are alternately made of high-

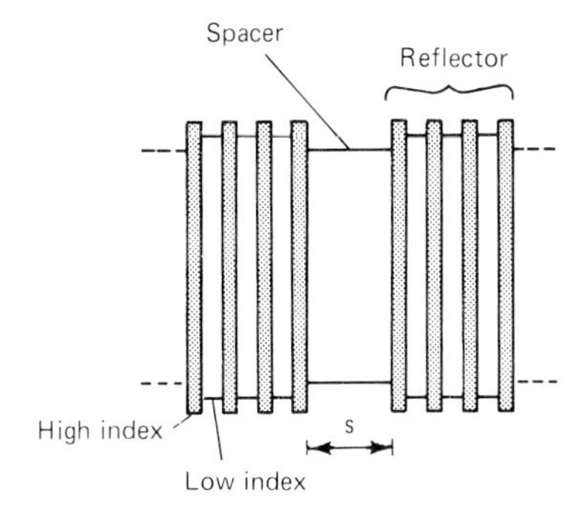


**FIGURE 3.78** Representative structure of an interference filter. The resonator is formed by two dielectric multilayer reflectors. Resonators can be cascaded.

refractive-index and low-refractive-index materials (e.g., $MgF_2$, with $n = 1.35$, or $SiO_2$, with $n = 1.46$, as low-index material and $TiO_2$, with $n = 2.3$, as high-index material).

By a proper choice of thickness, refractive index, and number of layers, and by cascading more than one resonator, it is possible to design a variety of spectral response shapes. To increase tolerance with respect to possible shifts of the source wavelength, bandpass filters often have a squared shape, as in Fig. 3.75, with a rather flat high-transmission region around the center wavelength, while, to reduce crosstalk between adjacent channels, the edges of the transmission band must be as sharp as possible.

Bandwidths as narrow as 5 to 10 nm (at −3 dB) can be achieved, but at the expense of a higher attenuation at the center wavelength, as a large number of layers (e.g., 30 to 40) are required. Usually the spacing between adjacent channels is at least 30 nm.

Basic configurations of 2-channel and $N$-channel demultiplexers, making use of interference filters, are shown in Fig. 3.79. Appropriate microlenses (i.e., ball, GRIN rod, rod) are used for collimating and focusing purposes. Interference filters operate at an incidence angle different from zero, thus separating the reflected beam from the incident beam. Tilting the filters also results in a shift of the center wavelength and can be used for fine-tuning the channels. The attenuation of practical interference filters, even at the center of the passband, is not negligible; therefore, the maximum number $N$ of filters in the cascaded configuration of Fig. 3.79*b* is usually limited to 4 or 5. A bifurcated tree configuration, by reducing the number of filters, could result in some advantage, from this point of view.

The schematic structure of a possible, practical implementation of the cascaded filter configuration is shown in Fig. 3.80 for a four-channel wavelength-division demultiplexer.[57]

Simple, compact, low-loss two-channel demultiplexers (or multiplexers) can be realized by means of GRIN-rod lenses,[50] as shown in Fig. 3.81. Two quarter-pitch lenses are used for collimating and focusing. The interference filter is deposited at the interface between the two lenses. Off-axis entry at the first lens allows an easy separation of the reflected beam. Additional coating on the lens surface at the output port for $\lambda_1$ can further reduce residual crosstalk. Multimode

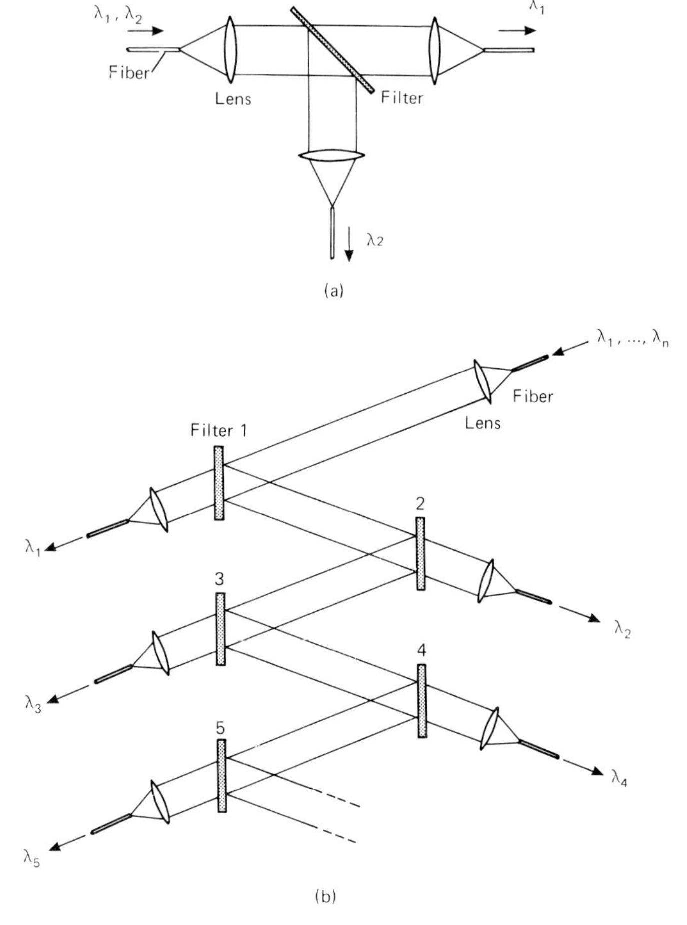


**FIGURE 3.79** Basic configurations of (*a*) 2-channel and (*b*) *N*-channel interference filter demultiplexers.

fiber devices operating at 850 and 1300 nm, or 1200 and 1300 nm, show typical insertion losses below 2 dB and crosstalk attenuation better than 30 dB.

Interference filters can be deposited directly on the end of properly cut and polished fibers, according to the configurations of Fig. 3.82. Low-loss, two- and three-channel demultiplexers can be obtained in this manner.[58]

### 3.5.3 Micro-Optical Devices for WDM Based on Angular Dispersion

As shown in Fig. 3.83, angular dispersion demultiplexers (or multiplexers) operate on a different principle: the input collimated beam impinges on a dispersive element, which, in turn, separates the different input wavelengths into output beams that exit at different angles. The angular separation depends on the dispersion relationship, $\theta = \theta(\lambda)$ that is characteristic of the dispersive element. The

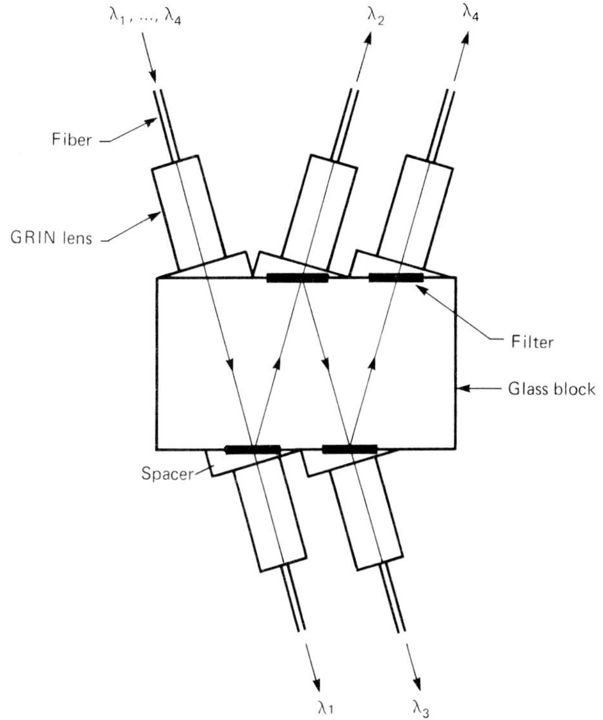


**FIGURE 3.80** Schematic structure of a practical four-channel wavelength-division demultiplexer, based on GRIN-rod lenses and interference filters. (*Adapted from Ref. 57.*)

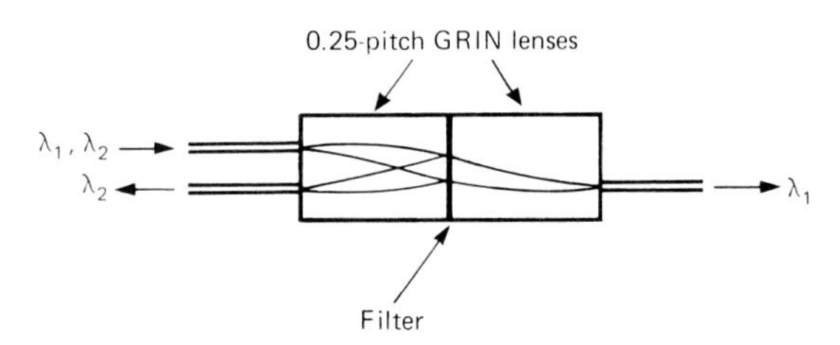


**FIGURE 3.81** Wavelength-division demultiplexer making use of GRIN-rod lenses, in a two-channel configuration. (*Adapted from Ref. 50.*)

separated output beams are then focused and are collected by separate optical fibers.

This kind of demultiplexer is quite effective with narrow-spectrum sources such as laser diodes; in this case, well-defined images of the input fiber are produced in the focal plane of the output lens, as shown in Fig. 3.84*a*. These images can be made to exactly match the core size of the output fiber. With broad (optical) bandwidth sources, such as LEDs, the images in the output focal plane are

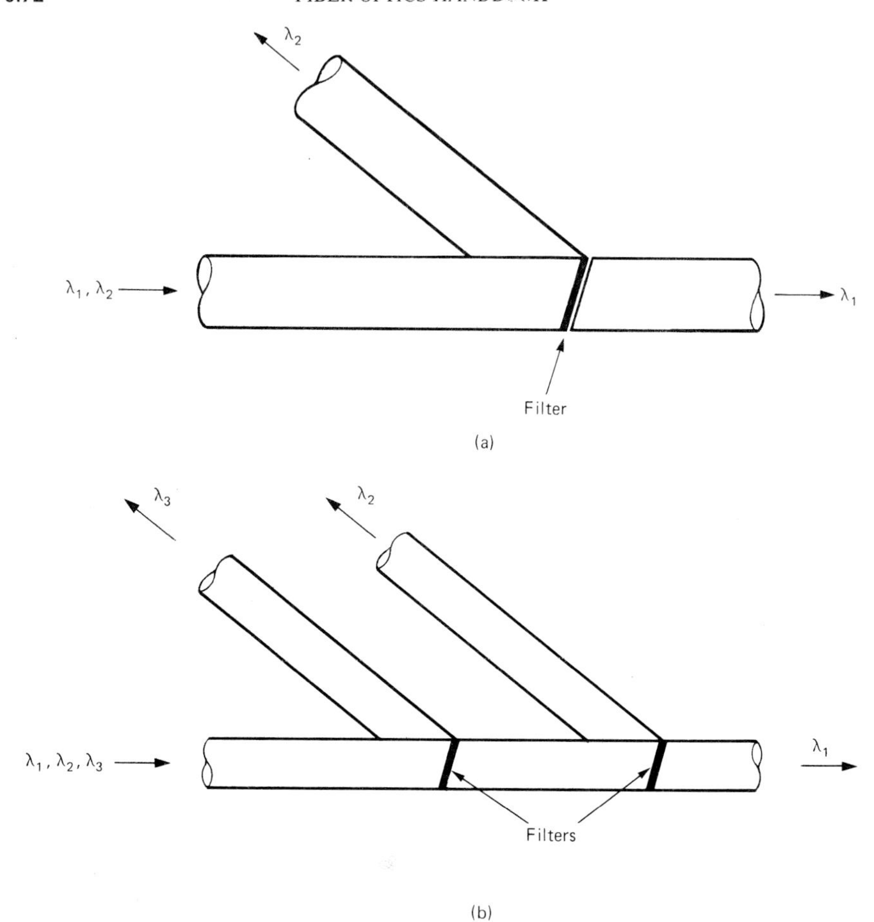


**FIGURE 3.82** Wavelength-division demultiplexers with interference filters directly deposited on fiber faces. (*a*) Two-channel and (*b*) three-channel. (*Adapted from Ref. 58.*)

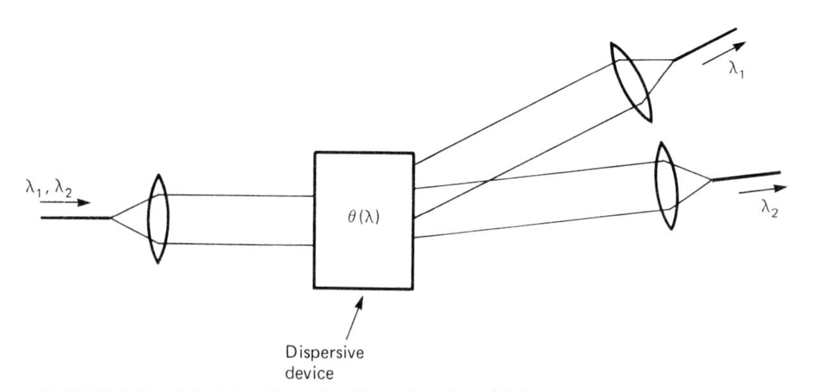


**FIGURE 3.83** Principle of angular dispersion demultiplexers.

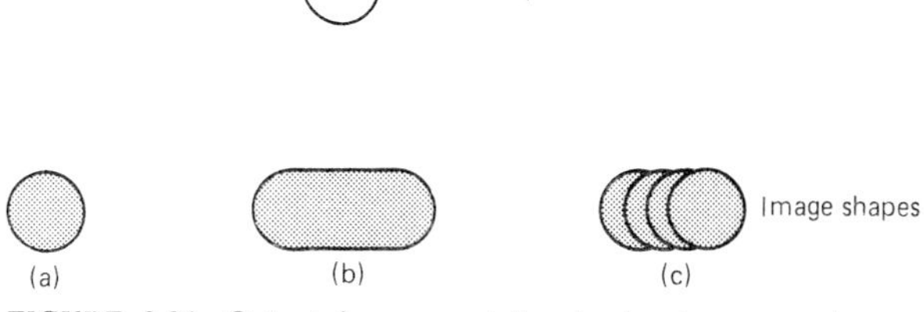


**FIGURE 3.84** Output images, relative to input source shape, from an angular dispersion demultiplexer, for (*a*) an ideal monochromatic source, (*b*) a broad-spectrum source, and (*c*) a monochromatic source affected by wavelength drift.

no longer circular spots, but assume an elongated shape (Fig. 3.84*b*), because of the combination of dispersion ($d\theta/d\lambda$) of the dispersive element and the broad spectral width of the source. Coupling efficiency is therefore reduced by the resultant mismatch between spot and output fiber core. In the case of laser sources, the presence of wavelength shifts (as produced, for instance, by temperature) can cause a spatial misalignment of the output spot with respect to the corresponding fiber core and, furthermore, can increase crosstalk into adjacent channels.

For these reasons, these demultiplexers are not used with broad-spectrum sources (LEDs). To reduce the effects of wavelength drifts of laser diodes, the core size of the output fiber can be larger than the input fiber (assuming 1:1 magnification), so that the output spots can be accepted with looser tolerances. Nonetheless, the use of fibers having different core sizes is not always convenient in the general case, and it is undesirable in bidirectional links.

To achieve a higher crosstalk rejection, the spatial separation between adjacent output ports should be increased. This would require either a dispersive element with higher dispersion or larger focal-length lenses, at the expense of larger size for the device.

The prism is a simple dispersive device which can be used either as in Fig. 3.85*a* or in the more compact Littrow configuration of Fig. 3.85*b*.[59] In this case, the output angle $\theta$ depends on the refractive index $n$, which, in turn, depends on wavelength $\lambda$ for a given material. Unfortunately, no available material can produce a value of $dn/d\theta$ high enough for practical applications in the 1.3- to 1.5-μm region.

Diffraction gratings are widely used dispersive devices. Blazed reflection gratings, as configured in Fig. 3.86, are commonly used for WDM applications.

Grating demultiplexers (or multiplexers) can be implemented according to two different basic constructions:

- Linear grating and focusing elements
- Self-focusing gratings

In the former case, the Littrow configuration (Fig. 3.87*a*) is often preferred[59] because it allows the use of only one lens, thus reducing device size, and because it results in lower astigmatism. The basic structure of a Littrow grating demultiplexer, making use of a 0.25-pitch GRIN-rod lens,[59] is shown in Fig.3.87*b*. In this case, the grating is placed—at the proper angle—on the back surface of the GRIN lens.

As an alternative, a concave mirror can be used as a focusing element,[60] as shown in Fig. 3.88. In such a device, the grating is still used in the Littrow configu-

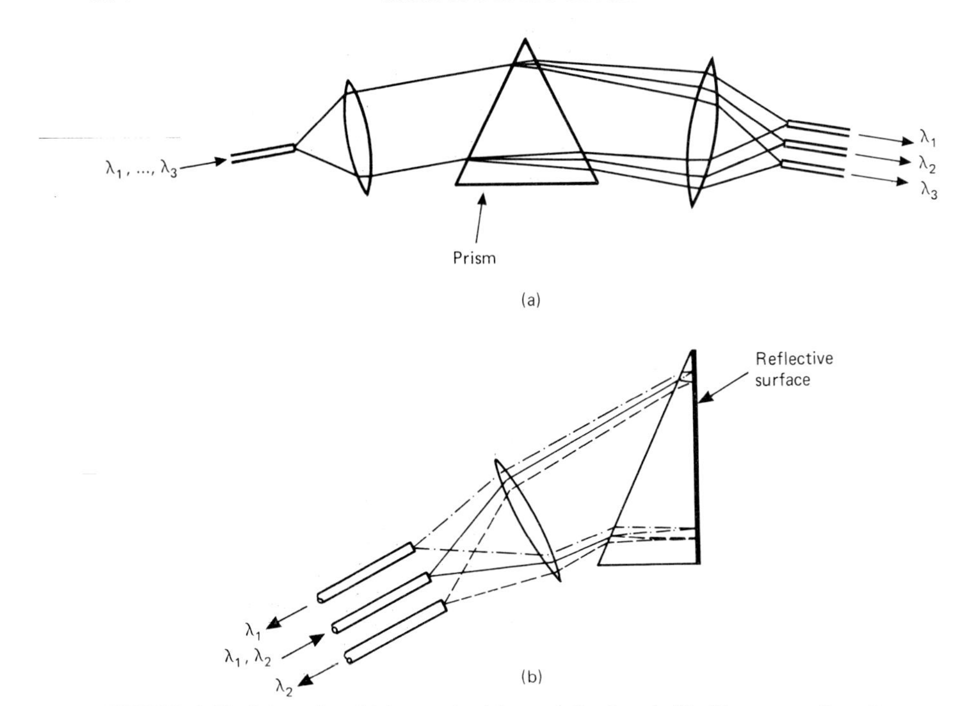


**FIGURE 3.85** Prism demultiplexers in (*a*) generalized and (*b*) Littrow configurations. (*Adapted from Ref. 59.*)

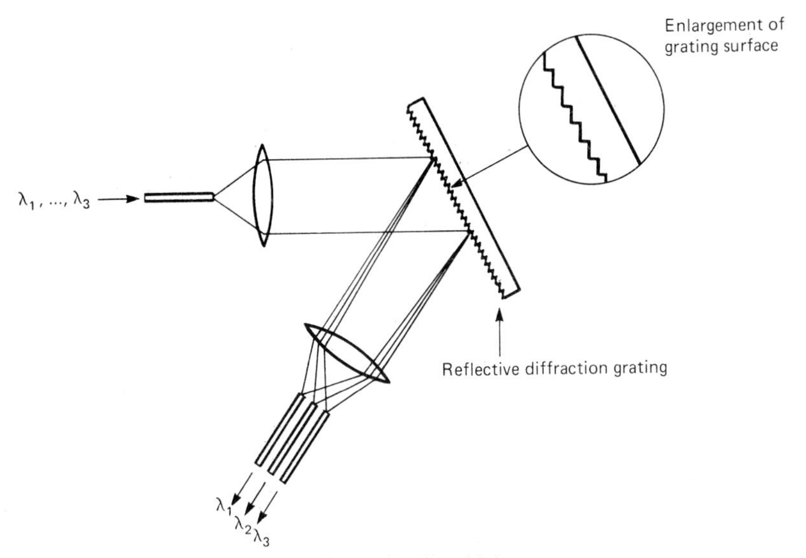


**FIGURE 3.86** Basic structure of a grating demultiplexer.

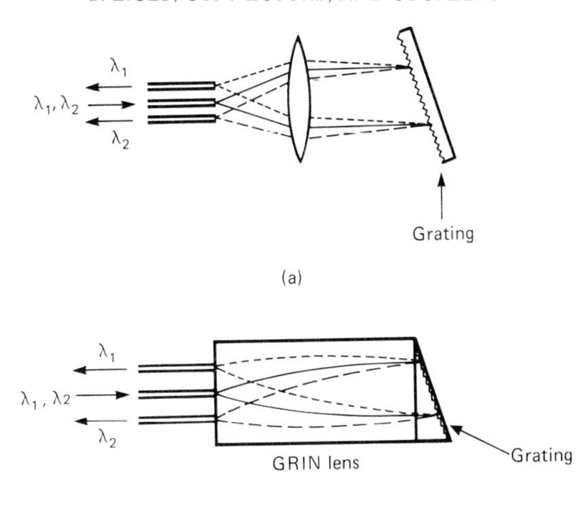


FIGURE 3.87 Littrow-configured grating demultiplexers. (*a*) Generalized form; (*b*) GRIN rod lens form. (*Adapted from Ref. 59.*)

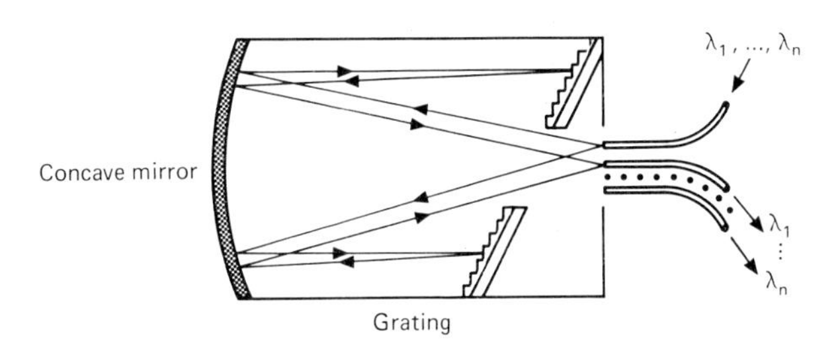


FIGURE 3.88 Basic structure of a grating demultiplexer using a concave mirror as both a collimating and a refocusing element. (*Adapted from Ref. 60.*)

ration. Here, the input and output fibers enter the device through a hole in the center of the grating. The cone emitted by the input fiber is reflected and collimated by the concave mirror, then undergoes angular dispersion by means of the diffraction grating, and, finally, is focused by the same concave mirror onto the output fibers.

If a properly shaped, concave grating is used, a self-focusing configuration can be obtained, where the beam-shaping effect is produced without the need of lenses or external mirrors. This results in a simple and compact device. In this case, the grating is used in the Rowland configuration, as shown in Fig. 3.89*a,* and simultaneously provides wavelength discrimination and focuses the output beam onto the output fibers.

A view of a practical device operating according to this principle[61] is shown in Fig. 3.89*b*. The choice of the core size of the output fibers determines the bandwidth of the output channels.

A device operating according to a similar principle can be implemented by means of a planar waveguide and a concave cylindrical grating (Fig. 3.90). In this case, a relatively simple fabrication technique is employed.[62]

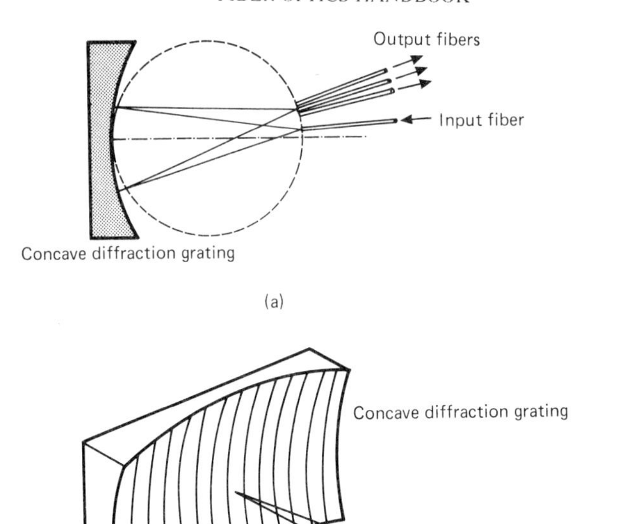


**FIGURE 3.89** Self-focusing, concave grating demultiplexer. (*a*) Schematic of the Rowland configuration; (*b*) view of a practical device. (*Adapted from Ref. 61.*)

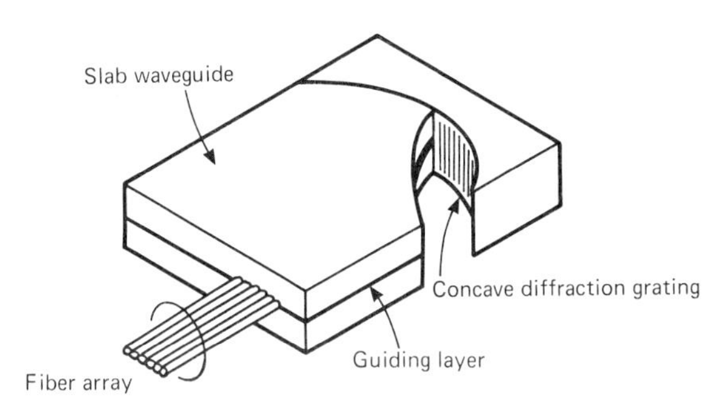


**FIGURE 3.90** Demultiplexer using a concave cylindrical grating and planar waveguide. (*Adapted from Ref. 62.*)

### 3.5.4 Coupled-Fiber WDM Devices

The techniques described apply mainly to multimode fibers, since, in this case, the complete processing of the input beam (i.e., collimation, dispersion, focusing, etc.) can be easily obtained with low insertion losses. Such processing is not easy to achieve, however, when the output fibers are single-mode, because of the small core size and the unavoidable image deterioration that results from aberrations, distortions, and other imperfections of practical optical elements.

Therefore, multiplexers and demultiplexers for single-mode fibers have been preferably developed with coupled-fiber technology, as discussed in Sec. 3.4.4. In the coupled-fiber technique, the cores of two (or more) fibers are placed close enough to allow overlap of the fields guided by adjacent fibers. In this coupling configuration, a spatially periodic transfer of guided power takes place between the adjacent cores, as in Fig. 3.65.

With reference to Fig. 3.65, the power $O_4$ that is present at a core section of the coupled fiber at a distance $z$ from the beginning of the coupling region, is related to the input power $I_1$ and to the coupling length $z$ by Eq. (3.51), which is recalled here for convenience:

$$\begin{aligned} O_3(z) &= I_1 \cos^2 (\Delta\beta z) \\ O_4(z) &= I_1 \sin^2 (\Delta\beta z) \end{aligned} \tag{3.51}$$

where $\Delta\beta$ is the coupling coefficient that determines the *coupling length* of the power transfer between the two cores and depends on the optical and geometrical parameters of the coupled guides, specifically on the wavelength. According to the simplified model[56] already mentioned in Sec. 3.4.4, which refers to fused biconical taper couplers, the coupling coefficient $\Delta\beta$ in Eq. (3.51), is given by Eq. (3.53), which clearly shows the dependence on the wavelength $\lambda$. Therefore, from Eq. (3.51), the amount of power that, after a length $z$ has transferred to the coupled fiber depends on wavelength.

This behavior can be explained in a heuristic manner: when the wavelength varies, the mode field diameter varies as well. This produces a different overlap of the fields in the core of the adjacent fiber, thus changing the coupling coefficient $\Delta\beta$ and, hence, the amount of power transferred to the other guide. This principle is illustrated in Fig. 3.91.

There is a point $z_0$ along the coupling region where all of the power at a given wavelength is contained within the core of the input fiber, while all of the power at another, different wavelength has been transferred to the core of the coupled fiber, as indicated in Fig. 3.92. Since the total power at each wavelength must be conserved (neglecting the excess losses), the power relationship at the two output ports is exactly complementary. This coupling behavior allows the discrimination of two different wavelengths that are simultaneously present at the input port of the device.

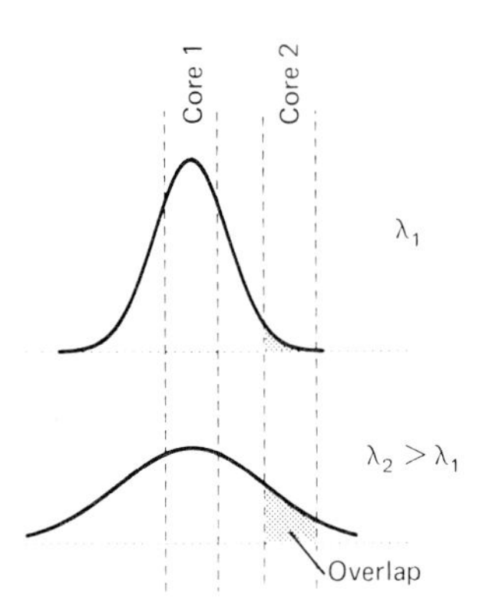


**FIGURE 3.91** Schematic illustration of the field overlap between two adjacent fiber cores in the coupling region of a fiber coupler, at two different wavelengths, $\lambda_1$ and $\lambda_2$.

Single-mode multiplexing and demultiplexing devices have been fabricated for wavelengths of 1.3 and 1.55 μm. The coupling region can be fabricated either by the widely used fused biconical taper technique or by the polished fiber approach (see Sec. 3.4.4).

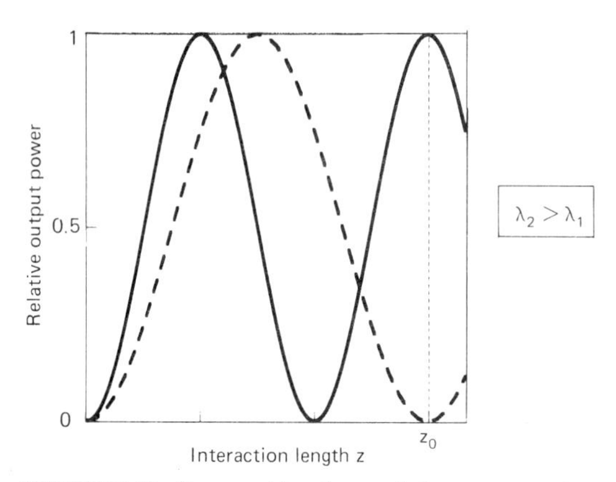


**FIGURE 3.92** Power exiting the coupled output port of a fiber coupler ($O_4$ in Fig. 3.65), as a function of length $z$ of the coupling region, for two different wavelengths, $\lambda_1$ and $\lambda_2$.

Excess losses on the order of 0.5 dB can be routinely achieved, while the crosstalk attenuation is usually on the order of 15 to 20 dB. Cascaded couplers can be used as additional filters to improve on these values. An important feature of coupled-fiber WDM devices is their inherently easy fabrication and, consequently, their relatively low cost.

## 3.6 MICROLENSES FOR OPTICAL-FIBER COUPLING DEVICES

Many optical functions, such as branching, multiplexing, demultiplexing, filtering, coupling, attenuating, switching, and isolating, are usually performed outside the fiber by means of so-called *micro-optics devices*. These devices usually operate in a collimated beam arrangement, which is simple to handle and yields low coupling losses. The basic configuration for this class of device is shown in Fig. 3.93. Two lenses are needed, to collimate the beam exiting the fiber to the left and to refocus it on the core of the fiber to the right, after the required processing (e.g., splitting, filtering, adding, and switching).

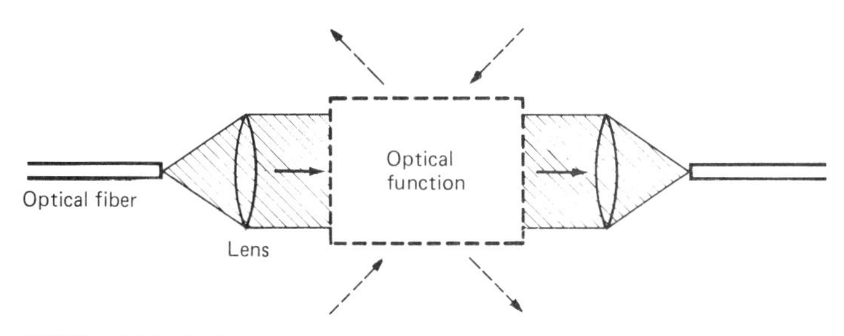


**FIGURE 3.93** Basic collimated beam configuration for fiber coupling to external optical processes.

Another important application calling for the use of lenses is efficient coupling between source and fiber. In this case, important parameters are optical quality, small size, easy incorporation into modules, and low cost.

Lenses to be used in micro-optics modules for optical fibers must be optimized to provide minimum insertion loss. The lens numerical aperture, therefore, should match that of the optical fiber. The focal length must be short enough to reduce lens aberrations and size, but not so short as to produce too narrow and divergent a beam. Typical lens sizes for these applications range from a few hundred micrometers to a few millimeters.

Lenses currently used fall into two general categories: homogeneous and nonhomogeneous. Homogeneous lenses have a uniform refractive index, whereas nonhomogeneous lenses have a graded-refractive-index distribution.

### 3.6.1 Homogeneous Lenses

The general characteristics of thick, homogeneous spherical lenses can be found in most general optics texts. The specific lens shapes shown in Fig. 3.94, which find application in micro-optics devices and in source-fiber coupling, can be characterized by the formulas that follow. These formulas, which are for paraxial rays only, determine focal length $f$ and the positions of principal planes $h_1$ and $h_2$ with respect to the corresponding lens vertices. In the following, $n$ is the refractive index of the lens material, and $R$ is the radius of the spherical surface.

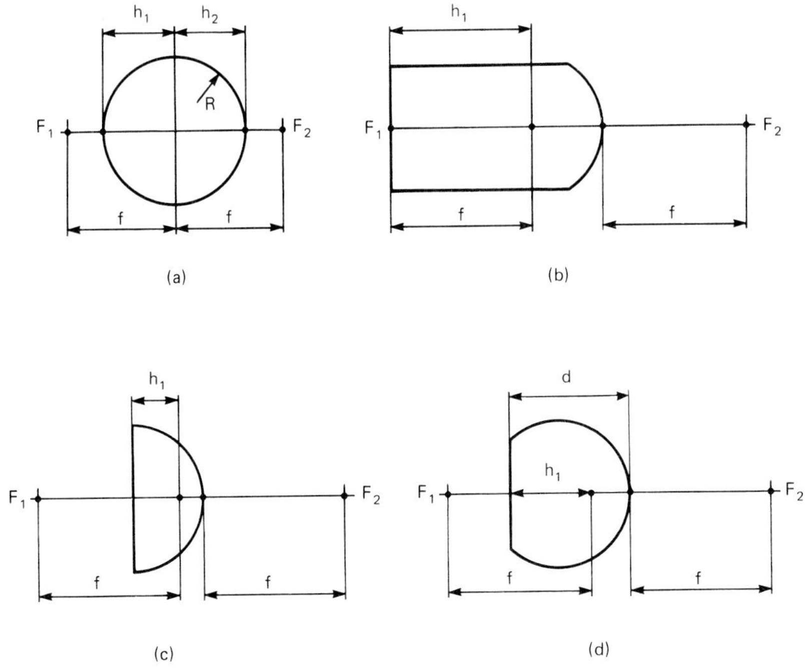


**FIGURE 3.94** Specific homogeneous lens shapes. (*a*) Ball; (*b*) rod; (*c*) hemisphere; (d) truncated sphere.

*Ball Lens.* See Fig. 3.94*a*.

$$f = \frac{nR}{2(n - 1)} \tag{3.57}$$

$$h_1 = -h_2 = R \tag{3.58}$$

*Rod Lens.* The lens thickness is designed so that the left focal point lies on the flat surface (i.e., $f = h_1$). (See Fig. 3.94*b*.)

$$f = \frac{R}{n - 1} \tag{3.59}$$

$$h_1 = \frac{R}{n - 1} \qquad h_2 = 0 \tag{3.60}$$

*Hemispherical Lens.* See Fig. 3.94*c*.

$$f = \frac{R}{n - 1} \tag{3.61}$$

$$h_1 = \frac{R}{n} \qquad h_2 = 0 \tag{3.62}$$

*Truncated Sphere.* See Fig. 3.94*d*.

$$f = \frac{R}{n - 1} \tag{3.63}$$

$$h_1 = \frac{d}{n} \qquad h_2 = 0 \tag{3.64}$$

where $d$ is lens thickness.

### 3.6.2 Graded-Refractive-Index Rod Lenses

The most widely used nonhomogeneous lenses are the graded-refractive-index rod lenses, or GRIN-rod lenses, consisting of a glass or plastic cylinder of length $L$ whose refractive index distribution has cylindrical symmetry (Fig. 3.95), as represented by the following relationship:

$$n^2(r) = n_1^2\,[1 - (gr)^2 + h_4(gr)^4 + h_6(gr)^6 + \cdots] \tag{3.65}$$

where $n_1$ = refractive index on the lens axis
$g$ = main focusing constant (sometimes indicated as A)
$h_4, h_6, \ldots$ = higher-order constants

This profile is usually obtained by an ion-exchange process. Actual lenses have diameters of 1 to 2 mm. For paraxial rays, higher-order terms ($h_4$, $h_6, \ldots$) can be ignored, and the index distribution simply becomes

$$n^2(r) = n_1^2\,[1 - (gr)^2] \tag{3.66}$$

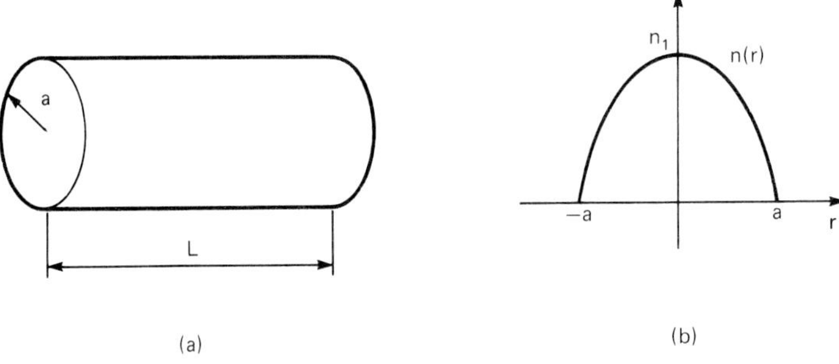


FIGURE 3.95 GRIN rod lens. (*a*) Lens geometry; (*b*) refractive index of GRIN-rod as a function of radius.

In such a lens, rays follow sinusoidal paths, refocusing periodically, as in Fig. 3.96. The length of the period, or pitch $P$, of the lens is

$$P = \frac{2\pi}{g} \tag{3.67}$$

After one-quarter-pitch length, all rays coming from a point source $Q$ on the input face become parallel. After one-half pitch, an inverted image of $Q$ is formed.

GRIN-rod lens focusing properties can be described by analogy with homogeneous lenses, in terms of focal length $f$ and principal plane positions $h_1$ and $h_2$:

$$f = [n_1 g \sin (gL)]^{-1} = \frac{P}{n_1 2\pi \sin (2\pi L/P)} \tag{3.68}$$

$$h_1 = -h_2 = \frac{\tan (gL/2)}{n_1 g} = \frac{P \tan (\pi L/P)}{n_1 2\pi} \tag{3.69}$$

For the quarter-pitch (sometimes indicated as $P/4$ or $0.25P$) lens shown in Fig. 3.97, which is of particular interest for optical-fiber devices, these become

$$f = \frac{1}{n_1 g} = \frac{P}{n_1 2\pi} \tag{3.70}$$

$$h_1 = -h_2 = \frac{1}{n_1 g} = \frac{P}{n_1 2\pi} \tag{3.71}$$

Numerical aperture (NA), at a point $Q$ on the input face of the lens (Fig. 3.98), is limited by the condition under which a ray strikes the surface of the lens and is given by

$$\text{NA} = \sin \theta_{\max} = g n_1 \left[ \frac{a^2 - r^2}{1 - (r^2/a^2) \sin^2 \phi} \right]^{1/2} \tag{3.72}$$

It depends on the position $r$ and azimuthal coordinate $\phi$ of the incident ray, as

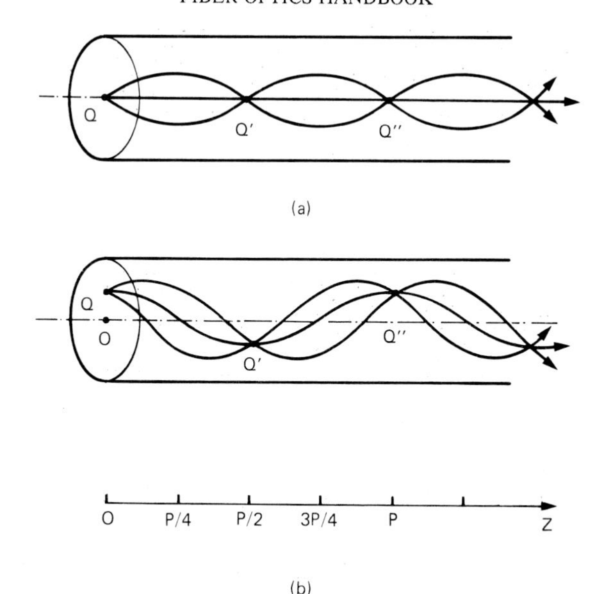


**FIGURE 3.96** Ray paths within a GRIN-rod lens: (*a*) Focusing of rays from an on-axis point; (*b*) focusing of rays from an off-axis point.

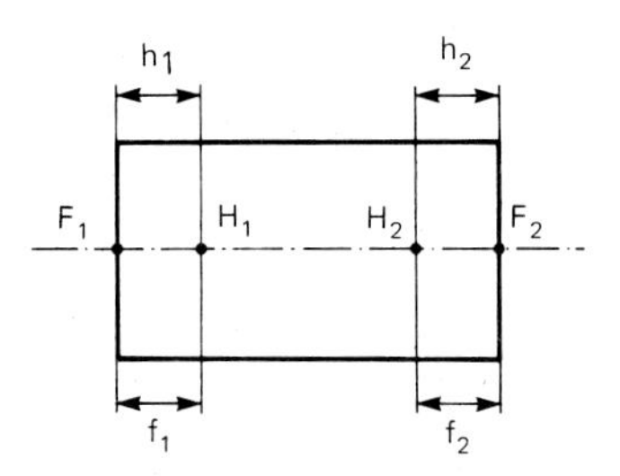


**FIGURE 3.97** Location of focal and principal planes for a quarter-pitch GRIN-rod lens.

defined in Fig. 3.98. For meridional rays (i.e., rays lying in a plane containing the lens axis), $\phi$ is zero and the local numerical aperture is

$$\mathrm{NA} = g n_1 (a^2 - r^2)^{1/2} \tag{3.73}$$

When the paraxial condition is removed, the index profile defined by Eq. (3.66) no longer assures that all rays are exactly focused in one point. In such a general case, index distributions exist,[63] being optimum for specific groups of rays, but not for all rays at the same time. The ideal profile for meridional rays has the form

$$n^2(r) = n_1^2 \operatorname{sech}^2(gr) = n_1^2 [1 - (gr)^2 + \tfrac{2}{3}(gr)^4 - \tfrac{17}{45}(gr)^6 + \cdots] \tag{3.74}$$

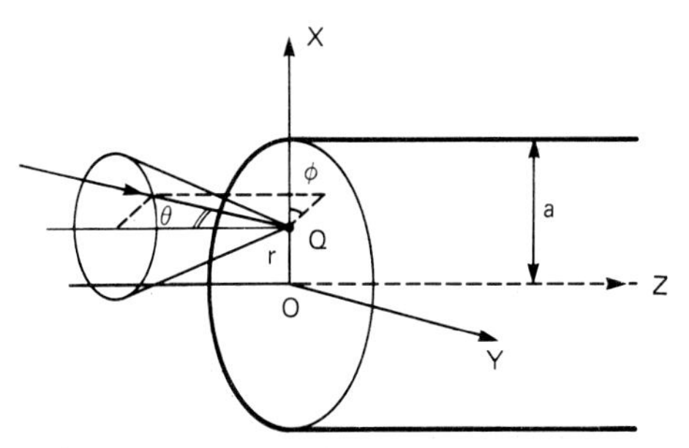


**FIGURE 3.98** Coordinate system that is used for the definition of the GRIN-rod numerical aperture.

while the ideal profile for helical rays (i.e., rays following helical paths with constant radius) is

$$n^2(r) = n_1^2\,[1 + (gr)^2]^{-1} = n_1^2\,[1 - (gr)^2 + (gr)^4 - (gr)^6 + \cdots] \qquad (3.75)$$

Actual profiles differ from these ideal profiles. In any case, for basic design purposes, index profiles corresponding to the simple quadratic relation of Eq. (3.66) can be assumed.

Since the application of lenses in micro-optic devices usually involves imaging to couple power from one fiber to another, or from a source to a fiber, lens aberrations can limit the resulting coupling efficiency.

Spherical aberration of homogeneous lenses with a low refractive index ($n \approx 1.5$) is about 2 times higher than comparable GRIN-rod lenses. If high-refractive-index materials ($n$ = 1.7 to 1.9) are used for homogeneous lenses, the resulting aberration is about the same as for GRIN-rod lenses.[41]

Sometimes GRIN-rod lenses are used in an off-axis condition, as in Fig. 3.96*b*. In this case, astigmatism can contribute significantly to coupling losses.

## ACKNOWLEDGMENTS

The authors are indebted to G. Coppa and M. Artiglia for their assistance with coupling theory and to B. Costa for his helpful suggestions about the overall content of the chapter.

## REFERENCES

1. R. Olshansky et al., "Differential Mode Attenuation Measurements in Graded-Index Fibers," *Applied Optics,* vol. 17, no. 11, pp. 1830–1835, 1978.
2. R. Olshansky, "Mode Coupling Effects in Graded Index Optical Fibers," *Applied Optics,* vol. 14, no. 4, pp. 935–945, 1975.
3. P. Di Vita et al., "Realistic Evaluation of Coupling Loss between Different Optical Fibres," *Journal of Optical Communications,* vol. 1, pp. 26–32, 1980.

4. G. Coppa et al., "Effects of Joints on Mode Power Distribution and Losses in Multimode Optical Fibres," *Optical and Quantum Electronics,* vol. 15, pp. 143–153, 1983.
5. P. Di Vita et al., "Evaluation of Coupling Efficiency in Joints between Optical Fibres," *Alta Frequenza,* vol. XLVII, no. 5, pp. 414–423, 1978.
6. D. Gloge,"Offset and Tilt Loss in Optical Fiber Splices," *Bell System Technical Journal,* vol. 55, no. 7, pp. 905–916, 1976.
7. D. Marcuse et al., "Guiding Properties of Fibers," Chap. 3 in S. E. Miller and A. G. Chynoweth (eds.), *Optical Fiber Telecommunications,* Academic Press, New York, 1979.
8. S. Seikai et al., "Structural Parameter Specifications of a Graded-Index Fiber on the Basis of Splice Loss," *Transactions of IECE of Japan,* vol. E65, no. 8, pp. 485–491, 1982.
9. J. Sakai et al., "Splice Loss Evaluation for Optical Fibers with Arbitrary-Index Profile," *Applied Optics,* vol. 17, no. 17, pp. 2848–2853, 1978.
10. G. Coppa et al., "Analysis of Joint Losses between Monomode Fibres," *CSELT Rapporti Tecnici,* vol. XI, no. 6, pp. 427–429, 1983.
11. D. Marcuse, "Loss Analysis of Single-Mode Fiber Splices," *Bell System Technical Journal,* vol. 56, no. 5, pp. 703–718, 1977.
12. S. Nemoto et al., "Analysis of Splice Loss in Single-Mode Fibres Using a Gaussian Field Approximation," *Optical and Quantum Electronics,* vol. 11, pp. 447–457, 1979.
13. C. M. Miller, "Loose Tube Splices for Optical Fibers," *Bell System Technical Journal,* vol. 54, no. 7, pp. 1215–1225, 1975.
14. G. Cocito et al., "COS2 Experiment in Turin: Field Test on an Optical Cable in Ducts," *IEEE Transactions on Communications,* vol. COM-26, no. 7, pp. 1028–1036, 1978.
15. T. Kimura et al., "Recent Advance on Optical Fibre Transmission Technology," *Japan Telecommunications Review,* pp. 3–10, Jan. 1976.
16. P. Melman, "Elastic Tube Splice Performance with Single Mode and Multimode Fibres," *Electronics Letters,* vol. 18, no. 8, pp. 320–321, 1982.
17. Y. Kato et al., "Arc-Fusion Splicing of Single Mode Fibers 1: Optimum Splice Conditions," *Applied Optics,* vol. 21, no. 7, pp. 1332–1336, 1982.
18. T. Katagiri et al., "Optical Microscope Observation Method for a Single-Mode Optical-Fiber Core for Precise Core-Axis Alignment," *Journal of Lightwave Technology,* vol. LT-2, no. 3, pp. 277–283, 1984.
19. M. Fujise et al., "Core Alignment by a Simple Local Monitoring Method," *Applied Optics,* vol. 23, no. 15, pp. 2643–2648, 1984.
20. C. M. Miller, "Local Detection Device for Single-Mode Fiber Splicing," *Technical Digest of Topical Meeting on Optical Fiber Communication,* Phoenix, Arizona, Optical Society of America, 1982, pp. 44–45.
21. Y. Kato et al., "Arc-Fusion Splicing of Single-Mode Fibers. 3: A Highly Efficient Splicing Technique," *Applied Optics,* vol. 23, no. 15, pp. 2654–2659, 1984.
22. M. Miyauchi, "New Reinforcement for Arc Fusion Spliced Fibre," *Electronics Letters,* vol. 17, no. 24, pp. 907–908, 1981.
23. C. M. Miller, "Fiber-Optic Array Splicing with Etched Silicon Chips," *Bell System Technical Journal,* vol. 57, no. 1, pp. 75–90, 1978.
24. M. de Vecchis et al., "Overview of the Cylindrical Grooved Structure," *Conference Record of Globecom '83,* San Diego, vol. 3, pp. 1165–1169, 1983.
25. R. E. Wagner et al., "Interference Effects in Optical Fiber Connections," *Applied Optics,* vol. 21, no. 8, pp. 1381–1385, 1982.
26. R. E. Epworth, "The Phenomenon of Modal Noise in Analogue and Digital Optical Fi-

bre Systems," *Proceedings of the Fourth European Conference on Optical Communication,* Genova, Sept. 12–15, 1978, pp. 492–501.

27. K. Petermann, "Nonlinear Distortions and Noise in Optical Communication Systems due to Fiber Connectors," *IEEE Journal of Quantum Electronics,* vol. QE-16, no. 7, pp. 761–770, 1980.
28. N. Suzuki et al., "A New Demountable Connector Developed for a Trial Optical Transmission System," *Technical Digest of the First Integrated Optics and Optical Fiber Communication Conference,* 1977, pp. 351–354.
29. Y. Morimoto et al., "Design and Fabrication of Low Loss Single-Mode Optical Fiber Connector," in *Fiber Optic Couplers, Connectors and Splice Technology, Proceedings of the Society of Photo-Optical Instrumentation Engineers,* vol. 479, 1984, pp. 36–41.
30. N. Kurochi et al., "A Development Study on Design and Fabrication of an Optical Fibre Connector," *Proceedings of the Third European Conference on Optical Communication,* München, Sept. 14–16, 1977, pp. 97–99.
31. P. Hensel, "Triple-Ball Connector for Optical Fibres," *Electronics Letters,* vol. 13, no. 24, pp. 734–735, 1977.
32. N. Suzuki et al., "Demountable Connectors for Optical Fiber Transmission Equipment," *Review of the Electrical Communication Laboratories,* vol. 27, no. 11-12, pp. 999–1009, 1979.
33. J. Minowa et al., "Optical Componentry Utilized in Field Trial of Single-Mode Fiber Long-Haul Transmission," *IEEE Transactions on Microwave Theory and Techniques,* vol. MTT-30, no. 4, pp. 551–563, 1982.
34. M. Saruwatari et al., "Active and Passive Optical Devices for Use in Single-Mode Fiber Transmission Systems," *Review of the Electrical Communication Laboratory,* vol. 31, no. 3, pp. 299–309, 1983.
35. W. Radermacher, "A High-Precision Optical Connector for Optical Test and Instrumentation," *Hewlett-Packard Journal,* vol. 38, no. 2, pp. 28–30, 1987.
36. T. Yoshizawa et al., "Optical Plastic Components for Fiber Interconnection," *Review of the Electrical Communication Laboratories,* vol. 34, no. 1, pp. 111–117, 1986.
37. P. K. Runge et al., "Demountable Single-Fiber Optic Connectors and Their Measurement on Location," *Bell System Technical Journal,* vol. 57, no. 6, pp. 1771–1790, 1978.
38. A. Jacques et al., "Connecteurs et Composants Passifs pour Fibres Multimodales," *Commutation et Transmission,* no. 2/3, pp. 55–68, 1982.
39. J. B. Despouys, "High Performance Low Cost Optical Connector," *Proceedings of the Eighth European Conference on Optical Communication,* Cannes 1982, pp. 331–336.
40. J. G. Woods "Single Mode Fiber Optic Connectors and Splices," in *Fiber Optic Couplers, Connectors and Splice Technology, Proceedings of the Society of Photo-Optical Instrumentation Engineers,* vol. 479, 1984, pp. 42–47.
41. A. Nicia, "Lens Coupling in Fiber-Optic Devices: Efficiency Limits," *Applied Optics,* vol. 20, no. 18, pp. 3136–3145, 1981.
42. S. Matsuda et al., "Low-Loss Lens Connector for Single-Mode Fibers," *Applied Optics,* vol. 21, no. 19, pp. 3475–3483, 1982.
43. J. P. Carrol et al., "Design Considerations of the Expanded Beam Lamdek Single-Mode Connector," *Proceedings of FOC/LAN,* San Francisco, September 1985.
44. N. Suzuki et al., "Low Insertion- and High Return-Loss Optical Connectors with Spherically Convex-Polished End," *Electronics Letters,* vol. 22, no. 2, pp. 110–112, 1986.
45. W. C. Young et al., "Low-Loss Field-Installable Biconic Connectors for Single-Mode Fibers," *Technical Digest of Topical Meeting on Optical Fiber Communication,* New Orleans, Optical Society of America, pp. 14–15, 1983.
46. N. Suzuki et al., "High-Performance Optical Connectors for Subscriber Loop

Systems," *Review of the Electrical Communication Laboratories,* vol. 32, no. 4, pp. 619–625, 1984.

47. R. Rao et al., "High Return Loss Connector Design without Using Fibre Contact or Index Matching," *Electronics Letters,* vol. 22, no. 14, pp. 731–732, 1986.
48. C. M. Miller, "A Fiber-Optic-Cable Connector," *Bell System Technical Journal,* vol. 54, no. 9, pp. 1547–1555, 1975.
49. T. Satake et al., "Loss Analysis of a Single-Mode 5-Fiber Plastic Connector," *Transactions of IECE of Japan,* vol. E69, no. 3, pp. 180–182, 1986.
50. K. Kobayashi et al., "Micro-Optic Devices for Branching, Coupling, Multiplexing and Demultiplexing," *Technical Digest of the First Integrated Optics and Optical Fiber Communication Conference,* Toyko, 1977, pp. 367–370.
51. K. Nosu et al., "Slab Waveguide Star Coupler for Multimode Optical Fibres," *Electronics Letters,* vol. 16, no. 15, pp. 608–609, 1980.
52. Y-F. Li et al., "Coupling Efficiency of a Multimode Biconical Taper Coupler," *Journal of the Optical Society of America,* Part A vol. 2, no. 8, pp. 1301–1306, 1985.
53. E. G. Rawson et al., "Bitaper Star Couplers with up to 100 Fibre Channels," *Electronics Letters,* vol. 15, no. 14., pp. 432–433, 1979.
54. D. Marcuse, *Light Transmission Optics,* Chap. 10, Van Nostrand, New York, 1972.
55. K. O. Hill et al., "Optical Fiber Directional Couplers: Biconical Taper Technology and Device Applications," *Proceedings of the Society of Photo-Optical Instrumentation Engineers,* vol. 574, 1985, pp. 92–99.
56. F. P. Payne et al., "Modelling Fused Single-Mode-Fibre Couplers," *Electronics Letters,* vol. 21, no. 11, pp. 461–462, 1985.
57. K. Nosu et al., "Multireflection Optical Multi/Demultiplexer Using Interference Filters," *Electronics Letters,* vol. 15, no. 14, pp. 414–415, 1979.
58. G. Winzer et al., "Single-Mode and Multimode All-Fiber Directional Couplers for WDM," *Applied Optics,* vol. 20, no. 18, pp. 3128–3135, 1981.
59. W. J. Tomlinson, "Wavelength Multiplexing in Multimode Optical Fibers," *Applied Optics,* vol. 16, no. 8, pp. 2180–2194, 1977.
60. J. P. Laude, "Diffraction-Limited Wavelength Multiplexers/Demultiplexers: A New Approach," *Technical Digest of the Third Integrated Optics and Optical Fiber Communication Conference,* San Francisco, 1981, p. 66–67.
61. R. Watanabe et al., "Optical Demultiplexer Using Concave Grating in 0.7–0.9 μm Wavelength Region," *Electronics Letters,* vol. 16, no. 3, pp. 106–107, 1980.
62. Y. Fujii et al., "Optical Demultiplexer Using a Silicon Concave Diffraction Grating," *Applied Optics,* vol. 22, no. 7, pp. 974–978, 1983.
63. W. J. Tomlinson, "Aberrations of GRIN-Rod Lenses in Multimode Optical Fiber Devices," *Applied Optics*, vol. 19, no. 7, pp. 1117–1126, 1980.

# CHAPTER 4
# FIBER-OPTIC TEST METHODS

**Felix P. Kapron**
*Bell Communications Research*

## 4.1 TEST METHODS COVERED

### 4.1.1 Test Method Strategies

The chronology of fiber-optic measurements begins even before the manufacturing of the fiber, connector, emitter, detector, etc., are complete. For example, before a glass fiber is drawn, the preform may be subjected to a geometry and refractive index profile measurement. If it fails here, no further expense is necessary to process it into what would turn out to be unacceptable fiber. During the draw, the manufacturer will monitor the thickness and concentricity of the organic jacketing applied to the fiber. Other optical components will similarly have on-line test sequences prior to formation of a final product. Fiber measurement parameters include attenuation, cutoff wavelength (applies to single-mode fiber), numerical aperture (applies to multimode fiber), mode field diameter (single-mode), bandwidth (multimode), chromatic dispersion (mostly single-mode), bending losses, polarization (single-mode), strength and fatigue, and environmental resistance (temperature, liquid immersion, hydrogen, nuclear radiation, torsion, flexing, abrasion).

Note that not all measurements are for the direct benefit of the *user*. Some are done only for the *manufacturer*, as for factory process control or quality assurance to a specification. Moreover, not all measurements are done all the time. Depending upon the importance and tolerance of the parameter, along with the degree of manufacturing process control, *measurement sampling* will run from 0 to 100 percent for any particular parameter. Measurements form a significant portion of production costs, and there is always pressure to reduce them.

Continuing our example, the fibers are tested during and after cabling with respect to attenuation, bandwidth (multimode), crosstalk, splice loss, cable strain vs. fiber strain, and cable environmental resistance. After installation in the field, measurements may be done for installation or repair verification or for upgrading. Examples of such measurements are attenuation, chromatic dispersion (single-mode), bandwidth (multimode), splice loss, polarization, transmitter-coupled power, receiver-coupled power, signal-to-noise ratio, and bit-error ratio.

Generally, this chapter will detail test methods that are of more interest to the user than to the product supplier. However, the distinction between the two is relative. For example, a cabler is a user of a product from a fiber supplier, but a cable installer is a user of a product from a cable supplier. A transmitter maker is a user of a product from a light-emitting diode (LED) or injection laser diode supplier, but the maker is a supplier to a user incorporating it in terminal equipment.

At any stage, this chapter will be more concerned with test methods designed to measure performance parameters for the user rather than those measuring design parameters for the manufacturer. The chapter will deal mainly with individual components, though at times these may be assembled into subsystem links.

The test sets for measuring fiber-optic parameters may be *fixed* in position, as is usual in the factory, or *portable,* as is required in the field. The latter will tend to be smaller and lighter and be self-powered or otherwise capable of remote operation. Measurement test sets may be *integrated,* i.e., capable of measuring several parameters simultaneously or sequentially, or *discrete,* i.e., devoted to separate parameters. Integrated test sets may be more compact, but they have superfluous features if not all parameters are measured to the same sampling level. Discrete sets may require more fiber handling, but they can be performance-optimized for each parameter and their number allocated according to sampling rate.

With the maturity of the fiber optics industry, test sets for virtually all measurements are available commercially. Performance-to-price ratios are continually improving; test sets are becoming faster and more automated, requiring less skill and knowledge on the part of the operator. The function of this chapter is therefore not to provide details of equipment assembly and usage procedures, as may have been required for "home-made" construction some years ago. Topics such as resolution or accuracy will not be treated in detail. Rather, the emphasis is on an understanding of the *principles involved* so that a user can better appreciate the physical phenomena underlying the parameters being measured, and can make more intelligent choices from the available equipment.

### 4.1.2 Standard Test Methods

There are a number of organizations prominent in fiber-optic technology standards. Several of these are at the system level, addressing particular applications, e.g., synchronous optical networks or computer data lines. More generally, at the component level, there are only a few standards groups.

In the United States, the Electronic Industries Association (EIA) and the Telecommunications Industry Association (TIA) publishes two types of publicly available fiber-optic documents. One is specifications, and at the top of the hierarchy are *generic specifications* (e.g., dealing with all fiber or cable types). This is followed by *sectional specifications* (e.g., multimode fiber, single-mode fiber, buried cable, aerial cable), and then a *blank detail specification* giving tables of parameters contained in the generic specification and appropriate sectional specification. The supplier and user "fill in the blanks" to form the *detail specification* of actual parameter values.

The EIA has over 200 *Fiber-Optic Test Procedures* (FOTPs),* documents numbered according to the scheme EIA/TIA-455-*UVW,* where *UVW* is the FOTP number, and *Optical Fiber System Test Procedures* (OFSTPs), numbered EIA-526-*XYZ,* where *XYZ* is the OFSTP number. Many of these documents have been published; others are currently in various stages of review and balloting. These

*Numbered and unnumbered FOTPs cited in this chapter are listed under "References" at the end of the chapter.

documents have been drafted by and voted on by both suppliers and users and cover measurement of the parameters in the specifications cited, mainly on fibers, cables, and connectors, but specifications for other active and passive components—tools, test equipment, and systems—are emerging. This chapter will frequently cite FOTPs (but not OFSTPs).

Another U.S. source is a series of documents called *Technical References* (TRs) published by Bellcore (Bell Communications Research), largely for regional Bell operating companies and other clients. These outline requirements and objectives for a number of fiber-optic components. Products are tested and analyzed in accordance with the requirements of the TRs.*

International bodies dealing with fiber-optic technology standards include the International Electrotechnical Commission (IEC). Like the EIA, the IEC generates a similar specification hierarchy and test method standards. The International Telephone and Telegraph Consultative Committee, known as the CCITT, the acronym for its name in French, considers fiber, subsystems, and networks, but not active or passive components. Test methods are divided into *reference test methods* (RTMs) that are performed strictly according to the definition, and are accurate and reproducible. *Alternate test methods* (ATMs) may be easier to implement and may deviate from the strict definition; however, there must be ways to relate results to those obtained from the reference test methods. Both EIA and Bellcore documents have influenced international standards, and they are generally in harmony with them, although some differences do appear.

The American National Standards Institute (ANSI) has formed a Fiber-Optic Coordinating Committee of domestic standards groups, from the component to the network level. Moreover, specific subject-matter experts on U.S. delegations to the international bodies assist in standards development to ensure product compatibility.

The emphasis in this chapter will be on *standardized* measurements. Little reference will be given to alternative procedures that have developed historically, or that are subjects of research only. Consequently no references to the voluminous research and development literature will be given. Alternative test methods will be discussed if they help the reader understand the standard test methods, or if standard methods do not yet exist.

## 4.2 POWER MEASUREMENTS

### 4.2.1 Power Attributes

The measurement of optical power is fundamental to a broad range of fiber-optic test methods. The power to be measured may vary with wavelength, i.e., as a continuous spectral scan or at discrete wavelengths. It can change in time, being either a steady (DC) level, or modulated as an analog continuous wave or as a digital pulse stream. Power can vary with position in a spatial or angular scan. Also, it can vary with polarization.

### 4.2.2 Sources and Detectors

A typical source used for measurements at many wavelengths over a broad wavelength range is a white light source, such as a tungsten halogen lamp. To obtain

*Selected Bellcore TRs are listed under "References."

the desired wavelengths, its output is filtered by a diffraction grating or Fabry-Perot cavity in a monochromator or optical spectrum analyzer or with interference filters on a rack or wheel. The detector is usually a semiconductor device and may be a silicon (Si) photodiode sensitive up to about 1050 nm; at longer wavelengths, indium gallium arsenide phosphide (InGaAsP) diodes are used, while germanium (Ge) photodiodes can cover both ranges, though with more noise and less sensitivity. When a detector is coupled to a fiber, a multimode fiber attached to the detector package can assist in providing high light-collection efficiency. Sometimes the source or detector may be cooled to increase stability and reduce noise. In this way, a wavelength scan from 600 nm to 1800 nm, which encompasses the range of interest for most communications and sensor applications with silica-based fiber, may be obtained.

Usually, fewer wavelengths or narrower wavelength ranges called for in the specific applications are sufficient for routine measurements. In the case of only a few discrete wavelengths, a filtered light-emitting diode (LED) or laser diode (LD) may be used as sources. Gallium aluminum arsenide phosphide (GaAlAsP) diodes are used in the 850-nm region, and indium gallium arsenide phosphide (InGaAsP) diodes are used around 1300 nm and beyond. Both types are used in commercially available stabilized light sources. The detectors may be found in commercial optical power meters; some are calibrated to absolute (rather than relative) power. The source may be modulated at low frequency and electronically phase-locked to the detector to increase system sensitivity. This phase-sensitive detection and lock-in amplification is convenient particularly if the source and detector are collocated. For other types of measurements, higher-speed modulation may be required. Complete optical loss test sets incorporating such components are commercially available.

### 4.2.3 Field Distributions

In many cases dealing with optical components, it is necessary to obtain what are called near-field or far-field optical power distributions. These are useful, for example, in analyzing source-to-fiber coupling, or source and fiber geometrical parameters.

The *near-field* power distribution is the light intensity as a function of position on the emitting surface. Such a surface can be, for example, the active output area of an LED or of a fiber end. Near-field measurement is covered in FOTP-43. As shown in Fig. 4.1, an objective lens, such as a microscope objective, is used

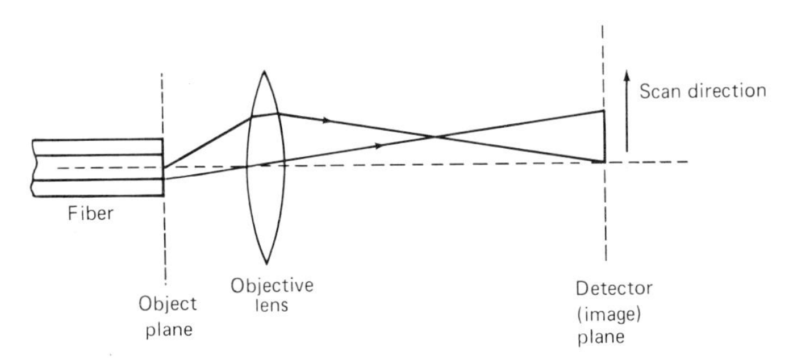


FIGURE 4.1 Near-field power distribution.

to magnify the small emitting area that is common in fiber optics. The image is then scanned in a plane by a movable detector, or in a line if the near-field pattern is circularly symmetric. Detector arrays are also used, usually to give a video display on a monitor. Care must be taken that the arrays do not vary in linearity or sensitivity, especially at longer wavelengths.

As shown in Fig. 4.2*a*, a *far-field* distribution is a record of the light intensity as a function of angle some distance away from the emitting surface. Far-field measurement is covered in FOTP-47. It is important that the distance $z$ between the detector scan or array and the emitting area of largest dimension $d$ be such that

$$z > \frac{(10d)^2}{\lambda} \tag{4.1}$$

where $\lambda$ is the wavelength of the light used. This ensures that the emitting area appears almost as a point to the detector. For a fiber, for example, dimension $d$

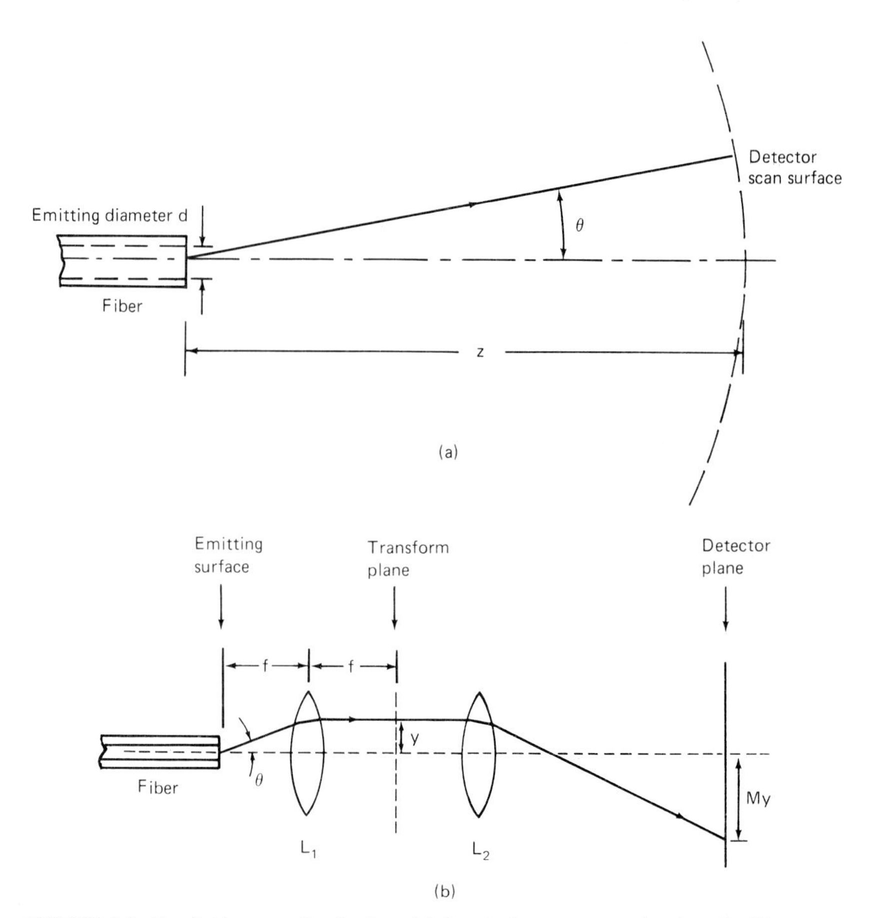


FIGURE 4.2 Far-field power distribution. (*a*) Standard arrangement (projected); (*b*) convenient alternate arrangement (transformed). (*FOTP-47*)

corresponds to the core diameter for a multimode fiber and to the mode field diameter for a single-mode fiber. The corresponding values of distance $z$ should exceed 0.2 to 1.2 m and 2 to 10 mm, respectively, depending on those diameters. Note that the detector scan surface is ideally spherical. If a planar surface is used, an arithmetic correction should be made, though this is often neglected at smaller angles.

To shrink these distances, especially with large source areas, and to allow for detector planar surfaces, the alternative arrangement of Fig. 4.2*b* is sometimes used. A Fourier-transforming lens $L_1$ and a relay lens $L_2$ are placed between the emitting area and the detector plane. The first lens must be one focal length, $f$, from the fiber-end face. The lens establishes the following relationship between the scan position $y$ in the Fourier transform (second focal) plane and far-field angle $\theta$:

$$y = f \sin \theta \tag{4.2}$$

The second lens $L_2$ produces a magnified image (magnification $M$) of the transform plane in the detector plane.

### 4.2.4 End Preparation

The preparation of good fiber ends is essential to both measurements and operating systems, and it is covered in FOTP-57. The fiber jacket is removed by mechanical stripping, chemical solvent (which should be well-vented), or possibly burning. Depending on the details of these methods, fiber strength near that point may be affected.

A flat end-face should generally be closer than 2° to perpendicularity (with respect to the fiber axis), and often closer to ½°. A variety of portable hand-held tools are commercially available. Fiber cleaving, cutting, or splicing tools generally abrade or scratch the fiber's cylindrical surface. Cleaving is induced with a longitudinal tensile force, applied to a straight or slightly bent fiber during and/or after the abrasion.

In addition, the ends may be further prepared by polishing. Typically, the fiber is mounted within a vertical holder and polished in a figure 8 pattern with two or three powders or sequentially decreasing grit size. Final cleanliness is critical. Sometimes the differing dopants in the core and cladding glasses may result in a slight depression of up to several micrometers at the core. Moreover, surface compaction may produce a slightly higher refractive index grading to a depth of less than 1 μm in the end face.

Visual inspection of an end-face can be conducted with a simple microscope to check for notches, scratches, shatter, lip, etc. An alternative optical bench setup in FOTP-57 involves mounting the fiber horizontally and shining a helium-neon (He-Ne) laser beam at an (arbitrary) angle to the vertical end-face. The reflection (at an equal angle) strikes a target card a distance $D$ away from the end face. Rotating the fiber one turn within the holder causes a light circle of radius $r$ to be described on the target. The end-face angle is then

$$\theta = \frac{1}{2} \arctan \frac{r}{D} \tag{4.3}$$

The use of an interference microscope is outlined in FOTP-179. An alternating pattern of light and dark fringes is formed because of differences between the fiber-end face and a flat mirror in the microscope. A pattern of concentric circular fringes indicates a concave or convex-end face, whereas uniformly spaced straight lines indicate a flat but nonperpendicular-end face. Counting fringes gives an accurate measure of end face curvature or end angle. More complex surface breaks can also be analyzed.

## 4.3 FIBER POWER LOSS

### 4.3.1 Attenuation

***Absorption and Scattering.*** *Attenuation* is the loss of useful signal power along a fiber. It is important in determining the optical power budget across a fiber link between the transmitter and receiver. The spectral dependence of attenuation is important for the supplier in determining the quality of a product. For the user, spectral attenuation affects the dispersion performance of spectrally broad sources and fiber upgradability for multiwavelength operation. In terms of its wavelength dependence, the spectral attenuation coefficient can be written as

$$\alpha(\lambda) = A\lambda^{-4} + B + Ce^{D\lambda} + E(\lambda) + Fe^{-G/\lambda} \tag{4.4}$$

where $A$ through $F$ are adjustable parameters that are fixed from curve fits to the measurement. Attenuation is by absorption and scattering.

*Absorption* is the conversion of light into heat. The $C$ term is ultraviolet (uv) absorption due to electronic transitions between the valence and conduction bands distributed around 140 nm. The $F$ term is infrared (ir) absorption due to molecular vibration states having a primary peak around 8 μm, but with higher-order combination and overtone bands at shorter wavelengths. The above are *intrinsic* absorptions.

*Extrinsic* absorption includes impurity absorption which is often caused by $OH^-$ ions in silica, which have a fundamental vibration around 2.7 μm, but with important overtones near 1385 and 1245 nm. This, with possible absorption due to hydrogen in-diffusion (which can be reversible unless chemical bonding occurs at higher temperatures), forms the $E$ term. Finally, absorption due to defects formed during the glass's thermal history (for example, during drawing) or to various nuclear radiations may be important in some applications. Impurity absorption due to transition metal ions is of no concern in modern fiber.

Absorption can be measured directly by microcalorimetry. Because absorption is so low, however, a rather long fiber has to be placed in a thermal flask while light passing through it causes the temperature to rise only a small fraction of a degree. Such testing is rare today.

*Scattering* is the conversion of light bound and propagating within the fiber into leaky, unbound, radiative, or back-scattered light. The wavelength-dependent $A$-term is called *Rayleigh scattering.* The $B$ term is wavelength-independent. *Intrinsic* scattering is attributable to refractive index variations on a molecular level, from either thermal or compositional fluctuations. *Extrinsic* scattering is attributable to waveguide or index profile errors or to environmental effects such as bending. The backward component of scattering (i.e., backscatter) can detrimentally contribute to bidirectional crosstalk, but is also useful for the optical time-domain reflectometry (OTDR) test instrument (Sec. 4.6). Finally, *nonlinear scattering* occurs at high optical power levels in fibers and with very coherent sources. This can usefully generate new wavelengths or be used in amplification (depending on the fiber dopant composition). On the other hand, it can limit the amount of power coupled into the fiber, or cause crosstalk between wavelength-division-multiplexed (WDM) or optical frequency-division-multiplexed (OFDM) channels.

Scattering can be measured directly by use of light-integrating spheres or cubes. As with absorption, low power levels make this difficult, and such testing is rare.

Almost always, it is the *total* attenuation—absorption plus scattering—that is measured. If $z$ (in kilometers) is the position along the fiber and $\alpha(z)$ is the local value (in decibels per kilometer) of the attenuation coefficient there, then the power decay between the interfaces $z_1$ and $z_2$ ( $> z_1$) is

$$P(z_2) = P(z_1) \cdot 10^{-A/10} \tag{4.5a}$$

Here the total attenuation is

$$A = \int_{z_1}^{z_2} \alpha(z)\, dz \qquad \text{dB} \tag{4.6a}$$

Practical units of $P$ are in watts (or sometimes in microwatts or nanowatts). If $p$ is logarithmic power in decibels relative to 1 microwatt (dBm, where 0 dBm $\equiv$ 1 mW), then

$$p(z_2) = p(z_1) - \frac{A}{10} \tag{4.5b}$$

The total attenuation between the interface points is

$$A = 10 \log \frac{P(z_1)}{P(z_2)} \qquad \text{dB} \tag{4.6b}$$

$$= 10\,[\,p(z_1) - p(z_2)] \tag{4.6c}$$

If between $z_1$ and $z_2$ the attenuation per unit length is a constant, it is defined as the attenuation coefficient

$$\alpha = \frac{A}{z_2 - z_1} \qquad \text{dB/km} \tag{4.7}$$

As shown in Sec. 4.3.1, "Modal Effects," this may not be true even if the fiber is structurally uniform.

***Cutback and Insertion.*** The *cutback* method is given in FOTP-46 for multimode fiber and in FOTP-78 for single-mode fiber. As shown in Fig. 4.3*a*, the power $P_L$ exiting a full length $L$ of fiber is measured. As already mentioned in Sec. 4.2.2, "Sources and Detectors," this may be at one or more discrete wavelengths, or as a wavelength (spectral) scan. Without disturbing the source-to-fiber power cou-

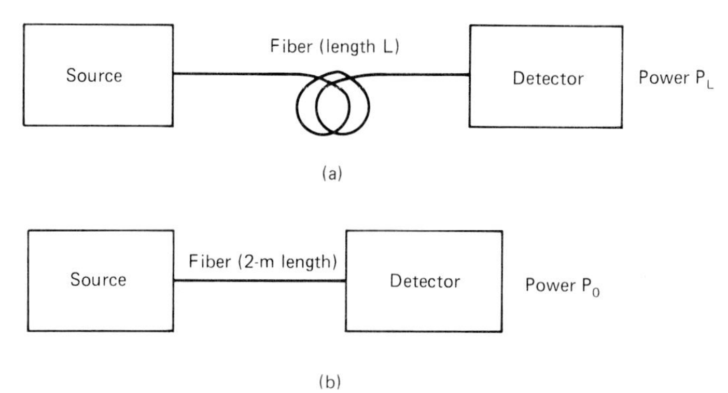


FIGURE 4.3 Attenuation by cutback. (*a*) Test fiber; (*b*) shortened test fiber. (*FOTP-46*)

pling, the fiber is cut back to a convenient (effectively zero) 2-m length as in Fig. 4.3*b*. The power measurement is repeated to give $P_0$. The attenuation is calculated as in Eq. (4.6*b*):

$$A = 10 \log \frac{P_L}{P_0} \quad \text{dB} \tag{4.8a}$$

The attenuation coefficient is

$$\alpha = \frac{A}{L} \quad \text{dB/km} \tag{4.8b}$$

If the detector or power meter measures in units of dBm, a form similar to Eq. (4.6*c*) is used.

Since only fiber-to-detector power coupling changes between measurements, and this variation can be made less than 0.01 dB, the cutback method is regarded as the reference test method for attenuation measurements. However, it does require an inconvenient cut, destruction of at least 2 m, and access to both fiber ends. It is therefore more suitable to the factory than the field. The insertion loss or attenuation by substitution method avoids this, though at the expense of some variability in the source-to-fiber coupling. This method is an *alternate test method* and is covered in FOTP-53.

Using the *insertion* or *substitution* method of this FOTP, the attenuation $A_R$ of a reference fiber is measured once by the cutback technique. (Note that this measurement is not normalized to length as is the attenuation coefficient.) This reference measurement can be done, if necessary, by a different test setup, although it should be similar to the one currently being used. The reference fiber is inserted and the detected power $P_R$ is obtained as in Fig. 4.4*a*. This need be done only once, but with care taken that the launch condition and wavelengths are similar to those of the original reference test setup. The test fiber (or fibers, in turn) is inserted, with care to maintain the same launch and detection conditions. In each case, the power $P_T$ is detected as in Fig. 4.4*b*. The attenuation of the test fiber is then

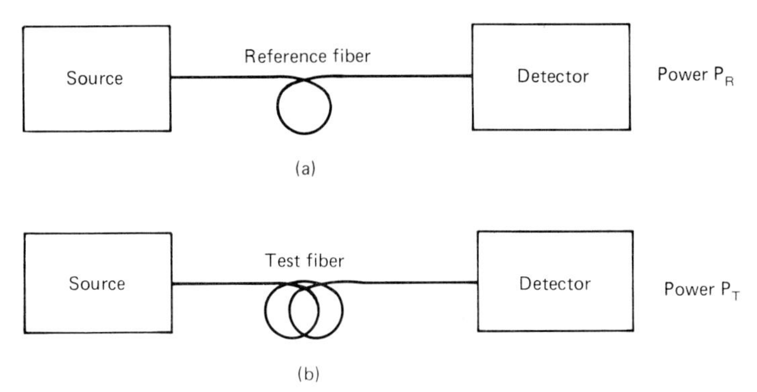


**FIGURE 4.4** Attenuation by insertion or substitution. (*a*) Reference fiber of known attenuation; (*b*) test fiber. (*FOTP-53*)

$$A_T = A_R + 10 \log \frac{P_R}{P_T} \tag{4.9}$$

Measurement of fiber loss can also be done using an optical time-domain reflectometer. This is covered in Sec. 4.6, "Optical Time-Domain Reflectometry."

***Modal Effects.*** This subsection will show how high-order modes may affect fiber attenuation results. In a multimode fiber, a particular ray pattern, or *mode,* has longitudinal periodicity along the fiber and angular periodicity within the cross section. A *bound mode* occupies a doughnut-shaped area within the core cross section; the mode is evanescently decaying beyond the area. With a *leaky mode,* light can tunnel from the doughnut to an outer region of the cladding. Higher-order leaky modes are increasingly lossy, since the outer surface of the cladding is in contact with a higher-index fiber jacket material (e.g., acrylate, silicone, or nylon) that serves to mechanically protect the fiber. *Unbound modes* occupy the whole cross section, except for a small central region of the core. As a result of the scattering described in this section under "Absorption and Scattering," some intermodal coupling can occur.

The fiber jacket acts as a *cladding-mode stripper.* If a large, diffuse light source excites all fiber modes, the cladding-mode stripper will ordinarily eliminate the highest-order modes within the first meter of fiber. If the fiber jacket has a refractive index lower than that of the cladding, 10 cm or so of the jacket may need to be removed and the fiber immersed in a higher-index fluid to effectively remove high-order modes.

Many leaky modes are not removed with a cladding-mode stripper. Figure 4.5*a* shows power propagation measured by multiple cutback along a length $L$ of multimode fiber excited by an 850-nm LED. The measured power decay can be empirically fit to the following equation, where $p$ is in decibels:

$$p(L) = p(0) - \alpha_s L - p_t(1 - e^{-L/L_t}) \tag{4.10}$$

Here $p(0)$ is the (absolute or relative) coupled power and $\alpha_s$ is a *steady-state* attenuation coefficient. The *transient* power $p_t$ can range from below 0.2 dB for a

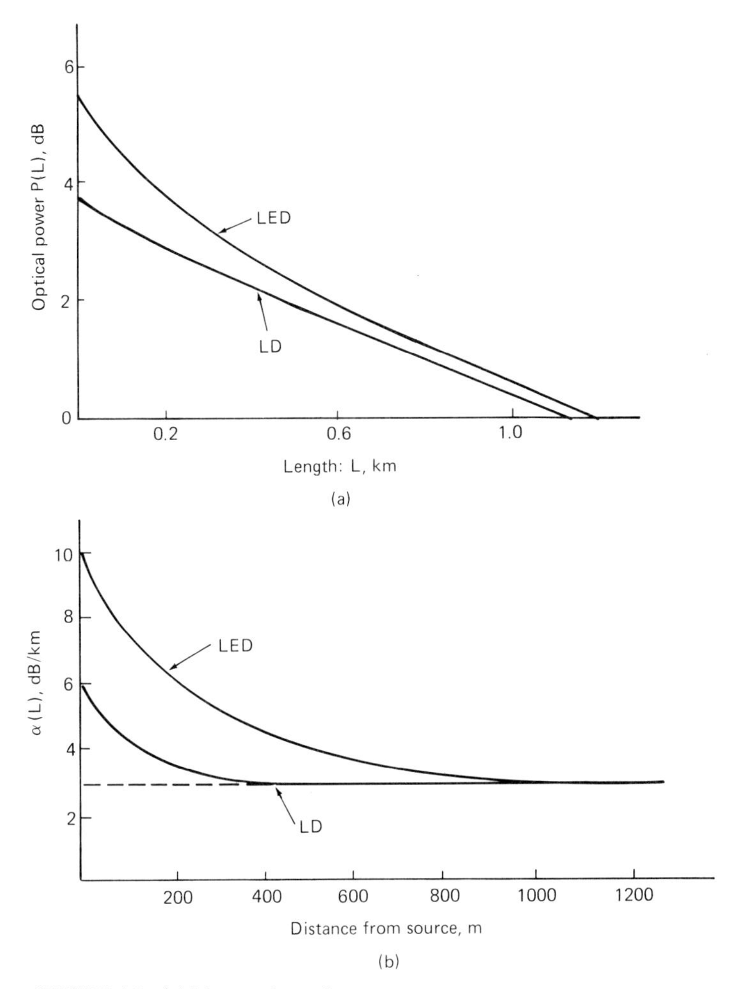


FIGURE 4.5 (*a*) Measured transient power due to LED and LD coupling to multimode fiber at 850 nm; (*b*) the resultant calculated transient attenuation coefficient.

laser diode to almost 2 dB for a surface-emitting LED (SLED). This power is lost over 4 or 5 times the transient length $L_t$, where $L_t$ ranges from below 100 m to several hundred meters. Beyond this length, $p(L)$ is linear (in decibels).

The corresponding attenuation coefficient from Fig. 4.5*b* is

$$\frac{dp}{dL} \equiv \alpha(L) = \alpha_s + \frac{p_t}{L_t} e^{-L/L_t} \tag{4.11}$$

The transient term decays exponentially with length $L$ and is important for fiber and splice and connector measurements. It can be made to disappear with use of a *mode filter,* which we shall discuss in the next section. Such behavior might be characteristic of longer spans of multimode fiber that are underfilled with laser diode excitation and extend many transient lengths. With shorter-distance, local-network fibers using LED sources, overfilled conditions may be more appropriate.

The effects just described are summed over all propagating modes. There are several ways of probing modal effects in detail. One can use a single-mode fiber to selectively launch one or a few modes onto the end face of a multimode fiber. Increasing the spot distance from the cross section center or the beam angle with the fiber axis increases the mode number. With such scans and cutback, the differential mode attenuation plot (attenuation vs. mode number) of Fig. 4.6 may be generated.

New methods are evolving in exciting two or three mode groups and describing their attenuation and dely characteristics by a mode-transfer matrix. In another technique, not yet standardized, one measures the near-field power profile $N(r)$ at the fiber-end face anywhere along a link using the method of Sec. 4.2.3, "Field Distributions," along with the refractive index profile $n(r)$. One method of obtaining $n(r)$ is from the near field of an overfilled fiber. The mode power distribution (MPD) is then given by

$$\text{MPD} = \frac{dN/dr}{dn/dr} \tag{4.12}$$

Figure 4.7 shows examples of MPD for LEDs.

With single-mode fiber, the second-order $LP_{11}$ mode group is a transient mode, and it can be a problem if the operating wavelength is too low. The first-order $LP_{01}$ mode group is tightly bound. Practical values of the cutoff wavelength of that mode depend upon fiber length and bend conditions, the measurement of which is discussed in Sec. 4.4.2 under "Launch Conditions." Here we show how to measure $\alpha_{11}(\lambda, R)$, the attenuation coefficient of the secondary mode, which depends on wavelength and on the bend radius $R$. If the fiber is spatially and an-

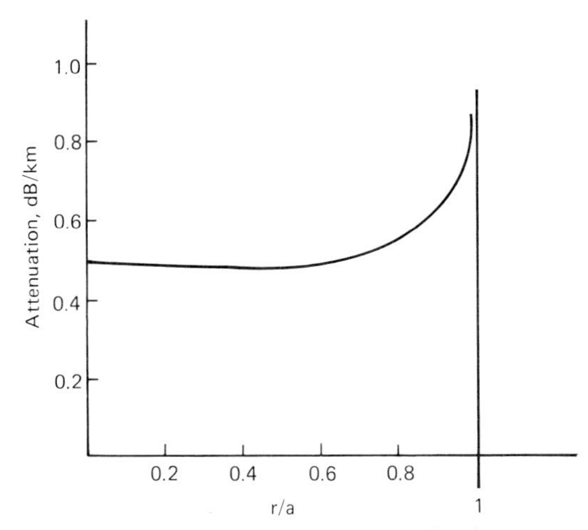


**FIGURE 4.6** Differential mode attenuation, i.e., attenuation coefficient vs. radial position *r* of a small spot incident on a multimode fiber-end face of radius *a*.

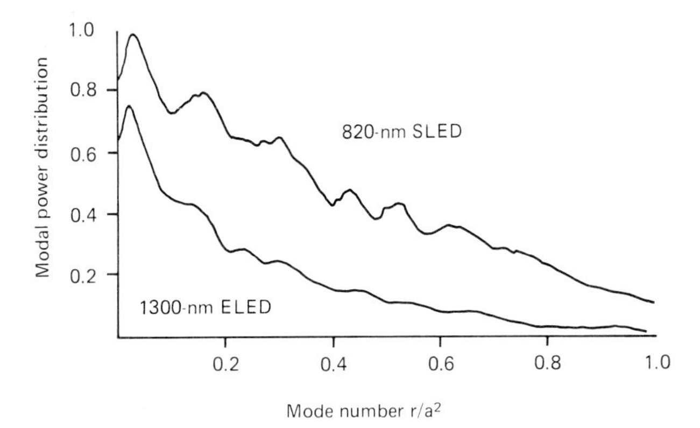


**FIGURE 4.7** Mode power distributions for a shortwave surface-emitting LED (SLED) and a longwave edge-emitting LED (ELED).

gularly overfilled (i.e., in the near field and far field), the power decay, analogous to Eq. (4.10) for the multimode fiber, is

$$P(\lambda, L) = P_{01}(\lambda)e^{-\alpha_{01}(\lambda)L} + P_{11}(\lambda)e^{-\alpha_{11}(\lambda)L} \tag{4.13}$$

where the $P$'s are absolute power levels (e.g., in units of milliwatts). Here $P_{01}$ and $P_{11}$ are the power levels at $L = 0$ launched into the primary ($LP_{01}$) and secondary ($LP_{11}$) modes, respectively, while $\alpha_{01}$ and $\alpha_{11}$ are the attenuation coefficients of those modes. Typically, $L$ is a few meters to tens of meters, and $\alpha_{01}$ is negligible compared to $\alpha_{11}$ in the wavelength region of interest, so the first spectral scan of the output power is just

$$P_1 = P_{01} + P_{11}e^{-\alpha_{11}L} \tag{4.14a}$$

Next, the secondary mode is removed by wrapping the fiber several times around a mandrel having a diameter of about 50 mm, and power level $P_2$ is measured:

$$P_2 = P_{01} \tag{4.14b}$$

Then the fiber is placed in a gentle bend of radius $R$, but cut back to a length $L_o$ of about a meter, and power level $P_3$ is measured:

$$P_3 = P_{01} + P_{11}e^{-\alpha_{11}L_o} \tag{4.14c}$$

From the three measurements of Eqs. (4.14), the $LP_{11}$ attenuation coefficient is calculated to be

$$\alpha_{11}(\lambda, R) = \frac{10}{L - L_o} \log \frac{P_3 - P_2}{P_1 - P_2} \tag{4.15a}$$

$$\approx a(R)e^{B\lambda} \tag{4.15b}$$

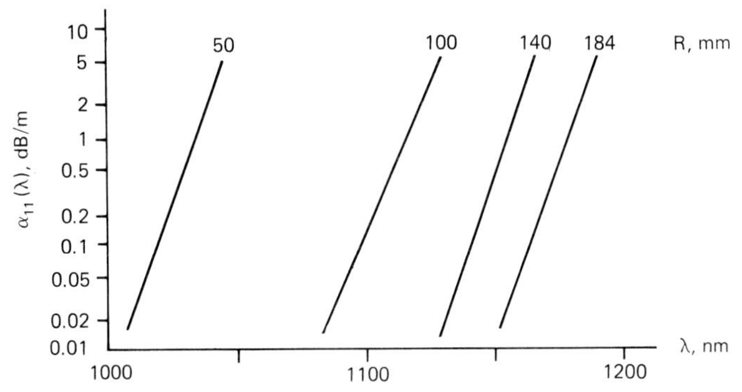


**FIGURE 4.8** Measured spectral attenuation coefficient $\alpha_{11}(\lambda, R)$ of the second-order mode group for various values of the bend radius $R$.

Here $a(R)$ and $B$ are fiber parameters. Equation (4.15*b*) produces straight lines on the log-linear plot of Fig. 4.8. Note that, measured at a fixed wavelength, the attenuation coefficient of the $LP_{11}$ mode increases greatly as the bend radius is reduced. Section 4.3.2, under "Definitions," describes how this relates to the cutoff wavelength.

***Mode Filters.*** For a multimode fiber, the topic of using a mode filter or equilibrium mode simulator to achieve an equilibrium mode distribution for steady-state attenuation values has long been controversial. The purpose of an equilibrium mode simulator is to ensure that the attenuation of a concatenated link equals the summed attenuations of the individual sections. The agreements resulting from standards discussions are covered in FOTP-50.

In one method of simulating equilibrium mode distribution, beam optics ($L_1$, $A_1$, $L_2$) is used to produce a light cone incident upon the fiber-end face as in Fig. 4.9*a*. The incident beam has a spot diameter, measured at the 5 percent intensity level in the near field, equal to 70 percent of the core diameter (measured in Sec. 4.5.3). The beam has a cone angle, measured at the 5 percent intensity level in the far field, equal to 70 percent of the numerical aperture (also measured in Sec. 4.5.3). Proper alignment of the incident beam, with respect to the fiber, is verified by means of a reflex viewer.

The other equilibrium mode simulation method involves a mode filter. A mode filter is qualified by first overfilling (spatially and angularly) the full fiber length of 1 or more kilometers, and measuring $\theta_5$, the far-field angle at the 5 percent intensity level. The mode filter is then applied to a 1- to 2-m cutback length, and a new $\theta_5$ is measured. The new value must lie between 94 and 100 percent of the earlier $\theta_5$ value to allow for slightly underfilled launch conditions. After this initial qualification, the mode filter may be repeatedly applied to the test fiber for attenuation measurements.

There are two types of mode filter for multimode fiber in general use. One type uses a "dummy" fiber similar to the test fiber, 0.5 to 1 km long, which is spliced before the test fiber, as shown in Fig. 4.9*b*. Perhaps more common is a mandrel wrap applied to the test fiber, consisting of four or five turns around a 20- to 30-mm-diameter mandrel, as shown in Fig. 4.9*c*. Then the cutback method (FOTP-46) or insertion and substitution method (FOTP-53) can be used.

With single-mode fiber, a 280-mm-diameter single turn has been defined as the

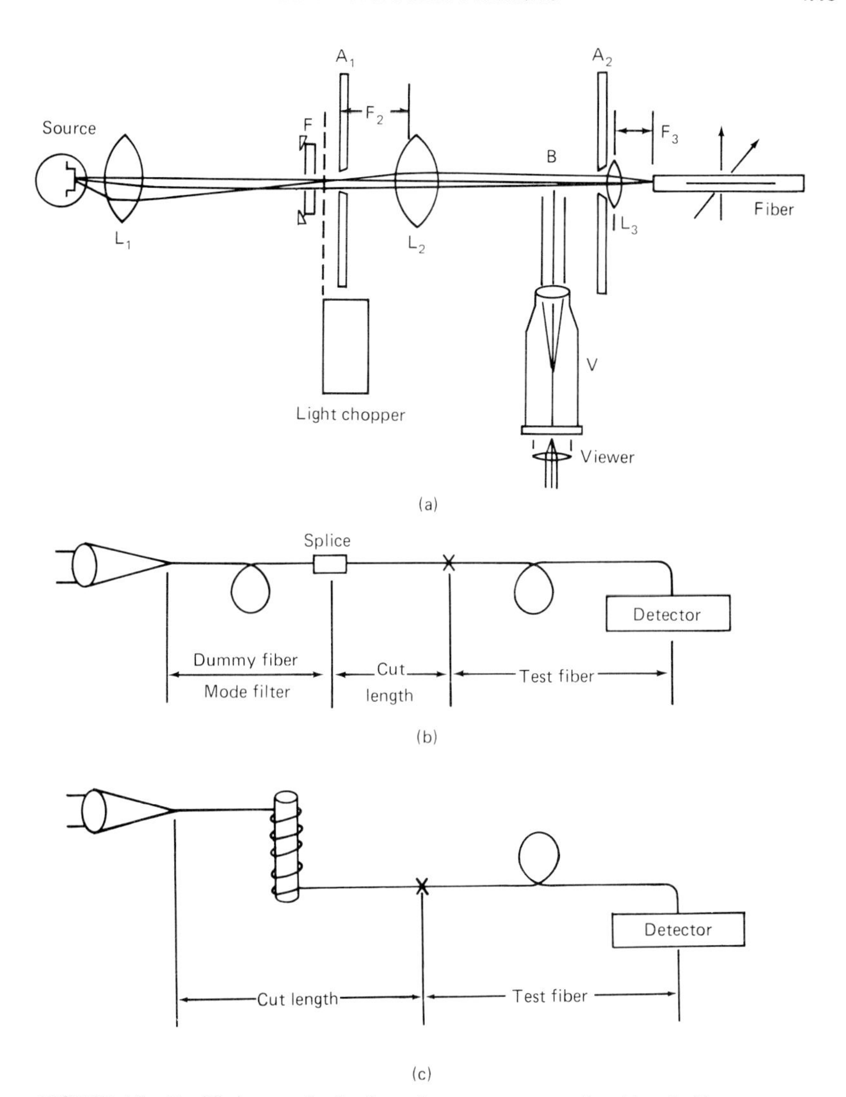


FIGURE 4.9 Equilibrium mode simulator for measurement of multimode fiber attenuation. (*a*) Limited phase space launch; (*b*) dummy fiber mode filter; (*c*) mandrel-wrapped mode filter. (*FOTP-50*)

test condition for a 2-mm length used in cutoff measurements (FOTP-80). The remainder of any reeled fiber should be at no smaller diameter. An FOTP-77, higher-order mode filter procedure is currently being defined for a number of single-mode measurements such as attenuation and mode field diameter.

***Bend Losses.*** Bending is generally of two types. *Macrobending* refers to bend diameters larger than the fiber diameter, but possibly as small as short-term fiber breakage levels. *Microbending* refers to deviations in the fiber axis that are only micrometers in extent over periods of millimeters.

From a ray viewpoint appropriate for multimode fibers, macrobend losses occur because the fiber axis reduces the acceptance or numerical aperture (NA) at any point on the cross section of the fiber core. Therefore, some bound modes strip off. From a mode viewpoint, the field seeks to preserve a uniform phase front around the bend, but this requires the portion on the outside of the bend to travel faster. When that part exceeds the speed of light in the cladding, it radiates away.

There are two kinds of macrobend losses. A *transition* loss occurs whenever the fiber axis radius of curvature changes, e.g., when a straight fiber goes into a bend or when a fiber changes from bent to straight, etc. A *steady-state* loss occurs with distance into a constant bend. The macrobend attenuation test procedure (FOTP-62) does not distinguish between the two, although the method can be configured to obtain such a distinction.

In FOTP-62, Method A measures the fiber attenuation increase due to a change from the straight condition (no bends less than 280-mm diameter) to a bent condition. First, a straight fiber is coupled between the source and detector; a power reading $P_o$ (either spectral or single wavelength) is taken. Then the fiber is wound on the outside or inside of a mandrel, without stress or fiber crossovers, for the specified diameters and numbers of turns. The power $P_i$ is determined for each of these configurations. These values and equations similar to Eq. (4.8*a*) give the attenuation increases.

Method B measures the total attenuation in the bent condition. Power $P_i$ is measured for each configuration as above. Following cutback, the power $P_f$ is measured. By a similar calculation, this gives the required total attenuation for each bend.

Figure 4.10*a* and *b* shows single-mode bend attenuation according to Method A. Note the larger attenuation at longer wavelengths (where the fundamental mode is farther from cutoff and has a larger mode diameter). Generally, fibers with smaller mode diameters and higher secondary-mode cutoff wavelengths are more bend-resistant. Bellcore TR-20 and CCITT G.650 call for 100 turns around a 76-mm- (3-in) diameter mandrel to simulate the total number of splice housing turns between long-distance repeaters. Note that this gives fewer sites for bend

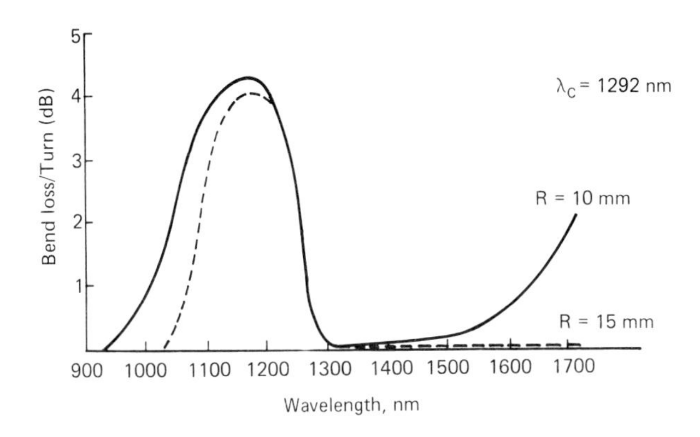


**FIGURE 4.10a** Spectral attenuation increase due to bending for a fiber of cutoff wavelength $\lambda_c$, equal to 1292 nm.

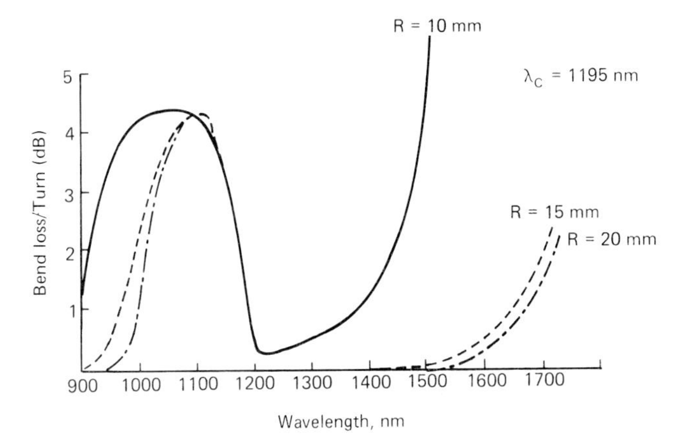


**FIGURE 4.10b** Spectral attenuation increase due to bending for a fiber of cutoff wavelength $\lambda_c$ equal to 1195 nm.

transition loss than would likely be experienced in the field. With both multimode and single-mode fiber, the induced loss should be under 0.5 dB at the wavelength of operation. A single-bend test at 32-mm diameter simulates an inadvertent tight bend; loss should not exceed 0.5 dB at 1550 nm for single-mode fiber (Bellcore TR-20).

A nonstandard basket-weave test attempts to come closer to the conditions a fiber might see in a cable—a combination of macrobending and microbending. The fiber is wound under tension onto a 150-mm-diameter drum. Figure 4.11 shows the resulting spectral attenuation curves for various winding tensions.

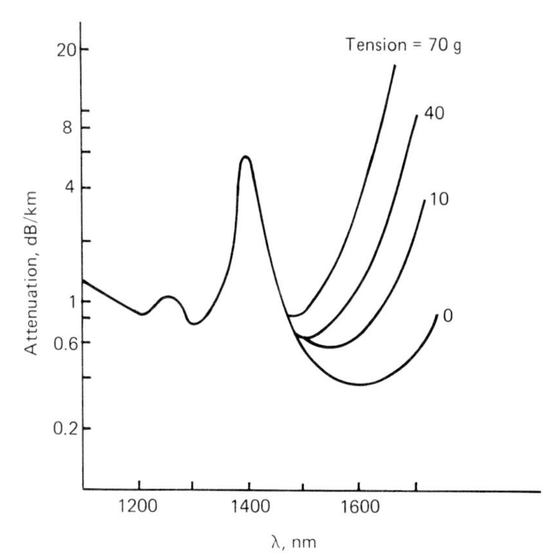


**FIGURE 4.11** Basket-weave attenuation at various tensions.

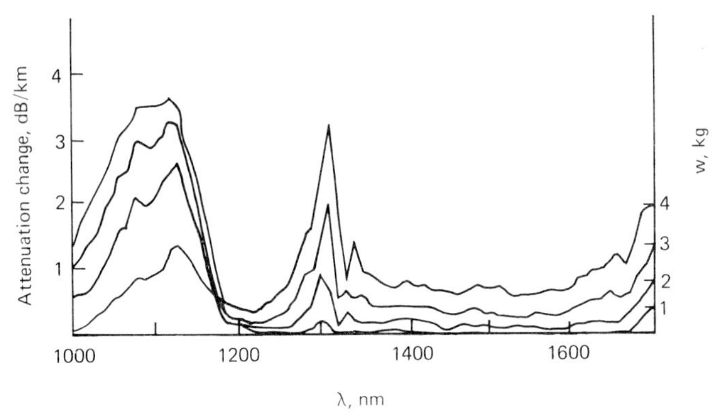


FIGURE 4.12 Microbend attenuation with various pressures due to applied weights of 1 to 4 kg.

Microbend attenuation results from numerous bends of very small radii. Microbends cause mode mixing in multimode fibers, and conversion of bound modes to unbound modes in both single-mode and multimode fibers. Measuring the attenuation due to microbends is dealt with in FOTP-68. Silicon carbide sandpaper is placed on two aluminum plates; the specific grit depends upon the fiber jacket thickness. The fiber is sandwiched between the plates along four straight paths of 380 mm each. The spectral attenuation change with increasing weights gives plots such as Fig. 4.12.

Bending measurements cannot accurately simulate features of the application environment of the fiber such as cables, jumpers, and splice housings. They provide general guidelines, however, and are useful for comparisons between fiber types.

### 4.3.2 Cutoff Wavelength

*Definitions.* A multimode fiber has many cutoff wavelengths. For a parabolic-index fiber, the number of bound propagating modes is

$$M = \left(\frac{\pi a}{\lambda}\right)^2 (n_1^2 - n_2^2) \tag{4.16}$$

where $a$ is the core radius and $n_1$ and $n_2$ are the core peak and cladding indexes, respectively. For a given fiber, operation around 1300 nm yields fewer modes than around 850 nm. As the wavelength is increased, an increasing number or modes are "cut off." Alternatively, as the core diameter is reduced from 50, 62.5, 85, or 100 μm down to 8 or 9 μm, and the core-cladding index difference is reduced from 1 or 2 percent down to about 0.35 percent, the number of mode groups reduces to one or two.

Simplistically, the *cutoff wavelength* is the shortest wavelength above which the fiber is single-mode. The value of the *theoretical* cutoff wavelength can be calculated from the fiber refractive index profile. However, the near field of the secondary mode group is that of a doughnut, somewhat less tightly bound than the primary fundamental mode peaked at the center. The practical cutoff wavelength, measured in fiber (or especially in cable), as discussed in the following text, is usually more than 100 nm below the theoretical value.

Cutoff wavelengths have been determined by double-mode/single-mode transitions established by means of near-field, far-field, pulse-splitting, polarization, coherence, and transmitted-power techniques. The last of these has been adopted as the standard method of FOTP-80. Recall from Sec. 4.3.1, under "Modal Effects," that the secondary mode has an attenuation coefficient that depends on curvature; hence cutoff corresponding to a certain decibel loss will depend upon the sample's radius of curvature $R$ and length $L$. The former dependency is shown in Fig. 4.13*a*, where the curves may be fit to

$$\lambda_c(R) = \lambda_c(\infty) - \frac{A}{R} \pm \frac{B}{R^2} \tag{4.17}$$

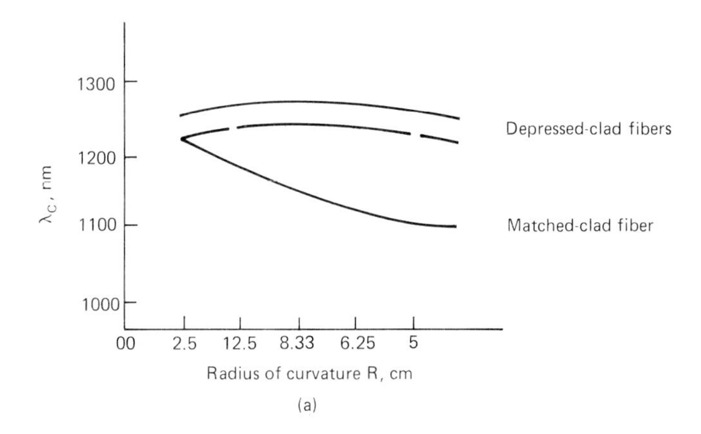


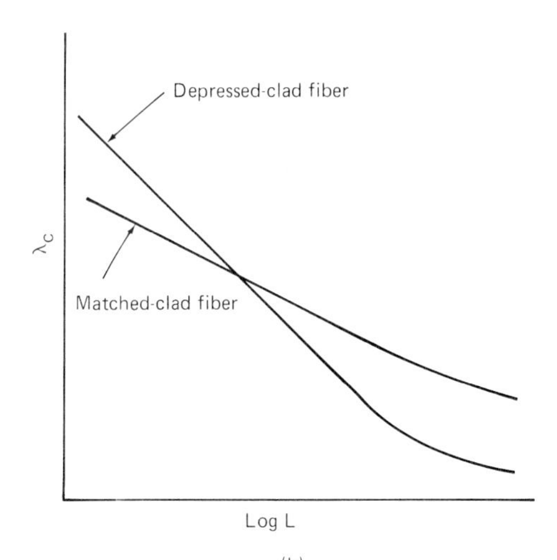


**FIGURE 4.13** Cutoff wavelength $\lambda_c$ dependence on (*a*) bend radius and (*b*) length.

where $A$ and $B$ are positive fiber parameters. Note that the matched-cladding cutoff is more bend sensitive than the depressed-cladding cutoff. However, the depressed-cladding cutoff is more length-sensitive than the matched-cladding cutoff. This is shown in Fig. 4.13*b*, which may be fit to the curve

$$\lambda_c(L) = \lambda_c(2) - A \log \frac{L}{2} \tag{4.18}$$

which follows from Eq. (4.15*b*). Slope parameter $A$ varies from 20 to 40 nm/decade for matched-cladding fibers to 55 to 75 nm/decade for depressed-cladding fibers.

***Fiber Cutoff.*** After considerable negotiation in standards groups, it was decided that the fiber sample would have a length $L = 2$ m, which conveniently corresponds to the cutback length in attenuation measurements. The radius of curvature is $R = 140$ mm, and can be applied in either of two ways. As in FOTP-80, Fig. 4.14*a* shows a single-turn configuration, while the split-mandrel configuration of Fig. 4.14*b* is convenient for fiber handling. First, the transmitted spectral power $P_s(\lambda)$ through the "straight" fiber sample around the expected cutoff wavelength is measured. Then there are two choices.

In the *bend-reference* technique, a section of the fiber is bent to a diameter of 60 mm or less to suppress the secondary mode. Then the corresponding power $P_b(\lambda)$ is measured. The bend attenuation is computed to be

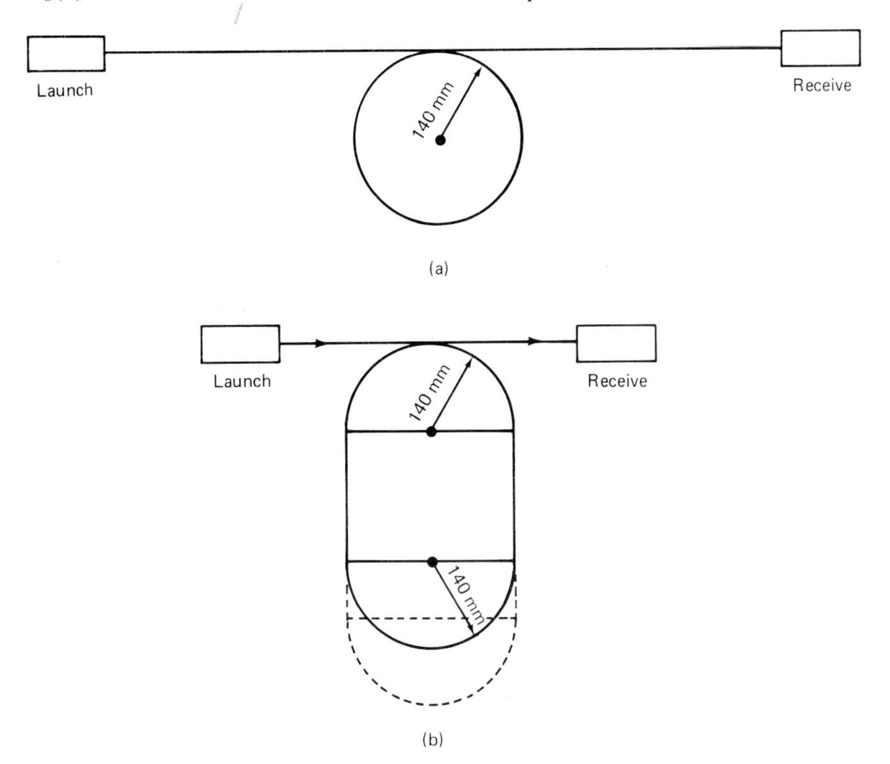


**FIGURE 4.14** Measurement configurations for determining cutoff wavelength of uncabled fiber. (*a*) Single turn; (*b*) split mandrel. (*FOTP-80.*)

$$A_b(\lambda) = 10 \log \frac{P_s(\lambda)}{P_b(\lambda)} \tag{4.19a}$$

as in Fig. 4.15*a*. The longest wavelength at which the attenuation equals 0.1 dB establishes the cutoff wavelength $\lambda_{cf}$.

In the *multimode-reference* technique, the single-mode fiber is replaced with a few meters of multimode fiber (for which the deployment condition is not critical). Then the spectral power $P_m(\lambda)$ is measured and the relative attenuation is computed:

$$A_m(\lambda) = 10 \log \frac{P_s(\lambda)}{P_m(\lambda)} \tag{4.19b}$$

as in Fig. 4.15*b*. The longest-wavelength portion of the curve is fitted by a straight line and the 0.1 dB $\lambda_{cf}$ wavelength is identified as shown. A round-robin study has determined that the two methods give equivalent results.

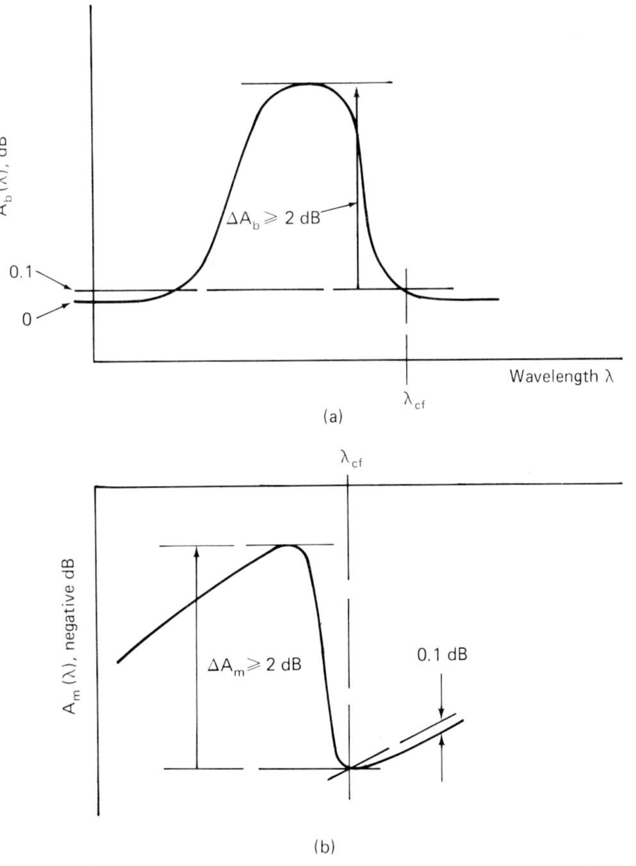


**FIGURE 4.15** Cutoff spectra for (*a*) bend reference technique and (*b*) multimode reference technique. (*FOTP-80*.)

Note that the measurements in Fig. 4.15 require an attenuation difference of no less than 2 dB. This is to ensure sufficient excitation of the secondary mode. Including polarization effects, there are two modes in the primary $LP_{01}$ mode group and four modes in the secondary $LP_{11}$ mode group. For equal excitation of both groups, the maximum attenuation difference is expected to be $\Delta A = 10 \log_{10} (6/2) = 4.77$ dB.

***Cable Cutoff.*** Compared to the previous measurements, the values of the cutoff wavelength of a cabled single-mode fiber will be smaller. This is because of the length and bend effects already described. Clearly, the value of cable cutoff $\lambda_{cc}$ depends on the nature of the fiber, the cable, and the deployment conditions. Except for the last, the procedure of FOTP-170 is similar to that of FOTP-80 for uncabled fiber.

Figure 4.16*a* shows an essentially straight 22-m cable length with a 1-m decabled fiber length at each end. A loop of 76-mm (3-in) diameter is applied to each end to simulate the effects of splice organizers. This deployment is intended to simulate a cable repair length, the situation that would result in the highest practical value of cutoff wavelength. There is usually a linear relationship between $\lambda_{cf}$ and $\lambda_{cc}$, but the effect is not readily predictable.

Figure 4.16*b* shows an alternative deployment condition that avoids the use of

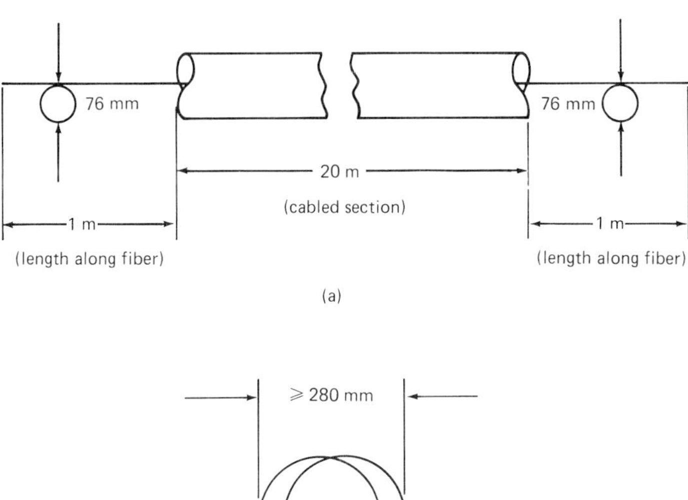


**FIGURE 4.16** Measurement configurations for determining cutoff wavelength for uncabled fiber. (*a*) Cabled form; (*b*) uncabled form. (*FOTP-170.*)

a cable. This is advantageous for both the fiber manufacturer selling to the cabler and to the user buying from the cabler. An uncabled fiber length of 22 m is used with a minimum diameter of 280 mm to conservatively simulate cabling effects; the end loops are also applied. The cutoff values measured are generally no smaller than those measured under the previous condition.

***Bimodal Effects.*** The practical consequence is that for a link to be truly single-mode, the transmitter central wavelength $\lambda_t$ should exceed the largest value of $\lambda_{cc}$. The latter is generally taken as 1250 to 1270 nm, allowing for possible LED spectral tails and wavelength-division multiplexing.

If bimodal transmission does occur, two consequences are differential mode delay and modal noise. Differential mode delay will cause a single narrow input pulse to exit the fiber as two output pulses. This pulse separation can be as large as $\Delta T = 3$ ps/m (or ns/km) around 850 nm, but can often be eliminated by loop filters near the receiver. Modal noise can occur with very coherent laser diode sources having a coherence time that exceeds this pulse separation between lossy joints. Such splices or connectors may launch some secondary modes which may not be able to decay away prior to the next joint. This results in loss fluctuations due to the bimodal speckle pattern there, combined with disturbances to either the fiber or joint. Bimodal transmission can be used for low bit-rate operation.

## 4.4 FIBER INFORMATION CAPACITY

### 4.4.1 Index Profile

The refractive index profile of a multimode fiber is very closely related to its modulation frequency roll-off and bandwidth. For a single-mode fiber, the profile determines its dispersion properties. Generally the user is interested in the *effects* of the profile design, so we will only briefly review how the index profile $n(r)$ itself is measured. The manufacturer generally tests the preform along its length prior to draw. In some commercially available instruments, the preform index profile is related to the changing deflection angle of a narrow beam that scans transversely to the preform. This is not practical for a small-diameter fiber.

The reference test method for both multimode and single-mode fibers is the refracted near-field method of FOTP-44. A short fiber is immersed in a cell containing a fluid of slightly higher refractive index. A light cone (typically from a 633-nm wavelength He-Ne laser for best resolution) with a large angle and small spot is scanned along a fiber cross-sectional diameter. The guided light is transmitted by the fiber out of the cell, while the leaky rays are blocked by a circular opaque stop. The refracted inbound rays form an annular core beyond the stop, and these are focused onto a detector. It turns out that the collected optical power, as a function of scan radius, equals the square of the refracted index. With care taken with oil cleanliness and index accuracy, very good refractive index profiles can be obtained, but the method is somewhat tedious.

Several near-field (Sec. 4.2.3) methods can be used. In the cladding near-field technique, cladding modes are uniformly excited by injection of a large-angle cone into the side of a bare fiber tip. The near-field scan can be related to $n(r)$. A more standard transmitted near-field method is given for multimode fiber in FOTP-43. An overfilled launch is used, typically with an 850-nm wavelength and 2-m fiber length. Cladding modes and leaky modes are stripped off. The near-field

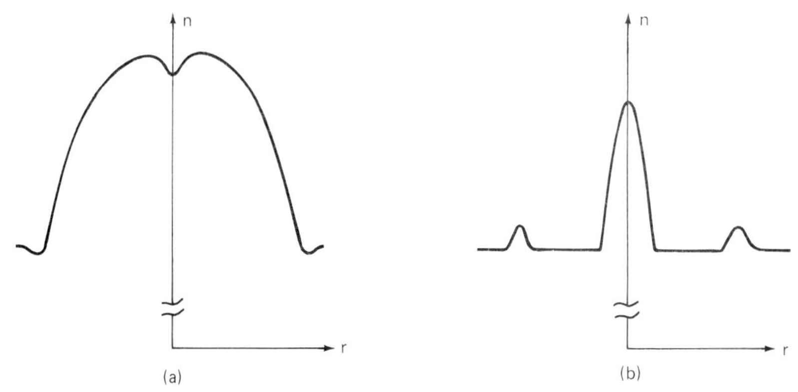


FIGURE 4.17 Refractive index profiles for (*a*) parabolic-index multimode fiber and (*b*) triangular-index single-mode fiber.

pattern is related to $n^2(r)$. For single-mode fiber, there is no FOTP for a transmitted near-field method. However, the procedure has been demonstrated using monochromatic light and an inversion of the scalar wave equation for fibers to yield the profile.

Finally, a transverse interference method is given in FOTP-29. It is not as often used since it requires an expensive parallel-path-transmission interference microscope. Viewing a short bare fiber on its side produces optical interference fringes which are digitized for computation of the refractive index profile.

Figure 4.17 shows some fiber refractive index profiles. The multimode profile has a near-parabolic profile characteristic of low-distortion fibers. Very slight profile errors can lead to severely reduced bandwidth. The single-mode profile has a near-triangular profile and a surrounding index ring characteristic of some fibers with minimum dispersion shifted to the 1550-nm region.

### 4.4.2 Multimode Distortion

***Differential Mode Delay.*** In Sec. 4.3.1, under "Modal Effects," we saw that in multimode fibers different modes have different attenuations and that differential mode attenuation can be measured with selective mode excitation. Analogously, differential mode delay can also be measured, with results as shown in Fig. 4.18. This is especially useful for the manufacturer who can use such information to optimize the index profile (Sec. 4.4.1). For example, a low-bandwidth, step-index multimode fiber has high-order modes that lag the low-order modes, and this would show on a differential mode delay measurement. The graded-index profile is designed to closely equalize the optical path lengths of both mode types. If insufficient profile correction is applied such that the step-index situation still applies, although to a much smaller degree, the profile is said to be *undercompensated.* Such optimization is very wavelength sensitive.

Profiles can readily be designed that are overcompensated so that low-order modes are relatively slower. Some fibers may even have mixed compensations along their profiles. Differential mode delay measurements can be useful also to the user since, when sections of multimode fiber are concatenated, the differential mode delay of the overall link will be minimized if the compensations of ad-

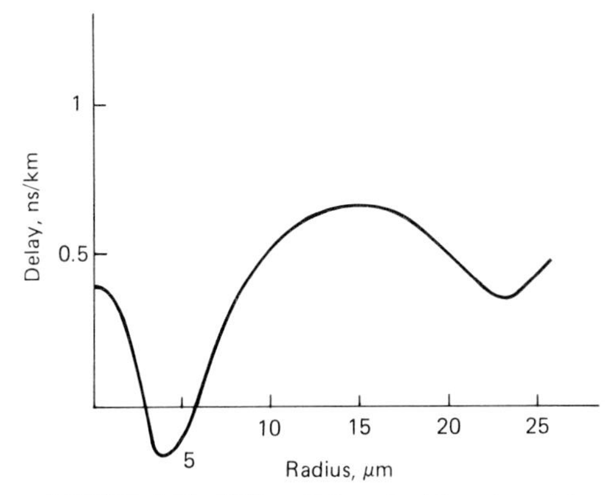


**FIGURE 4.18** Differential mode delay, i.e., delay vs. radial position of a small-diameter beam incident on a multimode fiber-end face.

jacent fibers are of opposite types. However, interest in such effects diminished as single-mode fibers became more prevalent.

*Launch Conditions.* Both differential mode attenuation and differential mode delay combine to affect the temporal distortion of a pulse. FOTP-54 specifies that, in the measurement of multimode distortion or bandwidth, a *mode scrambler* shall be positioned between the light source and the test fiber. Its function is to produce a uniform overfilled spatial and angular radiation distribution for the launch, independent of the radiation properties of the source and its alignment. It is not to assure length concatenation predictability as was the purpose of the equilibrium mode simulator ("Mode Filters," in Sec. 4.3.1) with attenuation. One scrambler example is a series combination of 1-m lengths of step-index, graded-index, and step-index fibers. Another example is introducing a series of serpentine microbends in the beginning of the test fiber.

As with mode filters or equilibrium mode simulator devices for attenuation, the mode scrambler must be qualified. The near-field intensity (as measured by FOTP-43) must vary less than 25 percent across the fiber core, while the far-field intensity numerical aperture (as measured by FOTP-177; see Sec. 4.5.3) must exceed the numerical aperture of the fiber.

In measurement, the mode scrambler is followed by a cladding-mode stripper. Such a launch has been as controversial as the equilibrium mode simulator for attenuation, since it may not represent the modal distribution of an installed link. The latter may range from underfill for laser excitation and fuller fill with LEDs ("Modal Effects" in Sec. 4.3.1). The difference in results between the two should not be too great, since, even with overfill, the higher-order modes usually contributing to distortion effects are naturally filtered out along the fiberlength.There is experimental evidence to confirm this, and both launch conditions are allowed by the EIA and CCITT, G.650.

*Time and Frequency Domains.* The terms *time domain* and *frequency domain* refer to the two ways in which multimode distortion may be measured and specified. The time-domain measurement is covered in FOTP-51. There the source is usually a repetitively pulsed laser diode with a spectral width specified to be sufficiently small so that chromatic effects are negligible. Signal averaging is often used at the detector end, and the source and detector combined response time $t_0$ is usually less than 0.5 ns root mean square (rms).

First, the output pulse from the full length $L$ of test fiber is measured. The rms pulse width may be calculated as $t_L$, but usually the Fourier transform is taken and the amplitude roll-off $dB_L(f)$, in decibels, is calculated. Here $f$ is the modulation frequency and the phase portion of the baseband response is usually neglected. Then the fiber is cut back to the lesser of 10 m or 1 percent of the original length. As an alternative to this cutback method, the insertion or substitution method may be used, as was done with attenuation in "Cutback and Insertion," in Sec. 4.3.1. This is nondestructive, need be done only once, and is only slightly less accurate than the measurement by cutback. In either case, an optical attenuator may be needed and the input pulse duration $t_0$ and roll-off $dB_0(f)$ are obtained. The instrument used in this test calculates the fiber baseband frequency response

$$dB_F(f) = dB_L(f) - dB_0(f) \qquad \text{MHz} \tag{4.20a}$$

as shown in Fig. 4.19, and the pulse broadening

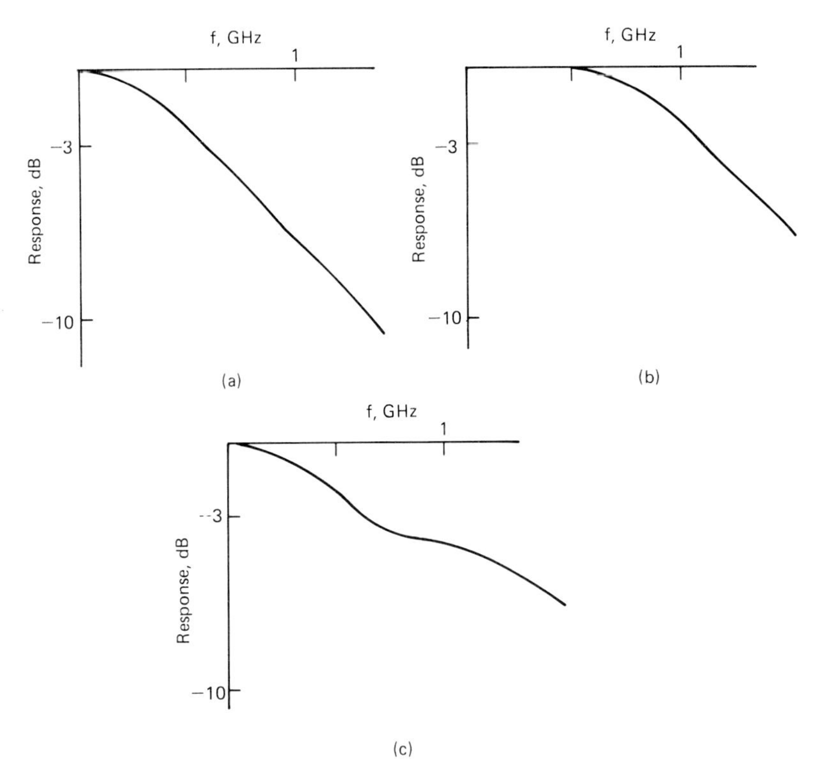


**FIGURE 4.19** Baseband response versus modulation frequency $f$ (including source and detector effects) for (*a*) long fiber, (*b*) shortened fiber, and (*c*) fiber without source and detector effects.

$$t_F = (t_L^2 - t_0^2)^{1/2} \qquad \text{ps} \tag{4.20b}$$

The 3-dB optical bandwidth is the lowest frequency $f_3$ for which the power has decayed in half, i.e., $dB_F(f_3) = -3$. (This is equivalent to the $-6$-dB electrical bandwidth, as the photodiode is a square-law detector. The electrical power across a load resistor is proportional to the square of the photocurrent, and that photocurrent is proportional to the optical power.) Some instruments may use the full-fiber transfer function $H_F(f)$ [of which $dB_F(f)$ is only the amplitude] necessary to compute the impulse response $h(t)$. This is convenient for observing "humps" in the temporal response corresponding to mode groups that may be arriving before or after the main pulse. Sometimes the frequency response is fit to a gaussian curve. Then the 3-dB reciprocal bandwidths follow a rule similar to Eq. (4.20*b*), and the time and frequency domains are related by the rule of thumb

$$f_3\ (\text{MHz}) \cdot t_F\ (\text{ns, FDHM}) = 440 \tag{4.21}$$

where FDHM indicates a full-duration, half-maximum measurement.

The frequency-domain measurement is covered in FOTP-30. There, a swept-frequency generator is used to drive the laser diode, and the radio-frequency (RF) amplitude after the detector is displayed. With cutback or insertion procedures similar to those for the time-domain, a result analogous to Eq. (4.20*a*) is obtained. Phase information may be retained with the use of a vector voltmeter or a network analyzer, but this is usually inconvenient; the results are not usually transformed into the time domain. The field version of the procedure is given in OFSTP-1.

***Wavelength and Length Effects.*** Because the refractive index profile varies slightly with wavelength and because multimode bandwidth is very sensitive to profile, the bandwidth varies with wavelength. Usually, bandwidth is designed to peak around 1300 nm (with values exceeding 3 GHz over a kilometer), or around 850 nm, or at both regions (in the last two cases, with lower bandwidths). Because of large core-cladding index differences, fibers with a larger NA (Sec. 4.5.3) tend to have smaller bandwidths.

Fiber bandwidth may vary with fiber length, even along a uniform fiber. This is because fiber scatter described in "Absorption and Scattering," Sec. 4.3.1, causes mode mixing that tends toward equalizing the differential mode delay, giving pulse spreading that is sublinear with length. The effect is more acute for higher NA and higher-attenuation fibers, especially if the operating wavelength is far from the maximum-bandwidth wavelength. Ideally the bandwidth and pulse spreading behave as

$$f_3(L) = f_3(1)L^{-\gamma} \tag{4.22a}$$

$$t_F(L) = t_F(1)L^{\gamma} \tag{4.22b}$$

where $L$ is in kilometers. One extreme value for $\gamma$ is unity, which indicates no mode mixing; in this case, bandwidth is given in units of megahertz-kilometers and broadening in units of nanoseconds per kilometer. The other extreme is $\gamma \rightarrow ½$ (units of $\text{MHz} \cdot \text{km}^{1/2}$ and $\text{ns/km}^{1/2}$), which indicates complete mode mixing. Figure 4.20 shows the straight lines that result from a log-log plot of Eq.

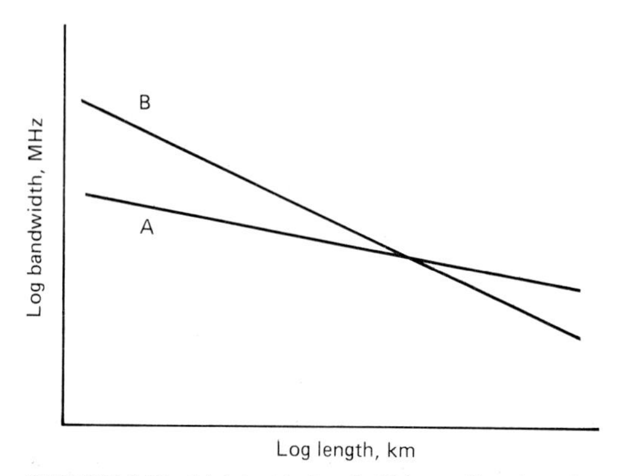


**FIGURE 4.20** Multimode bandwidth vs. fiber length. Compared to fiber *A*, fiber *B* has a higher NA and attenuation, and has a less than optimum profile.

(4.22*a*) for both types of fiber. Such measurements of length dependence are very tedious and require fiber destruction by multiple cutbacks.

The concatenation effects of "Differential Mode Delay," Sec. 4.4.2, are also length-dependent. In this case, the individual bandwidths $f_3(n)$ of the individual fibers (regardless of lengths $L_n$) are added to give the link bandwidth

$$f_3 = \left\{ \sum_n [f_3(n)]^{-1/\gamma} \right\}^{-\gamma} \tag{4.23}$$

Still another form for nearest-neighbor fibers is

$$f_3 = [f_3^{-2}(1) + f_3^{-2}(2) + 2r_{12} f_3^{-1}(1) f_3^{-1}(2)]^{-1/2} \tag{4.24}$$

where $r_{12}$ is a correlation coefficient. The coefficient is negative for fibers of opposite compensations at the joint, and positive for similar compensations. The absolute value approaches 1 without mode mixing, and 0 with mixing. Such multimode measurement exercises are of limited interest now with the emergence of single-mode fiber applications, but may find new relevance in low-capacity, short-haul, multimode networks.

Finally, these methods have application for single-mode fiber operated below the cutoff wavelength with two or three mode groups ("Bimodal Effects" in Sec. 4.3.2). Figure 4.21 shows the time-domain and frequency-domain behavior of single-mode fiber operated in bimodal transmission.

### 4.4.3 Chromatic Dispersion

The preceding multimode effects occur even with monochromatic sources. A real source, however, has some wavelength spread, and each wavelength has a different delay per unit length $\tau(\lambda)$ down the fiber. This occurs in both multimode and single-mode fiber.

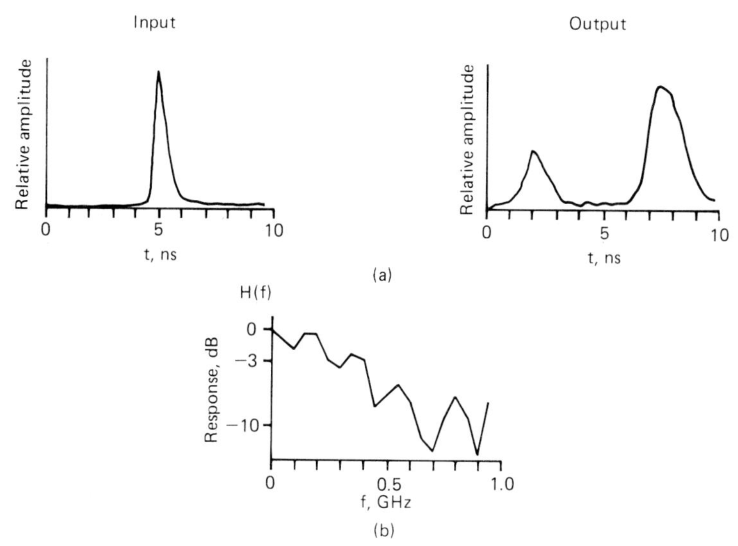


**FIGURE 4.21** Bimodal propagation at 825 nm over 1.5 km of fiber that is single-mode around 1300 nm. (*a*) Input pulse and output pulses; (*b*) fiber frequency response.

***Cause and Effect.*** A practical source has some wavelength (or optical frequency) spread $\Delta\lambda$ (in nanometers), and each wavelength (or frequency) component has a different group delay $\tau(\lambda)$ per unit length (about 9.8 μs/km or ns/m) down the fiber. This means that, analogously to Eq. (4.20*b*) for multimode distortion, a pulse down a fiber length $L$ spreads by

$$\sigma(\lambda) = L\,\Delta\lambda\,[D^2(\lambda) + \tfrac{1}{2}S^2(\lambda)\cdot(\Delta\lambda)^2]^{1/2} \qquad (4.25)$$

where $\sigma$ and $\Delta\lambda$ are rms quantities. Here

$$D(\lambda) = \frac{d\tau}{d\lambda} \qquad \text{ps/(nm}\cdot\text{km)} \qquad (4.26a)$$

is the chromatic dispersion coefficient, and

$$S(\lambda) = \frac{dD}{d\lambda} \qquad \text{ps/(nm}^2\cdot\text{km)} \qquad (4.26b)$$

is the dispersion slope.

Measurements of chromatic dispersion do not usually separate the interrelated mechanisms causing the effect. The mode has a phase index (or effective refractive index), $n(\lambda)$, from which

$$N(\lambda) = n - \lambda\frac{dn}{d\lambda} \qquad \tau(\lambda) = \frac{N}{c} \qquad (4.27)$$

where $N$ is the group index. One mechanism is the material dispersion of the refractive index variation for the base silica glass forming the fiber. The smaller effect of profile dispersion relates to the relative index variations of the doped silicas forming the index profile. The intermediate effect of waveguide dispersion relates to the ratio of profile dimensions to photon wavelength. The latter two effects can be manipulated to yield fibers with tailored dispersion properties.

It is important to realize that chromatic dispersion occurs in both single-mode and multimode fibers. For the latter, in the 1300-nm region, it may not be as important as multimode distortion, especially if the source spectral width is small. Nevertheless, modal effects such as launching conditions can influence measurement results.

***Measurement.*** Measurement methods for chromatic dispersion have included (1) direct pulse broadening with very short pulses and long fibers, (2) frequency sweeping, (3) heterodyning and coherence, and (4) interferometry over fiber only a few meters long (a measurement of interest more to fiber manufacturers). Standards and commercial equipment have evolved along the lines described in the following discussion.

Multiwavelength sources are required. For the time-domain measurements of FOTP-168, the source used to be a pulsed Q-switched, mode-locked neodymium-doped:yttrium aluminum garnet (Nd:YAG) laser and a nonlinear Raman fiber. Today, the source is usually composed of multiple laser diodes, possibly with wavelength tuning by temperature variation. For the frequency-domain measurements of FOTP-169 and FOTP-175, two or more LEDs with variable filters may be used. A complete scan can be done in less than a minute to minimize length errors due to temperature fluctuations.

In Fig. 4.22 (FOTP-168) the pulse delay vs. wavelength is measured for both the long and short fiber. If $\Delta T(\lambda)$ is the delay difference for the length difference $L$, then the group delay per unit length is

$$\tau(\lambda) = \frac{\Delta T(\lambda)}{L} \tag{4.28}$$

The chromatic dispersion coefficient in Eq. (4.26*a*) follows by differentiation.

To obtain the phase delay $\phi(\lambda)$ vs. wavelength (FOTP-169), the pulse generator is replaced by a high-frequency oscillator at a constant modulation frequency $f$ and the delay generator and oscilloscope are replaced by a phase meter or vec-

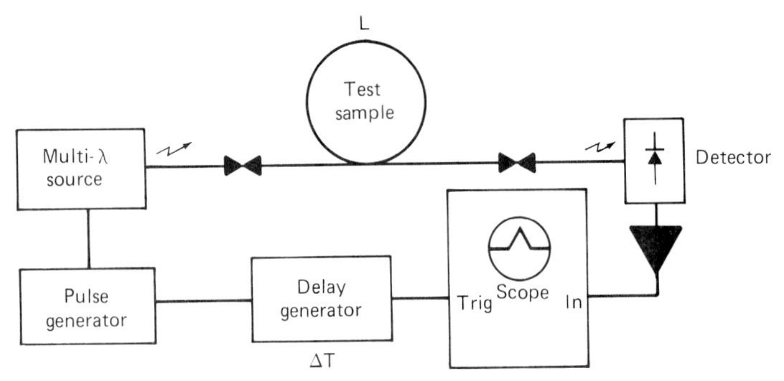


FIGURE 4.22 Configuration for measuring chromatic dispersion by time delay.

tor voltmeter. An electrical or optical reference channel is connected between the oscillator and meter. The group delay is obtained from

$$\phi(\lambda) = 2\pi f L \tau(\lambda) \tag{4.29}$$

and differentiated as before.

The wavelength scan may be chosen to use wavelength pairs closely spaced by $\delta\lambda$ around each central $\lambda$. If the corresponding differential delay $\delta\tau(\lambda)$ for each pair is recorded via Eq. (4.28) or (4.29), then the chromatic dispersion coefficient may be more directly given by

$$D(\lambda) \cong \frac{\delta\tau(\lambda)}{\delta\lambda} \tag{4.30}$$

Other implementations of direct chromatic dispersion measurements are covered by FOTP-175. One of these utilizes the simultaneous transmission of the wavelength pair to two detectors as in Fig. 4.23*a*. The other utilizes low-frequency modulation between the two wavelengths to a single detector, as in Fig. 4.23*b*.

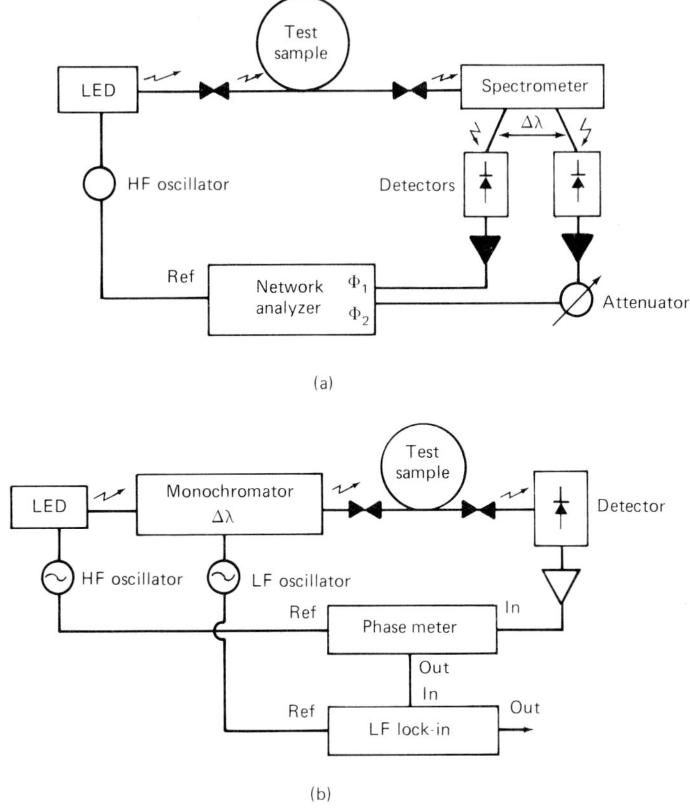


**FIGURE 4.23** Configuration for measuring direct chromatic dispersion by means of (*a*) dual wavelengths and (*b*) double modulation.

*Curve Fits.* For purposes of fiber specification and system calculation, analytical expressions for the dispersion coefficients are useful. For *dispersion-unshifted* (EIA Class IVa) fiber, the spectral group delay is fitted to the three-term Sellmeier expression

$$\tau(\lambda) = \tau(\lambda_0) + \frac{S_0}{8}\left(\lambda - \frac{\lambda_0^2}{\lambda}\right)^2 \tag{4.31a}$$

Here $\lambda_0$ is the zero-dispersion wavelength and $S_0$ is the zero-dispersion slope. The dispersion coefficient (from the preceding or by direct fit) is

$$D(\lambda) = \frac{S_0}{4}\left(\lambda - \frac{\lambda_0^4}{\lambda^3}\right) \tag{4.31b}$$

Figure 4.24 shows the spectral group delay and chromatic dispersion coefficient for a representative concatenated fiber. The fiber's chromatic dispersion is specified by

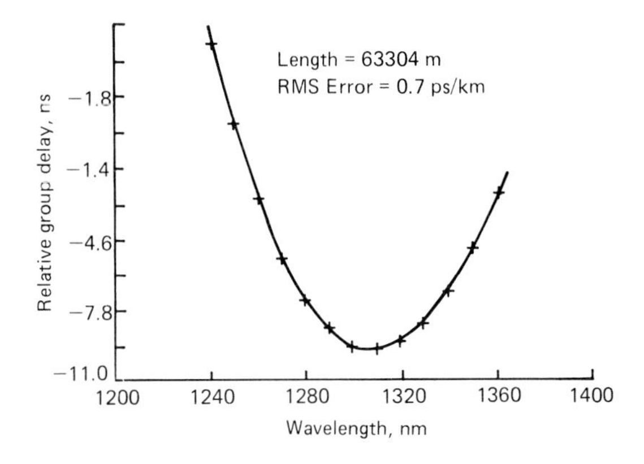


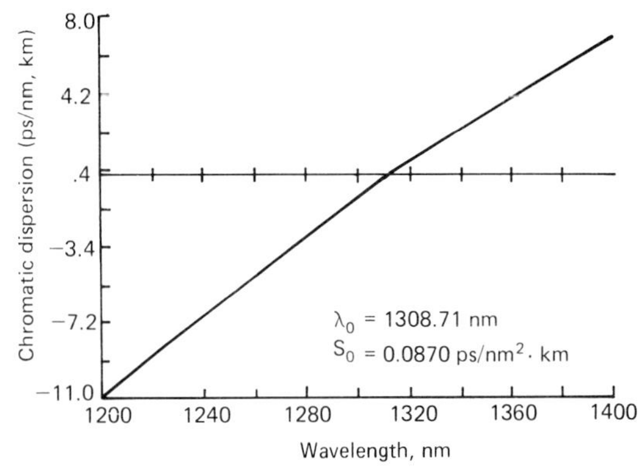


**FIGURE 4.24** Group delay and chromatic dispersion for a 63-km dispersion-unshifted fiber.

$$S_{0max} = 0.095 \text{ ps/nm}^2 \cdot \text{km} \tag{4.31c}$$

$$\lambda_{0min} = 1295 \text{ nm or } 1300 \text{ nm} \qquad \lambda_{0max} = 1322 \text{ nm}$$

For *dispersion-shifted* (EIA Class IVb) fiber, the corresponding expressions are

$$\tau(\lambda) = \tau(\lambda_0) + \frac{S_0}{2}(\lambda - \lambda_0)^2 \tag{4.32a}$$

$$D(\lambda) = (\lambda - \lambda_0)S_0 \tag{4.32b}$$

Typically, $S_0$ ranges from 0.055 to 0.085 ps/(nm$^2$ · km) and $\lambda_0$ nominal is 1550 nm, but specifications have not yet been standardized.

For *dispersion-flattened* (EIA Class IVb) fiber, two or more values of $\lambda_0$ may occur. A tentative curve fit is the five-term Sellmeier equation

$$\tau(\lambda) = A\lambda^4 + B\lambda^2 + C + D\lambda^{-2} + E\lambda^{-4} \tag{4.33}$$

and its derivative.

## 4.5 SPLICE AND CONNECTOR PROPERTIES

### 4.5.1 Intrinsic and Extrinsic Factors

The fiber joint (splice or connector) influences a fiber link in several ways. The insertion loss will affect the end-to-end power budget, while the return loss will affect transmitter and receiver performance. Distortion can affect end-to-end bandwidth or dispersion. The causes of these are classified into two general types. *Intrinsic* effects result from fiber imperfections, while *extrinsic* effects result from splice or connector imperfections. The relative importance of these effects depends upon the method used in jointing, i.e., fusion splicing, mechanical splicing or connectorizing, or inserting lensed connectors (see Chap. 3).

Examples of intrinsic effects are fiber mismatches on either side of the joint. Geometrical aspects include cladding noncircularity and core-cladding eccentricity. For multimode fibers, the important factors are core noncircularity and mismatches in index power profile, core diameter, and numerical aperture (producing losses that differ bidirectionally). For single-mode fibers, mode field noncircularity and mode field diameter mismatch are important (producing losses that are theoretically bidirectionally equal).

Examples of extrinsic effects are poor fiber ends (rough or angled), longitudinal end separation, transverse or lateral offset of the fiber axes, and angular misalignment of the axes.

We will now cover measurement of these various intrinsic and extrinsic factors and measurement of their effects on insertion loss, return loss, and distortion. Optical time-domain reflectometry techniques are treated in Sec. 4.6.

### 4.5.2 Fiber Geometry

The measurement of fiber geometry is treated in FOTP-176. For graded-index multimode and for single-mode fibers, this procedure covers cladding diameter, cladding noncircularity, and core-cladding concentricity. It also covers core noncircularity for graded multimode fiber only.

In this procedure, the fiber is typically a few meters long, without sharp bends, and is overfilled with light at the input end. At the output end, light silhouettes the cladding; for example, white light can be reflected off a mirror about 10 mm from the fiber end at about a 10° angle from the perpendicular to the fiber axis. An objective lens magnifies (20 to 80 ×) the output end and, in a fully automated setup necessary for the method, sends the image to a video camera, video monitor display (with cross hairs), video digitizer, and computer.

A master scan of the output image cross section is made to identify the edges of the core and cladding. In either cartesian coordinates or polar coordinates, numerical routines are performed to find the centers $\mathbf{r}_c$ and $\mathbf{r}_g$ of the circles giving the best fit to the core and cladding (glass), respectively. Iterative calculations determine the maximum, minimum, and average radii of the core [$R_{c\min}$, $R_{c\max}$, $R_c = \frac{1}{2}(R_{1\min} + R_{1\max})$] and of the cladding ($R_{g\min}$, $R_{g\max}$, $R_g$). The following are calculated:

$$\text{Cladding diameter} = 2R_g \tag{4.34a}$$

$$\text{Cladding noncircularity} = \frac{R_{g\max} - R_{g\min}}{R_g} \times 100\% \tag{4.34b}$$

$$\text{Core-cladding concentricity error} = |\mathbf{r}_g - \mathbf{r}_c| \quad \text{single mode} \tag{4.34c}$$

$$= |\mathbf{r}_g - \mathbf{r}_c| \times \frac{100\%}{2R_c} \quad \text{multimode} \tag{4.34d}$$

Core noncircularity (multimode fiber)—as in Eq. (4.34*b*)

Other, less sophisticated methods can also be used. In FOTP-45 a short (25-mm) fiber sample is mounted vertically to provide an end view through a microscope (100 to 400 ×). Manual measurements similar to those of FOTP-176 are performed on the image (using a calibrated reference scale) or on a photograph of the image. FOTP-27 allows for cladding diameter measurement by mechanical calipers or by side-view or end-view microscopy. More sophisticated methods, involving Michelson and Fizeau interferometry, are also being investigated. This is in response to user needs for tighter fiber geometrical tolerances (in splice and connector assemblies) and the means of measuring them.

Note that core radii are not fully examined in the preceding discussion. For single-mode fiber, core parameters (and numerical aperture) are not specified for the user, but are replaced by the mode field diameter of Sec. 4.5.4 (which is measured at the wavelength of operation). For multimode fiber, core diameter is obtained more accurately and repeatably using methods in the next section.

Coating geometry is covered in FOTP-55. A short fiber sample is examined through a microscope; the end is viewed directly, viewed on a screen, or photographed. Four-coating, outside diameter scans are taken 45° apart, and the jacket diameter is calculated as the average of these. Coating noncircularity is determined by using orthogonal diameters, and is determined as the larger of the two noncircularities so calculated. Coating-cladding concentricity is determined similarly.

Manufacturers use on-line methods for monitoring glass and jacket dimensions during the fiber draw and use the monitoring data for feedback control. Usually laser light beam scattering is employed. These measurements are not usually performed by the user, and are therefore not discussed here.

### 4.5.3 Core Diameter and Numerical Aperture

The parameters of core diameter and numerical aperture are useful for multimode fiber only. Both of these parameters are treated in "umbrella" FOTPs that allow several other FOTPs to be used for implementation. Core diameter and numerical aperture are useful for calculations involving source-to-fiber, fiber-to-fiber, and fiber-to-detector couplings.

FOTP-58 treats the core diameter derived from $n(r)$, as obtained in Sec. 4.4.1, "Index Profile," via transverse interference, refracted ray scanning, or near-field radiation. In either of the two index methods, the curve between the 10 and 80 percent intensity points is fitted to the index contrast

$$\begin{aligned}\Delta n(r) &= n(r) - n_2 \\ &= \Delta n(0)\left[1 - \left(\frac{r}{a}\right)^g\right]\end{aligned} \tag{4.35}$$

where $n_2$ is the cladding index as in Fig. 4.25*a*. The best-fit parameters are the on-axis value $\Delta n(0)$, the core "radius" $a$, and the particular value of the profile power-law parameter $g$ ($\approx 2$). The core diameter $D$ is calculated at the 2.5 percent value of $\Delta n(0)$, i.e.,

$$D = 2a(0.975)^{1/g} \approx 1.975a \tag{4.36}$$

For the near-field method of Fig. 4.25*b*, $D$ is simply taken equal to $2a$. Alternatively, no curve fit need be used and $0.025\Delta n(0)$ is found directly. The equivalence of these methods in giving closely equal values of core diameter for a variety of multimode fiber profiles has been established by an EIA/NIST (National Institute of Standards and Technology) measurements round-robin.

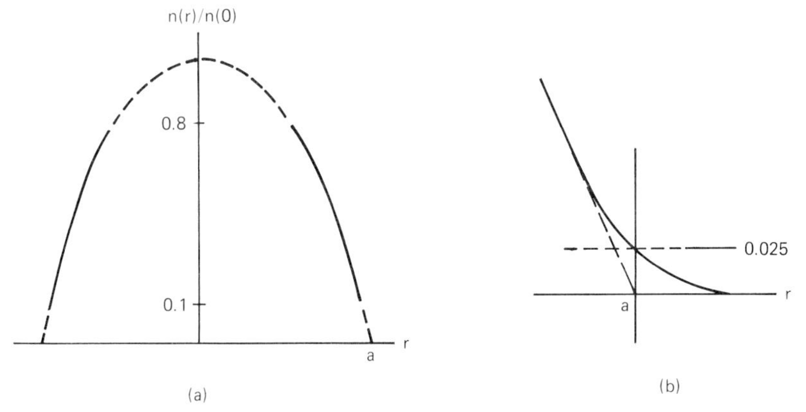


**FIGURE 4.25** Multimode fiber core diameter determination by (*a*) index profile fitting (limited region) and (*b*) near-field point (solid curve data).

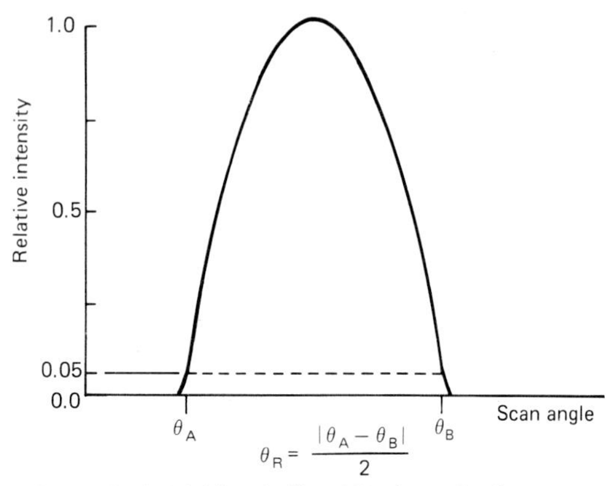


**FIGURE 4.26** Multimode fiber NA determination.

FOTP-177 treats numerical aperture derived from far-field radiation (Sec. 4.2.3) or from the refractive index profile (Sec. 4.4.1) obtained via transverse interference or refracted ray scanning. For the far-field radiation shown in Fig. 4.26,

$$NA = \sin \theta_5 \tag{4.37}$$

where $\theta_5$ is the 5 percent intensity half-angle; the measurement is taken at 850 nm. For the index profile methods, the power-law curve fit of Eq. (4.35) is used. Then

$$NA = K \frac{n_2 \sqrt{2\Delta(0)}}{1 - \Delta(0)} \tag{4.38}$$

where

$$\Delta(0) = \frac{\Delta n(0)}{n_2}$$

The silica cladding index $n_2$ is 1.460 at 563 nm and 1.457 at 633 nm. The corrective $K$ factor has been determined by another EIA/NIST round-robin.

Both core diameter (CD) and NA are important factors describing source-to-fiber coupling and fiber-to-fiber coupling. Such coupling is proportional to the *mode volume* defined by $(\text{CD} \times \text{NA})^2$. For fiber joints there will be no extrinsic losses (apart from reflection) in going from a fiber of CD = 50 μm, NA = 0.21 to a fiber of CD = 62.5 μm, NA = 0.27 (i.e., from one fiber to another of higher mode volume). However, there will be a loss in the reverse direction. For source-to-fiber coupling, the latter fiber will generally accept more light, especially from an LED that characteristically has a wide emission angle and sometimes a large spot size.

### 4.5.4 Mode Field Diameter

*Usefulness.* The mode field diameter (MFD) parameter is for single-mode fiber only; it replaces the CD and NA parameters of multimode fiber. It is usually given the symbol $2w_0$ and relates to several transmission properties. A small MFD generally means smaller bending losses ("Bend Losses" in Sec. 4.3.1), since less of the cross-sectional optical power resides in the outer, evanescent portion of the bend. Conversely, it generally means greater jointing difficulty. For example, intrinsic loss due to an offset $x$ of the fiber axes is

$$\text{Transverse loss} = 4.343\left(\frac{x}{w_0}\right)^2 \quad \text{dB} \tag{4.39}$$

Extrinsic loss due to a relative mismatch $R$ of the mode diameters $2w_1$ and $2w_2$ of the two fibers is

$$\text{Mismatch loss} = 4.343R^2 \quad \text{dB} \tag{4.40a}$$

where

$$R = \frac{2(w_1 - w_2)}{w_1 + w_2} \tag{4.40b}$$

Spectral MFD can be related to waveguide dispersion.

Values of MFD range from about 8½ to 10 μm for dispersion-unshifted fiber around 1310 nm. Values at 1550 nm are about 13 percent larger, and this relates to the greater bend sensitivity there. Values for dispersion-shifted fiber range from 7 to 8½ μm at 1550 nm. Since some optical power resides outside the core, these are larger than core diameter values (which are not relevant for determining splice loss or bending loss for single-mode fiber).

*Methods.* Especially for dispersion-unshifted fibers around 1300 nm, the optical mode field has the gaussian near-field intensity

$$N(r) = N(0)e^{-2r^2/w_0^2} \tag{4.41}$$

A straightforward method of measuring MFD, therefore, is the near-field scan technique of FOTP-165. Since this method is commonly performed with a vidicon (Sec. 4.2.3), it is less convenient than the far-field scan of FOTP-164. The far-field intensity

$$F(\theta) = F(0)e^{-2\theta^2/\theta_w^2} \tag{4.42a}$$

from which one obtains the MFD

$$2w_0 = \frac{2\lambda}{\pi \tan \theta_w} \tag{4.42b}$$

is the spatial Fourier transform (in field) of the near-field intensity. In both cases the gaussian fitting functions are obtained by maximizing the integral overlap of the function with the data as outlined in the FOTPs.

Another near-field method, in CCITT G.650, relies on transverse offset loss at a mechanical splice or connection. To avoid growth of the field from the near field into the Fresnel region, the fiber-end faces must be separated less than 5 μm and index-matching material is used to prevent reflection effects. The power transmittance is fitted to

$$T(x) = T(0)e^{-x^2/w_0^2} \tag{4.43}$$

which is a generalization of Eq. (4.39) to larger offsets.

There are two more far-field methods. In FOTP-167, a wheel of circular apertures of various sizes is placed between the fiber end and collection optics leading to a detector. The measured light power passed through each aperture vs. the NA is fit to the expression

$$P(\theta) = P_\infty(1 - e^{-2[(\pi w_0/\lambda)\tan\theta]^2}) \tag{4.44}$$

In a variation of this technique, an aperture of fixed diameter slides longitudinally between the fixed fiber and detector, effectively varying the aperture NA. Finally, FOTP-174 utilizes a knife edge in the far field. The edge is a distance $D$ from the fiber and is scanned. The power vs. edge position $x$ is fitted to the complementary error function

$$T(x) = T(x_0)\ \mathrm{erfc}\,[A(x - x_0)] \tag{4.45a}$$

This determines $x_0$, the center of the pattern, and parameter $A$. The MFD is then

$$2w_0 = \sqrt{2}A\lambda\frac{D}{\pi} \tag{4.45}$$

*Nongaussian Measurements.* The gaussian simplicity does not apply at 1550 nm, especially with dispersion-shifted fibers. In that case the rms far-field, or Petermann II, definition has been adopted by standards bodies and it will soon apply to all fibers. For the far-field scan method of FOTP-164, the MFD is

$$2w_0 = \frac{\sqrt{2}}{\pi}\left[\frac{\int_0^{\pi/2} P_m(\theta)\sin\theta\cos\theta\,d\theta}{\int_0^{\pi/2} P_m(\theta)\sin^3\theta\cos\theta\,d\theta}\right]^{1/2} \tag{4.46}$$

where $P_m(\theta)$ is the measured data. Typical data are shown in Fig. 4.27; for dispersion-shifted fiber, accurate results require scans below the −30-dB level relative to the center. Scan angles that are double those for dispersion-unshifted fibers are necessary.

For the integrated far-field methods, reflective optics are often utilized to collect all the light. For the variable aperture in FOTP-167, the nongaussian definition is

$$2w_0 = \pi\lambda\left[\int_0^{\pi/2} a(\theta)\sin\theta\cos\theta\,d\theta\right]^{-1/2} \tag{4.47a}$$

where the complementary aperture function analogous to Eq. (4.44) is

$$a(\theta) = 1 - \frac{P(\theta)}{P_\infty} \tag{4.47b}$$

For the knife-edge far field in FOTP-174, there is little commercial interest to date in developing a nongaussian procedure, although it has been used in the lit-

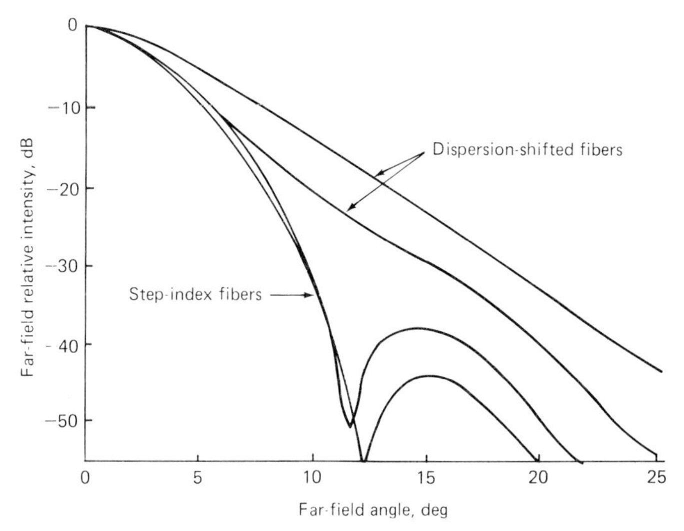


**FIGURE 4.27** Far-field scans for mode field diameter on two fiber types.

erature. The same can be said for the two near-field techniques, the scanning in FOTP-165 and the transverse offset.

### 4.5.5 Insertion Loss

The power across a joint or connector is affected by a number of intrinsic and extrinsic factors, as outlined in Sec. 4.5.1. A number of measurement conditions also affect the power, especially those related to modal conditions of multimode fiber or single-mode fiber near cutoff.

Fiber joints are sensitive to the modes that enter, and they affect the modes that leave. With multimode fibers, differential mode attenuation ensures that joint losses will generally be larger with full excitation than with underexcitation. This is even more true with attenuation. In a practical system, this sensitivity relates to the length of fiber preceding the joint to a source or earlier joint; differences in this length show up in both long-haul and local network systems. For measurement of insertion loss, either full excitation or an equilibrium mode simulator should be specified for repeatable results. Radiative modes are usually excited at a joint, and these contribute to the joint loss. This means that the insertion loss is composed of a local loss just across the joint, plus a transient loss ("Modal Effects" in Sec. 4.3.1) distributed along the length following the joint. With a long fiber between the joint and the receiver, a smaller apparent total loss will result than if a short fiber were used there.

Figure 4.28 illustrates the distinction between local and distributed joint losses and their dependence upon launch conditions. The effects of long and short transmitting fibers between the source and joint are shown in Fig. 4.29. Note that with the long fiber, there is much less sensitivity to the launch numerical aperture.

Also affecting splice and connector loss is source coherence, as already mentioned in "Bimodal Effects" in Sec. 4.3.2 for single-mode fibers. If the coherence

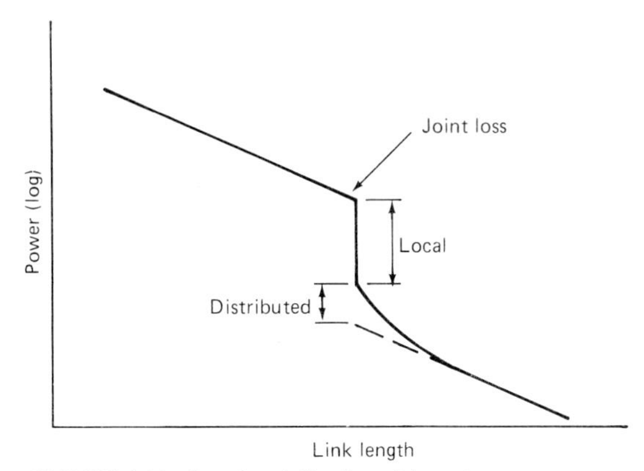


**FIGURE 4.28** Local and distributed (transient) losses and their launch dependence.

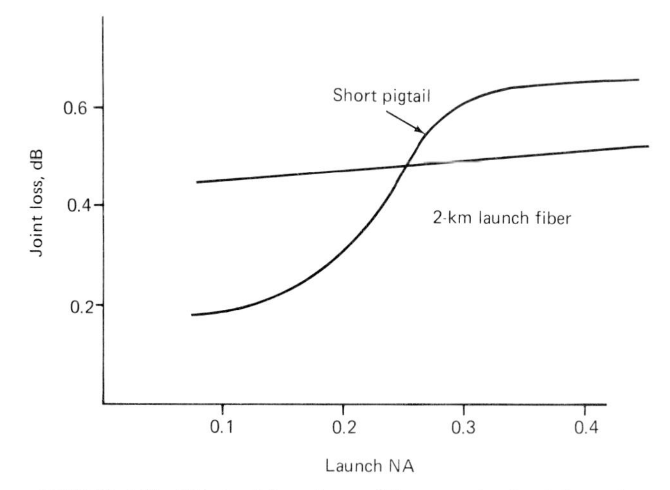


**FIGURE 4.29** Effect of launch conditions on the local loss of a multimode mechanical splice.

time of the source exceeds the multimodal pulse broadening discussed in "Time and Frequency Domains" in Sec. 4.4.2, then the input fiber will yield a beam with a speckle pattern at the joint. (This is more likely for a joint nearer the transmitter, within the coherence length of the source.) The output fiber will receive most of these speckles if the loss is low. However, the nature of the speckle pattern (and therefore the loss) can vary with source modulation or with mechanical or thermal fluctuations of the fiber or joint, leading to a phenomenon called *modal noise*.

With single-mode fiber, these modal effects can be avoided if the fiber or cable cutoff wavelength, as measured according to Sec. 4.3.2, is below (or only slightly above) the transmitter wavelength. Bimodal effects can be even more serious than multimode effects since the second-order mode can carry a significant fraction of total power. Because of this and because of the spectral variation of mode field diameter, joint losses are a more sensitive function of wavelength for single-mode fiber than for multimode fiber.

There are a variety of configurations for measuring insertion loss. With respect to input-output fiber lengths and generally decreasing joint loss, these are short-long, short-short or long-long, and long-short. Figure 4.30 shows a cutback technique combining the last two conditions. Here power $P_3$ is measured after the receiving fiber (optional), power $P_2$ is measured just after the splice, and power $P_1$ is measured just before the splice. Then, in decibels,

$$\text{Local loss} = 10 \log \frac{P_1}{P_2} \tag{4.48a}$$

$$\text{Transient loss (optional)} = 10 \log \frac{P_2}{P_3} \tag{4.48b}$$

The total splice loss is the sum of these.

The same variety of configurations can be used in insertion or substitution techniques. According to FOTP-34, a power measurement is taken before the joint is made, and another taken after the joint is made. The before/after power ratio then yields the loss.

With single-mode connectors, Bellcore TR-TSY-000326 (see References) prescribes a concatenation test in which the additional loss due to the insertion of an increasing number of connectors separated by short (2- to 3-m) jumper cables is measured. In another measurement, a jumper cable is prepared with each connector half. The source is coupled to the input launch jumper, and the power out of the connector half is measured. The other connector half is then put on, and the output test jumper gives a second power reading at the detector. The absolute power ratio (or logarithmic power difference in decibels) yields the connector insertion loss.

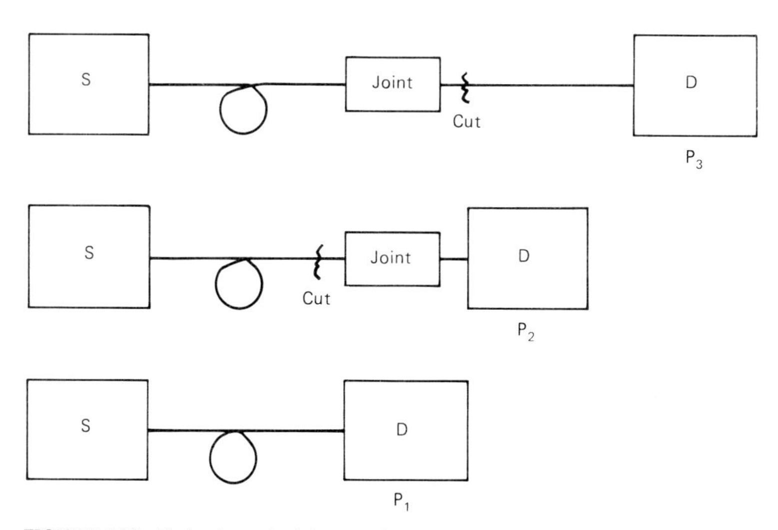


**FIGURE 4.30** Cutback method for transient and local losses.

Other noninvasive means of measuring joint loss have been developed, but these have not been standardized. With macrobend or microbend techniques, for example, light can be locally launched into the fiber just ahead of the joint. In like manner, light can be extracted just before and just after the joint. A comparison of the before and after powers can be used to optimize a splicing procedure. In another technique, scattered unguided power rather than guided power is observed. The scattered power is much more sensitive to joint loss. Calibration relating the scattered partial power to loss is necessary. Finally, OTDR methods are covered in Sec. 4.6.4.

### 4.5.6 Reflectance and Return Loss

Reflections along a fiber link can adversely affect laser stability, and multiple reflections can add signal noise to the detector. They can occur at a butt joint or lensed joint but are unlikely at a fusion joint. At a fiber-air interface, the Fresnel reflectance $R$, not to be confused with mode diameter mismatch in Eq. (4.40$b$), is about 3.5 percent, or $-14.6$ dB. The optical return loss is defined as

$$\mathrm{ORL} = -10 \log R \tag{4.49}$$

For the two fiber-end faces, this corresponds to an ORL of 14.6 dB. Reflectance is the term sometimes preferred for single components, whereas return loss applies to a series of components (including fiber) along a link.

Various mechanisms can cause the reflectance to be even larger. In one mechanism, interference is produced in the cavity formed between the two fiber endfaces, and this causes the reflectance to vary periodically with longitudinal displacement of the faces and with the transmitter wavelength. This alternately constructive and destructive interference has a period of $\lambda/2$. It is especially prominent with coherent laser sources and can add 3 dB to the reflectance (see Chap. 3). Another effect occurs if the cleaved ends are further polished. In this case, a thin higher-index layer forms on the end face to increase reflection. (Fusion may anneal this out.) The combined two effects can yield reflectances over 22 percent, or $-6.6$ dB, or optical return losses under 6.6 dB.

Reflectance values are low with fusion splicing and with carefully designed mechanical joints. Index-matching material can greatly reduce reflection, but some applications rule out index fluid or gel because of handling or lifetime degradation problems. Fiber-to-fiber physical contact joints also attempt to eliminate the index discontinuity, but end-face damage is a concern. Another method is to cut or polish one or both end faces at an angle to the fiber axis such that reflections are directed at angles away from the fiber axis. The tradeoff here is in slightly increased insertion loss, so index fluid is sometimes also used. By whatever method, optical return losses above 40 dB (reflectances below 0.01 percent) are readily achievable, and these levels appear to have no deleterious effects on fiber link performance.

Reflectance measurements are typically done in one of two ways. An optical continuous-wave reflectometer (OCWR) is described in FOTP-107 and TR-TSY-000326. As shown in Fig. 4.31, a 2 by 2 coupler is used. A continuous-wave LED or laser diode source is connected to input port $a$ while a detector is connected to input port $b$. A jumper cable with the reflecting components under test is spliced (with a high optical return loss) to output port $c$. The output port $d$ is made nonreflecting with an index-matching gel or tight fiber loop. The power $P_R$ at port

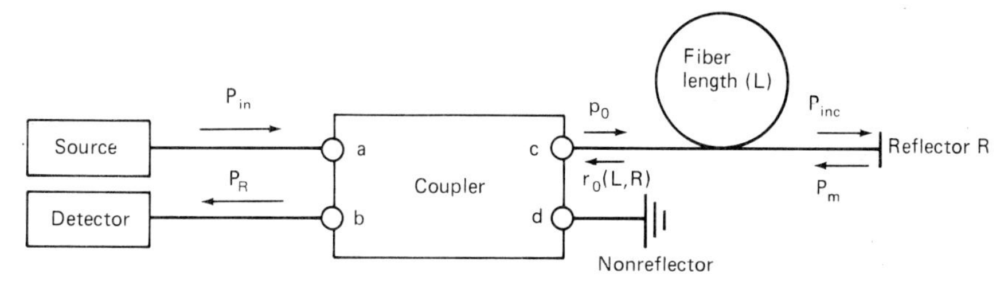


FIGURE 4.31 Optical return loss using a coupler.

*b*, resulting from reflections by the component and the coupler, is measured. The jumper is removed and replaced by a nonreflecting termination; power $P_0$ due only to the coupler is measured. Then the detector is moved to port *c* to measure the power $P_{inc}$ incident upon the reflector.

The fraction of reflected power would be $(P_R - P_0)/P_{inc}$ except that the loss in traveling from port *c* to port *b* must be accounted for. The loss in this direction is measured by connecting the source to port *c* and the detector to port *b* and measuring $P_{out}$. Then $P_{in}$ is measured by connecting the source directly to the detector so that $P_{out}/P_{in}$ is the fraction of transmitted power along the path from port *c* to port *b*. Finally, the optical return loss of the components (including the fiber) under test is

$$\text{ORL} = 10 \log \frac{P_{out}P_{in}}{P_{in}(P_R - P_0)} \tag{4.50}$$

The optical continuous-wave reflectometer is a DC instrument and cannot therefore provide information on the location or loss of possibly multiple reflecting components in a link. The OTDR can be used to do this ("Reflectance" in Sec. 4.6.4), but this is more complex and expensive.

## 4.6 OPTICAL TIME-DOMAIN REFLECTOMETRY

### 4.6.1 Instrument Operation

The optical time-domain reflectometer has proved to be a most useful laboratory and portable field instrument. It has the advantages of displaying information from along the fiber length and of requiring access to only one fiber end for most applications. Many OTDR requirements and objectives are given in Bellcore Technical Reference TR-TSY-000196 (see References).

The OTDR has a laser diode transmitter that is coupled to send rectangular light pulses into the fiber. The fiber scatters light ("Absorption and Scattering" in Sec. 4.3.1) continuously along its length, to a level proportional to the pulse energy. Part of this light is back-scattered in a reverse direction, along with light reflected from discrete points and having essentially the input pulse shape. The light heading back toward the transmitter is intercepted by a coupler or switch and directed to a photodiode receiver.

The pulse durations range from 0.1 ns to 10 μs in a tradeoff between resolution (corresponding to 1 cm to 1 km) and backscatter range, which is energy-dependent. The pulses are repetitive to assist in signal processing (and thereby enhance the signal-to-noise ratio), but the modulation period is no less than one round-trip time in the fiber (about 10 μs/km). Various coding techniques may also be employed. A display plots optical power as half the actual decibels vs. half-time or fiber distance; the halves correct for the round-trip effect.

### 4.6.2 Length Calibration

The OTDR instrument requires that one enter the group index $N$ ("Cause and Effect" in Sec. 4.4.3) or group delay per unit length or equivalent for the particular test sample fiber or cable being used. This value may be supplied by the manufacturer; otherwise it can be measured according to FOTP-60. It should not be calculated, since it depends upon the wavelength, the refractive index profile (for undoped silica, $n = 1.447$ and $N = 1.462$ around 1300 nm), modal conditions (in multimode fiber), and cable length (if used) vs. fiber length.

Fiber length measurement begins with a reference fiber or cable that is similar to the type of test sample but for which the length is known accurately by mechanical means. The reference length is connected to the OTDR, or to a "dead-zone fiber" between it and the OTDR. A trace, or "signature," for the latter condition is shown in Fig. 4.32. The dead-zone fiber may be used in many OTDR measurements to reduce the influence of the initial reflection at the OTDR connector; the reflection between the dead-zone fiber and the best fiber or cable can more easily be minimized. Manual or automatic cursors are used to locate the points $z_1$ and $z_2$ at the beginning of each reflection, corresponding to both ends of the reference length if the dead-zone fiber is used. Without a dead-zone fiber, only $z_2$ is located.

The precalibrated reference length is determined as

$$L = z_2 - z_1 \qquad \text{(length of dead-zone fiber not known)} \tag{4.51a}$$

or

$$L = z_2 - \text{known length of dead-zone fiber} \tag{4.51b}$$

or

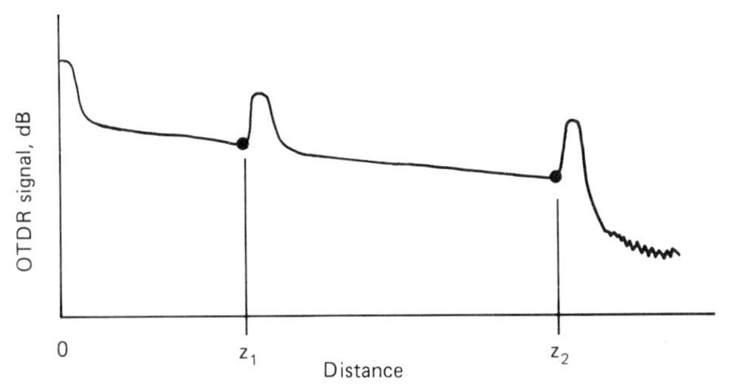


**FIGURE 4.32** OTDR trace of a test sample ($z_1$ to $z_2$) with a section of dead-zone fiber (0 to $z_1$) preceding it. (*FOTP-60.*)

$$L = z_2 \qquad \text{(dead-zone fiber not used)} \tag{4.51c}$$

The instrumental value of the group index $N$ and hence the precalibrated length $L$ is adjusted until the latter coincides with the known length of the test sample. This value is now the calibrated value of $N$ for the test sample, and can be used for those OTDR measurements in which accurate length information is required.

### 4.6.3 Attenuation

Attenuation between points along a fiber is covered in FOTP-61. These points are chosen to be on the "uniform" portions of an OTDR trace, away from abrupt changes. Figure 4.33, for a dead-zone fiber, shows the cursor readings taken at $z_0$

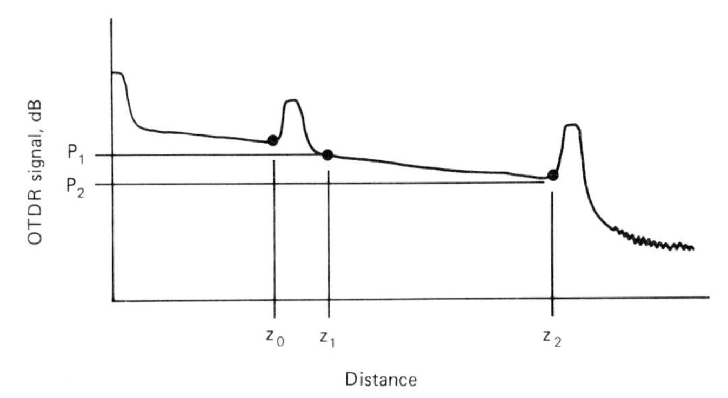


FIGURE 4.33 OTDR attenuation procedure using a dead-zone fiber. (*FOTP-61*.)

at the beginning and at $z_2$ at the end of the test sample, and at $z_1$ just after the reflection dead zone. Between fiber locations $z_1$ and $z_2$, the attenuation is $(P_1 - P_2)$ dB, while the attenuation coefficient is $(P_1 - P_2)/(z_2 - z_1)$ dB/km.

Usually, OTDRs are equipped with least-squares data-fitting routines to do these estimates. Extrapolated back to the beginning of the test sample at $z_0$, the overall attenuation is

$$A_T = \left(\frac{z_2 - z_0}{z_2 - z_1}\right)(P_1 - P_2) \qquad \text{dB} \tag{4.52}$$

If a dead-zone fiber is not used, then $z_0 = 0$.

The OTDR signal at a point depends on both the forward incident power there and the local backscatter capture coefficient. The latter can vary with length depending on the fiber index profile, NA (multimode), MFD (single-mode), etc. The effect of the backscatter variation can be eliminated with *bidirectional* readings (from both ends) that are averaged. (The differences can be used to plot backscatter vs. length.)

OTDR attenuation values agree closely with cutback attenuation values ("Cutback and Insertion" in Sec. 4.3.1) if suitable precautions are taken. The cursors must be placed away from nonuniform portions of the trace caused by local backscatter or reflection effects. The wavelengths used in both measurements should agree to within 5 nm. For multimode fibers, launch conditions become important.

Bidirectional averaging should be used if access to both fiber ends is possible. Finally, if the attenuation coefficient is desired, the length must be obtained via an accurate group index value *N*.

### 4.6.4 Faults and Joints

Manufacturers routinely check their fibers or cables for *point defects* that are indicated by temporary or permanent local deviations of the OTDR signal in the upward or downward direction. These have lengths limited usually by the pulse duration, and may vary with wavelength, direction, sample handling, etc. One usually wants these point defects to be minimal. There are a variety of types, and measurement is covered in FOTP-59.

Figure 4.34*a* shows a fault with apparent loss (as with some splices) and loss and reflection (as with some connectors). Figure 4.34*b* shows a "neutral" fault

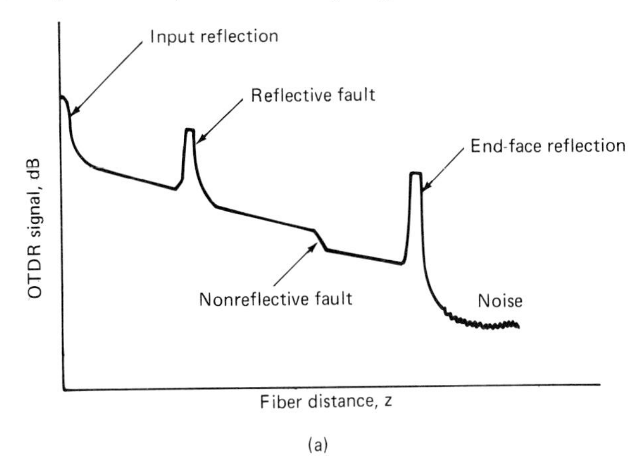


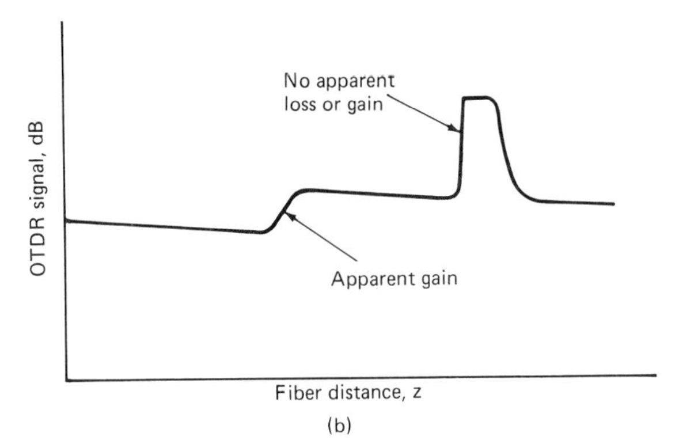


**FIGURE 4.34** OTDR traces of various types of faults. (*a*) Reflective and nonreflective; (*b*) with differing gain and reflection properties. (*FOTP-59.*)

and one with "gain." Quantitative measurement may not always be of concern in some applications, but we will consider it below. An FOTP specifically for the insertion loss of splices and connectors is now in preparation (see FOTPs not yet numbered in References).

*Insertion Loss.* The pulse duration tends to smear the fall (or rise) of the signal across a fault. Some OTDRs require placement of a pair of cursors on each side of the fault, away from reflection effects. The two lines defined this way are extrapolated to the beginning of the fall (or rise), and their vertical separation gives the apparent loss (or gain) in decibels. Some OTDRs do these calculations automatically.

One cause of an apparent gain at a joint is a receiving fiber that has a higher backscatter coefficient than the transmitting fiber. For example, in single-mode fiber that coefficient varies inversely as the square of the mode field diameter. Figure 4.35*a* shows a trace across a joint in which the second fiber has a smaller

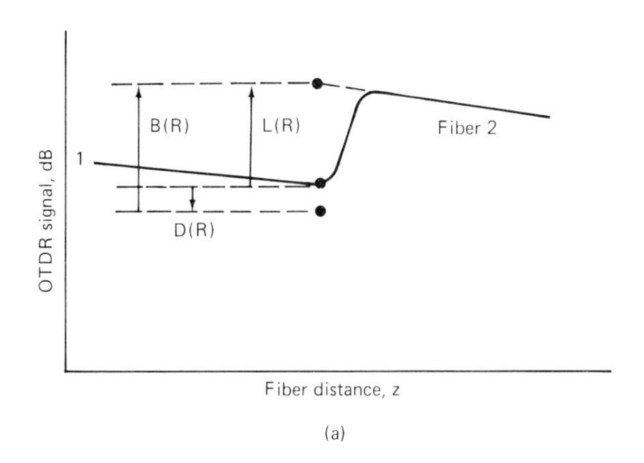


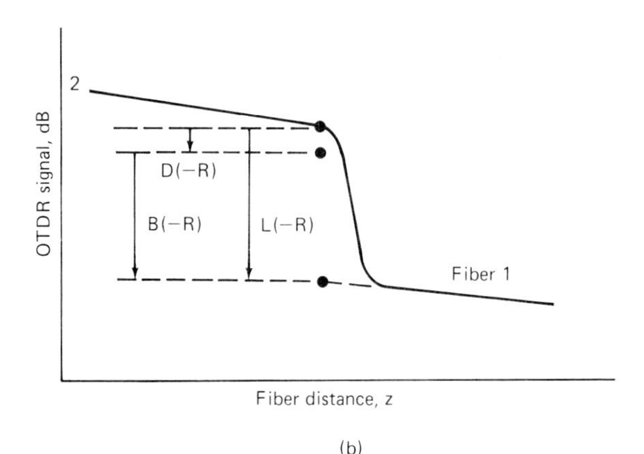


**FIGURE 4.35** OTDR trace across a joint with fibers of different mode field diameters. (*a*) $\omega_1 > \omega_2$; (*b*) $\omega_2 < \omega_1$.

MFD (larger backscatter). If $R$ is the MFD mismatch of Eq. (4.40*b*), the power loss (downward) is

$$D(R) = 4.343R^2 \tag{4.40a}$$

and the backscatter change (upward) is

$$B(R) = -4.343R \tag{4.53}$$

since $R$ is at most 0.1 to 0.2, the magnitude of $B(R)$ is always considerably larger than $D(R)$, and the sum $L(R) = D(R) + B(R)$ yields an apparent gain.

Figure 4.35*b* shows the trace taken in the opposite direction. Now only $B(R)$ changes sign and exaggerates the downward loss. Note that averaging the bidirectional OTDR readings cancels out the effect of the backscatter coefficient change and leaves only the true power loss.

***Reflectance.*** The optical continuous-wave reflectometer measurement in "Reflectance and Return Loss," Sec. 4.5.6, is tedious and best used with single reflectors. It is less useful for multiple reflections. For systems evaluation, knowledge of reflection magnitude and location, along with attenuation between reflectors, is important, and the OTDR can supply these. Because the reflected signal is proportional to pulse power, whereas the back-scattered signal is proportional to pulse power times pulse duration, short pulses characteristic of short-haul, high-resolution OTDRs are particularly effective in quantifying optical return loss. An FOTP is in preparation as this chapter goes to press.

If $T$ is the pulse duration and $H$ is the pulse peak height in decibels above the backscatter level, then

$$\text{Reflectance} = B + 10 \log [(10^{H/5} - 1)T] \quad \text{dB} \tag{4.54}$$

Here $B$ is the fiber backscatter level below the incident power. If $T$ is in nanoseconds, then $B$ is decibels below the power of a 1-ns pulse; it varies among fiber types from about $-78\frac{1}{2}$ to $-80\frac{1}{2}$ dB. The measurement should be calibrated with respect to the OTDR instrument and fiber type if accurate reflectance values are required. This is done usually with a calibrated reflector of known optical return loss. The OTDR must be linear for high reflections. Care must also be taken if precise location is desired, since location can vary with the reflectance by an amount approaching the rise time of the pulse.

## 4.7 RELIABILITY

### 4.7.1 Fiber and Joint Environments

Since fibers, cables, splices and connectors, and other devices will see a range of environmental severity, a number of FOTPs have been devised to test for survivability. There are a great variety of tests involving temperature and humidity, temperature cycling, mechanical shock, abrasion, fluid immersion, fungus, etc.; we will not detail all of these. Inspection occurs before, possibly during, and after the testing. Usually mechanical integrity, strength, and optical loss are monitored. We will outline only two specific environmental tests.

FOTP-49 examines the attenuation resulting from gamma irradiation of the fiber, but the procedure could be extended to other components. The gamma-ray source is cobalt 60 or cesium 137. The dose rates are low (less than 20 rad/h to more than 100 rad) for a 5-km fiber length, or high (5 to 250 rad/s for $3 \times 10^3$ to $10^6$ rad) for a 50-m fiber. The shorter length is used because of the higher induced attenuation coefficient. The light source coupling is kept below 1 μW to avoid the phenomenon of photobleaching. The temperature chamber is kept dark for the same reason and kept cold to enhance the effects of the damage and to prevent thermal bleaching.

The preceding test is for military and specialized commercial applications. Another test is for hydrogen effects that can cause additional fiber attenuation. Such attenuation can be due to indiffused molecular hydrogen at room temperature or lower; the effect is reversible with removal of the hydrogen source.

Permanent hydrogen bonding can occur at high temperatures. An FOTP addressing the potential sources of emitted hydrogen, such as fiber jacketing and cabling materials, and the metallic corrosion observed in underwater working links, is now being written.

### 4.7.2 Strength

Fiber strength depends upon the size of statistically distributed cracks in the fiber. There is an inverse relationship between the *inert* strength and crack size. An inert environment is one free of moisture and other deleterious elements, and is usually a vacuum with perhaps some dry inert gas.

***Proof Testing.*** To guarantee minimal strength of a fiber or of a splice or connector, a controlled tensile stress is applied for a given time period. Some fusion splicing machines, for example, have proof testers built in. FOTP-31 covers the fiber case in which the jacketed fiber passes (under tension) between rotating capstans. Typically 1 or 2 m of fiber is under stress at a given instant, and speeds may exceed 5 or 10 m/s. The testing may be done by the manufacturer on-line during the draw or later, or by the user to some enhanced level. If the fiber breaks, it will happen at its weakest points. The result of this testing process is to reduce the lengths of available fiber.

Figure 4.36 shows the stress-vs.-time plot for proof testing. The fiber experiences a load time $t_L$ to a proof-test level $\sigma_p$ (typically 50 to 200 kpsi) which is maintained for a dwell time $t_D$ (standardized to 1 s), followed by an unload time $t_U$. The effectiveness of the proof testing is proportional to

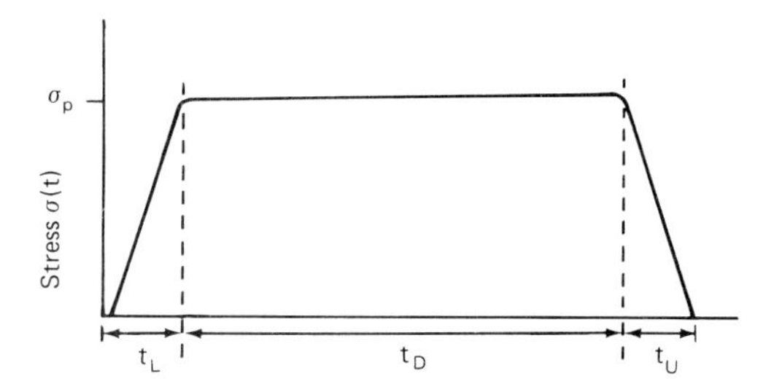


FIGURE 4.36 Stress history for proof testing.

$$\sigma_p\left(t_D + \frac{t_L + t_U}{n + 1}\right)^{1/n} \tag{4.55}$$

where $n$ is a stress-corrosion coefficient, typically around 20. With this normalization, alternative values of stress and times can be substituted to give the same effect.

***Fatigue.*** Unlike proof testing, fatigue lifetime testing is always destructive. We will talk in terms of stress $\sigma$, although for glass the strain $\varepsilon$ is related by

$$\sigma = E(1 + 3\varepsilon)\varepsilon \tag{4.56}$$

where $E \approx 10^4$ kpsi is Young's modulus. FOTPs for static and dynamic fatigue are now being written. The tests may be done in tensile form (to be discussed), but also in the form of a uniform bend, or as a two-point bend in which a few centimeters of fiber are wedged between two closing face plates. The theoretical interpretation of static or dynamic bend tests is somewhat more complex than tensile testing.

***Static Fatigue.*** The stress history for static fatigue testing is shown in Fig. 4.37*a*. A fiber sample of length $L$ is subjected to a *constant* applied stress $\sigma_a$ until it breaks at a failure time $t_f$. (This may take seconds to months, depending upon the stress and environmental conditions.) If $S_i$ is the initial inert strength of the fiber prior to testing, then the static lifetime equation is

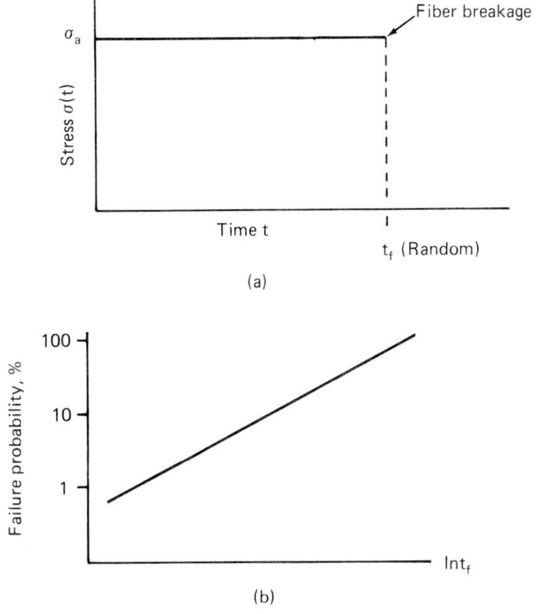


**FIGURE 4.37** Static fatigue plots. (*a*) Stress history; (*b*) Weibull failure time.

$$t_f = (BS_i^{n-2})\sigma_a^{-n} \tag{4.57}$$

where $B$ is a parameter. If the test is repeated usually for a number of samples at each of several applied stresses, a plot of log $t_f$ vs. log $\sigma_a$ should yield a straight line of slope $-n$ and intercept log $(BS_i^{n-2})$.

There is scatter in the failure times $t_f$ because of the random nature of the initial strength $S_i$. *In theory, the static survival probability distribution is*

$$P(t_f, \sigma_a, L) = \exp\left[-\left(\frac{t_f}{t_o}\right)^b \left(\frac{\sigma_a}{\sigma_o}\right)^{nb} \left(\frac{L}{L_o}\right)\right] \tag{4.58}$$

Here

$$b = \frac{m}{n-2} \tag{4.59}$$

$m$ is the Weibull $m$ value (see Chap. 1, Sec. 1.7.1), and $t_o$, $\sigma_o$, and $L_o$ are dimensional scaling parameters. Usually the applied stress and gauge length are held fixed so that a plot of ln ln $1/P$ vs. ln $t_f$ yields a straight line of slope $b$ as in Fig. 4.37*b*; one can also obtain the mean value of failure time at the 50 percent probability level. Repeating the test for other applied stresses leads to additional mean values that can be used in Eq. (4.57) with less data scatter.

To form the static Weibull plot above, the $N$ measured failure times (for a fixed fiber length and applied stress) are listed in order of increasing duration $t_1, t_2, \ldots, t_M, \ldots, t_N$. The cumulative failure probability is

$$F(t_M) = \frac{M - \frac{1}{2}}{N} \tag{4.60}$$

On special Weibull scale paper, a plot of ln ln $1/(1 - F)$ vs. ln $t$ yields a straight line.

*Dynamic Fatigue.* The stress history for dynamic fatigue testing is shown in Fig. 4.38. In this test, a fiber sample is subjected to a constant stress *rate*

$$\dot{\sigma}_a = \frac{\sigma_f}{t_f} \tag{4.61}$$

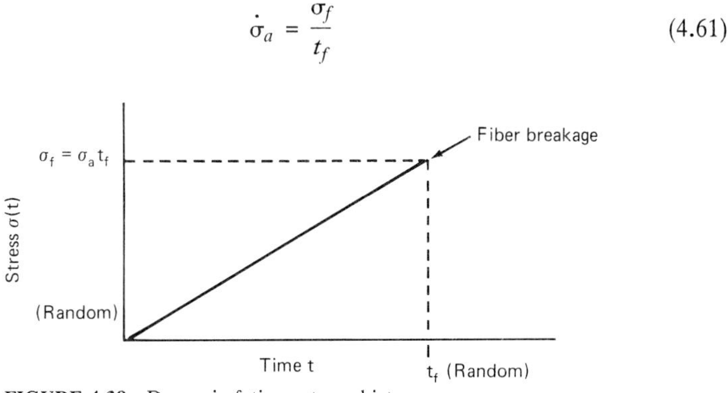


**FIGURE 4.38** Dynamic fatigue stress history.

until it breaks at a final failure stress $\sigma_f$ at a failure time $t_f$. (This can take less time than for static fatigue. For very low stress rates, some preloading at a fast rate is allowed.) Then the dynamic lifetime equation is

$$t_f = (n + 1)(BS_i^{n-2})\sigma_f^{-n} \tag{4.62a}$$

which may be compared to Eq. (4.57), or equivalently

$$\sigma_f^{n+1} = (n + 1)(BS_i^{n-2})\dot{\sigma}_a \tag{4.62b}$$

A plot of log $t_f$ vs. log $\sigma_f$ should yield a straight line of slope $-n$ and intercept log $(n + 1)(BS_i^{n-2})$. Alternatively, as per FOTP-76, a plot of log $\sigma_f$ vs. log $\dot{\sigma}_a$ yields the slope $1/(n + 1)$ and the same intercept.

The statistical dynamic survival probability distribution is

$$P(\sigma_f, \dot{\sigma}_a, L) = \exp\left[-\left(\frac{\sigma_f}{\sigma_o}\right)^{(n+1)b}\left(\frac{\dot{\sigma}_o}{\dot{\sigma}_a}\right)^{nb}\left(\frac{L}{L_o}\right)\right] \tag{4.63}$$

To form a dynamic Weibull plot, as in FOTP-28, the $N$ measured failure stresses (for a fixed fiber length and applied stress rate) are listed in order of increasing stresses. The cumulative failure probability is similar to that of Eq. (4.60). The static Weibull plot of ln ln $(1/P)$ vs. ln $\sigma_f$ yields a straight line of slope $(n + 1)b \approx m$. As with static fatigue, mean values of failure stresses for various rates may be used in Eq. (4.62*b*) to reduce data scatter.

## ACKNOWLEDGMENT

Where indicated, figures are adapted from EIA/TIA FOTPs with their permission.

## REFERENCES

### EIA/TIA Test Procedures

EIA *Fiber Optic Test Procedures, Optical Fiber System Test Procedures,* and fiber-optic specifications and standards are available from Electronic Industries Association, 2001 Pennsylvania Avenue, Washington, D.C. 20006, telephone 202-457-4900. FOTPs are identified by the code EIA/TIA 455-UVW, where UVW is the identifying number of the FOTP cited in this chapter.

FOTP-27. Methods for measuring outside (uncoated) diameter of optical waveguide fibers.

FOTP-28. Method for measuring dynamic tensile strength of optical fibers.

FOTP-29. Refractive index profile, transverse interference method.

FOTP-30. Frequency domain measurement of multimode optical fiber information transmission capacity.

FOTP-31. Fiber tensile, proof test method.

FOTP-34. Interconnection device total insertion loss test.

FOTP-43. Output near-field radiation pattern measurement of optical waveguide fibers.

FOTP-44. Refracted index profile, refracted ray method.

FOTP-45. Microscopic method for measuring fiber geometry of optical waveguide fibers.

FOTP-46. Spectral attenuation measurement for long-length graded-index optical fibers.

FOTP-47. Output far-field radiation pattern measurement.

FOTP-49. Procedure for measuring gamma irradiation effects in optical fibers and optical cables.

FOTP-50. Light launch conditions for long length graded-index optical fiber spectral attenuation measurements.

FOTP-51. Pulse distortion measurement of multimode glass fiber information transmission capacity.

FOTP-53. Attenuation by substitution measurement—for multimode graded-index optical fibers or fiber assemblies used in long length communications systems.

FOTP-54. Mode scrambler for overfilled launching conditions to multimode fibers.

FOTP-55. Methods for measuring the coating geometry of optical fibers.

FOTP-57. Optical-fiber-end preparation and examination.

FOTP-58. Core diameter measurement of graded-index optical fibers.

FOTP-59. Measurement of fiber point defects using an OTDR.

FOTP-60. Measurement of fiber or cable length using an OTDR.

FOTP-61. Measurement of fiber or cable attenuation using an OTDR.

FOTP-62. Optical fiber macrobend attenuation.

FOTP-68. Optical fiber microbend test procedure.

FOTP-76. Dynamic fatigue measurement of optical fiber by tension.

FOTP-77. Fabricating and qualifying a higher-order mode filter for single-mode measurements.

FOTP-78. Spectral attenuation cutback measurement for single-mode fibers.

FOTP-80. Cutoff wavelength of uncabled single-mode fiber by transmitted power.

FOTP-107. Return loss.

FOTP-164. Single-mode fiber, measurement of mode field diameter by far-field scanning.

FOTP-165. Single-mode fiber, measurement of mode field diameter by near-field scanning.

FOTP-167. Mode field diameter measurement—variable aperture method in the far field.

FOTP-168. Chromatic dispersion measurement of multimode graded-index and single-mode optical fibers by spectral group delay measurement in the time domain.

FOTP-169. Chromatic dispersion measurement of single-mode optical fibers by the phase-shift method.

FOTP-170. Cable cutoff wavelength of single-mode fiber by transmitted power.

FOTP-174. Mode field diameter of single-mode optical fiber by knife-edge scanning in the far field.

FOTP-175. Chromatic dispersion measurement of optical fibers by the differential phase shift method.

FOTP-176. Measurement method for optical fiber geometry by automated gray-scale analysis.

FOTP-177. Numerical aperture measurement of graded-index optical fiber.

FOTP-179. Inspection of cleaved fiber end faces by interferometry.

The following FOTPs are being written now. They are not yet numbered.

Measurement of splice or connector insertion loss using an OTDR.
Measurement of splice or connector return loss using an OTDR.
Static fatigue measurement of optical fiber by tension.
Dynamic fatigue measurement of optical fiber by two-point bending.
Static fatigue measurement of optical fiber by two-point bending.

### Bellcore Technical References

The following Bellcore generic requirements documents are relevant to this chapter. They are available from Bellcore, Customer Service, 60 New England Ave., Room 1B252, Piscataway, N.J. 08854-4196, telephone 201-699-5800.

TR-TSY-00020, Optical Fiber and Cable.
TR-TSY-000196, Optical Time Domain Reflectometers.
TR-TSY-000198, Optical Loss Test Sets.
TR-TSY-000264, Optical Fiber Cleaving Tools.
TR-TSY-000326, Fiber Optic Connectors for Single-Mode Optical Fibers.
TR-TSY-000468, Reliability Assurance Practices for Optoelectronic Devices.
TR-TSY-000761, Chromatic Dispersion Test Sets.
TR-TSY-000765, Splicing Systems for Single-Mode Optical Fibers.
TR-TSY-000886, Optical Power Meters.
TR-TSY-000887, Stabilized Light Sources.
TR-TSY-001028, Optical Continuous Wave Reflectometers.

### CCITT Recommendations

The following CCITT documents relate to specifications and test methods.

G.650 Test methods for single-mode optical fibre cable
G.651, Characteristics of a 50/125 μm Multimode Graded Index Optical Fibre Cable.
G.652, Characteristics of a Single-Mode Optical Fibre Cable.
G.653, Characteristics of a Dispersion-Shifted Single-Mode Optical Fibre Cable.

# CHAPTER 5
# OPTICAL SOURCES FOR FIBERS

**Paul Kit Lai Yu**
*University of California, San Diego*

**Kenneth Li**
*PCO, Inc.*

## 5.1 INTRODUCTION

The success of lightwave communications is largely the result of two technological breakthroughs. The first of these was the development of glass fibers with low optical attenuation. The second was the development of efficient semiconductor sources whose light output can be coupled to and transmitted in these fibers. This chapter deals solely with properties of semiconductor sources such as lasers and light-emitting diodes (LEDs), with special emphasis on those related to fiber communications.

## 5.2 WAVELENGTH CONSIDERATIONS

In optical transmission systems, the most important considerations related to fibers are their absorption and bandwidth. Signal attenuation in state-of-the-art fibers are attributed to various material effects. These include light absorption by impurities such as hydroxyl ($OH^-$) and transition-metal ions, inherent absorption by the ultraviolet edge of the electronic band, and the infrared edge of vibrational bands of constituent glass materials, scattering due to deformations formed during fiber drawing, Rayleigh scattering by refractive index inhomogeneities frozen into the glass lattice, and Mie scattering by foreign inclusions introduced during manufacturing. The technology for making fiber preforms has advanced to a stage wherein, basically, only intrinsic effects such as Rayleigh scattering and infrared absorption contribute to the losses. These two factors, in effect, define a transmission window in the near infrared spectrum, which extends in wavelength from 0.7 to 1.8 μm in current low-loss germanium dioxide- ($GeO_2$-) doped silica fibers. The $OH^-$ content in these fibers must be kept low, since an impurity level

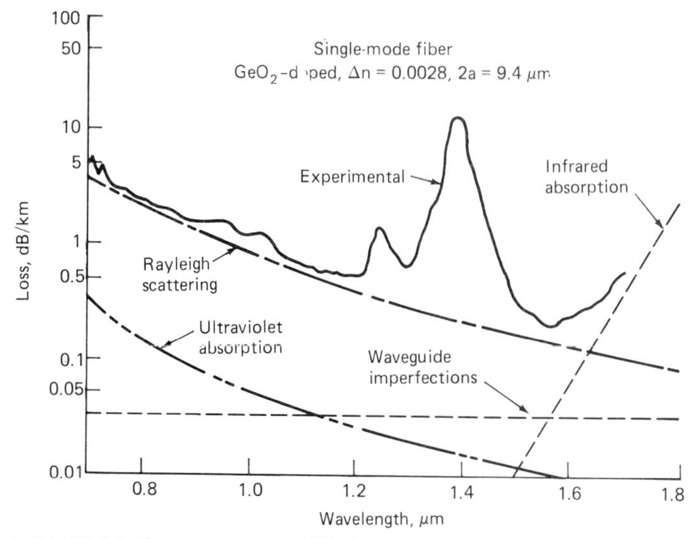


**FIGURE 5.1** Loss spectrum (solid line) for a contemporary germanium-doped silica, single-mode fiber. Dashed lines indicate the inherent limits imposed by Rayleigh scattering, ultraviolet and infrared absorption, and waveguide imperfections. *(From Ref. 1, with permission.)*

as low as 1 ppm can give rise to absorption peaks as high as 40 dB/km.[1a] Loss components associated with various intrinsic material effects are shown in Fig. 5.1.

The bandwidth of a fiber is determined by its structural and physical parameters as well as its material properties, as evidenced through chromatic and intermodal dispersion effects. Chromatic dispersion is the dominant effect that determines the bandwidth of single-mode fibers. However, at wavelengths where chromatic dispersion is minimum, birefringence effects due to core ellipticity and strain-induced anisotropy can limit the bandwidth. The specific material dispersion for $GeO_2$-doped silica passes through zero near $\lambda = 1.3$ μm. However, it is possible to control the dispersion in single-mode fibers by balancing the negative material dispersion against the positive waveguide dispersion. In this way, the wavelength which corresponds to minimum dispersion can be shifted to the range where absorption is lowest throughout the 1.2- to 1.6-μm region.

## 5.3 MATERIAL CONSIDERATIONS

For fiber optic applications, sources have been designed with direct bandgap semiconductor materials, in which the conduction band minimum coincides with the valence band maximum, to ensure direct radiative transition between these bands. Also, for emitter structures that employ more than one material, say, in heterostructures, lattice matching of these materials near the light-emitting region is needed to prevent excessive nonradiative transition via interface defects.

Emitters made with gallium arsenide/aluminum gallium arsenide ($GaAs/Al_xGa_{1-x}As$) compound semiconductors have been extensively characterized. In this material system, the bandgap of the ternary compound ($Al_xGa_{1-x}As$) can be tuned by changing the mole fraction $x$ of Al. The resultant ternary materials covering the bandgap range of 1.4 to 1.8 eV have a direct bandgap and are very close in lattice structure, since GaAs and AlAs have almost the same lattice constant. As a consequence, emitter devices with excellent heterointerface properties can be easily obtained from combinations of these materials. However, the wavelength range (0.7 to 0.9 μm) covered by them is outside that corresponding to lowest attenuation and zero total dispersion (material plus waveguide) in fibers. This greatly diminishes the interest in this material for emitters for long-distance, high-data-rate fiber communication systems

The quaternary indium gallium arsenide phosphide/indium phosphide ($In_xGa_{1-x}As_yP_{1-y}/InP$) system has emerged as a better alternative to GaAs/AlGaAs for fiber systems. With respect to the $In_xGa_{1-x}As$ ternary alloys, the addition of a group V element in the quaternary alloys eases the lattice matching to either InP or GaAs substrates. This extra degree of freedom allows one to tailor the desired device wavelength while maintaining the exact lattice-matching condition required for high-quality devices. For InGaAsP lattice-matched to InP, a wavelength range of 0.93 to 1.65 μm can be easily achieved. These quaternary alloys have a smaller bandgap and a higher refractive index than InP. The bandgap of quaternary $In_xGa_{1-x}As_yP_{1-y}$ lattice-matched to InP can be expressed as[2]

$$E_g = 1.34 - 0.72y + 0.12y^2 \tag{5.1}$$

where $x$ is approximately equal to $0.4526/(1 - 0.031y)$. Thus, emitter structures can be constructed by using quaternary materials for the active region and InP or quaternary alloys of larger bandgap for the cladding layers.

Traditionally, these quaternary alloys have been produced by means of liquid-phase epitaxy (LPE). LPE usually involves the growth of thin films on a substrate by passing over it solutions in which various constituents with fixed relative atomic ratios are dissolved. Crystal growth can be initiated by a lowering of temperature whereby the excess solutes are crystallized out. Because of its simplicity, LPE has been the primary method used in emitter manufacturing.

Other growth techniques, such as vapor-phase epitaxy (VPE),[3] molecular-beam epitaxy (MBE)[4] and metalorganic chemical vapor deposition (MOCVD),[5] have been extensively studied and have demonstrated growth of InGaAsP layers with good control of doping and layer thickness uniformity over large-diameter (2- to 3-in) wafers. In both VPE and MOCVD, constituent elements enter the growth chamber as gaseous compounds. VPE uses halide compounds, while MOCVD uses hydride and metalorganic compounds. In order to achieve uniform composition, it is essential that these gases be premixed in a manifold before they enter the growth chamber. Since the solid composition, dopant concentrations, and the growth rate of epilayers can be controlled by adjusting the gas flow rates, the temperature, and the pressure in the chamber, multilayer structures can be obtained in the same growth run. In molecular-beam epitaxy, an ultra-high vacuum chamber is used where various elements are vaporized and impinge on a heated substrate surface in the form of molecular beams. The correct stoichiometry and composition of epilayers can be achieved by adjusting the shutters that control the arrival rates (fluxes) of the various constituent beams.

Because MBE and MOCVD can support the growth of large wafers with good quality, it is expected that these techniques will become dominant in the manufacture of emitters for fiber communication systems.

In the rest of this chapter, various aspects of the two most common light emitters—LEDs and lasers—will be described with special attention to their applications in optical-fiber communication.

## 5.4 LIGHT-EMITTING DIODES (LEDs)

### 5.4.1 LED Structures

Emitter devices such as lasers and LEDs require a large population of electrons and holes in the active region of the device and a strong radiative recombination cross section between them inside the active region. This necessitates the use of a *pn* junction at the active region for carrier injection and the use of direct bandgap materials for efficient light generation in the active region. For efficient carrier injection and carrier confinement, a double heterojunction diode concept is used in which electrons and holes are injected into the active region from *n* and *p* sides, respectively, and are prevented from leaving the active region by the heterobarrier at the opposite side of the active region.

In the past, two basic LED structures in the GaAs/AlGaAs material system have been used extensively for short-haul, lightwave telecommunication links: the surface-emitting LED (SLED) and the edge-emitting LED (ELED). With the development of low-loss fibers in the 1.3-μm wavelength region, InGaAsP/InP ELEDs and SLEDs have become viable optical sources—even for medium-range, medium-data-rate fiber links. SLEDs have been used exclusively for multimode fiber links because they have a wide-emission far-field angle (which makes them very inefficient for coupling into small-core, single-mode fibers). On the other hand, ELEDs can be used for both single-mode and multimode fiber links and can be designed to operate in the superluminescent mode for higher bit-rate (greater than 400 Mb/s) transmission. In general, both SLEDs and ELEDs are reliable devices which have well-defined emission patterns. Because of their structural differences, ELEDs are more sensitive to temperature variations than SLEDs; nevertheless, ELEDs have significant advantages, such as coupled power and heat dissipation, over SLEDs.

***Surface-Emitting LEDs.*** The surface-emitting LED structure was first conceived by Burrus and Miller[6] in 1971. The SLED is also known as the *Burrus LED.* A double heterojunction is employed in this device, with the active layer sandwiched between layers of larger bandgap, to enhance the carrier confinement in the vertical direction. In Burrus' original design, a well was etched into the GaAs substrate to avoid reabsorption of light emitted through the substrate side.

A schematic diagram of a SLED is shown in Fig. 5.2. This structure has a large, planar *pn* junction at the interface between the active layer and one of the cladding layers. However, in order to increase the carrier density (as the recombination rate is proportional to it) inside the active region without drastically raising the injection current, the light-emitting area must be restricted to a small region. (Usually, the light-emitting region is circular and is typically 20 to 50 μm in diameter.) This can be achieved by confining the current to this region through methods such as proton bombarding the rest of the area to form highly resistive

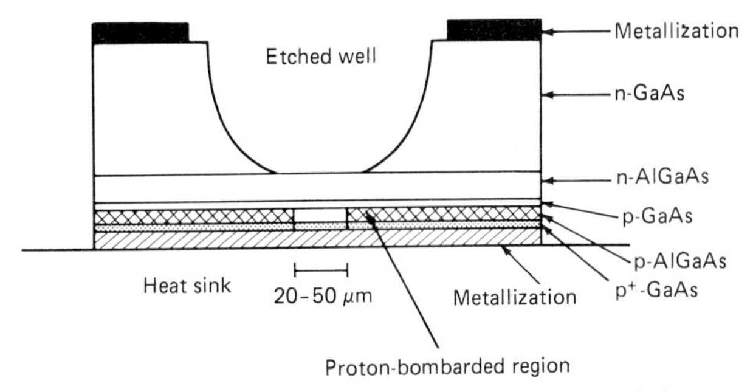


**FIGURE 5.2** Cross section of a GaAs/AlGaAs surface-emitting LED in which proton bombardment defines the current confinement region.

materials, dielectric (e.g., silicon dioxide) isolation of the injection current, mesa formation of the light-emitting region, and selective doping by diffusing zinc through a circular opening in the dielectric to reach the light-emitting region. All of these structures are shown in Figs. 5.2 to 5.5.

Light can be collected from SLEDs from either the substrate side or the epilayer side. The advantage of collecting light from the substrate side is that heat generated by the device near the surface can be properly conducted away by mounting the hot region near a good heat sink. In the case of GaAs/AlGaAs SLEDs (Fig. 5.2), where the active layer consists of either GaAs or $Al_xGa_{1-x}As$ (with little Al content), the GaAs substrate must be removed in order to avoid reabsorption of the emitted radiation. One way to achieve this is to use a selective etchant that forms a well in the GaAs substrate and leaves the AlGaAs cladding layer underneath the light-emitting region untouched. Alternatively, the whole GaAs substrate can be completely removed by etching to form planar devices.

As shown in Fig. 5.3, for 1.3-μm wavelength InGaAsP/InP SLEDs, four layers are epitaxially grown on an InP substrate, which can be: (1) a 2- to 5-μm-thick,

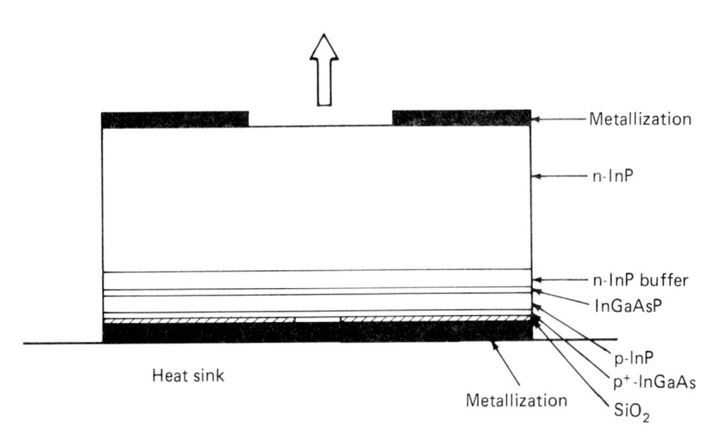


**FIGURE 5.3** Cross section of an InGaAsP/InP SLED with a dielectric layer for current confinement.

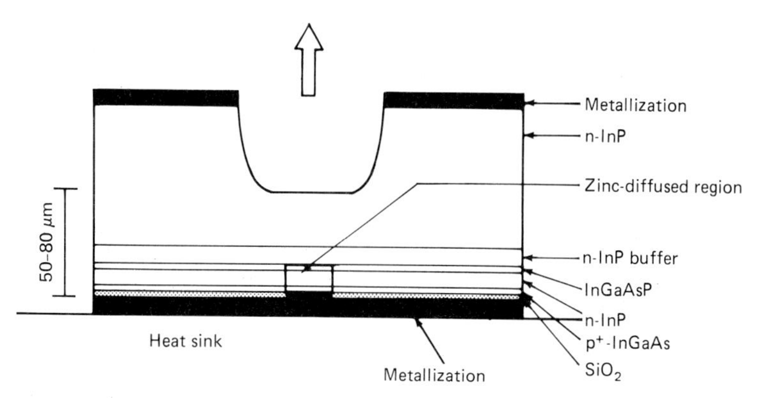


**FIGURE 5.4** Cross section of an InGaAsP/InP SLED in which zinc diffusion defines the current injection region.

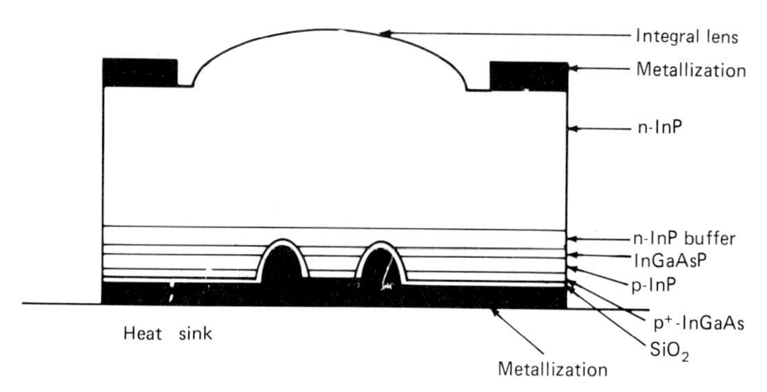


**FIGURE 5.5** Cross section of an InGaAsP/InP SLED using an etched mesa to achieve current confinement.

*n*-InP buffer layer doped with tin (Sn) to about $2 \times 10^{18}\text{cm}^{-3}$; (2) a 0.4- to 1.5-μm-thick, *p*-InGaAsP active layer ($\lambda_g$ = 1.3 μm) doped with zinc (Zn) ($5 \times 10^{17}$ to $2 \times 10^{18}$ $\text{cm}^{-3}$); (3) a 1- to 2-μm-thick, *p*-InP cladding layer doped with Zn or cadmium (Cd) ($0.5 \times 10^{18}$ to $5 \times 10^{18}$ $\text{cm}^{-3}$); or (4) a 0.2-μm-thick cap layer of small-bandgap InGaAs ($\lambda_g$ = 1.65 μm), which is heavily *p* doped ($1 \times 10^{19}$ $\text{cm}^{-3}$) to reduce contact resistance. The thickness and doping level of the active layer, as well as the compositions of other layers, are related to the power output and modulation speed requirements and can vary from structure to structure. Since InP is transparent to the emitted light at 1.3-μm wavelength, one can, in principle, leave a thick InP substrate in the completed device. However, the thinner the substrate is, the less the spreading of the emitted light will be due to light diffraction, making it easier to couple to fibers.

Diffraction effects can be further minimized, and thus more light can be coupled to the fiber, by etching an aligned well in the substrate, as shown in Fig. 5.4. A better approach, which can improve the light coupling significantly, is to form an integral lens directly on the substrate by etching, as shown in Fig. 5.5. This approach will be described later in this chapter.

*Edge-Emitting LEDs (ELEDs).* The edge-emitting LED (ELED) structure shown in Fig. 5.6*a* was first introduced by Ettenberg et al.[7] in 1976. The ELED is similar in structure to a stripe-geometry laser diode (see Sec. 5.5). However, unlike the laser diode, there is no lasing action in ELEDs because optical feedback is suppressed. This is done by intentionally destroying the facets or by using antireflection-coated facets. Alternatively, the contact to the active stripe can be partially metallized, as shown in Fig. 5.6*b*, so that light propagating toward the rear facet, as well as the feedback from the rear facet, is heavily attenuated in the unpumped region because of self-absorption.

As with the SLED, the InGaAsP/InP ELED is composed of four epitaxial layers of similar doping and thickness, with the exception that the active layer is much thinner (0.05 to 0.25 μm). Unlike the vertically emitting SLED, where waveguiding effects are absent, light from the ELED is guided by the active layer directionally along the stripe length (one of the benefits of using a double heterostructure) and then emitted through the front facet. This leads to a narrow vertical emission angle and better coupling to fibers. A thinner active region allows a larger portion of the optical field to spread into the cladding layers, thus minimizing self-absorption of the emitted light along the active stripe.

As in SLEDs, the injection level in ELEDs can be increased by restricting the injection region. This can be achieved by means of stripe injection, such as that in the oxide stripe structure shown in Fig. 5.6*a*. In this case, however, considerable current spreading can occur in the cladding and in the active layers underneath the contact stripe. This current spreading also affects the emission angle in the lateral direction. In order to improve the injection efficiency and to stabilize the lateral waveguiding, ELEDs with a buried heterostructure configuration or mesa structure, as shown in Fig. 5.6*c*, can be fabricated. As will be discussed later, these structures can produce optimal LED coupling to single-mode fibers and can provide a fast rise time under pulse modulation.

*Superluminescent Diodes (SLDs).* Superluminescent diodes (SLDs) are distinguished from both laser diodes and LEDs in that the emitted light consists of amplified spontaneous emission having a spectrum much narrower than that of LEDs but wider than that of lasers. Moreover—unlike laser diodes, which require optical feedback for stimulated emission to achieve lasing—SLDs have no built-in feedback mechanism.

The superluminescent diode was first investigated by Kurbatov et al.[9] in 1971. The SLD can have structural features similar to that of the ELED: namely, optical feedback may be prevented by using an oblique output facet, by having mirrors with zero reflectivity, or by partial metallization of the contact stripe. However, in contrast to ELEDs, a much higher injection level can be achieved in SLDs. This can be obtained by employing structures with good current and carrier confinement. Index-guided SLDs with a stripe width of 5 μm have been shown[10] to be capable of continuous-wave (CW) operation and a modulation rate as high as 600 MHz.

The SLD can be treated as a special kind of laser which has a much higher threshold current density and very "soft" light vs. current characteristics, such as shown in Fig. 5.7, because of its very low facet reflectivity. At low current level, the SLD operates like an ELED. At high injection current, the output power of an SLD increases superlinearly and the spectral width narrows as a result of the onset of optical gain. The output power, at a given current, depends strongly on the length of the pumped section. By reducing the stripe length while keeping the same current, the current density and the output power can be in-

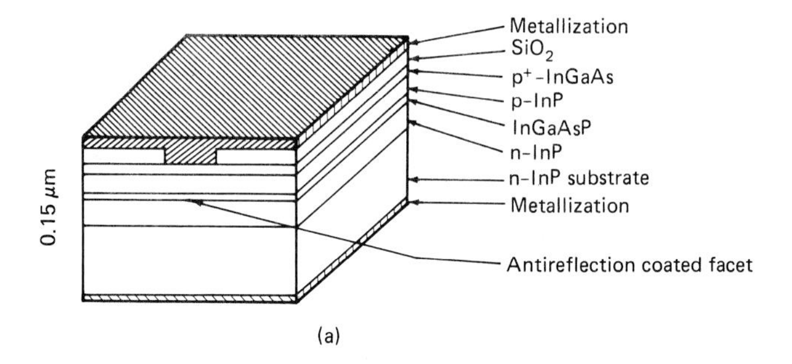


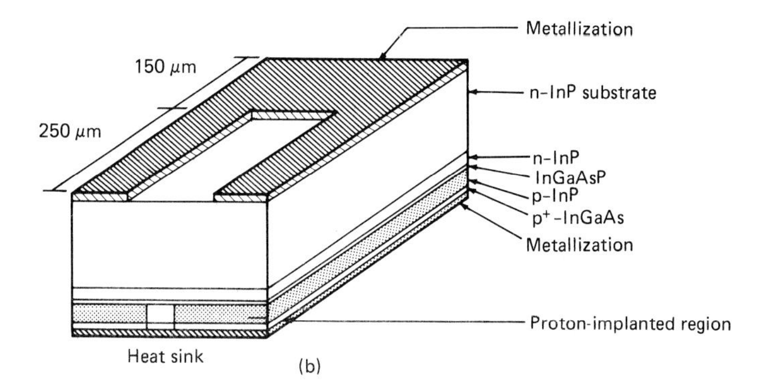


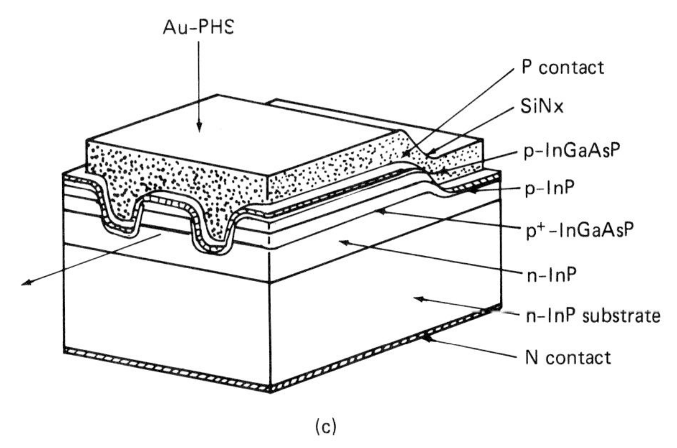


**FIGURE 5.6** Perspective views of (*a*) an InGaAsP/InP ELED with oxide stripe current confinement, (*b*) an InGaAsP/InP ELED with partially metallized *n* contact (the proton implantation is for current confinement), and (*c*) a high-speed, 1.3-μm, edge-emitting LED structure. (*From Ref. 8, with permission.*)

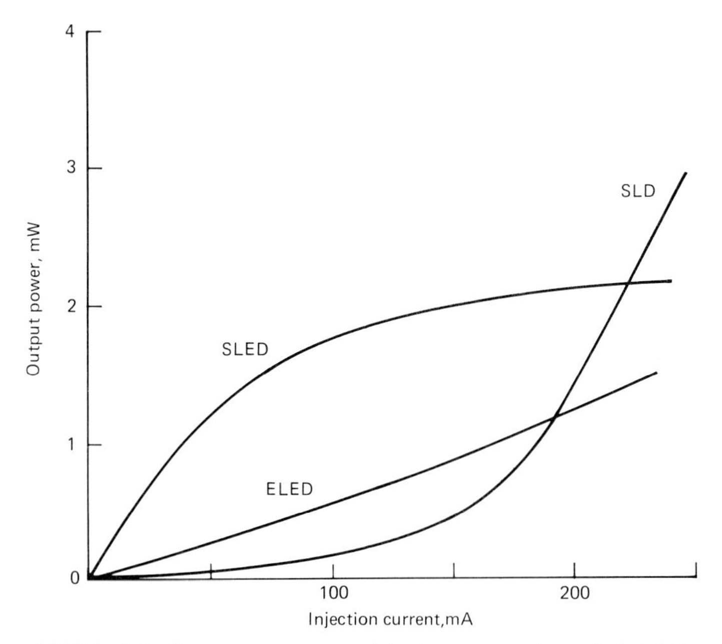


**FIGURE 5.7** Output power versus injection current for typical SLEDs, ELEDs, and SLDs. *(From Ref. 22, with permission.)*

creased. However, devices with short stripe lengths tend to heat up easily and this can result in a sublinear output dependence on temperature. A rate equation analysis[11] of the modulation performance indicates that a properly designed SLD can be useful for high bit-rate (greater than 1 Gb/s) transmission.

### 5.4.2 LED Spectral Linewidth and Radiation Pattern

*Spectral Linewidth.* Among the various characteristics of LEDs, the emission wavelength and the spectral width of source are the most important in the design of lightwave transmission systems. Both these properties affect the maximum data rate and distance of transmission because fiber loss and dispersion depend on wavelength. The peak emission wavelength $\lambda_p$ of an LED is determined mainly by the bandgap of the active layer. However, under pulsed operation, $\lambda_p$ can be shifted to a shorter wavelength by increasing the injection current, as a result of a band-filling effect. Alternatively, it can be shifted toward a longer wavelength (at a rate of about 0.6 nm/°C) by heating (which affects the bandgap and thus shifts the gain profile) or by increasing the active layer doping. The latter effect shifts the gain profile as a result of the formation of band tail states in highly doped materials.

The LED output spectrum is distinguished by its large, full-width at half-maximum (FWHM) spectral width $\Delta\lambda$. For an LED with an undoped active layer, the spectral width can vary as $\lambda_p^2$; this corresponds to a constant energy width of about 3 kT for $\lambda_p$ near 1.3 μm. There are, however, significant variations in $\Delta\lambda$

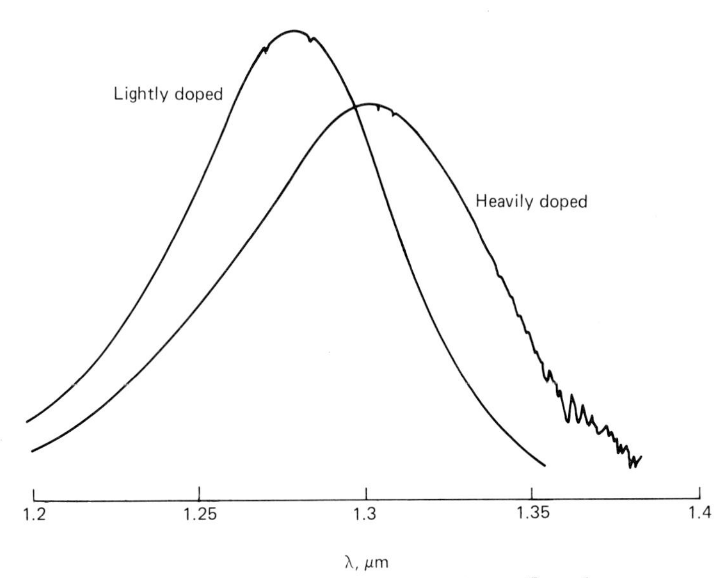


**FIGURE 5.8** Emission spectra for lightly doped ($5 \times 10^{17}$ $cm^{-3}$) and heavily doped ($10^{19}$ $cm^{-3}$) 1.3-μm-wavelength SLEDs. *(Courtesy of PCO, Inc.)*

among different LED structures. For an InGaAsP SLED emitting at 1.3 μm, the spectral width is typically about 100 nm. For an ELED, because of self-absorption along the active layer, the spectral width is about three-fifths that of the SLED and is typically 60 to 80 nm. Since the SLD operates in the region of amplified spontaneous emission, its spectral width is narrowed to about one-quarter that of the SLED and is typically 30 to 40 nm.

The LED spectral width can be broadened by heating or by operating at higher current. Both techniques are related to band filling. Alternatively, it can be broadened by increasing the active layer doping because of the formation of the band tail states. Figure 5.8 shows the spectra of two SLED devices having the same device parameters, except that the active layer doping levels are different. The spectral width increases from 75 nm to 95 nm when the Zn doping is changed from $5 \times 10^{17}$ to $10^{19}$ $cm^{-3}$. In this example, the LED rise time also improves from 15 ns to 1.5 ns with the increased doping level.

To narrow the spectral width of the SLED output, an $n$-doped InGaAsP filter layer with a bandgap larger than that of the active region can be grown between the $n$ InP buffer layer and another $n$ InP cladding layer,[12] as shown in Fig. 5.9.

***SLED Radiation Pattern.*** SLEDs have a large diameter-to-thickness ratio, with light emission being relatively uniform over the emitting area. Consequently, the radiation pattern is close to lambertian and the external radiant intensity (power per unit area per unit solid angle), $I(\theta)$, can be expressed as

$$I(\theta) \approx I_0 \cos \theta \qquad (5.2)$$

where $\theta$ is the angle between the emission direction and the normal to the emitting surface and $I_0$ is the axial radiance.

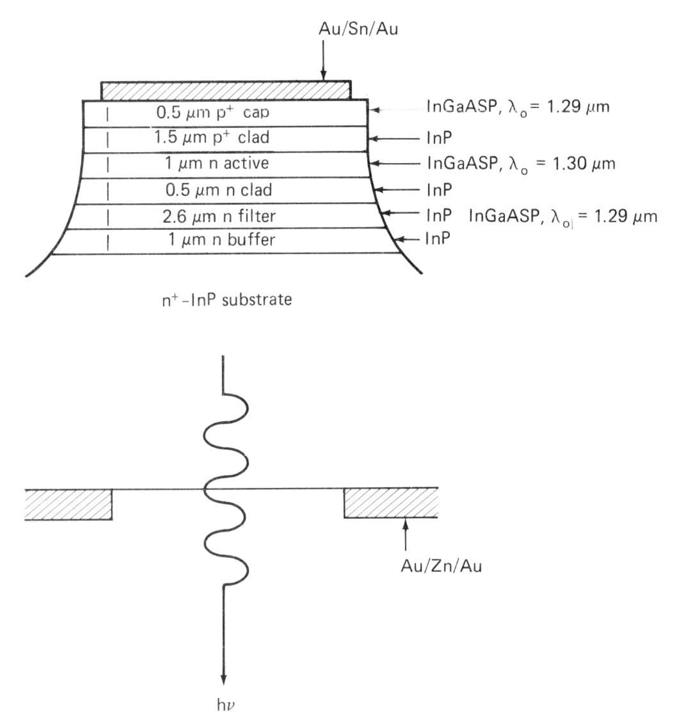


**FIGURE 5.9** Schematic cross section of a SLED with an InGaAsP filter layer. *(From Ref. 12, with permission.)*

In reality, because of imperfect current confinement near the edge of the active region, SLEDs usually have a nonuniform spatial emission profile which is more closely represented by a "stretched gaussian" function, with a gaussian tail near the edge. This has some effect on the coupling efficiency and will be discussed later.

The axial radiance $I_0$ of an SLED depends on the thickness of the radiative recombination region, the current density, and the internal quantum efficiency. It can be approximated by

$$I_0 \approx \frac{\eta_{\text{int}} T' J V}{4\pi n^2} \tag{5.3}$$

where $\eta_{\text{int}}$ = internal quantum efficiency of the LED
$T'$ = parameter which takes into account the absorption loss of emitted photons through the semiconductor and the transmission at the semiconductor-air interface
$J$ = current density
$V$ = junction voltage
$n$ = refractive index of the semiconductor

A high radiance can be obtained when the internal absorption is low and the back facet reflectivity is close to unity. A radiance as high as 200 W/(cm$^2$ · sr) has been achieved[13] for an SLED with an active region 2.5 μm thick and a doping level of $5 \times 10^{17}$ cm$^{-3}$.

***ELED Radiation Pattern.*** Because of the waveguiding effect of the stripe geometry, the ELED far-field radiation pattern is asymmetric and is more directional than that of the SLED. The output beam in the direction perpendicular to the junction plane is similar to that of the laser diode, with a far-field angle $\theta_\perp$ of about 30° (FWHM). The far-field pattern in this direction depends on the active region thickness and the index difference between the active and cladding layers. Typically, a far-field pattern with a strong main lobe, as shown in Fig. 5.10, can be obtained with an active layer thickness of about 0.12 to 0.15 μm in 1.3-μm-wavelength InGaAsP/InP ELEDs.

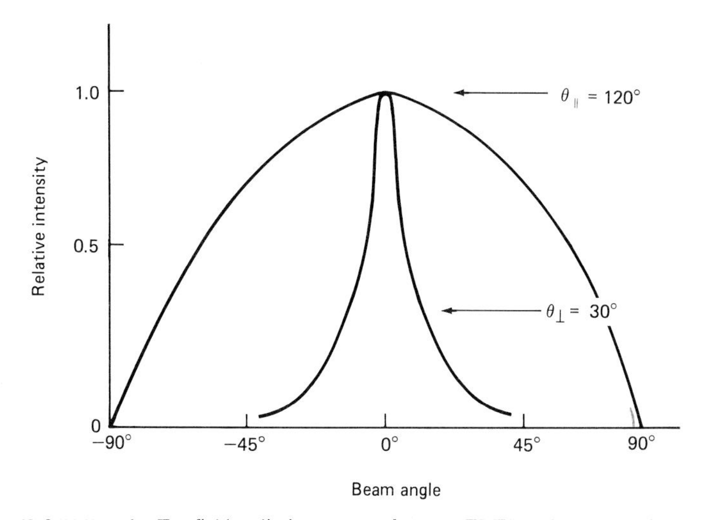


**FIGURE 5.10** Far-field radiation pattern from an ELED, where $\theta_\perp$ and $\theta_\parallel$ are the output beam angles in the directions perpendicular and parallel to the junction plane, respectively.

However, if an ELED is compared to a laser with the same stripe width, the output beam in the junction plane of the ELED will be much wider than that of the laser because of the lack of optical feedback. The distribution of the emitted radiation in this plane is approximated by a cos θ function. The emission angle can be reduced by using lateral index guiding or strong gain guiding, as in SLDs. High radiance values in the 1000 W/(cm$^2$ · sr) range at a 100-mA current level can be obtained for ELEDs.

### 5.4.3 LED-Fiber Coupling

Although the electroluminescence due to the recombination of the injected carriers is a very efficient process, the measured LED external quantum efficiency (light output/current) is very low. This can be explained by the fact that only a small fraction of the light generated within the active region strikes the semiconductor-air interface at an angle less than the critical angle $\theta_c$ and the rest is trapped inside the semiconductor by total internal reflection. For instance, the power emitted, $P_{out}$, into a medium of refractive index $n'$ from the planar surface of an SLED with a higher index $n$ can be approximated by

$$P_{out} = P_{in} T \frac{n'^2}{4n^2} \tag{5.4}$$

where $P_{in}$ = optical power generated inside SLED and $T$ = transmittance at semiconductor-air interface.

For an InP-air interface, $P_{out}$ is thus only a few percent of $P_{in}$.

On the other hand, the maximum power $P_{max}$ that can be coupled from an LED with uniform external radiance $I_0$ ($P_{out}$/emitting area) into a fiber with an acceptance angle $\theta_{c'}$ (and for small values of fiber numerical aperture, NA) can be approximated by

$$P_{max} = I_0 A \theta_{c'} \approx I_0 A \pi (\text{NA})^2 \tag{5.5}$$

where $A$ is the smaller of the cross-sectional areas of the emitter and the fiber core. Thus, in order to achieve high coupled powers, the product $I_0A$ must be maximized.

***SLED-Fiber Coupling.*** As discussed, it is clear that only those light rays that impinge on the fiber core at an angle (as measured from the fiber axis) less than the acceptance angle $\theta_{c'}$ will undergo total internal reflection at the core-cladding interface and can thus be guided along the fiber. Therefore, the fraction (or coupling efficiency) of LED light launched into a fiber strongly depends on its angular distribution. Because its radiation pattern is roughly lambertian (i.e., with a very wide angular intensity distribution), it becomes necessary to carefully consider ways to efficiently couple light from an SLED into a fiber.

The most common SLED-fiber coupling methods are butt coupling and lens coupling. Direct butt coupling is the simplest method to couple SLED radiation into a fiber. In order to achieve large coupled power in this approach, the LED emitting area should not be larger than the fiber core area, or a portion of the LED emission will be coupled into the fiber cladding. This populates evanescent modes which are quickly attenuated in the fiber. The butt-coupling efficiency $\eta_c$ for a lambertian light source (such as an SLED) can be obtained by integrating the radiant intensity over the LED emitting area and over the fiber solid angle $\theta_{c'}$. For step-index and graded-index fibers, $\eta_c$ is given by[14]

$$\eta_c = T_1 (\text{NA})^2 \qquad \text{(step index)} \tag{5.6a}$$

and

$$\eta_c = T_1 (\text{NA})^2 [1 - (2D^2)^{-1}] \qquad \text{(graded-index)} \tag{5.6b}$$

where $T_1$ = medium (e.g., air) transmissivity and includes LED-fiber coupling loss

NA = numerical aperture of fiber

$D$ = ratio of diameters of fiber core $d_f$ and emitting surface $d_s$

Parts *a* and *b* of Eq. (5.6) show that the coupling efficiency of a graded-index fiber approaches that of a step-index fiber when fiber core diameter $d_f$ is much larger than the emitting surface diameter $d_s$, but is only half that of the step-index fiber when these diameters are equal.

For small-core fibers with low NA, the observed $\eta_c$ is appreciably less than that predicted by Eq. (5.6). This is attributed to the actual LED emission, which deviates from the ideal lambertian emission profile. As mentioned earlier, a gaussianlike tail profile is more representative near the edge of the light-emitting region in view of the current spreading underneath the contact region.[15] The effect of this gaussian tail on $\eta_c$, for an SLED butt-coupled to a graded-index fiber with a 50-μm-diameter core, has been studied.[16]

Therefore, $\eta_c$ is far from optimal with butt coupling, as a large fraction of the LED emission cannot be coupled into the relatively narrow fiber acceptance angle. Much more efficient coupling can be achieved if lenses are used to collimate the emission from the SLED, especially when the source area is smaller than the fiber core. The coupling gain relative to butt coupling has been analyzed for a number of lens schemes such as lenses placed between the SLED and fiber, fibers with integral lenses, truncated lenses cemented to the LED structure, and lenses that are integrated with the LED structure.[17] All of these schemes are aimed at magnifying the emission area so that the size of the image of the LED emitting area matches the fiber core and, at the same time, the radiance of the image over the fiber acceptance solid angle is not significantly reduced. Figure 5.11 illustrates the coupling gain over butt coupling when truncated spheres of titanium trioxide:silicon dioxide ($Ti_2O_3$:$SiO_2$) are used as microlenses.

In general, the coupling efficiency reaches a maximum when the lens magnification factor $M$ equals the parameter $D$ in Eq. (5.6). For instance, for a spher-

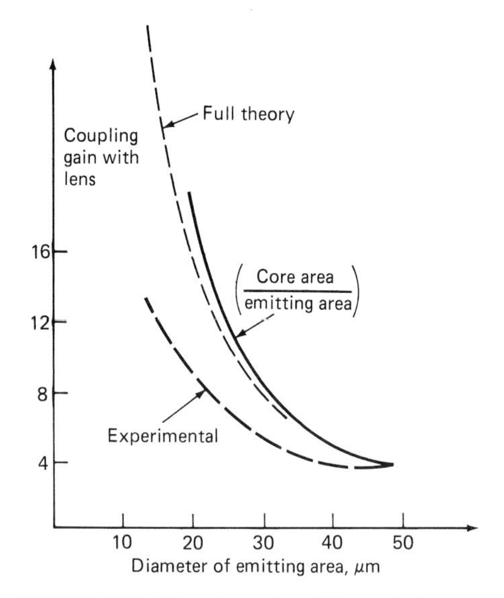


**FIGURE 5.11** Experimental and theoretical coupling gain due to lensing (with microlenses) as a function of source radius. *(From Ref. 18, with permission.)*

ical lens of radius $r$, the coupling gain $g$ over butt-coupling is the same for both graded-index fiber and step-index fiber and can be expressed as

$$g = T_L\left(\frac{d_f}{d_c}\right)^2 \tag{5.7}$$

where $T_L$ = lens transmissivity and the parameter $d_c$ is determined by the fiber numerical aperture and lens focal length $f$ (i.e., $d_c = 2f\text{NA}$).

Several lens-coupling configurations which have been adopted are shown in Fig. 5.12. In all of these schemes, light within a larger emission solid angle is transferred by a lens into a smaller solid angle which can be accepted by the fiber. The maximum coupling efficiency attainable with any lens scheme will ultimately be limited by the fraction of the emitted light that can be collected by the lens. With the simple lens and bulb-ended fiber configuration shown in Fig. 5.12, the lenses can collect only a fraction of the light emitted from the front face of the SLED, because of the air gap which separates the lens and the LED.

On the other hand, in the microsphere lens configurations, the adhesives used to hold the lens in place can also act as an immersion medium. Because the adhesive has an index higher than that of air, the lens can collect a relatively larger portion of the emitted light.

The highest coupling efficiencies can be obtained when the LED-lens interface is completely eliminated, as in the case of the integral-lens LED. In this case, the lens is made of the same material as the LED, so that $T_L$ of Eq. (5.7) becomes unity. Moreover, the integral lens has a large curved surface, which helps the radiative dissipation of heat generated within the LED. For the GaAs/AlGaAs material system—since the GaAs substrate underneath the emitting region will eventually be removed—the integral lenses can also be obtained by etching an array of circular recesses on a GaAs substrate (to define the lens shape) and then growing LED layers on this substrate. For the InGaAsP/InP system, the integral lens can be fabricated on the back side of the InP substrate after epitaxy. Methods for integral lens fabrication include ion-beam milling,[19] photoelectrochemical (PEC) etching,[20] and masked chemical etching[21] of the InP substrate.

A factor of 3 to 5 improvement in coupling efficiency, as compared with the butt-coupling case (with 0.16 NA fibers), can be obtained with bulb-ended fibers, and as much as 18 to 20 times improvement is obtainable with sphere-lens coupling. In principle, several hundredfold gains are feasible with integral-lens geometry if the emission region can be made as small as a few micrometers in diameter. Figure 5.13 shows that an integral-lensed LED with a 15-μm-diameter emitting region can couple 2.7 times more power into a step-index fiber than can

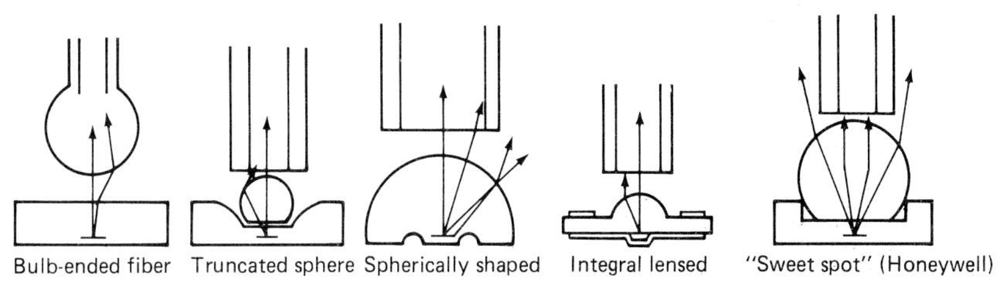


FIGURE 5.12 SLED-fiber lens-coupling schemes. The short, horizontal line indicates the active area of the SLED.

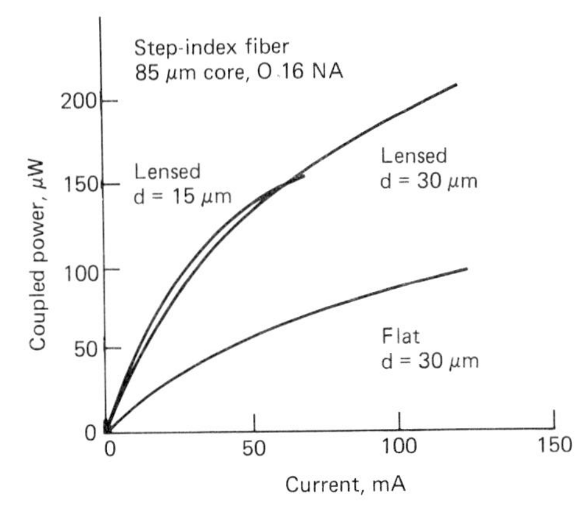


**FIGURE 5.13** Coupled power vs. current characteristics for two integral-lensed LEDs and a flat-surface LED. *(From Ref. 19, with permission.)*

a flat-surface LED with twice the diameter. In general, lens coupling is more advantageous than butt coupling when the fiber core diameter is much larger than the source diameter.

***ELED-Fiber Coupling.*** Because of the asymmetric far-field pattern of ELEDs, the analysis of ELED-fiber coupling is more complicated than that of SLED-fiber coupling. Generally speaking, the narrow beam divergence of ELEDs can result in a higher coupling efficiency to fibers with a small acceptance angle $\theta_c$ (i.e., low NA) than that achievable with SLEDs. This is especially true with graded-index fibers which have a small effective $\theta_c$. The coupling can be improved further by using real-index-guiding or gain-guiding ELED structures which have narrower beam divergence in the junction plane.

As in the case of SLED-fiber coupling, various lens-coupling methods can be employed in ELED-fiber coupling to improve the coupling efficiency over the butt-coupling scheme. However, special optical systems must be used because of the asymmetry of the far-field pattern of the ELED. The ELED has a narrower beam divergence in the plane perpendicular to the junction, $\theta_\perp$ (which is still larger than fiber $\theta_c$), than that in the junction plane, $\theta_{||}$, similar to that described in Fig. 5.2. For fiber coupling, the divergence in the junction plane, $\theta_{||}$, can be reduced by using a cylindrical lens whose axis is perpendicular to the junction plane. A second cylindrical lens, orthogonal to the first, can be used to narrow $\theta_\perp$.

Furthermore, the ELED near-field pattern is also asymmetric and is widest in the junction plane, and this limits the useful lens magnification. In practice, bulb-ended or tapered fibers are widely used to increase coupling efficiency. However, a truncated spherical microlens glued onto the emitting facet of the ELED has been used to improve coupling. A ridge-loaded ELED waveguide can also be used to improve the lateral near field. In the case of ELED to single-mode fiber, coupled power greater than 10 μW is possible, in contrast to the average power of 2.5 μW that is coupled from an SLED into a single-mode fiber.

Index-guided superluminescent diodes (SLDs) usually provide more output power than SLEDs or ELEDs because of the higher internal gain available in

SLDs. The narrow output far-field angle of SLDs results in higher coupled power into single-mode fibers. In the case of a 50-μm core, 0.2-NA fiber, a coupling efficiency as high as 26 percent has been realized in 1.3-μm SLDs. Coupled powers as high as 250 μW have been achieved at an injection current of 100 mA.[22]

***Heat Sinking and Alignment Tolerance Considerations.*** For both the SLED and ELED, fiber-coupling efficiency increases as the emitting spot size is reduced. However, as the spot size is reduced, the LED output power also decreases. This is attributed to the fact that the active emitting volume is proportional to the spot size. For the SLED at constant injection level, a smaller spot size can increase device heating, which reduces the internal quantum efficiency. Consequently, the design for maximum coupled power will not, in general, correspond to that for maximum coupling efficiency. For a given operating current and for a fiber with specified dimensions, the spot size and magnification of an LED can be optimized to provide the highest coupled power. As mentioned earlier, LEDs are usually mounted with the epilayer facing the heat sink to facilitate the removal of heat generated in the active region. Alternatively, at the cost of giving up some efficiency, they can be mounted substrate-side down. This is desirable from a manufacturing viewpoint for two reasons: assembly is greatly simplified and cost is reduced (because bad devices can be eliminated by testing in wafer form, before dicing and final assembly).

In ideal situations, the fiber and the LED are held at optimal positions relative to each other. In the butt-coupling case, for instance, the fiber is butted against the LED. In the lens-coupling case, the LED and the fiber are aligned and fixed in accordance with focusing properties of the lens. In practice, however, the relative position of components can be different from the optimum configuration, in accordance with alignment tolerances. These misalignments can cause a reduction in coupling efficiency but can result in a higher stability of the coupled power vs. displacement. For the SLED, the sensitivity of coupling efficiency $\eta_c$ to lateral, axial, and angular misalignment has been calculated numerically and verified experimentally.[22a] For a fiber with a core diameter larger than that of the source, $\eta_c$ increases as lens diameter decreases (because of the increased magnification); however, $\eta_c$ then also becomes increasingly sensitive to axial and lateral separation.

For the microlens coupling scheme, a lens with a diameter close to that of the fiber core can usually achieve a stable, optimum coupling, as shown in Fig. 5.14. By comparison, the Honeywell Sweet Spot configuration shown in Fig. 5.12 employs a relatively large lens and provides an output beam which is larger than the fiber core. This scheme has the advantage of reduced sensitivity to axial separation, at the cost of lower coupling efficiency. For the ELED, the asymmetry of the far-field pattern results in greater sensitivity to lateral displacement parallel (vs. perpendicular) to the junction plane. Sensitivity to axial separation also increases with lens coupling of the ELED.

### 5.4.4 LED Bandwidth

*LED Response.* One of the main reasons for using LEDs for fiber links is the fact that LED output intensity can be modulated directly with current. The maximum modulation rate of an LED will ultimately be limited by the carrier recombination lifetime. Parasitic elements such as space-charge capacitance and spreading diffusive capacitance can also cause a delay of carrier injection into the junction and

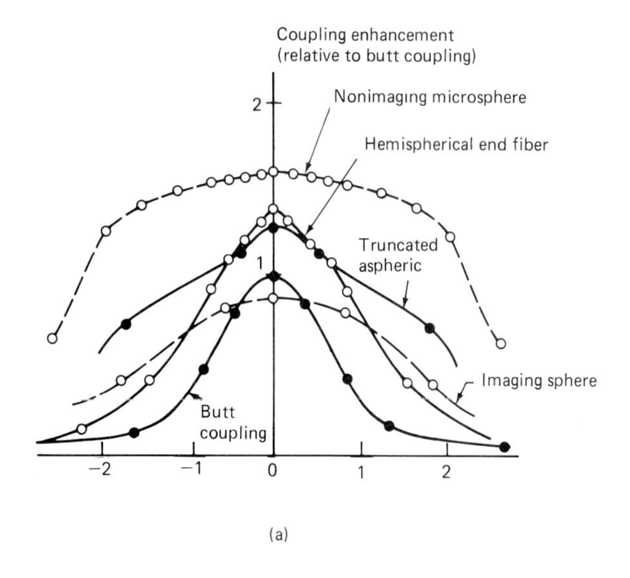


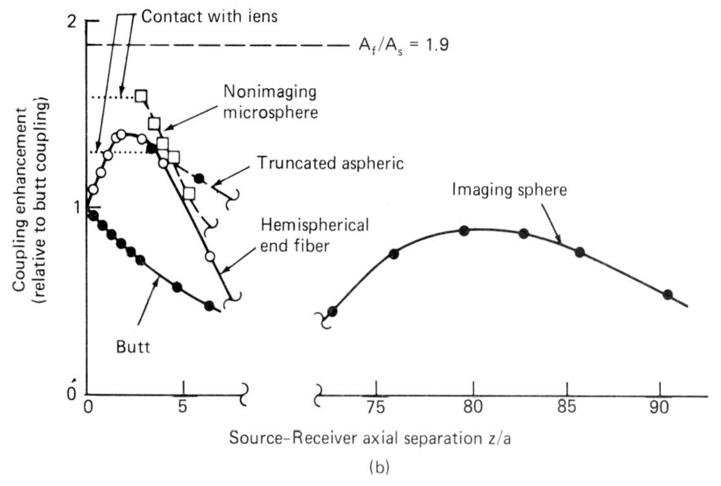


**FIGURE 5.14** Coupling gain (enhancement) for various lensing schemes as a function of (*a*) lateral separation and (*b*) axial separation. (*From Ref. 17, with permission.*)

thus cause an increase in the light output rise time.[23] The modulation bandwidth $\Delta f$ is defined as the frequency at which the detected power becomes half of its low-frequency value:

$$\Delta f = \frac{\Delta \omega}{2\pi} = \frac{1}{2\pi\tau} \tag{5.8}$$

where the carrier lifetime $\tau$ includes contributions from interfacial recombination as well as from radiative and nonradiative recombinations in the active region. In the limit, when the active layer thickness $d$ is much less than the carrier diffusion length and when the interfacial recombination velocity $s$ is small, $\tau$ can be expressed in terms of a reciprocal relationship

$$\frac{1}{\tau} = \frac{1}{\tau_r} + \frac{1}{\tau_{nr}} + \frac{2s}{d} \tag{5.9}$$

where $\tau_r$ and $\tau_{nr}$ are, respectively, bulk carrier radiative and nonradiative lifetimes. The carrier radiative lifetime is given by[24]

$$\tau_r = \frac{edN}{2J}\left(\sqrt{1 + \frac{4J}{edN^2B}} - 1\right) \tag{5.10}$$

where $B$ = radiative recombination coefficient
$N$ = active layer doping
$J$ = injection current density

Therefore, for cases where $\tau_r$ is dominant in Eq. (5.9), the bandwidth can be enhanced, at low injection level, by increasing the dopant concentration in the active region. Alternatively, the bandwidth can be enhanced by increasing the current density per unit active-layer thickness, since $\tau_r$ is proportional to $(d/J)^{1/2}$ at high injection level. A comparison of two LEDs with doped and undoped active regions at different injection levels is shown in Fig. 5.15.

Figure 5.16*a* and *b* shows the general effect of injection level and active-layer doping, respectively, on modulation bandwidth. In direct bandgap materials, the radiative lifetime is usually much shorter than the nonradiative lifetime. As the doping level is increased, however, impurity-assisted nonradiative processes become dominant and thereby reduce $\tau_{nr}$. At high injection levels, $\tau_{nr}$ can be further reduced as the nonradiative Auger recombinations (as discussed in Sec. 5.4.5) become significant. This is the case when the carrier concentration is larger than $2 \times 10^{18}$ cm$^{-3}$. This can cause an overall reduction in the carrier lifetime and lead to a larger modulation bandwidth. However, the increase in nonradiative processes also reduces the LED output as the radiation conversion efficiency (i.e., quantum efficiency) becomes smaller.

For an SLED with a doping level of approximately $1 \times 10^{17}$ to $5 \times 10^{17}$ cm$^{-3}$, a nominal bandwidth of 50 to 100 MHz is obtained. When the doping level is increased to $2 \times 10^{18}$ to $4 \times 10^{18}$ cm$^{-3}$, the bandwidth increases to more than 170 MHz. When the doping level exceeds $10^{19}$ cm$^{-3}$, for instance, with magnesium (Mg) as dopant, bandwidths as high as 1.2 GHz are obtained.

When LEDs are used as light sources in digital communication links, their rise times and fall times become critical parameters because they determine the data rate. LED response can be strongly affected by extrinsic parameters such as resistance-capacitance ($RC$) time constant, which depends on the series resistance of the diode, the space charge capacitance of the depletion region, and the diffusion capacitance at the edge of the light-emitting region. The series resistance is dominated by the contact resistance and can be improved by employing a low-contact-resistance metallization scheme.[25] The depletion region capacitance depends on device parameters such as area and thickness and must be optimized with respect to output power and speed. The diffusion capacitance is mainly caused by lateral current spreading. It can be reduced by raising the bias

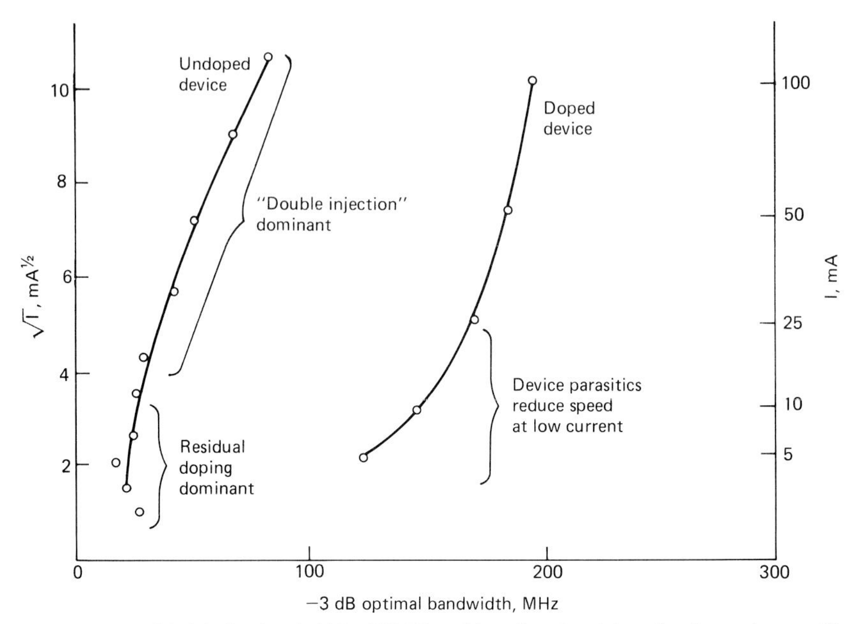


**FIGURE 5.15** Modulation bandwidth of SLEDs with undoped and doped active regions at different injection current levels. For an undoped device, at low injection levels, the $N$ in Eq. (5.10) is dominant, while at high injection, the dependence of bandwidth on current is evident. (*Courtesy of Plessey Research, Ltd.*)

current level (which causes a strong current-crowding effect in the emitting region), by adding a small DC prebias, or by employing current spikes.[23] Alternatively, the lateral current spreading can be avoided by employing structural designs with built-in lateral current confinement mechanisms as described earlier. A low-current (20-mA), high-coupled-power (7 μW into single-mode fiber), 1.3-μm ELED has been demonstrated with a modulation bandwidth greater than 500 MHz. This ELED structure employs the mass-transport technique to form a buried mesa structure for current confinement. Short stripe lengths (100 μm) are usually used in ELEDs for achieving higher current density.

In designing high-speed drive circuits, the LED capacitance should be taken into consideration, as described in Sec. 5.4.6. When the LED is in the off state, carriers (due to charged junction capacitance) remain in the active region and must be discharged through recombination, with a fall time $\tau$. This fall time can be improved by designing the circuit with a small forward bias to the LED even in the off state, thus allowing the LED capacitance to discharge. Alternatively, an intentional momentary reverse bias, which rapidly drains the carriers in the active region, can be applied to shorten the fall time.

***LED-Fiber Combined Response.*** The frequency response of an SLED—as coupled to a fiber—has specific features to be noted. Since the radiative lifetime is proportional to the square root of current density [see Eq. (5.10)] because of the lateral current spreading under the $p$ contact, the light generated at the center of the light-emitting region has higher response speed than that generated at the edge. Consequently, the effective modulation speed can be improved by prefer-

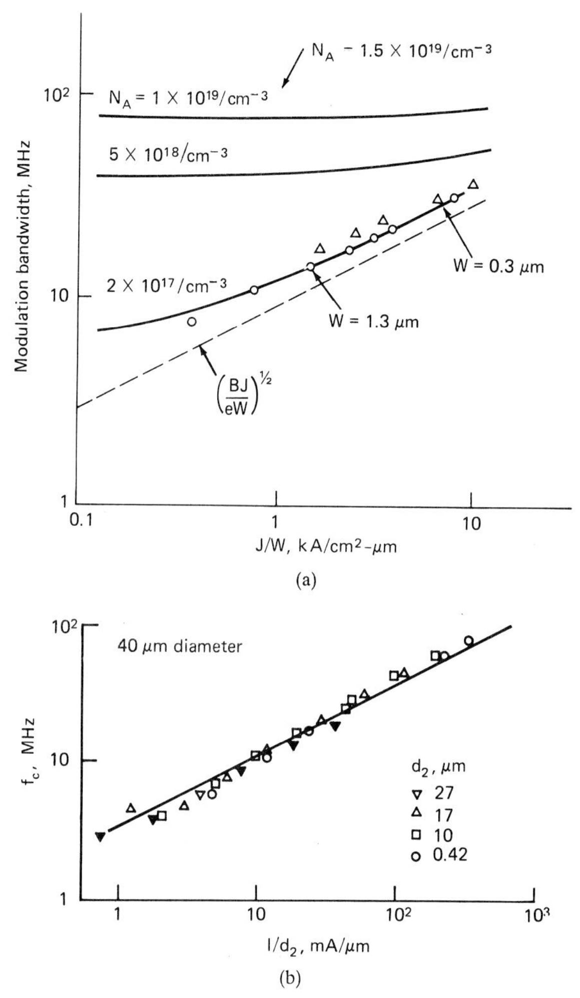


**FIGURE 5.16** Bandwidth considerations for SLEDs. (*a*) The modulation bandwidth of an SLED as a function of the normalized current density $J/w$, where $w$ is the thickness of the active region. (*From Ref. 13, with permission.*) (*b*) Cutoff frequency as a function of normalized current $I/d_2$ for four devices having various active layer thicknesses $d_2$. (*From Ref. 19, with permission.*)

entially focusing light from the center of the SLED onto a fiber by means of a lens. The fastest fiber-coupled SLED response occurs when the critical diameter $d_c$ of the lens is equal to the diameter of the uniform portion of the light spot, which also corresponds to a maximum in coupling efficiency. The integral SLED lens, combined with a lensed fiber, has achieved high coupled power as well as optimum modulation bandwidth, as shown in Fig. 5.17.

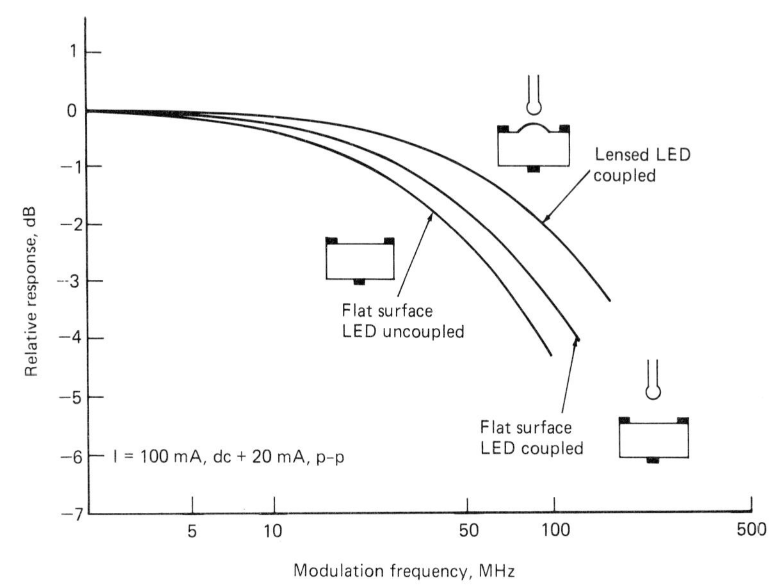


**FIGURE 5.17** Relative optical response vs. modulation frequency for three types of SLED-fiber configurations. *(From Ref. 15, with permission.)*

### 5.4.5 Linearity

First of all, it is observed that the temperature dependence of LED light output at constant drive current can be described by an empirical expression,

$$L = L_0 e^{T/T_0} \tag{5.11}$$

where $L_0$ is a constant. The parameter $T_0$ ranges from 105 to 130°K for an InGaAsP/InP SLED and is independent of the drive current. $T_0$ ranges from 70 to 80°K for an ELED, but its value can be reduced substantially at high current.

Second, the output power of an SLED initially increases linearly with current, but becomes sublinear at higher currents, as shown in Fig. 5.7. This sublinear behavior at high current can be partially explained by a self-heating effect which increases with injection level. However, this heating effect can be reduced by using short-duration current pulses and good heat sinking. The residual output saturation under short-pulsed drive, which is called radiance saturation, is observed in the InGaAsP/InP SLED.

Numerous models have been proposed to explain both the output temperature dependence and the output saturation in LEDs. In-plane superluminescence at high injection has been postulated to account for the output sublinearity in SLEDs.[26] In this model, the emission in the junction plane is enhanced at the onset of stimulated emission because of the high injection level, but this will be at the expense of emission in other directions, since the total gain remains fixed. As the injection level is increased further, however, the fraction of the in-plane emission increases and the surface emission gradually saturates. This model is supported by the light vs. current characteristic for the faceted SLED and by observed facet and side emissions in the ELED. However, the in-plane emission should not be significant in SLEDs which have a diameter less than 30 μm, and, furthermore, this mechanism alone cannot explain the observed independence of $T_0$ relative to the drive current.

Auger recombination, which is significant at high carrier density, has been proposed to explain the sublinear behavior and the temperature dependence of the LED output.[27] The enhanced saturation with increased injected carrier density (achieved by reducing the thickness and diameter of the emitting region) has been successfully modeled by assuming a nonradiative Auger component that depends on injection level.[28]

However, since Auger recombination is insignificant for slightly-doped materials, it cannot fully explain the sublinear behavior observed in LEDs which have an active layer doping of about $5 \times 10^{17}$ $cm^{-3}$. Instead, a drift-leakage model, which predicts that the leakage current from the active layer into the cladding layer increases with temperature, has been proposed. The leakage of injected electrons over the active-layer/*p*-cladding heterointerface has been verified in LEDs and in lasers. Carrier leakage is also expected to become significant when the injected carrier temperature far exceeds the lattice temperature.

None of the above-mentioned mechanisms alone can completely explain observed LED behavior. However, these mechanisms may be related and the temperature dependence may be due a combined effect. For instance, hot carriers resulting from Auger recombinations may be responsible for sizable leakage currents and may cause current saturation and $T_0$ behavior.

### 5.4.6 Drive Circuits

As mentioned earlier, an important class of LED drive circuits includes those designed for medium data-rate (less than 1-Gbit) transmission. In this case, the optimal design is one in which high-speed performance is limited only by the rise time and fall time of the LED and not by the drive circuit. In many drive circuits, a DC forward bias is added to the signal current to speed up charging of the diode capacitance, so that the rise time of the LED can be reduced. Similarly, a reverse-bias pulse can be used to shorten the fall time by accelerating the removal of carriers. However, whenever power dissipation and reliability are considered, return-to-zero (RZ) operation is preferred. Thus the basic requirements for this LED driver are that it can:

- Deliver current in the 50- to 100-mA range
- Switch the LED output on and off at high speed in response to a low-level input signal

One popular circuit is the common-emitter bipolar transistor switch shown in Fig. 5.18. This circuit has the merits of high gain, low input voltage, and an independently biased LED. The speed of this switch is limited by the forward current, the LED junction capacitance, and the minority charge stored at the collector-base junction of the transistor. The rise time of this driver can be further improved by using the speedup capacitor $C_1$ or by forward-prebiasing the LED. Also, the fall time can be reduced by using a Schottky diode clamped between the base and the collector of the transistor, which removes the minority carriers stored at the collector-base junction.

Another variation of the collector-loaded common-emitter switch is the emitter-coupled circuit, shown in Fig. 5.19. This circuit operates like a linear dif-

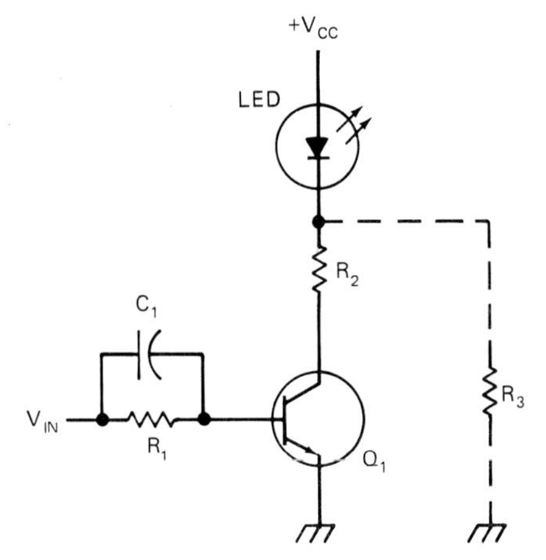


**FIGURE 5.18** A common-emitter switch-type LED driver. $R_2$ is a current-limiting resistor and $C_1$ is a speed up capacitor. *(From Ref. 29, with permission.)*

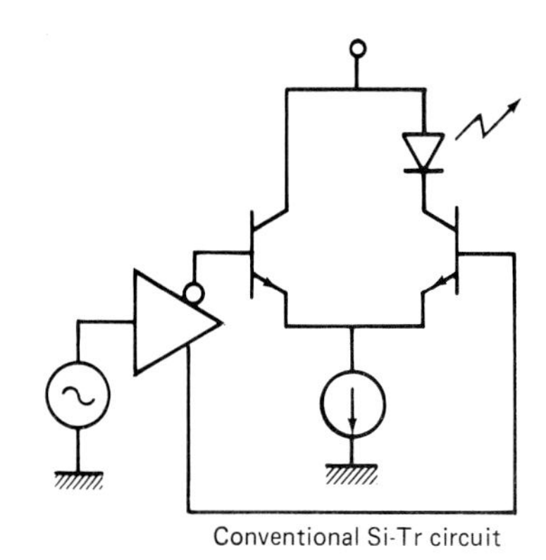


**FIGURE 5.19** An emitter-coupled silicon transistor driver for an LED. *(From Ref. 30, with permission.)*

ferential amplifier, except that the switching range is outside the linear region of the amplifier. This circuit has been used at speeds beyond 300 Mb/s and is compatible with commercial emitter-coupled logic (ECL) integrated circuits.

Low-impedance drivers which supply a low-impedance voltage step to the LED can also be used for high-speed operation. One such driver is the emitter-follower shown in Fig. 5.20. This circuit is capable of generating an optical pulse having a 2.5-ns rise time (equivalent to 100 Mb/s) from an LED having a capacitance of 180 pF.

A second type of low-impedance LED driver is the shunt configuration shown in Fig. 5.21, where the LED is in parallel with the switching element. Again, in

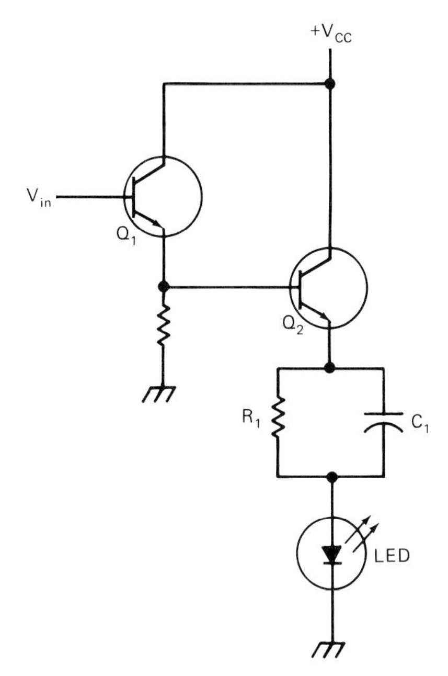


**FIGURE 5.20** A low-impedance, emitter-follower LED driver. The combination of $Q_1$, $Q_2$, $R_1$, and $C_1$ serves to optimize the speed of the circuit. *(From Ref. 29, with permission.)*

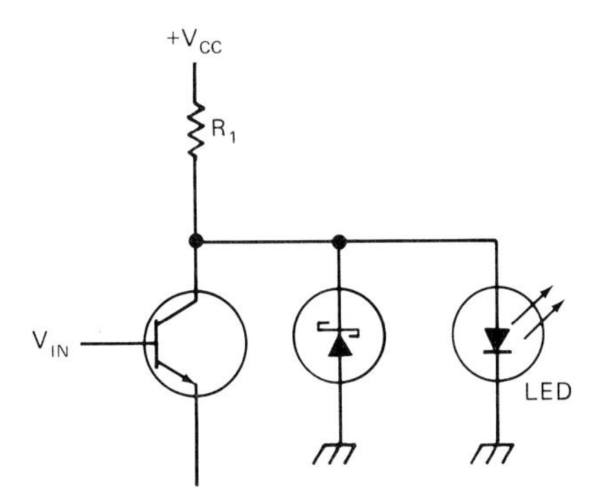


**FIGURE 5.21** A simple shunt driver where a low-impedance path is provided for LED turnoff. *(From Ref. 29, with permission.)*

this configuration, the LED can be slightly reverse-biased at turnoff by means of a Schottky diode clamp, which speeds up the removal of the stored charge.

An ECL-compatible shunt driver can provide a low-noise, constant-current source. A shunt driver with low output impedance can also be obtained in GaAs, which has the potential of very high speed because of the high electron mobility in GaAs material. One such circuit, shown in Fig. 5.22, consists of a buffer amplifier and a switching field-effect transistor (FET). An LED transmitter employing this driver has been shown to have a high data rate [400 Mb/s for non-return-to-zero (NRZ) coding] and exhibits neither overshoot nor undershoot in the LED output.

Because of the temperature dependence of the LED output, as described in Sec. 5.4.5, the LED can suffer a few decibels variation in optical output over normal operating temperatures, which typically range from 0 to 70°C. Since the absolute level of the LED output coupled into the fiber is usually small, and since many transmission systems are limited by received power, the designer has to compensate for this temperature variation to assure consistent performance. Compensation can be provided in the circuit design, so that constant LED output power is maintained over a wide temperature range. The usual approach is to make use of the fact that the cut-in voltage of silicon *pn* junctions is sensitive to temperature. Thus a silicon diode can be used in a temperature-compensating LED driver that raises or lowers the voltage applied to the LED in response to temperature changes.

Another problem occurs when SLDs are substituted for ELEDs, to couple

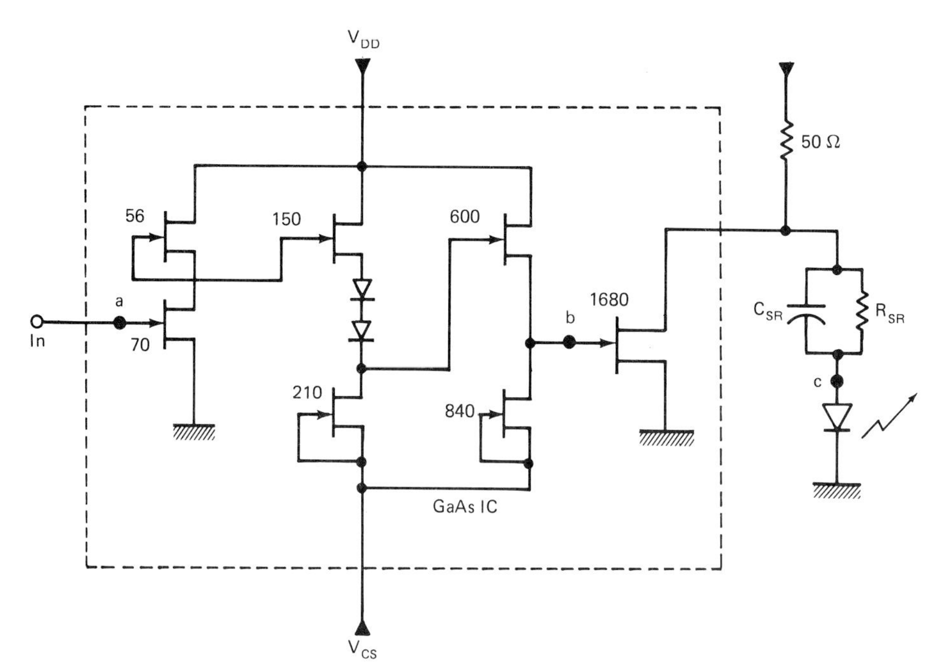


**FIGURE 5.22** Circuit diagram of an LED transmitter using a GaAs integrated circuit. Numbers denoted are FET gate widths in micrometers. A 400-Mb/s NRZ data rate has been obtained with this circuit. *(From Ref. 30, with permission.)*

more power into single-mode fibers. As the ambient temperature drops, the SLDs may lase at low current, making the light output characteristics very nonlinear near the threshold current. A temperature-regulating circuit can be used in this case.

### 5.4.7 Aging Effects

The reliability of the light source is very important to the success of an optical-fiber communication system. So far, LEDs are expected to have a longer life expectancy than laser diodes, as they are immune from problems associated with facet damage, threshold current degradation, transverse-mode switching, and so on, which plague lasers. However, in order to increase the light coupled into fibers, there is a tendency to drive LEDs at a high current density. As a result, LEDs may be operated at high temperature but, for the sake of economy, may not be provided with thermoelectric cooling.

The definition of LED lifetime varies among manufacturers and users. Generally speaking, the *end of life* of an LED is reached when its output drops to a certain percentage (say, 50 to 70 percent) of the initial value. The mean time to failure (MTTF) is defined as the average time interval between initial operation and the end of life in a large ensemble.

Like other device failures, LED failures can be divided into three kinds: infant failure, freak failure, and gradual degradation (for the long-lived devices). Infant failures result mainly from localized crystal defects and manufacturing defects. They are screened out in a high-temperature burn-in test which lasts about 100 to 200 h. Freak failures consist of those LEDs that pass the burn-in test but fail earlier than the majority of the population. Freak failures are caused by an extreme combination of similar processes that eventually cause failure in the main population. These processes usually have a well-defined time dependence, and they lead to a gradual reduction in quantum efficiency in the main population. For instance, it is observed that, as LEDs age, nonradiative defect centers move gradually into the active region by climbs or by diffusion. For the InGaAsP/InP LED, precipitate-like, dark-spot defects are observed during the gradual degradation.

***Life Tests for SLEDs and ELEDs.*** Most LED life tests are accelerated by applying temperature and current stresses to the device. In these tests, the light output is monitored and the MTTF is determined at constant DC and constant elevated temperature. The MTTF of devices operating at room temperature can be extrapolated from these results. In order to ascertain a good value for the MTTF, LEDs can be screened by a 100-h, 150-mA burn-in test to eliminate leaky or unstable devices such as those caused by indium inclusions in the device layers.

Usually, well-designed 1.3-μm Burrus-type SLEDs show no sign of degradation after operating at 5 kA/$cm^2$ and at 35°C for more than 1000 h. For life testing, double-heterojunction LEDs can be bonded with gold/germanium (Au/Ge) solder and operated at 100 mA dc (8 kA/$cm^2$). At 20°C operation, no dark-line defects are observed, even though the substrate may have a high dislocation density ($10^5$ $cm^{-2}$). At elevated temperature, $\langle 110 \rangle$ dark-line defects and dark-spot defects are observed, but with little decrease in light output.

It appears that defects in the dielectric isolation layer at the *p* contact are responsible for many LED failures. LEDs with Schottky barrier isolation at the *p* contact have only a few small dark-spot defects and have a projected MTTF (at

70°C) of more than $10^7$ h, which is independent of injection current density. By contrast, devices with dielectric isolation have a dense dark-spot-defect ring around the dielectric opening and fail catastrophically because of ⟨110⟩ dark-line defects and crystal damage at these defects. This limits the MTTF at 70°C to $2 \times 10^5$ h.

The reliability of ELEDs grown by liquid-phase epitaxy (LPE) has been compared to that of ELEDs grown by vapor-phase epitaxy (VPE) by operating them at a current density of 10 kA/cm$^2$. The VPE devices show a relatively faster degradation rate, independent of lattice mismatch, with a small activation energy (0.45 eV) and a projected MTTF (at 25°C) of about $10^6$ h. An order-of-magnitude improvement in MTTF ($2 \times 10^8$ h) for LPE-grown LEDs is obtained by mounting them on a heat sink with gold/tin (Au/Sn) solder rather than In solder (which leads to In in-diffusion and void formation, and consequently increases the thermal resistance of the diode).

*Aging Mechanisms.* From examinations of failed devices, several mechanisms responsible for LED aging have been identified. The major mechanisms are indium inclusion, dark-spot defects, dark-line defects, and misfits.

*Indium Inclusion.* Indium inclusion can originate from an incomplete melt wipe-off during liquid-phase epitaxy when the InP substrate is transferred from one solution to another, or from the substrate thermal decomposition during the soaking period of the growth. These inclusions can form an InGa-rich phase with low melting temperature and can move under thermal or strain gradients, thereby causing instabilities in the current-voltage (*I-V*) characteristics. The InP substrate is usually protected against thermal decomposition by exerting a phosphorus overpressure above it. This can be achieved by adding phosphine ($PH_3$) to hydrogen flow in the LPE, or by using a cover wafer and/or a tin solution saturated with phosphorus, or by a combination of these methods. In addition, in the LPE growth, the substrate can first pass through an indium melt to remove the thermally degraded surface, at the cost of reduced surface flatness.

*Dark-Spot-Defect Formation.* It is noted that InGaAsP/InP LEDs can have many dark-spot defects and still show little degradation in light output. The origins of dark-spot defects are debatable. Some studies suggest that they are a result of indium precipitation at the nucleation sites, while others suggest gallium and arsenic precipitation. However, it has been determined that the generation time for dark-spot defect formation is weakly dependent on temperature but strongly dependent on current density. It is believed that small dark-spot defects will not absorb light initially, but, when they grow in size, they can absorb light strongly. Another opinion is that the electrothermomigration of gold from the *p* contact can cause dark-spot defects in InGaAsP/InP LEDs, since gold can interact strongly and nonuniformly with semiconductor materials. By comparison, dark-spot defects do not develop during accelerated aging of LEDs having a platinum *p* contact.

*Dark-Line-Defect Formation.* Unlike GaAs LEDs, InGaAsP/InP LEDs only occasionally show dark-line defects. This may be attributed to the fact that the energy released in the electron-hole recombination in InGaAsP is only two-thirds of that in GaAs and is too small for creating point defects or for the migration of point defects to a dislocation. Like the dark-spot defects, the ⟨110⟩ dark-line defects can be produced by dielectric stress over the *p* InP confining layer and they can cause rapid degradation of the LED light output.

*Misfit Dislocations.* It has been observed that misfit dislocations are related to the lattice mismatch in epilayers. For devices without misfit, the lattice mismatch ($\Delta a/a$) of the InGaAsP layer with respect to the InP must be less than 0.05 percent at room temperature. These grown-in misfit dislocations act as nonradiative recombination centers and can be identified as crisscross patterns in the ⟨110⟩ directions in electroluminescence measurements. As LEDs with an intentional misfit of $\Delta a/a = 0.1\%$ age, the electroluminescence contrast of the ⟨110⟩ misfit dislocation lines increases, and the light output degrades in time as $e^{-t/\tau}$. Here $\tau$ is strongly dependent on current density (approximately as $J^{-3}$), but is only weakly dependent on temperature. In heavily degraded LEDs, misfit dislocations are found in both the confining and active layers, and act as strong nonradiative recombination centers, thereby causing device degradation.

## 5.5 LASER DIODES

### 5.5.1 Structures

Among the different laser structures developed so far, only a few have been selected for long-wavelength (i.e., 1.3 and 1.5 μm) fiber communication system applications. These selections are based on stringent emitter requirements such as high performance, low cost, and excellent reliability. In the past, laser structures were loosely grouped into gain-guided and index-guided structures, according to their means of lateral mode confinement. By proper design, both of these laser types can give rise to either single-lateral-mode or multilateral-mode optical output under certain conditions. However, as applications grow, there appear to be increasing demands on laser performance. In regard to coherent communication systems, for instance, the focus of laser research has been on the attainment of stable, single-mode, tunable-emission-wavelength lasers.

As in double heterojunction LEDs, optical modes and injected carriers in laser diodes are confined in the (transverse) direction perpendicular to the *pn* junction by two heterojunctions. With the use of double-heterojunction techniques alone, the threshold current of broad-area laser diodes can be drastically reduced. In order to achieve room-temperature, continuous-wave (CW) operation, the threshold current must be reduced further. This is possible when the junction is limited to a narrow stripe region less than 10 μm in width. The resultant lasers are referred to as stripe-geometry lasers.

*Multiple- and Single-Lateral-Mode Lasers.* The simplest multilateral-mode lasers are gain-guided lasers. In these lasers, the carrier injection is limited to the stripe contact area, which ranges from 5 to 20 μm in width and 150 to 300 μm in length along the laser cavity. As a result, the portion of the active region directly underneath the contact has gain, whereas neighboring regions are quite lossy. The presence of gain can increase the real part of the refractive index slightly and thus give rise to a weak lateral waveguiding effect.

Some examples of this type of gain-guided structure are shown in Fig. 5.23. They include the oxide-stripe laser,[31] the proton-bombarded laser,[32] and the V-groove stripe laser.[33] Since the active region under the stripe is wide enough to support the oscillation of several lateral modes, the output tends to be multimode. This is further aggravated by the fact that, although the contact area is restricted, the actual injected carrier profile can be considerably wider because of current spreading, diffusion, and drift of carriers in the *p*-type cladding layer as

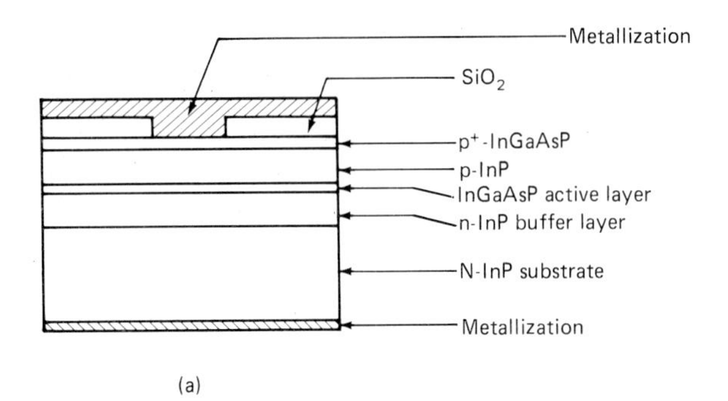


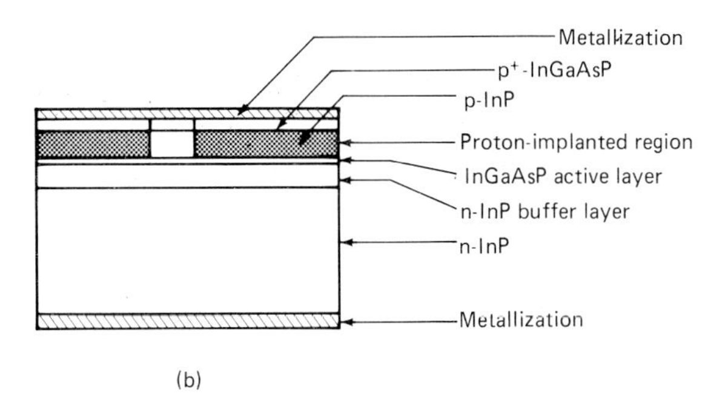


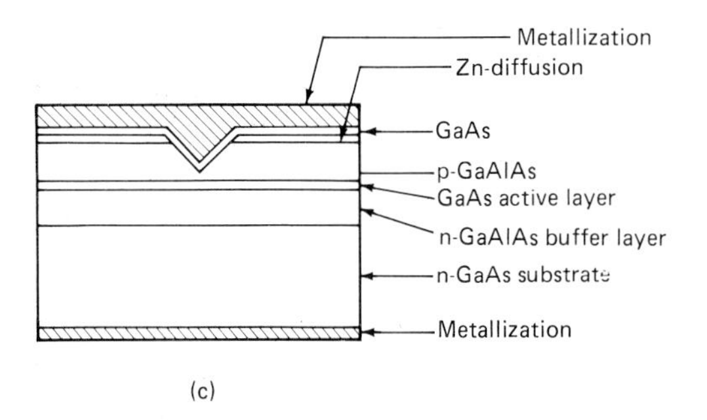


**FIGURE 5.23** Schematic cross sections of (*a*) an InGaAsP/InP oxide stripe geometry laser, (*b*) an InGaAsP/InP proton-bombarded stripe laser, and (*c*) a GaAs/AlGaAs V-groove stripe laser.

well as in the active layer. In the case of the oxide-stripe laser, the lateral diffusion[34] leads to an effective active width $w_{\text{eff}}$ of

$$w_{\text{eff}} = 0.7w + 1.95l_{\text{diff}} + \frac{7.47sl_{\text{diff}}}{w + 2.8l_{\text{diff}}} \tag{5.12}$$

where $l_{\text{diff}}$ = lateral diffusion length
$w$ = contact stripe width
$s$ = parameter describing the gain-guiding width

During lasing, the injected carrier profile can exhibit additional lateral spatial variation because of the high stimulated rate and the nonuniform lateral-mode profile. This can cause two opposite effects. First, at high injection levels, the fundamental lateral mode can cause excessive depletion of carriers in the stripe center through stimulated emission. This effect, called spatial hole burning, creates a dip in the gain profile, lowering net gain of the fundamental mode and allowing a higher-order mode to reach threshold conditions. Second, as a result of the free-carrier plasma effect and the abnormal dispersion of the band edge, the reduction in injected carriers in the stripe center can increase the index of the active layer there and thus stabilize the fundamental mode. The dependence of refractive index on carrier density is expressed as[35]

$$\frac{dn_r}{dn} \approx -2 \times 10^{-20}\ \text{cm}^3 \tag{5.13}$$

Depending on the relative strength of these two effects, antiguiding or guiding of the fundamental lateral mode can result.

In contrast to gain-guided lasers, index-guided lasers have sufficiently large dielectric steps (see Fig. 5.24) in the lateral direction to offset the influence of gain and loss. This results in a higher degree of lateral optical confinement. By controlling the width of the active region, a stable, fundamental lateral mode can be obtained. This dielectric step can be very abrupt and can be obtained by using lateral heterobarriers (buried heterostructures). The resulting lasers have an active region surrounded by material of higher bandgap and lower refractive index on all sides, and thus strong lateral carrier confinement can also be achieved. The most common buried-heterostructure lasers, as shown in Figs. 5.24 and 5.25, are planar buried-heterostructure lasers,[36] buried channel substrate lasers,[37] and double-channel planar buried-heterostructure lasers.[38]

The dielectric index change can also be as gradual as that obtained by gradually tapering the active layer in the lateral direction. The resultant index profile still favors lateral waveguiding. In this case, however, carriers may not be well-

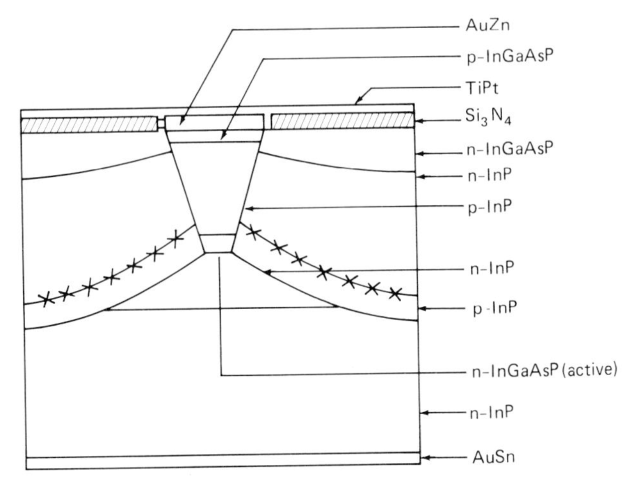


**FIGURE 5.24** Schematic diagram of an InGaAsP/InP buried-heterostructure laser on *n*-InP.

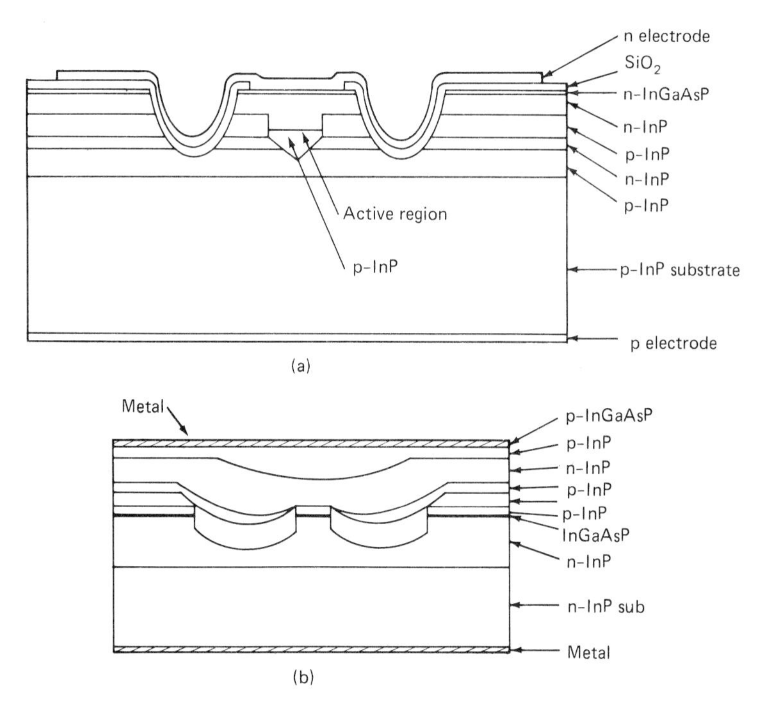


**FIGURE 5.25** Schematic diagrams of (*a*) an InGaAsP/InP buried channel substrate laser on *p*-InP and (*b*) an InGaAsP/InP double-channel planar, buried-heterostructure laser. (*From Ref. 38a, with permission.*)

confined laterally at the same time. Common structures of this kind are channel substrate lasers[39] and ridge waveguide lasers,[40] both of which are shown in Fig. 5.26.

Index-guided lasers, in general, possess lower threshold current and higher lateral-mode stability and mode selectivity than gain-guided lasers. Moreover, as will be noted in Sec. 5.5.2, index-guided lasers also have fewer longitudinal modes than gain-guided lasers.

***Single-Mode Laser Diodes.*** Single-mode lasers are those lasers that oscillate with a stable single longitudinal mode and a stable single transverse mode (in both directions). In contrast to the output of single-mode lasers, the optical output of multilongitudinal-mode laser diodes usually has higher relative intensity noise (RIN) because of the random switching among different modes. This noise can seriously degrade the quality of the transmitted signals. However, for multimode fiber communication systems, this multimode property can be an advantage, since the less coherent the laser emission is, the less are the optical-feedback-induced noise and the fiber modal noise. On the other hand, for single-mode fiber communication systems, it is highly desirable to use stable, single-longitudinal-mode laser sources with narrow linewidth (especially for long-haul applications).

Besides controlling the lateral mode structure, the lateral guiding mechanisms of laser diodes can also affect the longitudinal mode structure. Strongly index-

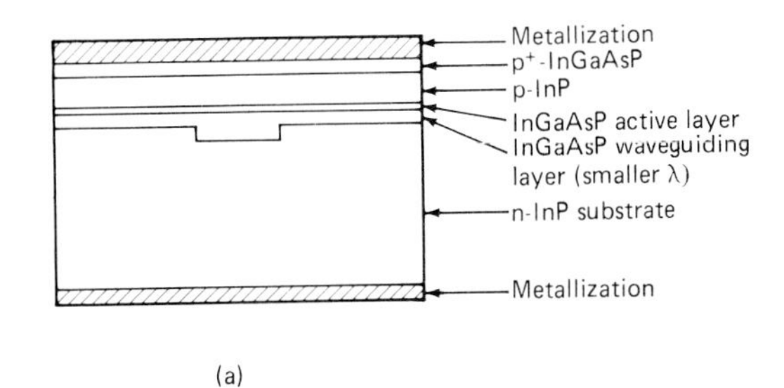


(a)

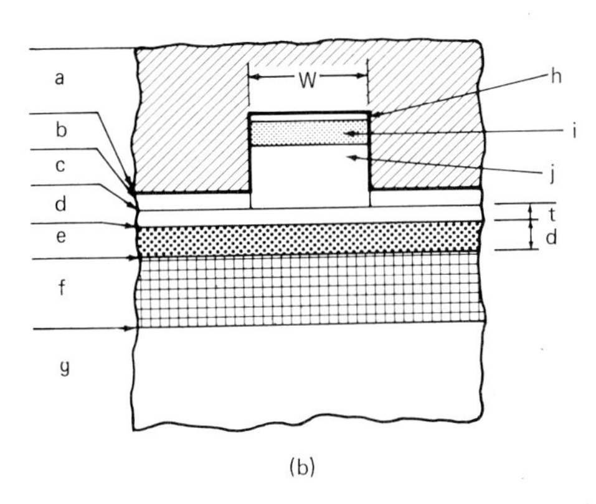


(b)

**FIGURE 5.26** Schematic diagrams of (*a*) an InGaAsP/InP channel substrate laser and (*b*) an InGaAsP/InP ridge guide laser: (*a*) Au pad; (*b*) Ti/Ag, Au contact; (*c*) $Si_3N_4$ insulator; (*d*) *p*-type InGaAsP (1.1-μm) buffer layer; (*e*) InGa AsP (1.3-μm) active layer; (*f*) *n*-type InGaAsP (1.1-μm) substrate layer; (*g*) *n*-type InP substrate; (*h*) Au/Zn alloyed contact; (*i*) *p*-type InGaAsP (1.3-μm) cap layer; (*j*) *p*-type InP ridge. (*From Ref. 40, with permission.*)

guided lasers usually have a single-longitudinal mode when operated slightly above threshold. Gain-guided lasers, on the other hand, usually have many longitudinal modes. Such behavior can be attributed to the large spontaneous emission factor β in the lasing modes. The spontaneous emission factor is the ratio of the spontaneous emission power in the lasing mode to the total spontaneous emission power and is given as

$$\beta = K \frac{\lambda^4}{4\pi^2 n^3 V \, \Delta\lambda} \tag{5.14}$$

where $\lambda$ = lasing wavelength
$\Delta\lambda$ = spectral width of spontaneous emission
$V$ = active volume
$n$ = refractive index
$K$ = astigmatism parameter

It is found that $K$ is close to unity for index-guided lasers and can be very large (greater than 10) for gain-guided lasers.

It is not totally clear why some index-guided lasers can operate in a single mode while others operate in multiple longitudinal modes. However, it is found that even index-guided lasers which operate in a single mode at DC can become unstable in emission wavelength when subjected to disturbances such as temperature change, optical feedback, or high carrier fluctuation during modulation. The resultant emission linewidth broadening and mode switching can lead to temporal pulse broadening in fibers, because of fiber dispersion, and degrade the quality of the transmitted signals for long-haul system applications. In order to maintain dynamic single-mode operation, extra means are needed to stabilize the lasing frequency. These can include:

1. Short cavity length
2. Mode selection by means of a composite cavity
3. Optical feedback by means of a grating incorporated inside the laser cavity

Short-cavity lasers are useful for achieving single-mode operation because, for a gain profile with a finite spectral width, the number of longitudinal modes within the positive gain region is inversely proportional to the length $L$ of the Fabry-Perot cavity. The longitudinal mode spacing $\Delta\lambda$ can be expressed as

$$\Delta\lambda = \frac{\lambda^2}{\left(n_{\text{eff}} - \lambda\frac{dn_{\text{eff}}}{d\lambda}\right)2L} \tag{5.15}$$

where $n_{\text{eff}}$ is the effective refractive index.

Thus, by reducing the cavity length, the Fabry-Perot mode spacing can be increased to a point where only a single longitudinal mode within the laser gain spectrum has a much higher preferential gain than the next mode, so that other modes cannot achieve threshold. For instance, by means of a microcleaving technique, as shown in Fig. 5.27, cavity lengths as short as 30 μm can be obtained. Also, by forming the facets with dry or wet chemical etching techniques, cavities as short as 23 μm can be achieved. However, even for these short-cavity lasers, multimode emission still appears under pulsed modulation, and is caused by a transient effect on the gain spectrum and the refractive index as the carrier density changes.

To further stabilize the longitudinal-mode structure, the composite cavity approach can be used. This approach incorporates additional mode selection techniques to stabilize the oscillation frequency. One such scheme places an external mirror close to one of the laser facets to create a multicavity effect. In another scheme, with a combination of a short cavity and an external mirror, a stable, single longitudinal mode is obtained even under pulsed modulation. In the same manner, a single longitudinal mode can be obtained with cleaved coupled-cavity lasers and interferometric lasers. These laser structures are shown in Fig. 5.28. Single-mode operation under modulation can also be obtained by external injection, in which the wavelength of one laser (the slave) is locked in value by light injected from a master laser. Laser linewidth narrowing can be achieved with this scheme.

Frequency-stable, single-mode lasers can also be achieved by adding a wavelength-selective grating in the active region. Distributed feedback, as shown in Fig. 5.29, and distributed Bragg reflector lasers are examples of this scheme.

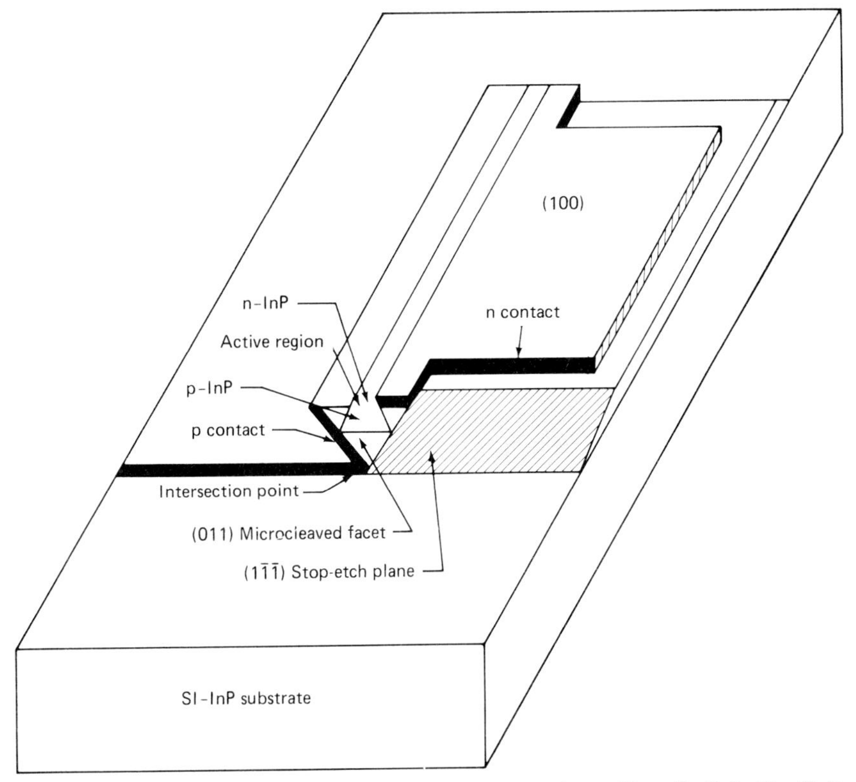


**FIGURE 5.27** A grooved InGaAsP/InP laser after microcleaving. (*From P. K. L. Yu, Ph.D. Thesis, California Institute of Technology, 1983.*)

The former utilizes a grating throughout the laser cavity and the latter uses gratings as reflectors outside the pumped region. With the incorporation of a grating structure, the oscillation wavelengths of distributed-feedback lasers become:

$$\lambda_o = \lambda_B \pm \frac{\lambda_B^2}{4n_r L} \tag{5.16}$$

which corresponds to two degenerate modes with the lowest threshold. The parameter $\lambda_B$ is the Bragg wavelength and is related to the grating period $\Lambda$ by

$$\lambda_B = \frac{2n_r \Lambda}{m} \tag{5.17}$$

where $m$ is an integer.

Low-threshold, distributed-feedback lasers with single-mode output can be obtained when the degeneracy is split by introducing a difference in the threshold gain of the two modes in Eq. (5.16). This can be done by means of end-facet reflectivity discrimination, a chirped grating, or an asymmetrical device structure

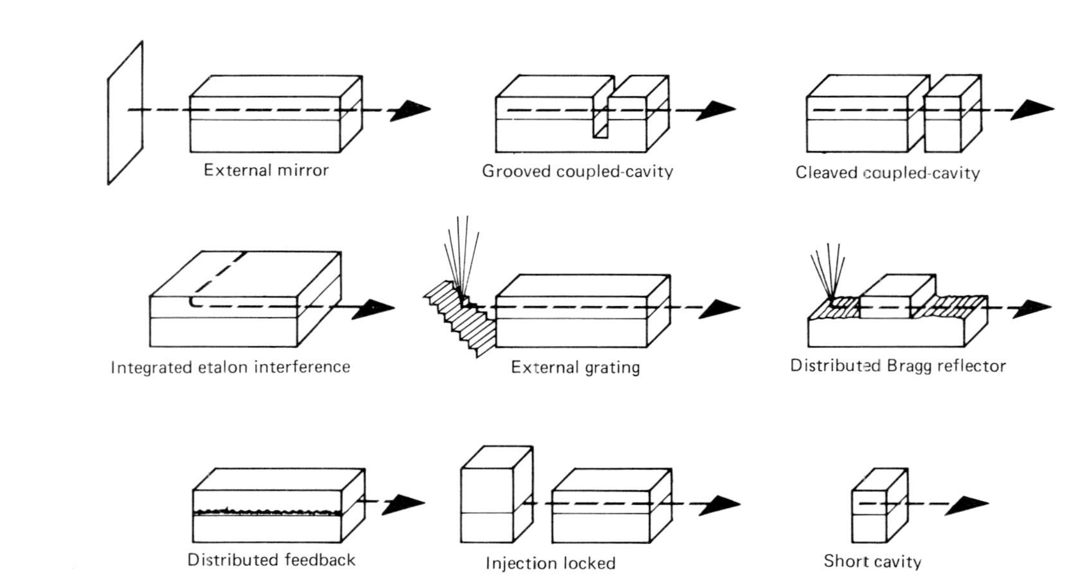


**FIGURE 5.28** Various schemes for laser longitudinal-mode stabilization. *(From Ref. 40a, with permission.)*

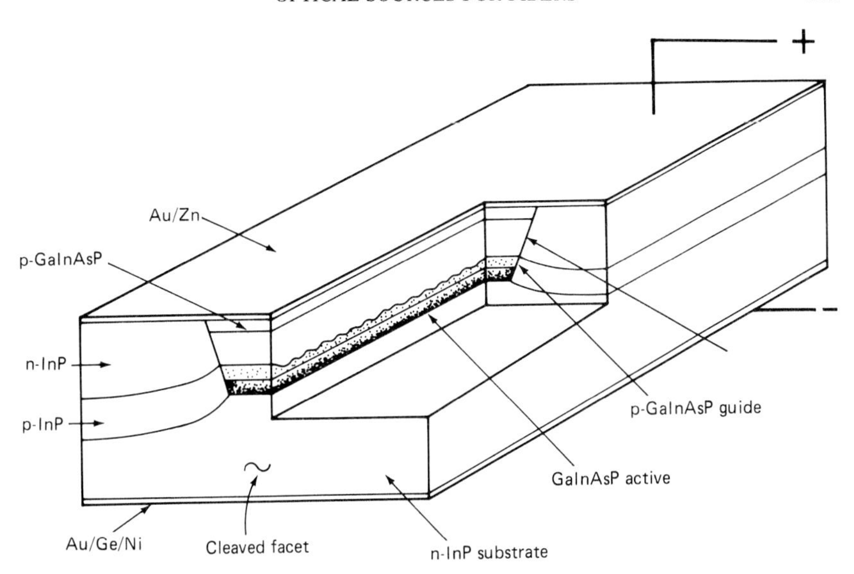


**FIGURE 5.29** Schematic diagram of a 1.55-μm InGaAsP/InP distributed-feedback laser.

(including those incorporating a λ/4 phase-shifter section in the grating). Wavelength stability can be maintained over a wide temperature range (20 to 70°C) and with a moderately high pulse modulation rate (greater than 500 Mb/s).

In contrast, distributed Bragg reflector lasers can oscillate only at the Bragg wavelength $\lambda_B$ because the grating is outside the pumped region. On the other hand, they usually have a higher threshold current than distributed-feedback lasers because of the weak coupling between the pumped and grating regions. Although a distributed Bragg reflector laser can be maintained in single-mode operation, its emission wavelength is more susceptible to temperature variation. In addition, because of the weaker coupling between the pumped and grating regions, the linewidth broadening under modulation due to carrier-induced refractive index variation is expected to be larger than that of distributed-feedback lasers. It is observed that the modal linewidth reaches a maximum when modulated near the laser resonant frequency, which is discussed in Sec. 5.5.4.

*High-Power Laser Diodes.* The two major factors which limit laser output power are:

- Catastrophic failure due to facet damage at high intensity
- Junction temperature, which can cause power saturation at high injection

It is generally believed that facet damage is caused by surface recombination of carriers at the facet, which makes the region near the facet absorbing. At high power levels, this absorption causes local heating and the region becomes more absorbing, eventually causing thermal runaway, and material meltdown occurs.

Since InGaAsP/InP materials have a much lower surface recombination velocity than GaAs/AlGaAs materials, it is expected that the InGaAsP/InP laser diodes can operate with higher power at the facets. Even so, a protective facet

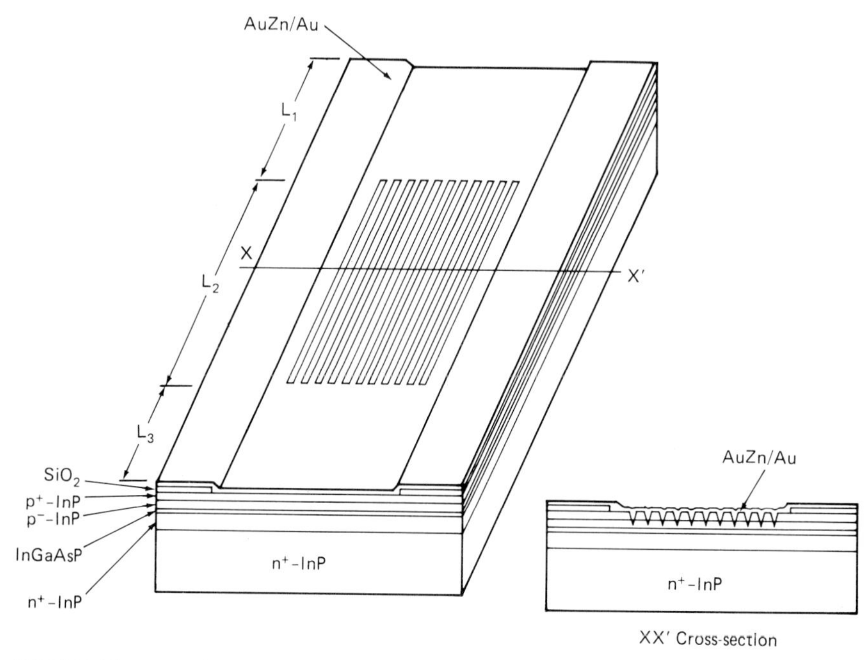


FIGURE 5.30 An InGaAsP/InP diffraction-coupled laser array. *(From Ref. 40b, with permission.)*

passivation coating should be applied to further enhance the threshold power density before catastrophic optical damage occurs. To increase output from the front facet, a highly reflective multilayer can be applied to the rear facet.

Higher output power allows longer distances between fiber-optic repeater stations and larger fan-outs for short-distance, fiber-optic links. Also, for analog modulation systems, higher output power implies higher optical power density inside the cavity, which, in turn, implies larger modulation bandwidth (see Sec. 5.5.4).

For ordinary buried-heterostructure lasers, a CW output greater than 100 mW is possible. While facet coating can help to increase the output power, additional approaches are needed if more power is required. As with GaAs/AlGaAs lasers, high-power InGaAsP/InP lasers can be achieved with a large optical cavity structure. In addition, high power can be produced by combining output power from closely spaced, multiple-stripe lasers—i.e., a laser array, as shown in Fig. 5.30.

### 5.5.2 Laser Diode Radiation Patterns and Spectral Characteristics

The radiation pattern of a laser diode is closely related to the structure of the laser waveguide. The lateral modal pattern of laser structures can be understood in terms of waveguiding theory.[34] For gain-guided lasers, the active stripe width is usually wide enough to support several lateral modes in the dimension parallel to the heterojunction. However, the width of the gain region can increase with injection level, thus making it difficult to control the radiation pattern.

For index-guided laser structures, the active region is surrounded by materials of lower refractive index. Such a waveguide can be tailored to provide fundamental transverse-mode operation in directions both parallel and perpendicular to the heterojunction.[41] For planar buried-heterostructure lasers with abrupt lateral heterointerfaces surrounding the active region, the resulting waveguide has a strong waveguiding effect on the propagating mode. For 1- to 3-μm wavelength lasers, the active region width should be less than about 2 μm, to avoiding higher-order-mode propagation. For buried-heterostructure lasers with a crescent-shaped active region (due to the gradual tapering of the active region), single fundamental transverse-mode operation can be maintained for a wider stripe.[42] Higher transverse-modes can appear at higher injection levels in buried-heterostructure lasers as a result of the spatial hole burning mentioned earlier. Beam divergence in the lateral direction is typically 20 to 25°.

Because of the thin active region (on the order of 0.15 to 0.2 μm), the beam emitted at the output face of a heterostructure laser diverges considerably in the plane perpendicular to the heterojunctions, resulting in a far-field angle ranging from 35 to 45°. By increasing the active region thickness, the divergence angle can be reduced until the first-order mode cutoff is reached. However, increasing the active layer thickness inevitably increases the threshold current. A better approach is to use a large optical cavity which is composed of a layer of InGaAsP with a slightly larger bandgap and lower index next to the active layer. The smaller index step allows the optical mode to spread out in the direction perpendicular to the heterojunction, thus narrowing the output beam.

***Longitudinal-Mode Spectral Characteristics.*** As mentioned earlier, gain-guided lasers are quite different from index-guided lasers in terms of their respective longitudinal-mode spectra. Gain-guided lasers have many longitudinal modes even at low power (as shown in Fig. 5.31*a*), while index-guided lasers are usually single mode at low power (see Fig. 5.31*b*). This is attributed to the fact that the rate of spontaneous emission into each longitudinal mode in a gain-guided laser is greater than that of an index-guided laser of the same dimensions. The curved wavefront in the junction plane of gain-guided lasers can enhance the $K$ factor in Eq. (5.14).

Index-guided lasers can be designed to oscillate mainly in a longitudinal mode. However, because of the strong coupling between the photon and carrier densities and the delicate balance between gain and loss, even index-guided lasers—which are highly single mode under CW operation—can become multimode in transient or pulsed operation. Also, during pulsed operation, the individual mode linewidth is broadened (i.e., chirping occurs) because of the change in refractive index (resulting from the changing carrier density), which shifts the mode frequency.

***Spectral Line Width.*** Frequency-stabilized, single-mode (both longitudinally and laterally) lasers are very attractive for long-distance, coherent communication systems which accommodate very high information capacity. Successful operation of these lasers, however, depends on the control of the spectral linewidth of the lasing mode, which is related to the noise characteristics.

The fundamental double-heterojunction laser noise characteristics were first investigated by Fleming and Mooradian,[43] who found the variation in linewidth, $\Delta\nu$, was inversely related to the output power, as predicted by the modified

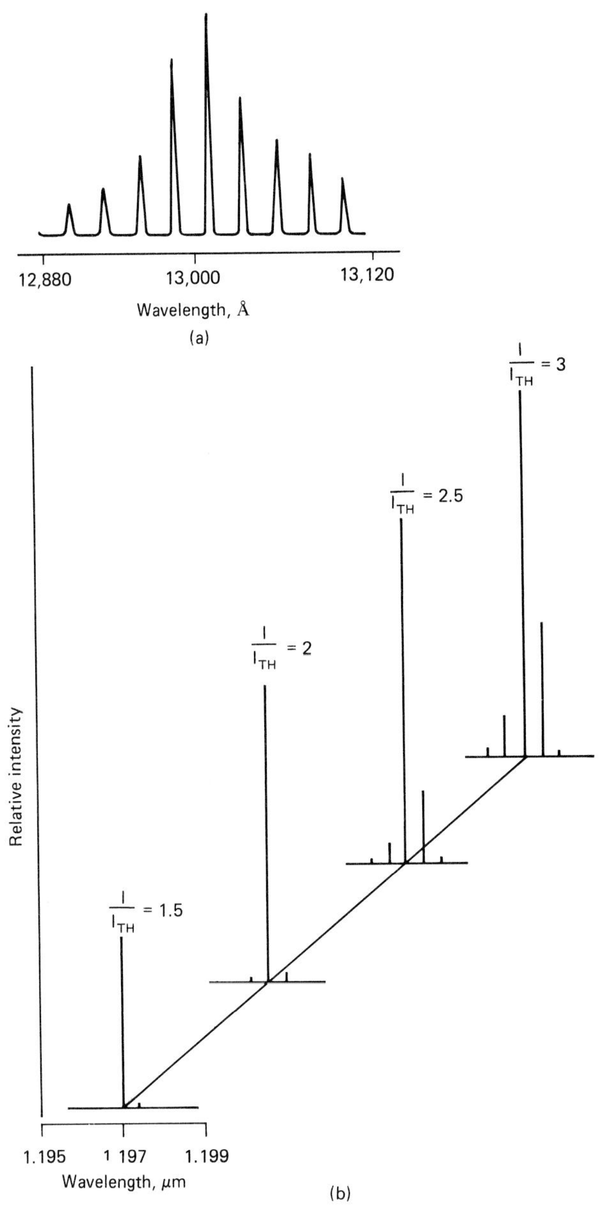


**FIGURE 5.31** Longitudinal mode spectra of (*a*) an InGaAsP/InP oxide stripe laser at an injection current level 1.3 times the threshold current (i.e., $I/I_{th}$ = 1.3), where the cavity length of the laser is about 250 μm, and (*b*) a buried crescent InGaAsP ($\lambda_g$ = 1.2 μm) laser at different injection levels. (*From Ref. 42, with permission.*)

Schawlow-Townes formula. Furthermore, the linewidth of laser diodes can be broadened by an extra factor, $(1 + \alpha^2)$, where $\alpha$ is the linewidth enhancement factor.[44,45] These linewidth factors can be expressed in the following terms:

$$\Delta\nu = \frac{V_g h\nu \Gamma G R \beta}{\pi P}(1 + \alpha^2) \tag{5.18}$$

and

$$\alpha = \frac{\partial \chi_R / \partial n}{\partial \chi_I / \partial n} \tag{5.19}$$

where $V_g$ = group velocity
$h\nu$ = photon energy
$\Gamma$ = mode confinement factor
$G$ = threshold gain
$R$ = reflectivity
$\beta$ = spontaneous emission factor
$P$ = laser output power
$n$ = carrier density
$\chi_R$ and $\chi_I$ = real and imaginary parts of the complex susceptibility $\chi$

The factor $\alpha$ reflects the strong amplitude-phase coupling of the optical field in the laser resulting from the highly detuned gain spectrum. Attempts to reduce this linewidth include the use of an external mirror, a coupled cavity, distributed feedback, or quantum wells in the active region.

### 5.5.3 Laser Diode-Fiber Coupling

In order to launch stable, maximum optical power into a fiber, effective schemes to couple light from lasers to fibers are necessary. Low coupling loss is difficult to achieve because of the small fiber spot size. For a single-mode fiber whose spot size can be as small as 5.5 μm, the laser-fiber coupling is even more difficult to optimize. Usually, a lens or combination of lenses is employed to shrink the fiber waist ($2w_f$) to match the laser waists ($2w_{\parallel}$ and $2w_{\perp}$), as shown in Fig. 5.32*a*, and thus bring about a high coupling efficiency. However, this can reduce the tolerance for offset misalignment. With waists coplanar and aligned, the coupling efficiency $\eta_a$, which is the ratio of the differential change in the coupled power to that of the laser diode output power (as a result of injection level change), can be expressed as[46]

$$\eta_a = \frac{4}{(w_f/w_{\perp} + w_{\perp}/w_f)(w_f/w_{\parallel} + w_{\parallel}/w_f)} \tag{5.20}$$

For simplicity, one can assume the laser beam is gaussian, therefore $w_{\parallel}$ equals $w_{\perp}$. When the laser and the fiber are offset by a radius $r$, the actual coupling efficiency $\eta$ becomes

$$\eta = \eta_a e^{(r/r_e)^2} \tag{5.21}$$

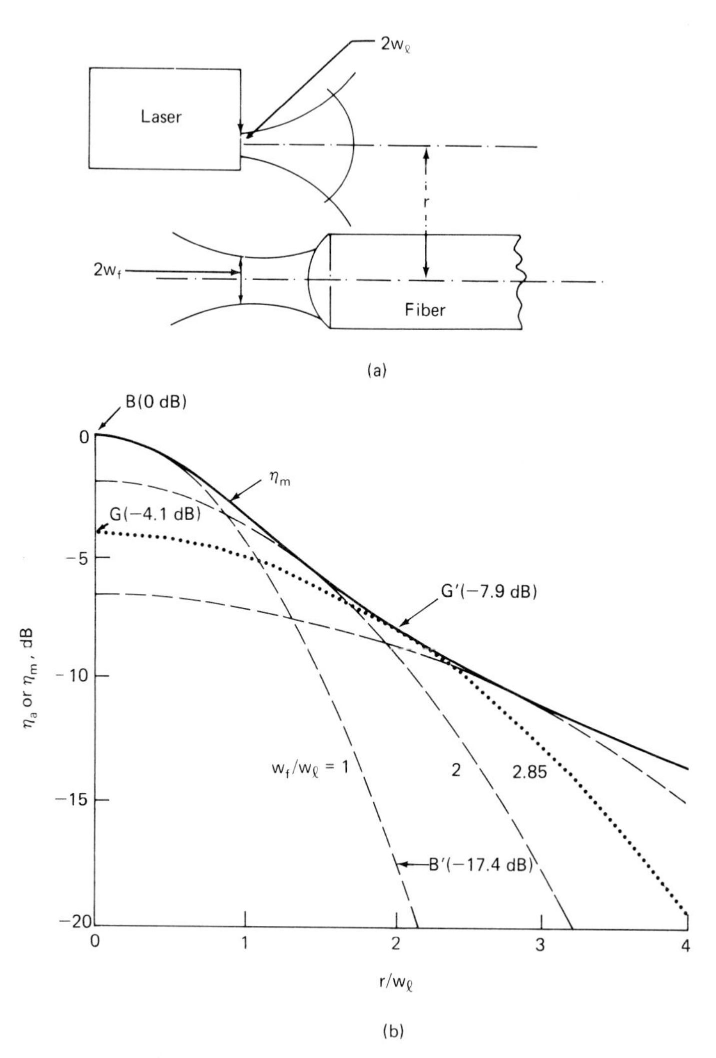


**FIGURE 5.32** Laser-fiber coupling efficiency. (*a*) Coupling geometry for a laser and a fiber having parallel axes and coplanar waists $2w_l$ (where $w_l = w_{\|} = w_{\perp}$) and $2w_f$, which are misaligned by offset distance $r$; (*b*) coupling efficiency $\eta_a$ (broken lines) and envelope $\eta_m$ (solid line) vs. $r$ for various $w_f/w_l$ ratios. (*From Ref. 46a, with permission.*)

where the lateral offset tolerance $r_e$ is given by

$$r_e = \sqrt{\frac{w_{||}^2 + w_f^2}{2}} \tag{5.22a}$$

and the angular offset tolerance $\theta_b$ is

$$\theta_b = 60 \frac{\lambda}{\pi^2} \left( \frac{1}{w_{||}^2} + \frac{1}{w_f^2} \right)^{1/2} \tag{5.22b}$$

The coupling efficiency as a function of $r/w_{||}$ is shown in Fig. 5.32*b*; it is concluded that a higher tolerance can be obtained for $w_f/w_{||}$ values greater than unity.

In general, methods using short-focal-length lenses lead to tighter tolerances, which can also cause difficulties in fabricating laser diode modules. In addition, there exist problems concerning temperature characteristics and hermeticity of the package.

Efficient coupling between a hermetically packaged channeled-substrate planar laser and a standard graded-index fiber has been achieved by a method involving the deformation (at room temperature) of an indium layer at the cap of the package, thereby providing a means to fix the position of a microlens relative to the laser and the fiber. With this method, an average coupling loss of 0.9 dB was obtained. The coupling efficiency varied by less than 0.1 dB during a test in which temperature was cycled 500 times between 20 and 70°C. This stability is attributed to the small size and the cylindrical shape of the package.

For reasons of stability and reliability, most methods developed for coupling a laser to a single-mode fiber require a hermetic fiber seal. Coupling can be accomplished in a number of ways. For instance, a cylindrical lens can be placed between the laser and the fiber, inside a fiber seal, to increase the coupling efficiency. To facilitate final alignment, a hemispherical glass droplet can be attached and aligned to the fiber core end before it is inserted into the package. Alternatively, the end face of the fiber can be polished into a quadrangular, pyramid-shaped hemicylindrical lens.

However, there are other methods which, though requiring a hermetically sealed fiber, have the advantage of automatic alignment between the lens and the fiber core and are thus attractive for low-cost transmission systems. For instance, in the method shown in Fig. 5.33, a conical fiber-end section (for shrinking the fiber waist) is obtained by etching the fiber end, with a resultant average coupling loss of 3 dB. In the high-index, tapered-end method, the fiber is heated and pulled to form a tapered end section, which is then dipped into molten glass of high index and pulled out to form a high-index lens by surface tension. With this technique, an average coupling loss of −2 dB is obtained. A simpler approach is to circumferentially etch the fiber end to form a lens. This requires the core to be concentric to the cladding for efficient coupling. In a similar method, the tapered end section is rounded by heating. The last two methods produce an average coupling loss of 3 dB.

Other methods employing separate single lenses—but without a hermetic fiber seal—can produce high coupling efficiency. However, the lenses must be tightly aligned individually in both axial and lateral directions. For instance, with a high index microbead, coupling losses as low as 1.6 dB can be achieved. Similarly, the ball and graded-refractive-index (GRIN) lens method, as shown in Fig. 5.34, and

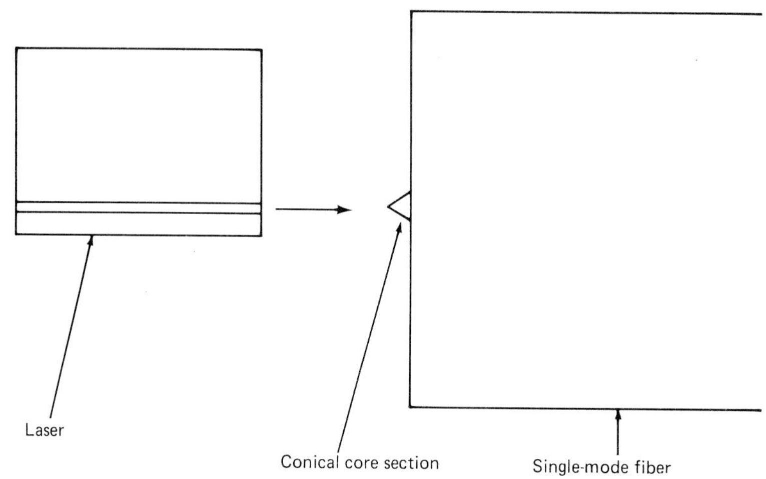


FIGURE 5.33 Laser to single-mode fiber coupling technique, utilizing etching to produce a conical core section at the fiber end.

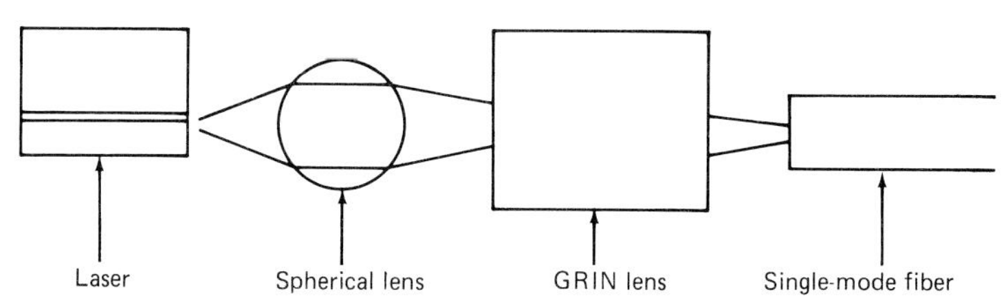


**FIGURE 5.34** Ball and GRIN lens scheme for laser to single-mode fiber coupling.

the convex GRIN lens method also produce coupling losses of around 3 dB. In all of these cases, the coupling loss has been dominated by lens aberration.

Coupling efficiencies greater than 50 percent can be achieved for a single-mode laser to single-mode fiber coupling, by means of a single-lens scheme or with fabricated, single-mode, fiber-end faces. However, the lateral and angular tolerances can be very tight. Looser tolerances can be obtained with the confocal, two-lens method, in which a large spherical lens is used to expand the laser waist and a GRIN-rod lens is used to enlarge the fiber waist. An extension of this method, as shown in Fig. 5.35, uses a spherical lens plus *two* GRIN-rod lenses to further ease tolerances. Besides high coupling efficiency and high misalignment tolerance to mechanical vibration and temperature variation, the coupling should result in low optical feedback into lasers, as lasers are very sensitive to both coherent and incoherent reflected light from the end fiber facet.[47,48] Noncoherent feedback leads to increased laser intensity noise, while coherent feedback alters the light amplitude and also shifts the spectrum. With a short, point-tapered (radius $< 10$ μm) fiber, high coupling efficiencies (75 percent), as well as very little feedback (less than $-30$ dB), can be obtained. In another scheme, the optical feedback is isolated by a nonreciprocal optical path in which

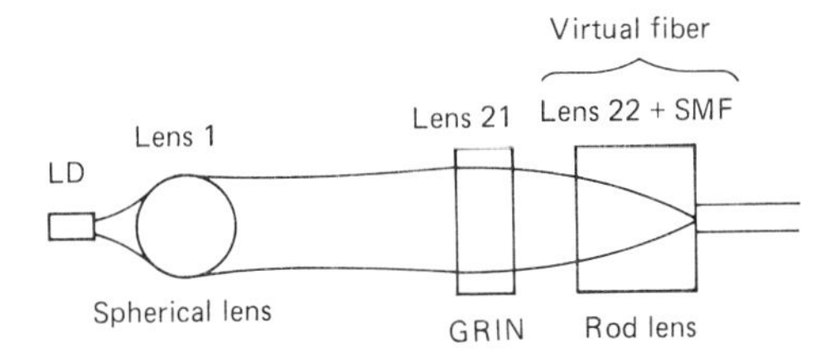


**FIGURE 5.35** Laser to single-mode-fiber coupling using a spherical lens and two GRIN-rod lenses. *(From Ref. 48a, with permission.)*

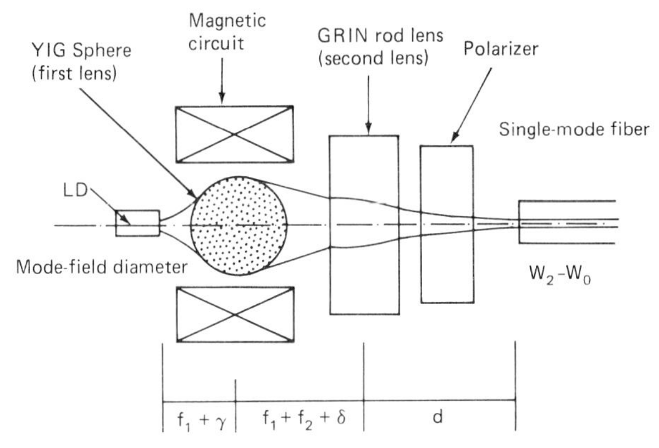


**FIGURE 5.36** Coupling of laser to single-mode fiber employing a YIG sphere for isolation. *(From Ref. 48b, with permission.)*

an yttrium iron garnet (YIG) GRIN-rod lens is inserted between the laser and a polarizer. The YIG lens shown in Fig. 5.36 serves both as a Faraday rotator for optical isolation and as a lens for effective coupling.

### 5.5.4 Laser Diode Bandwidth

***Laser Diode Direct-Modulation Response.*** For many present and future transmission systems, it is necessary to modulate the laser output at high frequency or high data rates. For reasons of simplicity, DC modulation has been the most popular method. At small-signal modulation, the output intensity spectra of laser diodes are relatively flat from low frequency, as shown in Fig. 5.37, until a resonance frequency—the relaxation oscillation frequency—is reached. This resonance is caused by the intrinsic coupling between photons and injected carriers inside the laser cavity. The output response rolls off quickly beyond this frequency. Coincidentally, the intensity noise spectrum of lasers, whose intensity varies as the square root of the injection current above the laser threshold cur-

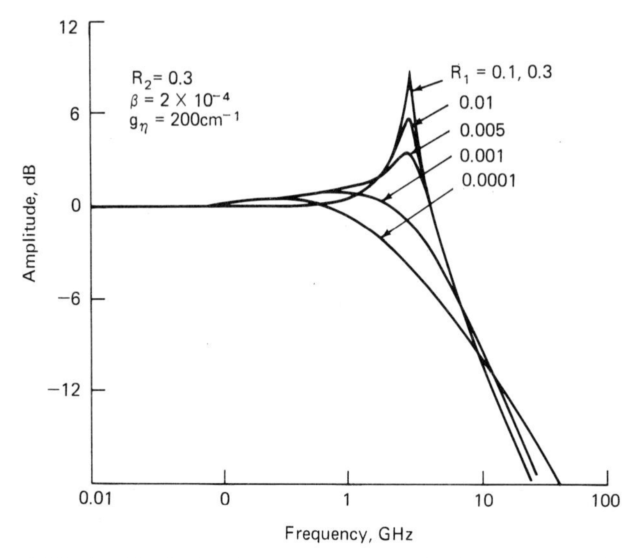


**FIGURE 5.37** Typical frequency responses for laser diodes having different front mirror reflectivities $R_1$. *(From Ref. 50, with permission.)*

rent, also peaks at this frequency. The existence of this resonance peak often limits the maximum direct modulation bandwidth of laser diodes to the multigigahertz range.

The dynamics of photons and carriers in lasers can be described with the help of the following one-dimensional rate equations:[49]

$$\frac{\partial I^+}{\partial t} + c\frac{\partial I^+}{\partial Z} = ANI^+ + \beta\frac{N}{\tau_s} \tag{5.23}$$

$$\frac{\partial I^-}{\partial t} - c\frac{\partial I^-}{\partial Z} = ANI^- + \beta\frac{N}{\tau_s} \tag{5.24}$$

$$\frac{dN}{dt} = \frac{J}{ed} - \frac{N}{\tau_s} - AN(I^+ + I^-) - \alpha N \tag{5.25}$$

where $I^+$ and $I^-$ = photon intensities of traveling waves in positive and negative $z$ directions, respectively, along laser waveguide
$N$ = carrier density
$A$ = differential optical gain constant (strongly dependent on material and temperature)
$\beta$ = fraction of spontaneous emission entering lasing mode
$\tau_s$ = carrier lifetime
$d$ = active layer thickness
$\alpha$ = distributed cavity loss
$J$ = injected current density
$c$ = velocity of light inside the cavity

By using Eqs. (5.23) and (5.24) in a small-signal analysis and by spatially averaging the photon and carrier densities inside the laser cavity, the relaxation oscillation frequency $f_r$ can be approximated as[50]

$$f_r = \frac{1}{2\pi}\sqrt{\frac{AP_0}{\tau_p}} \qquad (5.26)$$

where $P_0$ = steady-state photon density
$\tau_p$ = photon cavity lifetime (related to device parameters such as cavity length, facet reflectivity, and distributed loss $\alpha$)

$P_0$ is believed to be strongly related to other laser parameters such as linewidth (see Sec. 5.4.2), relative intensity noise, and catastrophic facet damage.

As can be seen from Eq. (5.26), the intrinsic laser modulation bandwidth can be enhanced by increasing the optical gain coefficient or the photon density or by decreasing the photon lifetime. These parameters can be separately adjusted to a certain degree through various means. For example, the gain coefficient $A$ can be increased by about a factor of 5 by cooling the laser from room temperature down to 77°K. Photon density in the active region can be increased by raising the forward bias level of the laser, but the optical output intensity at the laser facet must stay within the catastrophic mirror damage limit of mirrors. For GaAs/AlGaAs materials, this limits the maximum power density to about 1 MW/cm$^2$ before permanently damaging the mirrors. One method that has been used to alleviate this problem incorporates a window region near the facets to expand the optical mode volume there or to make window regions out of a higher-bandgap (i.e., nonabsorbing) material. With windowed buried-heterostructure lasers, a modulation bandwidth in excess of 10 GHz has been demonstrated in the GaAs/AlGaAs system. The modulation response of these lasers shows a reduction in—and sometimes an absence of—the resonance peak. This is attributed to superluminescence damping,[51] which reduces the facet reflectivity. For InGaAsP/InP materials, catastrophic mirror damage occurs at a much higher power density, and thus the output power is more likely to be limited by ohmic heating of the device at high injection levels.

Modulation bandwidth can also be enhanced by reducing the photon lifetime, which can be effected by decreasing the laser cavity length. However, as the cavity length shortens, the current density within the active stripe will rise, until a point is reached beyond which long-term reliability is compromised. In addition, the maximum photon density for lasers with short cavities may be smaller because of heating effects.

In addition to intrinsic modulation limitations, laser bandwidth limitations can be produced by the presence of parasitic capacitances, which come from the laser structure itself and from the associated wire bonding and package. Laser structures with a small junction area usually have low capacitance. However, in these lasers, the main parasitic capacitance comes from the area between the top and bottom contacts outside the lasing region. For example, the reverse blocking junctions in the buried heterostructure laser contribute much of the capacitance. Lasers fabricated on semi-insulating substrates have been shown to have low parasitic capacitance because of the distributed network nature of the parasitic reactances.

InGaAsP/InP lasers operating at 1.3-μm wavelength, with a constricted active

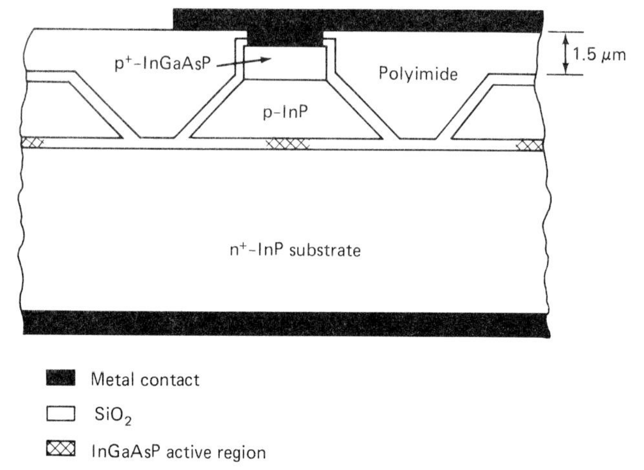


**FIGURE 5.38** Cross-sectional view of a constricted mesa laser. (*From Ref. 51a, with permission.*)

region to limit the area, as shown in Fig. 5.38, have demonstrated a modulation bandwidth in excess of 20 GHz.

***External Modulation of Laser Diodes.*** As discussed, in the direct pulse modulation of the laser diode output, the response is characterized by ringing at turn-on and turnoff because of relaxation oscillation. Associated with this ringing is chirping in the emission wavelength. These effects ultimately limit the maximum data rate and distance of transmission. An alternative means for light modulation is external modulation, where lasers act only as stable light sources whose output is coupled to an external modulator, which produces signal modulation. This modulation scheme has the potential for very high data rate transmission for two reasons. First, the photon and electron densities are not strongly coupled in the external modulator and therefore little relaxation oscillation will be detected. Second, both intensity as well as phase modulation of the laser output can be produced. Phase modulation is important in coherent communication systems because of its low insertion loss.

In external intensity modulators, the absorption of the laser output can be voltage-controlled by means of the electroabsorption effect. In bulk semiconductors, electroabsorption is caused by the spatial tilting of the band edges in response to an applied electric field, which enhances photon-assisted tunneling of electrons from the valence band to the conduction band.[52,53] In two-dimensional, quantum-well semiconductor materials, electroabsorption can be effected by shifting the exciton absorption line of the confined states in response to an electric field. This is commonly known as the *quantum confined Stark effect.*

High-speed intensity modulation (at greater than 10 GHz) of 1.55-μm-wavelength laser light has been demonstrated with ridge-loaded InGaAsP/InP electroabsorption modulators.

Intensity modulation can also be achieved by waveguide electro-optic modulators such as directional couplers, Mach-Zehnder interferometers, or total internal reflectors.[54] These devices depend on the linear electro-optic (Pockels) effect in materials such as lithium niobate ($LiNbO_3$) and lithium tantalate ($LiTaO_3$), where the anisotropic refractive index change is proportional to the electric field[55] and the polarization state of the laser light can thus be controlled and filtered by polarizers. Using external modulation, a high bit-rate/distance (more than 400 Gb · km/s) product has been demonstrated with cleaved, coupled-cavity laser diodes and $LiNbO_3$ external modulators.

Since the refractive index change can be controlled in these materials, they can be used for phase modulation of the laser output.

### 5.5.5 Linearity

A major difference between GaAs/AlGaAs and InGaAsP/InP laser diodes is the higher temperature sensitivity of the threshold current of InGaAsP/InP lasers. Experimentally, it is found that the threshold current $J_{\text{th}}$ of lasers can be approximately expressed as a function of temperature $T$ by

$$J_{\text{th}} = J_o e^{T/T_o} \tag{5.27}$$

where $T_o$ is a parameter describing the temperature sensitivity. For near room-temperature operation, the intrinsic value of $T_o$ for InGaAsP/InP lasers is approximately 60 to 70°K for 1.3-μm operation and 50 to 70°K for 1.55-μm operation. For GaAs/AlGaAs laser diodes, $T_o$ is about 110 to 140°K.

It has been suggested that the small values of $T_o$ observed in InGaAsP/InP lasers can be a result of a combination of mechanisms such as the carrier leakage over the heterobarrier, intervalence-band absorption, and interband Auger recombination.

Nonradiative interband Auger recombination processes can be significant in low-bandgap semiconductors and at high temperature. Among the 16 possible processes, only two of them, the CHCC and CHSH processes (C, H, and S denote conduction band, heavy hole band, and split-off hole band, respectively), shown in Fig. 5.39, are believed to be dominant in InGaAsP materials, especially when the carrier concentration exceeds $10^{18}$ $\text{cm}^{-3}$. In InGaAsP materials, which range from 1.0 to 1.6 μm in bandgap, it is estimated that the Auger lifetime of the carrier $\tau_A$ at high injection can be shorter than the radiative recombination lifetime $\tau_A$. This results in a higher threshold current and a reduced quantum efficiency as temperature increases. The threshold current density can be expressed as

$$J = ed\,\Delta n\left(\frac{1}{\tau_r} + \frac{1}{\tau_A}\right) + J_{\text{leak}} \tag{5.28}$$

where $\Delta n$ = injected carrier density
$J$ = density of total current injected into active region of thickness $d$
$J_{\text{leak}}$ = current-density component resulting from carriers which leak over heterobarrier and drift away under influence of electric field at heterointerface

Since the electric field at the heterojunction increases with bias, the leakage current continues to increase at levels above threshold.

Before carriers can effectively escape over the potential step at the heterointerface, they have to gain sufficient energy. In other words, the carrier

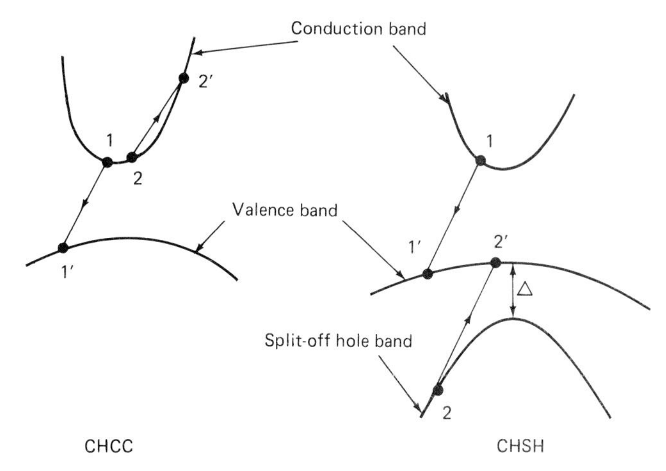


**FIGURE 5.39** Schematic diagrams of the normal CHCC and CHSH Auger processes, where C, H, and S stand for conduction-band electron, heavy-hole, valence-band electron, and split-off-hole, valence-band electron, respectively.

temperature has to exceed the lattice temperature. However, this extra kinetic energy can be derived from Auger recombination processes, and thus these two processes may be closely related. It has been observed that the rise in carrier temperature for 1.55-μm-wavelength materials during optical pumping is about 3 times faster than that for 1.3-μm materials. The $T_o$ value is therefore expected to be smaller in 1.55-μm material. For gain-guided InGaAsP/InP lasers, carriers are poorly confined in the junction plane. This, together with the temperature sensitivity of the threshold current of InGaAsP/InP lasers, can result in thermal runaway at high current levels and thus limit their use for long-haul communication systems. As shown in Fig. 5.40*a*, the light output vs. current (*L-I*) characteristic of 1.3-μm gain-guided lasers typically exhibits kinks as a result of lateral-mode instabilities.

For index-guided lasers, threshold currents are much lower (i.e., 10 to 30 mA) than those for gain-guided lasers (60 to 150 mA). The threshold currents and $T_o$ for various buried-heterostructure lasers are tabulated in Table 5.1. The *L-I*

**TABLE 5.1** Laser Structure Performance Comparison

| Laser structure | Threshold current at 23°C (mA) | $T_o$, °K |
|---|---|---|
| Planar BH | 10–20 | 70–106 |
| Nonplanar BH | 10–20 | 60–80 |
| Continuous active index-guided | 50–70 | 60–70 |

curves (see Fig. 5.40*b*) for these lasers are usually quite linear and kink-free, indicating stable, single-lateral-mode operation. The occurrence of kinks in the *L-I* curves of some index-guided lasers can be caused by mode stabilities in the laser

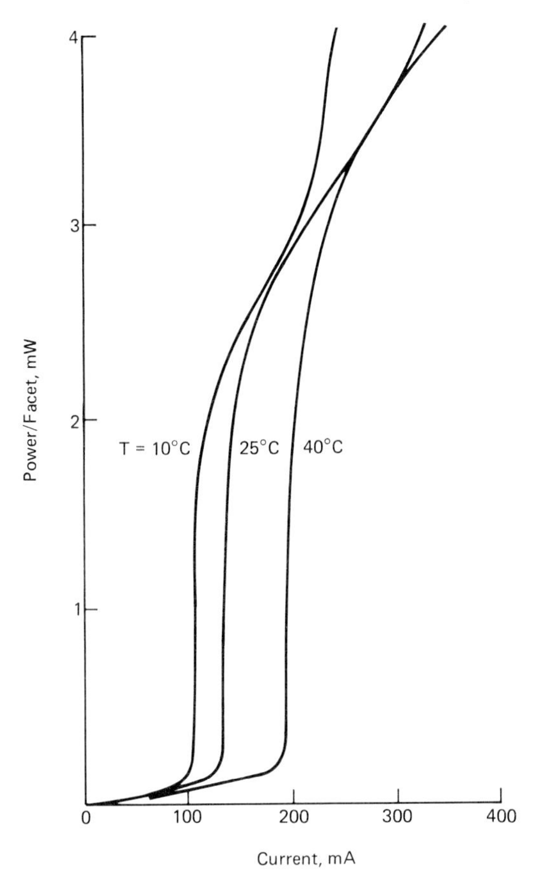


**FIGURE 5.40** Output power vs. current characteristics. (*a*) An InGaAsP/InP ridge waveguide laser for various temperatures. (*From Ref. 40, with permission.*)

waveguide at high injection. As noted before, the spatial hole burning in the active region along the junction is responsible for the higher-order-mode oscillation. The gradual decrease of external differential quantum efficiency at higher injection levels can be explained by a combination of the hot-carrier-induced leakage effect, the heating effect, and the effect of enhanced, reverse-biased junction leakage. The maximum operating temperature of index-guided lasers is thus limited by both the higher threshold current and the lower quantum efficiency at high temperature.

In linearity studies of lasers, electrical derivative techniques, which involve taking the $I\ dV/dI$ versus $I$ or the $dV/dI$ versus $I$ characteristics ($V$ and $I$ stand for voltage and current, respectively) have been very successful in determining the leakage current in the reverse-biased junction, to correlate the lateral-mode behavior and to identify processes affecting laser reliability. Furthermore, these results can be correlated with the $dL/dI$ characteristics ($L$ stands for light intensity) as a means of screening unreliable lasers, as described in Sec. 5.5.7.

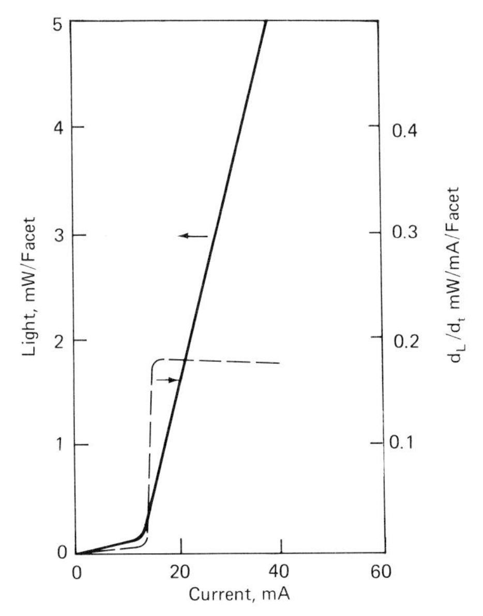


**FIGURE 5.40** Output power vs. current characteristics. (*b*) An InGaAsP/InP buried-heterostructure laser, with the *dL/dI* characteristic shown by the dashed line. (*From Ref. 55a, with permission.*)

### 5.5.6 Drive Circuits

The simplest driver to exploit the high-speed capability of lasers is a modified shunt driver using a GaAs metal-semiconductor, field-effect transistor (MESFET). The off-state current is usually set to a value close to and above threshold to minimize the turn-on delay, although this compromises the extinction (on/off) ratio of the light output. However, in practical transmitter designs with lasers as emitting sources, special attention must be paid to temperature control and the effects of laser aging. Since the threshold current and external quantum efficiency of InGaAsP/InP lasers are sensitive to temperature changes, separate circuits must be used to monitor and regulate the operating temperature. Also, as lasers age, the threshold current increases, and thus less light will be coupled to the fiber at the same current. The drive circuit, therefore, should have a feedback mechanism to adjust the current.

A feedback-stabilized laser driver circuit is shown in Fig. 5.41. In this circuit, the feedback from a photodiode is compared with the data pattern and a fixed reference, in order to control laser bias. This emitter-coupled circuit can provide output stability to within 1 percent at bit rates up to 274 Mb/s. The photocurrent detected by the photodiode, which is coupled to the back mirror, can be used to monitor the emission at the front mirror. However, this scheme can suffer from front-to-back mirror mistracking due to aging effects or to slight misalignment of a single-mode fiber during operation. Alternatively, light can be tapped from the fiber to the photodiode through a low-loss coupler.

Another laser control circuit, shown in Fig. 5.42, employs automatic bias con-

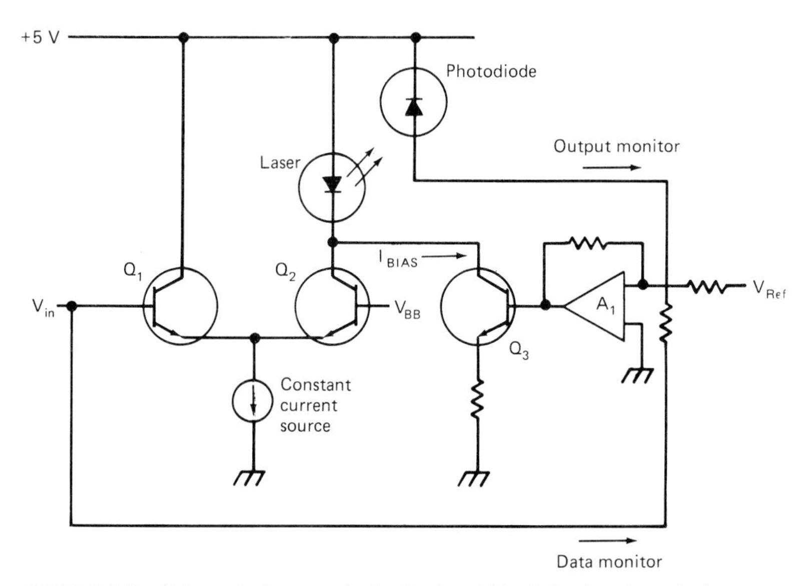


**FIGURE 5.41** Schematic diagram of a feedback-stabilized circuit, where the laser output from either a rear facet or a fiber tap is compared with the data pattern and a fixed reference to control the laser bias. *(From Ref. 55b, with permission.)*

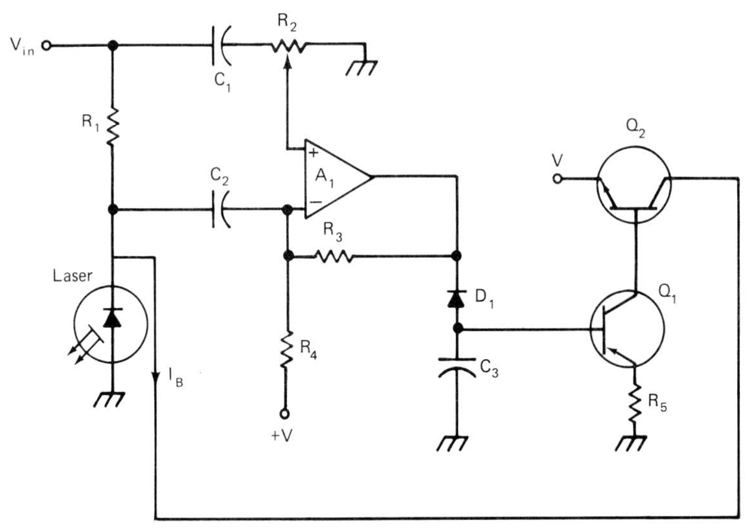


**FIGURE 5.42** Schematic diagram of an automatic bias control circuit that senses the laser threshold electrically. *(From Ref. 55c, with permission.)*

trol. This circuit makes use of the fact that the laser junction voltage becomes pinned at the onset of stimulated emission. By raising the bias current until the junction voltage is sensed to have reached a pinned value, the bias can be adjusted in response to temperature or aging.

Monolithic integration of lasers and electronic driving circuits on the same substrate, as initially proposed by Yariv,[56] offers such potential advantages as high-speed operation due to reduction of parasitic reactances, compactness, and ease of packaging. In GaAs/AlGaAs systems, the availability of high-performance laser sources and FETs has permitted the monolithic integration of source and driver. Figure 5.43 shows the laser driver portion of one such circuit, which is fabricated on a semi-insulating GaAs substrate and is designed to accept an input signal voltage compatible with silicon ECL levels. It consists of a differential current switch and a buffer amplifier. The differential circuits have a pulse reshaping function and are suitable for high-speed applications. Bit rates as high as 2 Gb/s have been obtained with this circuit when it is integrated with a laser diode and photomonitor.

The major breakthrough with these circuits is fabricating them on a common substrate with laser diodes. However, this requires the formation of laser facets by novel approaches such as reactive ion etching.[57] In the InGaAsP/InP system, the challenge is obtaining reliable, high-speed FETs or bipolar transistors on an InP substrate.

### 5.5.7 Aging

The study of laser and driver aging has been difficult because of the incomplete information gained from life test results. In the past, most laser aging studies were concentrated on the GaAs/AlGaAs systems, where standard experiments and criteria have been established. The median lifetime of GaAs/AlGaAs multimode lasers operated under CW conditions has exceeded $10^5$ h at room temperature. By comparison, the median lifetime of long-wavelength InGaAsP/InP lasers is much greater.

Laser diode lifetimes have been prolonged by improvements in the device structure and fabrication processes. However, for undersea optical transmission applications, the predicted laser reliability has not yet been confirmed, as no actual data for 25-year operation (which has been accepted as the mean service period of undersea systems) has been obtained. At present, most reliability studies have been focused on how to determine, from preliminary lifetime data, whether a particular batch of diodes should be used for a certain application or should be rejected.

Laser lifetime is usually measured by two criteria: changes in the lasing threshold $I_{th}$ and laser efficiency $\eta_c$. Both require the measurement of the optical output power from a laser facet by a large-area detector. Often the measurement is performed at elevated temperatures and results are extrapolated by an Arrhenius relationship to room temperature for an estimate of the laser lifetime.

In addition to testing individual-packaged laser diodes, lifetime tests have been extended to laser transmitter subsystems. In this case, a fiber-pigtailed laser is tested with its companion circuit elements. Part of the light output is tapped from the fiber and coupled to a photodiode, which adjusts the bias current to the laser, as described in Sec. 5.5.6.

Like LEDs, laser diode samples usually must pass several screening tests to eliminate infant failures before they are put into reliability tests. The screening

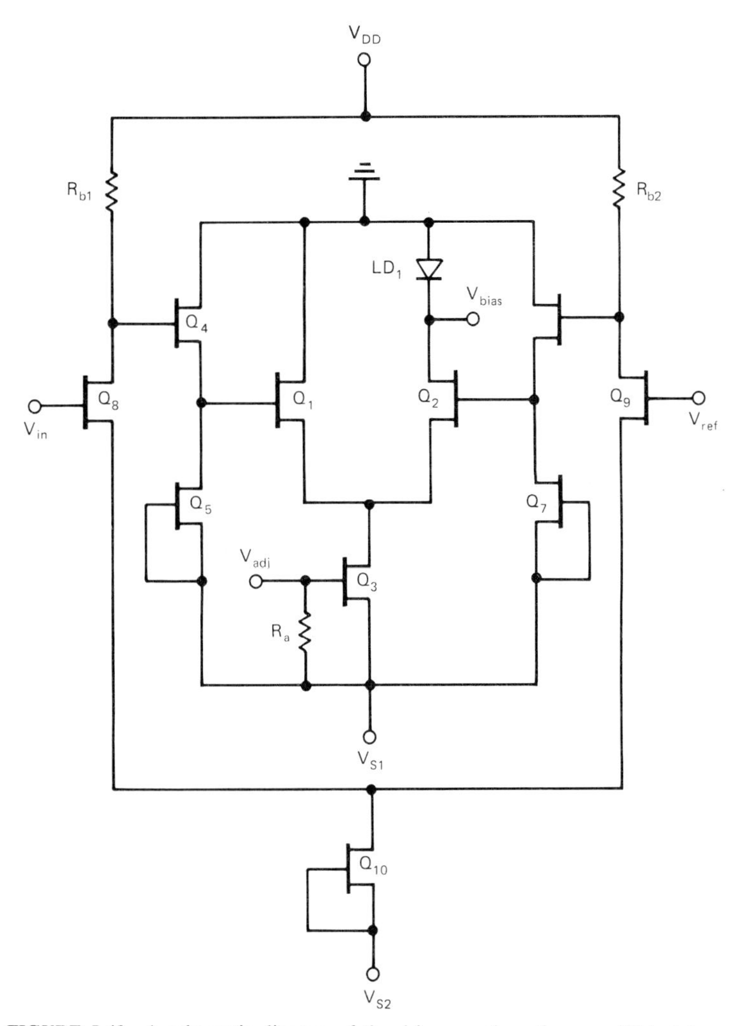


**FIGURE 5.43** A schematic diagram of the driver portion of a monolithic integrated circuit, incorporating source and driver on a semi-insulating GaAs substrate. The FETs $Q_1$, $Q_2$, $Q_3$ relate to the current-mode-logic switching circuit, while the other FETs relate to a buffer amplifier. The input signal is at gate $Q_8$. *(From Ref. 57a, with permission.)*

tests may involve operation at high injection levels (electrical stress) and/or high output level under high temperature (thermal stress) for several days. Those devices which show kinks in the *L-I* characteristics are eliminated from the life test. The remaining samples are operated at constant drive current, while under constant temperature. Since the InGaAsP/InP laser threshold current is quite sensitive to temperature changes, life test results taken at higher temperature (e.g., 60 to 70°C) are used to reveal information about structure and manufacturing process imperfections.

For a statistical study of semiconductor device aging, it is useful to divide failure modes into random and wear-out failure. Wear-out failures are mainly a result of gradual degradation of electrical and optical characteristics. The failure rate $f_w$ can be written as follows:[58]

$$f_w(t) = \frac{\sqrt{2}\,\exp\,[-\,(\ln t/t_m)^2/2\sigma^2]}{\sqrt{\pi}\,t\,\sigma\,\mathrm{erfc}\,[(\ln t/t_m)/\sqrt{2}\sigma]} \tag{5.29}$$

where $t$ = service time
$t_m$ = median lifetime
$\sigma$ = standard deviation in logarithmic time scale for logarithmic failure distribution

The longer $t_m$ and the smaller $\sigma$ are, the lower the wear-out failure rate will be. On the other hand, random failures $f_r$ are due to sudden failures and can be regarded as constant with time:

$$f_r(t) = \frac{r}{Nt} \tag{5.30}$$

where $r$ = number of failed devices
$N$ = sample size
$t$ = testing time, h

Lasers of the buried-heterostructure, double-channel planar buried-heterostructure, and V-groove stripe types all possess low threshold, high efficiency, and high stability characteristics and are predicted to have long lifetimes, i.e., in excess of $10^6$ h.

*Failure Mechanisms.* In the case of GaAs/AlGaAs lasers, similar gradual degradation behavior is observed when they are operated in both the lasing mode (above threshold) and the LED mode (below threshold). The independence of degradation activation energy with output power levels suggests that similar mechanisms are responsible for the gradual degradation of both lasers and LEDs, namely, those discussed in the section on aging mechanisms. In particular, dark-line defects are found in InGaAsP/InP lasers, and they are associated with lattice dislocations caused mainly by lattice mismatch. For lasers fabricated from wafers with an active-layer lattice mismatch less than $4 \times 10^{-4}$, no signs of dark-line defects are observed during life tests. Dark-spot defects are observed in some degraded lasers in the stripe region and they are associated with segregation of Ga and As atoms, or with mass transport during regrowth.

However, the rapid degradation of InGaAsP/InP lasers could be caused by a combination of other mechanisms. For the buried crescent lasers, it has been found that the rapid degradation at the high temperature is attributed to a decrease in the built-in potential of the InP *pn* junction formed between the *n*-InP cladding layer and the *p*-InP current-blocking layer. As a result of this degradation, the laser threshold will increase without affecting the external differential quantum efficiency, as shown in Fig. 5.44. This degradation can be eliminated by displacing the InP *pn* junction from the metallurgical interface by means of Zn out-diffusion. Similarly, in investigations of double-channel planar buried-heterostructure lasers under high electrical stress, it has been shown that, as la-

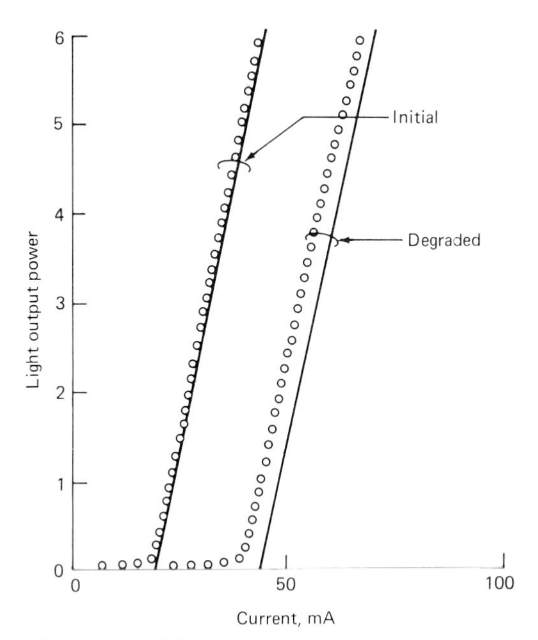


FIGURE 5.44 Calculated (solid line) and the measured (circles) light output vs. current of a 1.3-μm buried crescent laser before and after degradation. *(From Ref. 58a, with permission.)*

sers age, the leakage current gradually increases. This can be partially explained by the leakage around the mesa caused by thermal etching during regrowth.

Sudden failure is frequently a result of complications at the metal-semiconductor contact and the heat sink. Some system applications requiring highly reliable lasers favor the use of a laser chip mounted with the emission junction away from the heat sink. This technique avoids the shunt-path formation due to solder materials and also separates the metallization materials from the heat sink. Use of more stable solder materials, such as Au-Sn alloys, can also ease the problem of solder influx. Thermal characteristics of the substrate can sometimes account for the sudden failure of lasers, as demonstrated in the case where the *p*-InP substrate is used and the laser is mounted with the substrate side down on the heat sink. Of course, some laser failures can also be caused by hybrid circuit failures (loose bonds, transistor failure, etc.).

With careful attention to laser structures, metal-semiconductor contacts, and heat sinks, InGaAsP/InP lasers can meet the requirements imposed by transoceanic optical transmission systems.

## REFERENCES


1. T. Miya, Y. Terunuma, T. Hosaka, and T. Miyashita, "Ultimate Low-Loss Single-Mode Fibre at 1.55 μm," *Electronics Letters,* vol. 15, 1979, p. 106.

1a. S. E. Miller and A. G. Chynoweth (eds.), *Optical Fiber Telecommunication*, Academic Press, New York, 1979.

2. J. J. Hsieh, "Room Temperature Operation of GaInAsP/InP Double Heterostructure Diode Lasers Emitting at 1.1 μm," *Applied Physics Letters*, vol. 28, 1976, p. 283.

3. G. H. Olsen, "Vapor-Phase Epitaxy of GaInAsP," in T. P. Pearsall (ed.), *GaInAsP Alloy Semiconductors*, Wiley, New York, 1982, p. 11.

4. A. Y. Cho and J. R. Arthur, "Molecular Beam Epitaxy," *Progress in Solid State Chemistry*, vol. 10, pt. 3, 1975, p. 157.

5. R. D. Dupuis and P. D. Dapkus, "Continuous Room Temperature Operation of $Ga_{1-x}Al_xAs$-GaAs Double Heterostructure Lasers Grown by Metalorganic Chemical Vapor Deposition," *Applied Physics Letters*, vol. 32, 1978, p. 406.

6. C. A. Burrus and B. I. Miller, "Small Area, Double Heterostructure Aluminum-Gallium Arsenide Electroluminescent Diode Sources for Optical Fiber Transmission Lines," *Journal of Optical Communication*, vol. 4, 1971, p. 307.

7. M. Ettenberg, H. Kressel, and J. P. Wittke, "Very High Radiance Edge-Emitting LED," *IEEE Journal of Quantum Electronics*, vol. QE-12, 1976, p. 360.

8. J. Hayashi et al., "2 Gb/s and 600 Mb/s Single-Mode Fibre Transmission Using a High Speed Zn-Doped 1.3 μm Edge Emitting LED," postdeadline paper, *Optical Fiber Conference*, Reno, 1987.

9. L. N. Kurbatov et al., "Investigation of Superluminescence Emitted by Gallium Arsenide Diode," *Soviet Physics Semiconductor*, vol. 4, 1971, p. 1739.

10. R. Schimpe and J. Boeck, "Modulation Characteristic and Intensity Noise of CW Superluminescent Diodes," *Electronics Letters*, vol. 17, 1981, p. 715.

11. P. K. L. Yu, K. K. Li, and S. C. Lin, "High Frequency Modulation of Superluminescent Diodes," *Proceedings of the Society of Photo-Optical Instrumentation Engineers*, vol. 559, 1985, p. 149.

12. S. T. Forrest, P. P. Deimel, J-Y. Glacet, and R. A. Logan, "Narrow Spectral Width Surface Emitting LED for Long Wavelength Multiplexing Applications," *IEEE Journal of Quantum Electronics*, vol. QE-20, 1984, p. 906.

13. T. P. Lee and A. G. Dentai, "Power and Modulation Bandwidth of GaAs-AlGaAs High-Radiance LED's for Optical Communication Systems," *IEEE Journal of Quantum Electronics*, vol. QE-14, 1978, p. 150.

14. R. H. Saul, T. P. Lee, and C. A. Burrus, "Light-emitting Diode Device Design," in W. T. Tsang (ed.), *Semiconductors and Semimetals*, vol. 22, pt. C, Academic Press, New York, 1985.

15. O. Wada, H. Hamaguchi, Y. Nishitani, and T. Sakurai, "High Speed Response InGaAsP/InP DH LED's in the 1 μm Wavelength Region," *IEEE Electron Device Letters*, vol. EDL-3, 1982, p. 129.

16. J. A. Borsuk, "Light Intensity Profiles of Surface-Emitting InGaAsP LED's Impact on Coupling to Optical Fibers," *IEEE Transactions on Electron Devices*, vol. ED-30, 1983, p. 296.

17. J. G. Ackenhusen, "Microlens to Improve LED-to-Fiber Optical Coupling and Alignment Tolerance," *Applied Optics*, vol. 18, 1979, p. 3694.

18. R. C. Goodfellow, A. C. Carter, I. Griffith, and R. R. Bradley, "GaInAsP/InP Fast, High-Radiance, 1.05–1.3 μm Wavelength LED's with Efficient Lens Coupling to Small Numerical Aperture Silica Optical Fibers," *IEEE Transactions on Electronic Devices*, vol. ED-26, 1979, p. 1215.

19. O. Wada et al., "High Radiance InGaAsP/InP Lensed LED's for Optical Communication Systems at 1.2–1.3 μm," *IEEE Journal of Quantum Electronics*, vol. QE-17, 1981, p. 174.

20. F. W. Ostermayer, Jr., P. A. Kohl, and R. H. Burton, "Photochemical Etching of Integral Lenses on InGaAsP/InP LED's," *Applied Physics Letters*, vol. 43, 1982, p. 642.

21. J. Heinen, "Preparation and Properties of Monolithically Integrated Lenses on InGaAsP/InP Light-Emitting Diodes," *Electronics Letters,* vol. 18, 1982, p. 831.

21a. J. S. Escher et al., "Junction-Current-Confinement Planar Light-Emitting Diodes and Optical Coupling into Large-Core Diameter Fibers Using Lenses," *IEEE Trans. Electron. Devices*, vol. ED-29, 1983, p. 1463.

22. G. Arnold et al., "1.3 μm Edge-Emitting Diodes Launching 250 μW into a Single-Mode Fiber at 100 mA," *Electronics Letters,* vol. 21, 1985, p. 993.

23. T. P. Lee, "Effect of Junction Capacitance on the Rise Time of LED's and on the Turn-on Delay of Injection Lasers," *Bell System Technical Journal,* vol. 54, 1975, p. 53.

24. H. Namizaki, H. Kan, M. Ishii, and A. Ito, "Current Dependence of Spontaneous Carrier Lifetimes in $GaAs\text{-}Ga_{1-x}Al_xAs$ Double Heterojunction Lasers," *Applied Physics Letters,* vol. 24, 1974, p. 486.

25. W. X. Chen, S. C. Hsueh, P. K. L. Yu, and S. S. Lau, "Solid-Phase Epitaxial Pd/Ge ohmic contacts to $In_{1-x}Ga_xAs_yP_{1-y}$/InP," *IEEE Electron Devices Letters,* vol. EDL-7, 1986, p.471.

26. R. C. Goodfellow, A. C. Carter, G. J. Rees, and R. Davis, "Radiance Saturation in Small-Area GaInAsP/InP and GaAlAs/GaAs LED's," *IEEE Transactions on Electron Devices,* vol. ED-28, 1981, p. 365.

27. T. Uji, K. Iwamoto, and R. Lang, "Nonradiative Recombination in InGaAsP/InP Light Sources Causing Light Emitting Diode Output Saturation and Strong Laser-Threshold-Current Temperature Sensitivity," *Applied Physics Letters,* vol. 38, 1981, p. 193.

28. T. Uji et al., *Proceedings of the Electronical Communication Society National Conference,* paper 4-34, Tokyo, 1982.

29. P. W. Shumate and M DiDomenico, Jr., "Lightwave Transmitters," in H. Kressel (ed.), *Semiconductor Devices for Optical Communication,* 2d ed., chap. 5, Topics of Applied Physics, vol. 39, Springer-Verlag, 1980, p. 161.

30. T. Suzuki et al., "High Speed 1.3 μm LED Transmitter Using GaAs Driver IC," *Journal of Lightwave Technology,* vol. LT-4, 1986, p. 790.

31. H. D. Wolf, K. Mettler, and K. H. Zschauer, "High Performance 880 nm (GaAl)As/GaAs Oxide Stripe Lasers with Very Low Degradation Rates at Temperatures up to 120°C," *Japan Journal of Applied Physics,* vol. 20, 1981, p. L693.

32. R. W. Dixon, F. R. Nash, R. L. Hartman, and R. J. Hepplewhite, "Improved Light-Output Linearity in Stripe-Geometry Double Heterostructure (Al, Ga) As Lasers," *Applied Physics Letters,* vol. 29, 1976, p. 372.

33. P. Marschall, E. Schlosser, and C. Wolk, "New Diffusion-Type Stripe-Geometry Injection Laser," *Electronics Letters,* vol. 15, 1979, p. 38.

34. G. H. B. Thompson, *Physics of Semiconductor Laser Devices,* Wiley, New York, 1980.

35. C. H. Henry, R. A. Logan, and K. A. Bertness, "Spectral Dependence of the Change in Refractive Index Due to Carrier Injection in GaAs Lasers," *Journal of Applied Physics,* vol. 52, 1981, p. 4457.

36. M. Hirao et al., "Fabrication and Characterization of Narrow Stripe InGaAsP/InP Buried Heterostructure Lasers," *Journal of Applied Physics,* vol. 51, 1980, p. 4539.

37. H. Ishikawa et al., "V-Grooved Substrate Buried Heterostructure InGaAsP/InP Laser Emitting at 1.3 μm Wavelength," *IEEE Journal Quantum Electronics,* vol. QE-18, 1982, p. 1704.

38. T. Yanase et al., "VPE-Grown 1.3 μm InGaAsP/InP Double-Channel Planar Buried-Heterostructure Laser Diode with LPE-Burying Layers," *Japan Journal of Applied Physics Letters,* vol. 22, 1983, p. L415.

38a. I Mito et al., "InGaAsP Double-Channel-Planar-Buried Heterostructure Laser Diode [DCPBH LD] With Effective Current Confinement," *J. Lightwave Technology*, vol. LT-1, 1983, p. 195.

39. M. Ueno et al., "Optimum Designs for InGaAsP/InP ($\lambda$ = 1.3 $\mu$m) Planoconvex Waveguide Lasers under Lasing Conditions," *IEEE Proceedings—I, Solid State and Electron Devices,* vol. 129, 1982, p. 218.

40. I. P. Kaminow, R. E. Nahory, L. W. Stulz, and J. C. Dewinter, "Performance of an Improved InGaAsP Ridge Waveguide Laser at 1.3 $\mu$m," *Electronics Letters,* vol. 17, 1981, p. 318.

40a. T. E. Bell, "Single-Frequency Semiconductor Lasers," *IEEE Spectrum*, Dec. 1983, p. 38.

40b.. T. R. Chen et al., "Phase-locked InGaAsP Laser Array With Diffraction Coupling," *Appl. Phys. Lett.*, vol. 43, 1983, p. 136.

41. H. K. Kressel and J. K. Butler, *Semiconductor Lasers and Heterojunction LEDs,* Academic Press, New York, 1977.

42. K. L. Yu et al., "Mode Stabilization Mechanism of Buried Waveguide Lasers with Lateral Diffused Junctions," *IEEE Journal of Quantum Electronics,* vol. QE-19, 1983, p. 426.

43. M. W. Fleming and A. Mooradian, "Fundamental Line Broadening of Single-Mode (GaAl) Diode Lasers," *Applied Physics Letters,* vol. 38, 1981, p. 511.

44. C. H. Henry, "Theory of the Linewidth of Semiconductor Lasers," *IEEE Journal of Quantum Electronics,* vol. QE-18, 1982, p. 259.

45. K. Vahala and A. Yariv, "Semiclassical Theory of Noise in Semiconductor Lasers," *IEEE Journal of Quantum Electronics,* vol. QE-19, 1983, p. 1096.

46. M. Sumida and K. Takemoto, "Lens Coupling of Laser Diodes to Single-Mode Fibers," *Journal of Lightwave Technology,* vol. LT-2, 1984, p. 305.

46a. W. B. Joyce and B. C. De Loach, "Alignment-Tolerant Optical-Fiber Tips for Laser Transmitter," *J. Lightwave Technol.*, vol LT-3, 1985, p. 755.

47. R. Lang and K. Kobayashi, "External Optical Feedback Effects on Semiconductor Injection Laser Properties," *IEEE Journal of Quantum Electronics,* vol. QE-16, 1980, p. 347.

48. K. G. Elze, "Investigation of Optical Feedback Effects on Laser Diodes in Broad-Band Optical Transmission Systems," *Journal of Optical Communication,* vol. 2, 1981, p. 128.

48a. K. Kawano, M. Saruwatari, and O. Mitomi, "A New Confocal Combination Lens Method for a Laser-Diode Module Using a Single-Mode Fiber," *J. Lightwave Technology*, vol. LT-3, 1985, p. 739.

48b. T. Sugie and M. Saruwatari, "Effective Non-Reciprocal Circuit for Semiconductor Laser-to-Fiber Coupling Using a YIG Sphere," *J. Lightwave Technol.*, vol. LT-1, 1983, p. 122.

49. H. Statz and G. deMars, in C. H. Towns (ed.), *Quantum Electronics,* Columbia University Press, New York, 1960, p. 530.

50. K. Y. Lau and A. Yariv, "Ultra-High Speed Semiconductor Lasers," *IEEE Journal of Quantum Electronics,* vol. QE-21, 1985, p. 121.

51. K. Y. Lau and A. Yariv, "Effect of Superluminescence on the Modulation Response of Semiconductor Lasers," *Applied Physics Letters,* vol. 40, 1982, p. 452.

51a. R. S. Tucker, "High-Speed Modulation of Semiconductor Lasers," *J. Lightwave Technol.*, vol. LT-3, 1985, p. 1183.

52. W. Franz, *Zeitschrift für Naturforschung,* vol. A-13, 1958, p. 484.

53. B. O. Seraphin and N. Bottka, "Franz-Keldysh Effect of the Refractive Index in Semiconductors," *Physical Review,* vol. 139, 1965, p. A560.

54. R. C. Alferness, "Waveguide Electro-Optic Modulators," *IEEE Transactions on Microwave Theory Techniques,* vol. MTT-30, 1982, p. 1121.

55. A. Yariv, *Quantum Electronics,* 2d ed., Wiley, New York, 1975.

55a. N. K. Dutta, R. J. Nelson, P. D. Wright, and D. C. Craft, "Criterion for Improved Linearity of 1.3 μm InGaAsP-InP Buried Heterostructures," *J. Lightwave Technology*, vol. LT-2, 1984, p. 160.

55b. P. W. Shumate, F. S. Chen, and P. W. Dorman, "GaAlAs Laser Transmitter for Lightwave Transmission System," *Bell Syst. Tech. J.*, vol. 57, 1978, p. 1826.

55c. A. Albanese, "An Automatic Bias Control (ABC) for Injection Lasers," *Bell Syst. Tech. J.*, vol. 57, 1978, p. 1534.

56. A. Yariv, "Fundamental and Applied Laser Physics," in M. S. Feld, A. Javan, and N. A. Kurnit (eds.), *Proceedings of the 1971 Esfahan Symposium,* Wiley, New York, 1971, p. 897.

57. L. A. Coldren, K. Iga, B. I. Miller, and J. A. Rentschler, "GaInAsP/InP Stripe-Geometry Laser with a Reactive-Ion-Etched Facet," *Applied Physics Letters,* vol. 37, 1980, p. 681.

57a. H. Nakano et al., "Monolithic Integration of Laser Diodes, Photomonitors, and Laser Driving Circuits on a Semi-Conducting GaAs," *J. Lightwave Technol.*, vol. LT-4, 1986, p. 576.

58. A. S. Jordon, "A Comprehensive Review of the Log-Normal Failure Distribution with Application to LED Reliability," *Microelectronics Reliability,* vol. 18, 1978, p. 267.

58a. E. Oomura et al., "Degradation Mechanism in 1.3μm INGaAsP/InP Buried Crescent Laser Diode at a Higher Temperature," *Electron. Lett.*, vol. 19, 1983, p. 407.

# CHAPTER 6
# OPTICAL DETECTORS FOR FIBERS

**Kenneth Li**
*PCO Inc.*

**Paul Kit Lai Yu**
*University of California, San Diego*

## 6.1 RECEIVERS

The receiver is the most critical part of a fiber-optic communication system, as it determines the overall system performance in most situations. The function of the receiver is to detect the optical power transmitted through the fiber to the photodetector and to extract the transmitted information. In an analog fiber-optic system, the receiver demodulates the detected signal to obtain the transmitted information. In a digital fiber-optic system, the receiver output consists of the regenerated data and, normally, the recovered clock signal as well.

A key system performance parameter is the receiver sensitivity, which determines the minimum incident optical power required at the receiver to satisfy a specified value of bit error rate, for digital systems, or signal-to-noise ratio and signal-to-distortion ratio, for analog systems. A good receiver typically needs to have a large input dynamic range capability to accept unrestricted data format, fast acquisition time, multiple-bit-rate operation, low power consumption, and low cost. On many occasions, these requirements conflict with each other. A complete understanding of the optical receiver and system operation is necessary in order to examine all the tradeoffs required to optimize the overall system performance to meet a specific requirement.

Conversion of the received light to an electronic signal is accomplished by means of a photodetector. The commonly used photodetectors in fiber-optic communication systems are PIN photodiodes and avalanche photodiodes (APDs). Photoconductors, phototransistors, and other types of detectors have been used for special applications. A classic theoretical sensitivity analysis[1] for digital optical receivers, which has been used extensively,[2,3] features a high-impedance optical receiver with an integrated front end. When this receiver is implemented with a properly designed equalizing filter and amplifier, high sensitivity is achieved, but its dynamic range is limited. The transimpedance receiver design, which was developed[4,5,6] to overcome this shortcoming, provides improved dynamic range, allows simple system design, and has been widely used for various applications.

### 6.1.1 Characteristics of Optical Receivers

The functional block diagram of a typical digital receiver is shown in Fig. 6.1. The incident optical power can be detected by a PIN diode, an APD, a photoconductor, or a phototransistor. The photodetector is followed by a low-noise preamplifier, an automatic gain control (AGC) main amplifier, and a shaping filter. The shaping filter minimizes the effect of noise and intersymbol interference at the input of the regenerator, which samples the detected signal and regenerates the data that was originally transmitted. The clock required for sampling is recovered by the timing extraction circuitry, which can include either a phase-locked loop, a surface acoustic wave (SAW) filter or an inductance-capacitance ($LC$) tank circuit.

Of all the functional blocks, the most important are the photodetector and the preamplifier. These two elements determine the major receiver characteristics and overall receiver performance. Moreover, they distinguish the optical receiver from the traditional receiver for coaxial cable and microwave transmission systems. In comparison, the rest of the circuitry is common to many operational communication systems.

### 6.1.2 Receiver Sensitivity

As mentioned above, the receiver sensitivity is a measure of the minimum optical power level required at the receiver input so that it operates reliably with a bit error rate less than a predetermined value. It is defined as the average incident optical power $\langle P \rangle$ required for a certain defined bit error rate and is often expressed in units of dBm (0 dBm = 1 mW). In most cases, the defined bit error rate is $10^{-9}$. As fiber-optic technology continues to advance, it is not uncommon to encounter a requirement for bit error rate in the $10^{-12}$ to $10^{-15}$ range.

In the receiver circuit designs, the sensitivity of the circuitry is often defined as the average detected optical power, which is equal to $\eta\langle P \rangle$, where $\eta$ is the quantum efficiency of the photodetector. It is normal practice to complement the optical measurements, which are usually obtained with less accuracy, with the receiver's electrical measurements. The photocurrent can be measured very ac-

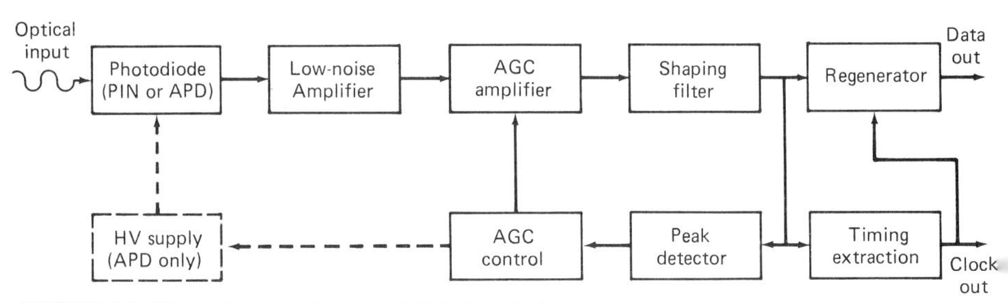


**FIGURE 6.1** Block diagram of a typical digital optical receiver. (*After S. Personick, Optical Fiber Transmission System, Plenum, New York, 1983.*)

curately and the sensitivity of the circuit can be evaluated without reference to the photodiode quantum efficiency. The receiver sensitivity is often defined to a certain performance level (bit error rate), and, in practical systems, a margin of 3 to 6 dBm must be allowed for system degradation due to temperature variation, tolerances, component aging, etc. Since the signal under consideration is measured as power, the sensitivity is given by

$$\text{Sensitivity} = 10 \log \frac{\langle P \rangle}{1 \text{ mW}} \qquad \text{dBm} \tag{6.1}$$

Consider now an ideal noiseless receiver and an ideal noiseless transmitter which sends an optical pulse if a 1 bit is transmitted and nothing if a 0 bit is transmitted. The receiver detects the 1 bit if one or more electrons are generated in that bit interval and detects the 0 bit if no electrons are generated. An error will be made if a pulse is transmitted and no electron is generated.

For any receiver, there exist various noise sources which determine the ultimate receiver sensitivity. For instance, consider a simple PIN photodiode in an optical receiver, where the signal-dependent shot noise is caused by the random generation of electron-hole pairs by the incident optical signal. In the bit interval of time, $T$, the number of electrons generated, $m$, is a random number. The probability density function of $m$ follows the Poisson distribution $P[\{m\}]$:

$$P[n = m] = \frac{N^m e^{-N}}{m!} \tag{6.2}$$

where $P[n = m]$ refers to the probability that exactly $m$ electrons will be generated within the bit interval and $N$ is the average number of electrons generated in this interval.

For an error to occur (i.e., no electron is generated), $n = 0$ and the probability is given by

$$P[n = 0] = e^{-N} \tag{6.3}$$

If a bit error rate of $10^{-9}$ is desired, $P[n = 0] = 10^{-9}$. From the above equation, $N = 21$; i.e., on the average, 21 photons must be transmitted for each optical pulse (for 100 percent quantum efficiency). The required optical power to satisfy this condition is referred to as the quantum limit.

However, in a practical receiver employing a PIN photodiode as detector, the major contribution to error comes from the thermal noise generated by the amplifier. In comparison, the error based on the Poisson statistics of photon detection is much smaller and can be neglected. With the help of standard communication theory[7] and by assuming that the amplifier thermal noise takes a gaussian distribution, it can be shown that, to achieve a $10^{-9}$ error rate, the receiver output response to an incident optical pulse must be 12 times as large as the root-mean-square value of the noise at the receiver output. By measuring the output noise and the amplifier gain, the sensitivity of the receiver can be determined.

### 6.1.3 Bandwidth

Depending on their configuration, front-end amplifiers for optical receivers can be classified into two types: high impedance and transimpedance.

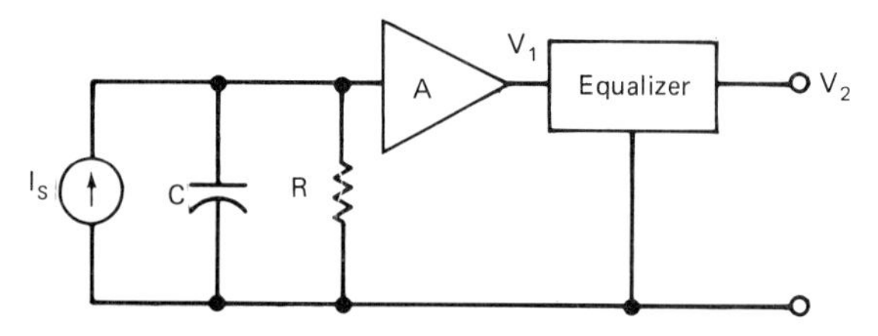


**FIGURE 6.2** High-impedance receiver amplifier. [*After T. V. Muoi, in E. E. Basch (ed.), Optical Fiber Transmission, Sams, Indianapolis, 1986.*]

***High-Impedance Amplifier Design.*** The block diagram of a high-impedance receiver amplifier is shown in Fig. 6.2. $R_L$ is the equivalent load resistor, which includes the bias resistor of the photodiode and the front-end transistor; $C$ is the total input capacitance of the amplifier including photodiode capacitance and stray capacitance. Assuming the amplifier has a gain $A$ which is flat over the frequency range of interest, the frequency response of the front-end amplifier shown in Fig. 6.3*a* is given by

$$\frac{V_I(f)}{I_s(f)} = \frac{AR_L}{1 + j\omega R_L C} \tag{6.4}$$

where $V_I$ is the output voltage of the preamplifier and $I_s$ is the photocurrent generated in the detector.

$R_L$ is chosen to be large to reduce thermal noise. As a result, the front-end roll-off frequency $1/2\pi R_L C$ is much smaller than the operating bandwidth. An equalizer is necessary to extend the receiver bandwidth out to the desired value, as shown in Fig. 6.3*b*. The ratio of $f_2/f_1$ is often referred to as the equalization ratio and can be as high as several decades.

***Transimpedance Amplifier Design.*** The transimpedance amplifier is a feedback amplifier and is shown in Fig. 6.4. $R_f$ is the feedback resistor and $C_f$ is the stray capacitance. $R_b$ is the photodiode bias resistance and $C$ is the total input capacitance. The transfer function of this amplifier is

$$\frac{V_2(f)}{I_s(f)} = -\frac{R_f}{1 + R_f/AR_b + j2\pi f R_f(C_f + C/A)} \tag{6.5}$$

For the case of $R_b \gg R_f$, $A \gg 1$, and $C_f \ll C$, the frequency response is shown in Fig. 6.5, where

$$f_2 = \frac{A}{2\pi R_f C} \tag{6.6}$$

The transimpedance amplifier is normally designed so that its bandwidth is high enough to accommodate the operating bit rate. It should be noted that the transimpedance receiver amplifier offers a larger dynamic range than the high-impedance design because of the increased gain of low-frequency components,

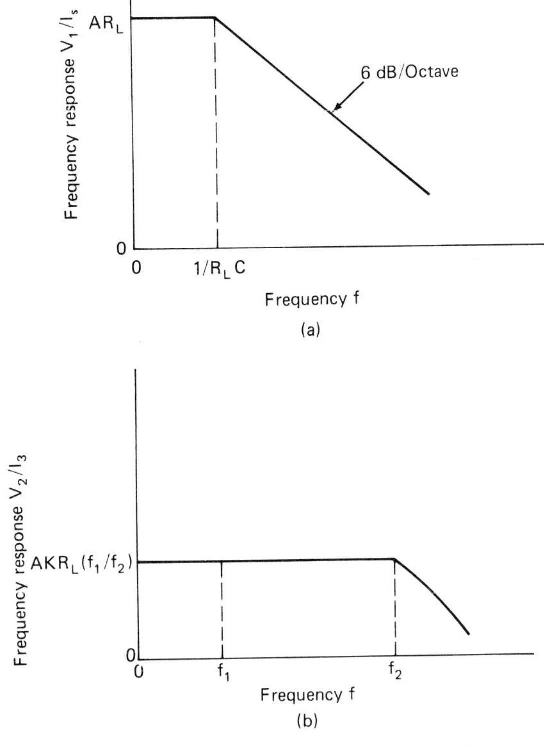


**FIGURE 6.3** Frequency response of a high-impedance receiver amplifier at (*a*) the front end and (*b*) the equalizer output. [*After T. V. Muoi, in E. E. Basch (ed.), Optical Fiber Transmission, Sams, Indianapolis, 1986.*]

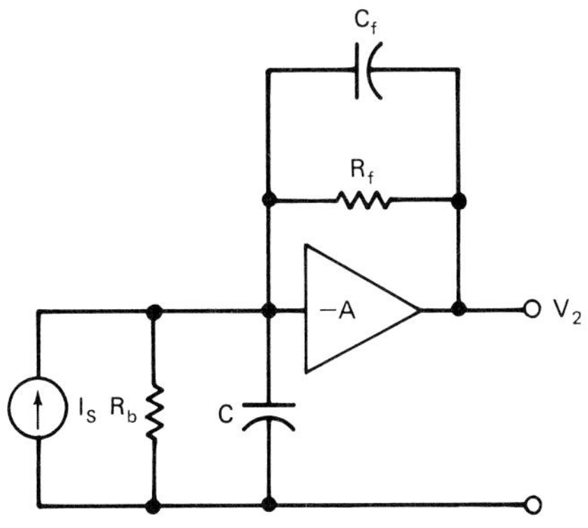


**FIGURE 6.4** Transimpedance receiver amplifier. [*After T. V. Muoi, in E. E. Basch (ed.), Optical Fiber Transmission, Sams, Indianapolis, 1986.*]

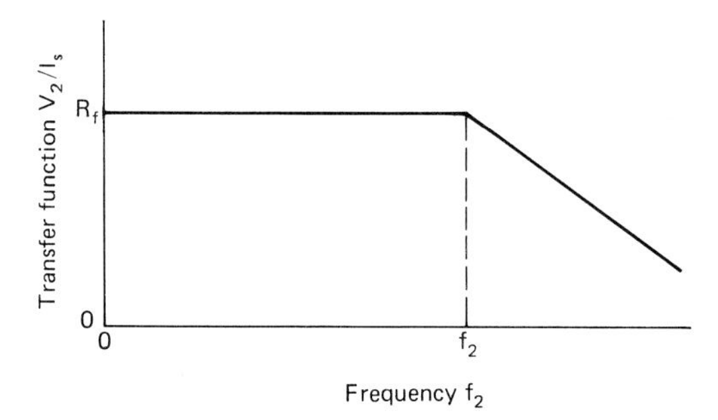


**FIGURE 6.5** Transimpedance receiver frequency response. [*After T. V. Muoi, in E. E. Basch (ed.), Optical Fiber Transmission, Sams, Indianapolis, 1986.*]

which causes saturation in the amplifier before equalization in the high-impedance design. The improvement in dynamic range is equal to the equalization ratio.

### 6.1.4 Dynamic Range

In fiber-optic communication systems, the receiver has to operate not only at the minimum allowable optical power level, but also at levels which can be significantly higher. The minimum required optical power is the receiver sensitivity, and the maximum usable optical power is limited by nonlinear distortion and saturation of the optical receiver. The ratio between the maximum and minimum optical power (in decibels) is the dynamic range of the receiver. The wide range of optical power incident at the receiver can be caused by transmitter output power changes, temperature fluctuations, losses due to fiber aging, variable connector losses, and repeater spacing. A wide receiver dynamic range is desirable since it allows flexibility and convenience in system configuration. The ability to accommodate a wide range of optical power levels means that the same receiver can be used for both long and short repeater spacings. In particular, a wide dynamic range requirement is critical in local-area network applications, where transmitter sources may be at different distances from the receivers and the signal may have to go through various numbers of splices, couplers, and splitters. In system testing, the transmitter is often jumpered directly to the receiver. As a result, the full output power from the transmitter is incident on the receiver. This often appears in system specifications as the maximum optical power requirement.

## *6.2 NOISE*

### 6.2.1 Detector Noise

The photodetector used in fiber-optic communication systems is often either a PIN or an avalanche photodiode. However, in monolithic detector-amplifier integration, a photoconductive detector is more compatible with the transistors in an integrated structure and therefore is worth consideration.[8]

Avalanche photodiodes (APDs) are more sensitive than PIN photodiodes because of their internal avalanche gain. Much effort has been devoted to APD signal and noise analysis, since these determine the ultimate limit of receiver sensitivity. The avalanche excess noise power has been theoretically derived and experimentally confirmed.[9–12] However, a knowledge of the noise power is not enough because APD avalanche gain distribution is highly nongaussian. The complete APD characterization, both theoretical and empirical, can be found in the literature.[13–16]

The receiver sensitivity analysis of optical receivers is more involved than that based on traditional communication theory (which mainly treats signal detection with additional gaussian noise). In optical communication, the quantum nature of photodetection creates a signal-dependent and time-variant Poisson noise process. This problem is further complicated by the nongaussian gain distribution of APDs. APDs are further discussed in Sec. 6.3.2 and in Chap. 7, Sec. 7.4.4.

In general, the noise of the photodetector consists of two components: (1) a signal-dependent noise component, which is the shot (or quantum) noise associated with the incident optical power and (2) a signal-independent noise component associated with the dark current of the photodetector. The second component is related to device structure and material parameters and is further discussed in Sec. 6.3.

For the simple PIN photodiode, the signal-dependent shot noise is caused by the random generation of electron-hole pairs by the incident optical radiation, as indicated in Eq. (6.2). This random nature of carrier generation sets a fundamental limit to the ultimate sensitivity of digital optical receivers. As noted earlier, even in the case of an ideal receiver, at least a minimum optical energy must be sent in a pulse so that it can be detected with a desired probability of error. The minimum limit on the required transmitted optical power for a certain probability of error is referred to as the *quantum limit.* For APDs, determination of the required transmitter output power is further complicated by the statistics of the avalanche multiplication process. Since the avalanche gain is a random process, "excess" noise is created. Furthermore, the probability density function of the noise current no longer follows the Poisson distribution. Because of the random nature of the avalanche multiplication process, the mean-squared value of the avalanche gain, $\langle g^2 \rangle$, is greater than the square of the mean avalanche gain value, $\langle g \rangle^2$. The ratio is referred to as the *avalanche excess noise factor* and is given by[9]

$$F(G) = \frac{\langle g^2 \rangle}{\langle g \rangle^2} = kG + \left(2 - \frac{1}{G}\right)(1 - k) \tag{6.7}$$

where $k = \beta/\alpha$, the ratio of the ionization coefficients of holes and electrons ($\beta$ and $\alpha$, respectively) of the APD ($k$ is defined such that $k \leq 1$).

For typical silicon (Si) APDs, $\beta/\alpha$ ranges from 0.02 to 0.04. For indium gallium arsenide/indium phosphide (InGaAs/InP) APDs, $\beta/\alpha$ ranges from 0.3 to 0.5; as a result, InGaAs APDs are noisier than Si APDs. For germanium (Ge) APDs, $\beta/\alpha$ is approximately equal to 2, and Ge APDs are much noisier than APDs fabricated in the other two material systems.

### 6.2.2 Noise in Signal

Noise in signal can be attributed to two major sources: (1) the light source itself and (2) the interaction of the light source with the fiber.

*Noise from Light Sources.* In a forward-biased *pn* junction, the carriers crossing the potential barrier exhibit shot noise. It has been shown that in light-emitting diodes (LEDs) the emitted photons also exhibit shot noise that is independent of the emission quantum efficiency.[17] The same phenomenon is observed when optical feedback is applied to the laser operating in a single longitudinal mode, since the emitted photons still obey Poisson statistics. For a multilongitudinal-mode laser, there is excess noise caused by the modal competition between various lasing modes. The noise spectrum of this excess noise depends on the number of modes and their relative amplitudes. Low-frequency, nonstationary noise occurs when a weak mode is competing with a strong mode for lasing, and broadband stationary noise occurs when two modes are about equal in intensity.[18] As the number of longitudinal lasing modes increases, the amount of excess noise is reduced because of an averaging effect among the modes.

***Noise from Light-Fiber Interaction.*** The noise produced by light-fiber interaction is mainly a result of the long coherence length of the source and the fluctuation of the laser modes. These effects are negligible in LED systems.

*Mode-Partition Noise.* Although the total power output of a laser remains constant, the distribution of power among different lasing modes fluctuates randomly. Because of this fluctuation, material dispersion in the fiber causes simultaneously emitted longitudinal modes to propagate at different speeds. As a result, it produces intensity fluctuation at the output of the fiber. This is often referred to as laser mode-partition noise.[19,20] In a multimode fiber system, the loss at the connectors and splices is wavelength-dependent. The noise caused by this mechanism is referred to as modal mode-partition noise.

*Modal Noise.* When the coherence length of the laser output is longer than the fiber in a multimode fiber system, a speckle pattern will be formed at the end face. For a misaligned connector or splices, the transmission will be dependent on the number of speckles captured by the receiving fiber. Modal noise is caused by the random fluctuation of the speckle pattern and is very sensitive to the coupling efficiency.[21]

*Feedback-Induced Intercavity Noise.* When a portion of the laser light is reflected back into the laser cavity by external optical components or discontinuity in fibers, two kinds of noise are generated.[22,23] One kind is the low-frequency noise induced by reflected waves at the laser-fiber coupling. This can be minimized by an antireflection coating or spherical lenses. The other kind is the noise caused by distant reflection. The frequency $f_m$ of the $m$th harmonic component of the noise spectrum is related to the round-trip time by

$$f_m = \frac{mc}{2nL} \tag{6.8}$$

where $n$ is the refractive index of the fiber, $c$ is the speed of light, and $L$ is the fiber length between the laser and the point of reflection.

## 6.3 PHOTODETECTORS

The appropriate wavelength for optical-fiber telecommunication is determined primarily by the absorption spectrum of the fiber. Dispersion of the fiber is an

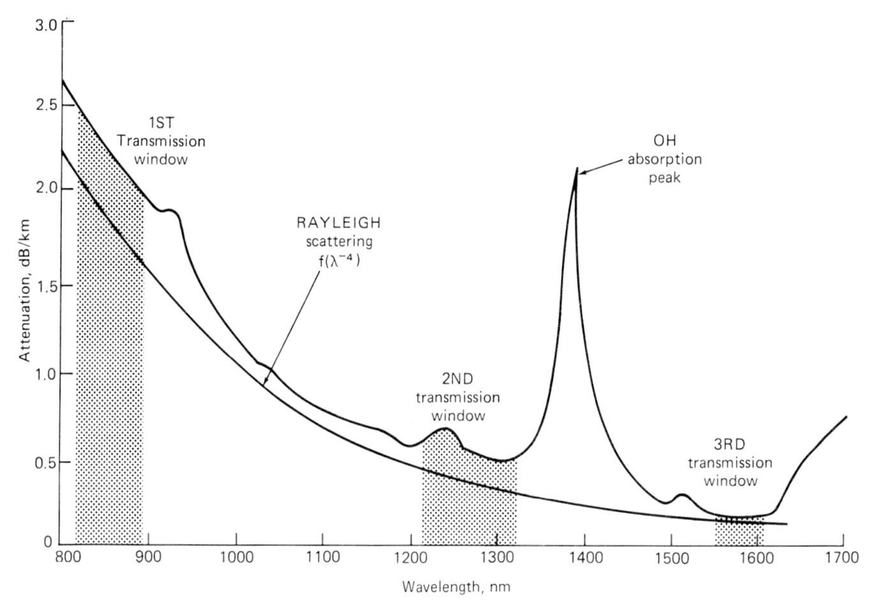


**FIGURE 6.6** Optical-fiber attenuation as a function of wavelength. [*After E. E. Basch et al., in J. C. Daly (ed.), Fiber Optics, CRC Press, 1984.*]

additional consideration for very wideband or high speed or long distance transmission. For single-mode fiber transmission, it is possible to minimize the dispersive effect in the 1.3- to 1.6-μm wavelength range. This effect is obtained by balancing mode dispersion and material dispersion. Dispersion-free transmission is possible around 1.3 μm. Figures 6.6 and 6.7 show the attenuation and material dispersion spectrum for some state-of-the-art multimode and monomode optical fibers. At a wavelength of 1.55 μm, the fiber loss is 0.2 dB/km, which is 1 order of magnitude below that at 0.82 μm. As a result, photodetectors and light sources in the 1.0- to 1.6-μm region are essential for long-distance, fiber-optic systems.

InP-based material is used for both emitters and photodetectors in this long-wavelength region. Germanium photodiodes and avalanche photodiodes have also been used for photodetection, as Ge absorption properties are quite uniform in this wavelength range. More recently, indium gallium arsenide/indium phosphide (InGaAs/InP) heterojunction photodetectors have been developed, and very high speed, high-quantum-efficiency, and low-noise photodiodes have been obtained. These detectors are essential for high-speed, long-haul fiber-optic communications. For 0.8- to 0.9-μm-wavelength photodetection, GaAs-based material is suitable for emitters, whereas silicon photodetectors, especially silicon avalanche photodiodes, are widely used in receivers. This is attributable to the well-established technology of silicon and its material characteristics, which lead to low-noise and high-sensitivity photodetectors. However, because of the long absorption length of 0.8-μm radiation in silicon and the resulting carrier diffusion effect, very high speed silicon photodetectors cannot be achieved easily. Instead, photodetectors with a GaAs/AlGaAs heterojunction have become more popular for high-speed, fiber-optic transmission systems, since these materials provide larger absorption coefficients and higher carrier-saturated velocities.

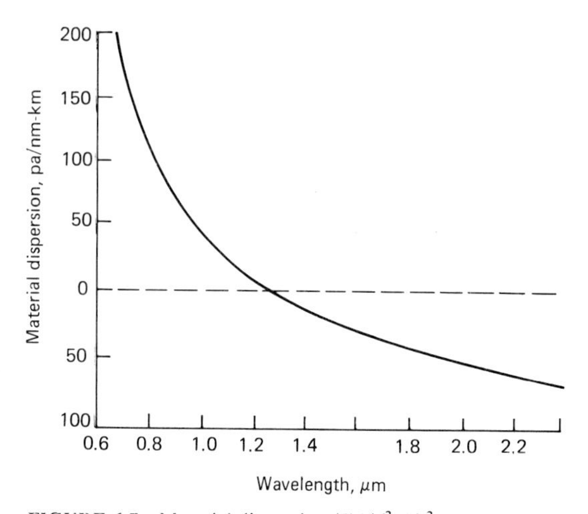


**FIGURE 6.7** Material dispersion $(l/c)(d^2n/dl^2)$ of silica glass. The group delay spread for 1-nm spectral width and 1-km fiber length is given as a function of wavelength. *(From Payne and Gambling, Elect. Lett., vol. 11, 1975, p. 8. © IEE.)*

### 6.3.1 pn and PIN Photodetectors

Silicon photodetectors which respond to radiation in the visible and near-infrared spectrum have been commercially available for many years. The structure and fabrication of these devices have reached a mature stage of development. Three structures are shown in Fig. 6.8: the *pn* junction photodiode, the PIN photodiode, and the metal-semiconductor contact Schottky photodiode. Ge-*pn* junction photodiodes, which respond from 0.8 μm to 1.7 μm, have been used in the longer wavelength region.

With advances in epitaxial growth techniques for various semiconductor compounds, heterojunction photodetectors have become feasible. Heterojunction photodetectors can be designed to achieve lower capacitance and higher quantum efficiency than can be achieved with homojunction photodetectors (those made of silicon or germanium). Heterojunction PIN photodiodes made in the GaAs/GaAlAs material system, for 0.8- to 0.9-μm-wavelength operation, and in the InGaAs/InP material system, for 1.0- to 1.6-μm-wavelength operation, have become readily available.[24]

Both *pn* junction and PIN photodetectors are normally operated in reverse-biased or short-circuit mode. A strong electric field exists in the junction where the photogenerated electron-hole pairs are separated and produce a photocurrent in the external circuits. Figure 6.9 shows the structure of a reverse-biased PIN photodetector. The carriers are generated from radiation incident (from the left) on the semiconductor material. The exposed surface of the diode has a reflectivity $r$ and the semiconductor has an absorption coefficient $\alpha$ at the energy $h\nu$ of the incident radiation. The $p^+$ and $n^+$ regions are designed to reduce the ohmic resistance of the photodiode. At normal reverse bias, the intrinsic $i$ region (with width $W_i$) is depleted, with the depletion region extending a distance $W_p$ into the $p$-type material and a distance $W_n$ into the $n$-type material. The total depletion width $W$ is thus equal to the sum of $W_i$, $W_p$, and $W_n$. Those electron-hole

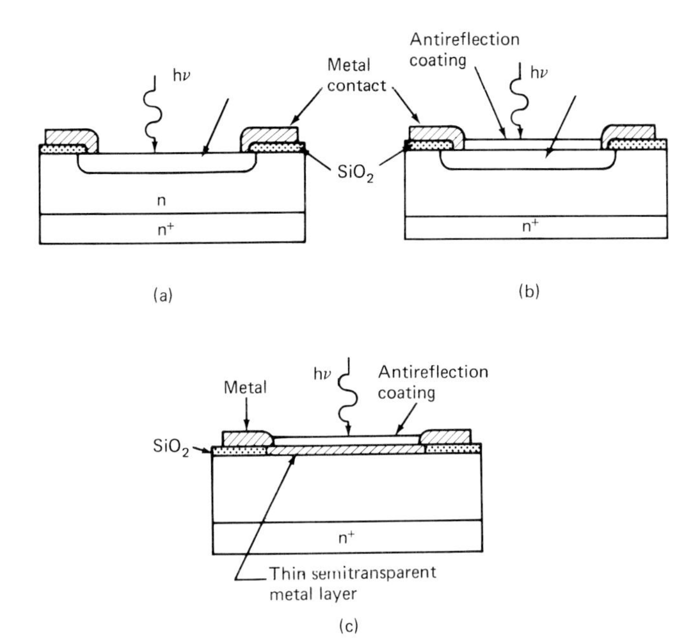


**FIGURE 6.8** Device configurations of some high-speed photodiodes: (*a*) *pn* diode, (*b*) PIN diode, and (*c*) metal-semiconductor diode. (*After H. Melchior, Laser Handbook, vol. 1, North Holland. © 1972 Elsevier Publ. Co.*)

pairs that are generated within this depletion region are separated by the electric field and contribute to the photocurrent directly. Outside the depletion region, some of the photogenerated electron-hole pairs recombine, but minority carriers, generated within a diffusion length $L_n$ or $L_p$ from the edges of the depletion region, are collected. They are subsequently swept across the depletion region by the electric field and contribute to the photocurrent. Since this portion of the photocurrent depends on the slow carrier diffusion process, it contributes to low frequency response and is to be avoided in high-speed applications.

***Responsivity.*** Responsivity $R$ is one of the critical parameters in a photodiode. It is defined as the amount of photocurrent generated per unit incident light power. It is related to the quantum efficiency $\eta$ by

$$R = \eta \frac{e}{h\nu} \qquad \text{A/W} \tag{6.9}$$

where $h\nu$ is the photon energy and $e$ is the electronic charge. The quantum efficiency of a photodiode can be estimated as follows. If the depletion region has an absorption coefficient of $\alpha_1$, a fraction $(1 - e^{-\alpha_1 W})$ of the incident light is captured at the reverse-biased junction. Assuming a reflectivity $r$ at the entrance facet and accounting for the optical power absorbed ($\alpha_2$) in the material between the front contact and the depletion region (of thickness $d$) gives a quantum efficiency of

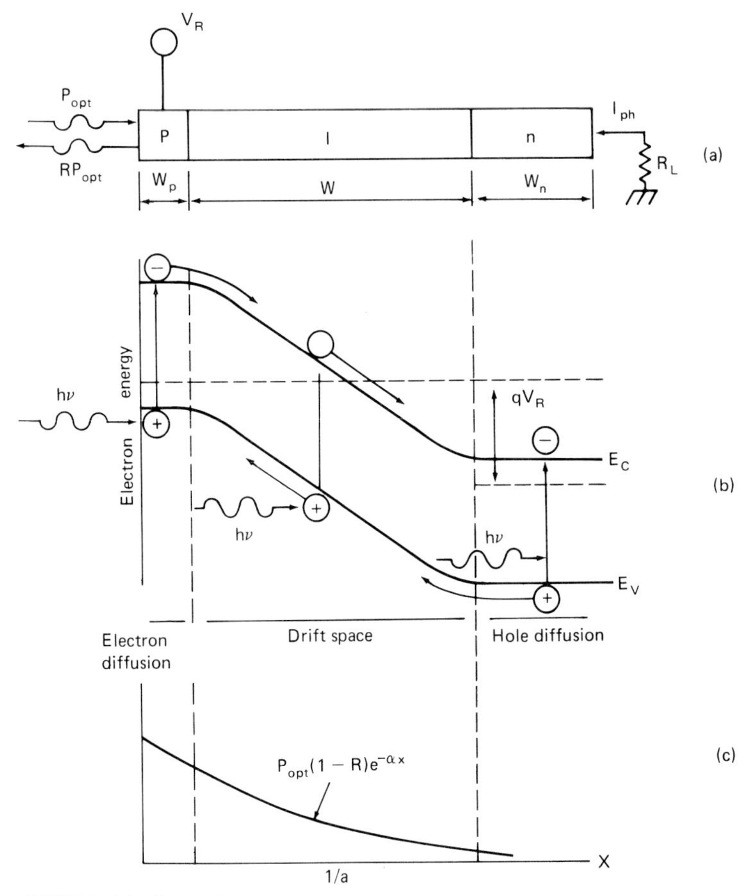


**FIGURE 6.9** Operation of a photodiode. (*a*) Cross-sectional view of PIN diode; (*b*) energy-band diagram under reverse bias; (*c*) carrier generation characteristics. (*After H. Melchior, J. Luminescence, vol. 7, 1973, p. 390.*)

$$\eta = (1 - r)e^{-\alpha_2 d}(1 - e^{-\alpha_1 W}) \tag{6.10}$$

For a typical silicon photodetector (where $\alpha_1 = \alpha_2$) that is fabricated from a silicon wafer with a resistivity $\rho \leq 100\ \Omega \cdot \text{cm}$, at a reverse-bias voltage of 10 V, the depletion width $W$ is about 10 μm. Responsivity in the 0.8- to 0.9-μm-wavelength region is 0.5 to 0.6 A/W. For the case of GaAs/GaAlAs and InGaAs/InP heterojunction photodetectors for use in the 0.8- to 0.9-μm- and 1.0- to 1.6-μm-wavelength regions, respectively, the contact layer material can be chosen so that it is transparent to the incident wavelength (or i.e., $\alpha_2 = 0$) and the material of the absorption regions can be chosen to provide a suitable absorption coefficient such that most of the incident light is absorbed in the depletion region. Figure 6.10 shows the typical responsivity spectra for Si, Ge, and InGaAs. A front entry and a back entry photodiode for 1.3-μm-wavelength detection are shown in Fig. 6.11. The InP contact layers are transparent to the 1.3-μm radiation. All of

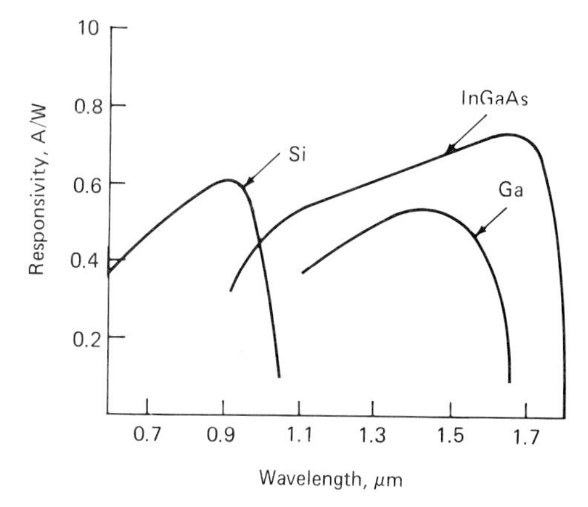


**FIGURE 6.10** Spectral responsivity of a PIN photodiode for different semiconductor materials. [*After W. Albrecht et al., in Clemens and Baack (eds.), Optical Wideband Transmission Systems, CRC Press, 1986.*]

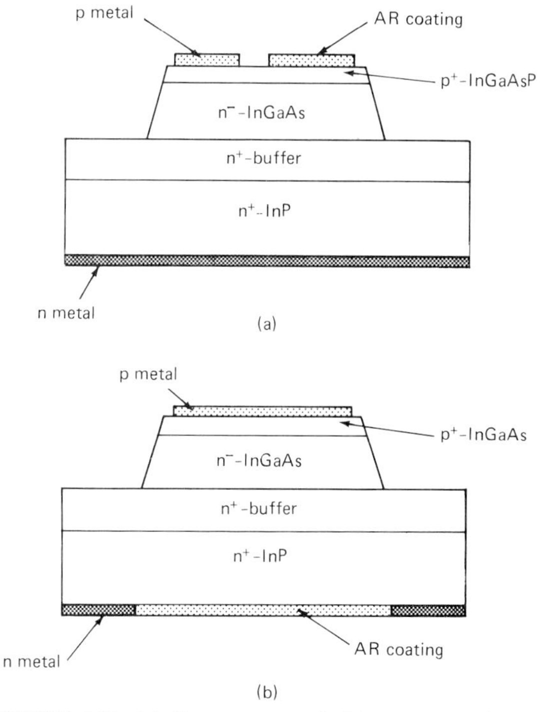


**FIGURE 6.11** (*a*) Front entry and (*b*) rear entry in an InGaAs/InP PIN photodiode for operation at a wavelength of 1.3 μm.

the incident light is absorbed in the InGaAs layer, whicn is about 3 to 4 μm thick. This InGaAs layer is grown undoped and has a background carrier concentration around $<5 \times 10^{14}$ $cm^{-3}$. At a relatively low reverse bias of less than 5 V, the layer is completely depleted. This provides high quantum efficiency and eliminates the slow-diffusion component. A maximum quantum efficiency of more than 95 percent can be achieved.

***Reverse Leakage Current (Dark Current).*** The reverse leakage current (dark current) is the current flowing through the device without illumination. This current, $I_D$, is composed mainly of (1) generation-recombination currents in the depletion regions and (2) the surface leakage current. Currents generated in bulk silicon can be kept low by using high-purity silicon and extreme care in processing and assembly to avoid introducing crystalline defects. Bulk currents as low as $2 \times 10^{-11}$ $A/m^3$ can be achieved. Surface leakage on silicon detectors can be reduced by passivation techniques such as depositing a layer of silicon dioxide. Surface leakage currents as low as $2 \times 10^{-11}$ A per millimeter of periphery at 10-V bias have been achieved. InGaAs/InP photodiodes are used in the 1.0- to 1.6-μm-wavelength region. Because of the small bandgap energy of InGaAs, both the bulk leakage and surface leakage are orders of magnitude higher than those of silicon diodes. Passivation of InGaAs, however, is a nontrivial task since the native oxide of this material is conductive. Polyimide and low-temperature chemical-vapor-deposition (CVD) silicon nitride have been used to passivate the InGaAs junction with good results.

To eliminate the surface effect of the InGaAs junction, various structures, such as planarized PIN photodiodes, can be used. The net effect is to surround the InGaAs junction with larger bandgap InP. Since dark current $I_D$ produces shot noise ($\langle i^2 \rangle = 2qI_DB$, where $B$ is the effective bandwidth of the receiver), $I_D$ should be kept as low as possible. In practical applications, dark current should not be the major noise-limiting element in a fiber-optic receiver. Nevertheless, high dark current has been observed to be associated with those devices which fail during burn-in or accelerated life testing. One should also note that the dark current increases with temperature. The exact temperature dependence is determined by the device structure, material purity, and processing parameters. For silicon diodes, the dark current at 70°C is about 10 times that at 25°C, and for InGaAs/InP diodes, the factor is about 20 times.[25]

***Capacitance.*** The resistance-capacitance ($RC$) time constant limitation arises from the small but not insignificant capacitance $C$ of the photodiode and the resistance $R$ of the load, which is usually taken to be 50 Ω. The photodiode capacitance consists mainly of the junction capacitance $C_j$ of the depletion region and the other parasitic capacitance $C_s$ related to the contact pad and the package. To reduce the junction capacitance, the area of the photodiode should be reduced or the depletion region width should be increased. However, the minimal area depends on the manner in which the light is coupled to the photodiode. For stable butt coupling, the fiber core will, in principle, limit the area to a few micrometers in diameter. The depletion region width is also related to the quantum efficiency and the carrier transit time across the depletion region, so an optimization is usually needed. In practice, as the junction capacitance of the photodiode is reduced, the parasitic capacitance becomes a more dominant factor in the total capacitance. Typical InGaAs/InP PIN photodiodes with a junction diameter of 55 μm and a depletion region width of 3 μm have a capacitance of 0.2 to 0.3 pF.

***Speed.*** The speed of a *pn* junction photodetector depends on the capacitance, the transit time of carriers through the depletion region, and the diffusion of carriers generated within a diffusion length from the edge of the depletion region. In a fully depleted PIN diode, the depletion width $W$ is approximated by the width of the intrinsic layer. The drift velocity of carriers is linearly proportional to the electric field for low field levels ($E < 10^4$ V/cm). At higher field levels, the electrons and holes approach a saturation velocity $V_{sat}$ of about $1 \times 10^7$ cm/s. This is true for both silicon and InGaAs. The transit time is

$$t_{\text{transit}} = \frac{W}{V_{\text{sat}}} \tag{6.11}$$

For a typical silicon photodetector, where $W = 50$ μm at $-50$ V reverse bias, the transit time is about 0.5 ns. For a typical InGaAs photodetector, $W$ equals 3 μm, which can be fully depleted at a reverse bias of around 5 V. This implies a transit time of less than 30 ps, which is significantly faster than that of the silicon photodetector. As these examples show, it is necessary to have a thin depletion layer to achieve high speed. On the other hand, the capacitance increases with reduced depletion width, and this contributes to the *RC* time constant. As a result, the thickness of the active region of a device is a tradeoff between competing effects of fast transit time (requiring a narrow depletion region) and the combination of high quantum efficiency and low capacitance (which require a wide depletion region). In addition, a device with low dark current requires a small area and minimum depletion volume, but as noted earlier, the effective coupling of light from an optical fiber imposes a minimum limit on the area.

***Linearity.*** Linearity in detection means that the output signal of the detector is linearly proportional to the input optical power. Linearity ensures a constant responsivity over a specified range of inputs. Reverse-biased photodetectors have a highly linear current output which is proportional to input optical power over 6 to 8 decades or more of photocurrent within a tolerance of a few tenths of a percent. The output saturation of the photodiode at high input optical power is caused by the collapse in electric field because of excessive photogenerated carriers, which also causes the reduction of the carrier velocities. Specific saturation behaviors depend upon the individual device structure, carrier concentration of the active layer, contact resistance, and illumination condition. In fiber-optic communication systems, saturation is not expected to be a problem because the power available is usually less than 1 mW.

### 6.3.2 Avalanche Photodetectors

The avalanche photodetector is a photodetector which contains a region of high electric field and exhibits avalanche multiplication of photogenerated carriers. The multiplication mechanism is based on the impact ionization effect in semiconductors, where free carriers created by photoabsorption are accelerated by a strong electric field until they gain sufficient energy, and, upon collision with other atoms, produce more electron-hole pairs, thus giving rise to gain in the photodetector. APDs require a higher bias voltage than PIN diodes to maintain a high electric field. The internal current gain is not a linear function of the applied voltage and is very sensitive to temperature. The avalanche process introduces extra noise (in addition to shot noise) because of the current multiplication in the device. This excess noise depends upon the material, device structure, gain, and illumination conditions and is the factor which ultimately limits the useful gain of the APD. This excess noise has been studied for the case of an arbitrary impact-ionization ratio $\alpha/\beta$ for both uniform and arbitrary electric field profiles, where $\alpha$ and $\beta$ are the ionization rates for electrons and holes, respectively.[26] In general, a large $\alpha/\beta$ ratio is necessary for low excess noise.

Figure 6.12 shows a schematic diagram of a silicon avalanche photodetector

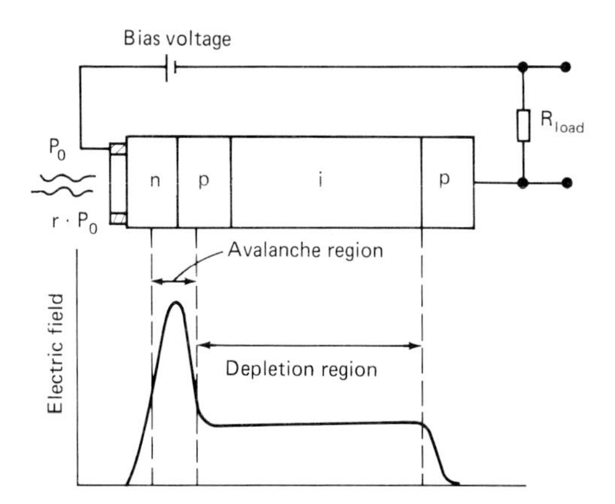


**FIGURE 6.12** Structure of an APD and the electric field distribution in the avalanche and depletion regions. [*After W. Albrecht et al., in Clemens and Baack (eds.), Optical Wideband Transmission Systems, CRC Press, 1986.*]

which consists of an $n^+$ contact, a $p$-type multiplying region, a drift (intrinsic) region, and a $p^+$ contact. In operation, under the large reverse-bias voltage, the depletion region extends completely from the $n^+$ to the $p^+$ contact. Avalanche multiplication occurs in the high-field region and this can be initiated by either electrons or holes. An electron-initiated process occurs if electrons are injected into the multiplying region because of light absorption at the right side of the region. Similarly, hole injection occurs for light absorbed to the left. Mixed injection of both carriers occurs when light is absorbed within the avalanche region. In general, low-noise operation occurs when only the carrier with the largest ionization coefficient is injected into the multiplying region. The shape and position of that region, as well as the absorption of light, must be optimized for each material system to achieve the highest performance. Silicon APDs are generally designed so that electrons are the principal carriers to undergo multiplication. Figure 6.13 shows several common structures for silicon APDs. Note that the APDs include a wide drift region for collecting photocarriers and a small avalanche region arranged for maximum injection of electrons rather than holes. A guard ring is incorporated in the planar structures to prevent edge breakdown at the perimeter of the multiplying region. A reverse-bias voltage of up to 400 V is not uncommon for this kind of diode.

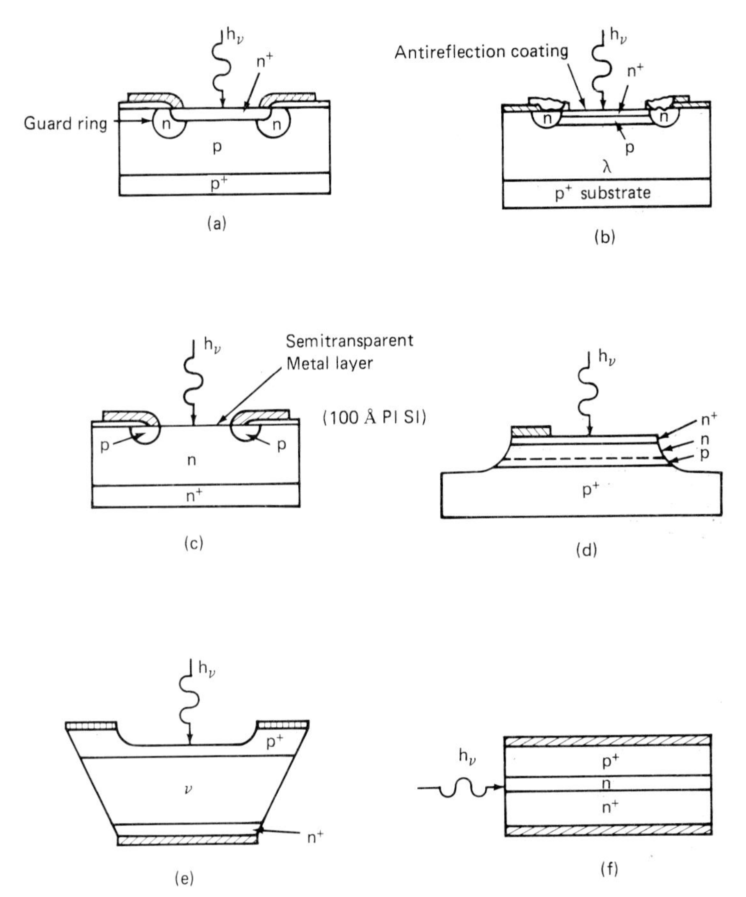


**FIGURE 6.13** Device configurations of some avalanche photodiodes. (*a*) Guard-ring structure; (*b*) guard-ring $n^+ppp^+$ structure; (*c*) metal-semiconductor structure; (*d*) mesa structure; (*e*) beveled PIN structure; (*f*) side-illuminated $p^+nn^+$ *mesa structure. (After H. Melchior, Laser Handbook, vol. 1, North Holland. © 1972 Elsevier Publ. Co.)*

InP lattice-matched materials (InGaAs and InGaAsP) are used for APD applications in the 1.0- to 1.6-μm-wavelength region. Figure 6.14 shows two simple APDs fabricated on layers lattice-matched to the InP substrate. The undoped InGaAs layer constitutes the absorption layer and the low-doped InP layer is the high-field layer for avalanche multiplication. Since the incident light is absorbed only in the InGaAs layer and not in the multiplication layer, this structure is referred to as a SAM (separate absorption and multiplication) APD.[27] There has been a lot of work on APDs based on the InGaAsP/InP material system and reasonably low leakage currents have been obtained with them at low voltages. However, the leakage current at high reverse-bias voltage became excessive because of tunneling in the narrow-bandgap materials. In addition, preliminary measurements of the ratio of impact ionization coefficients indicate that $\alpha/\beta \approx 0.5$, resulting in substantial excess noise.

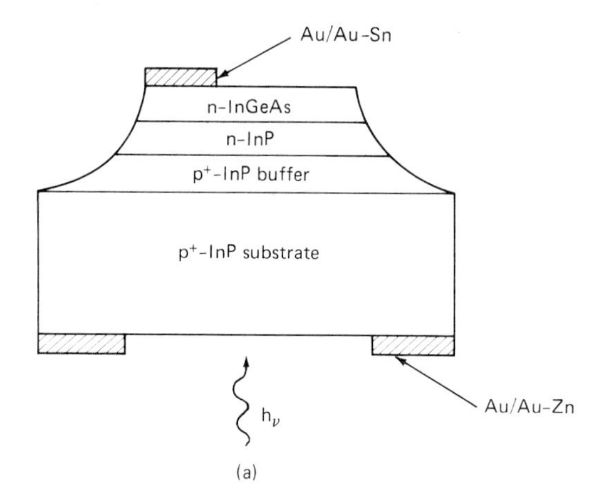


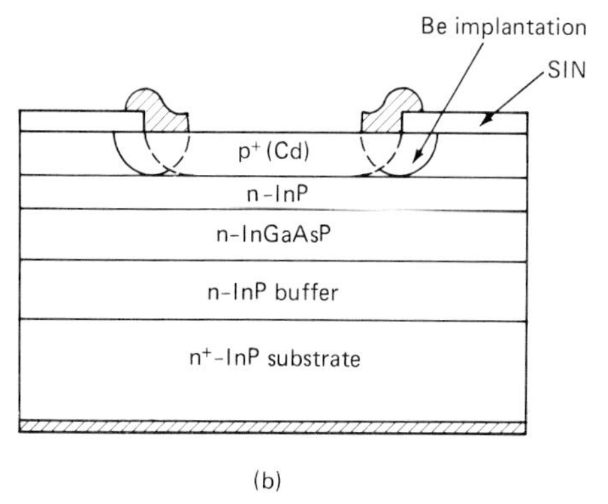


**FIGURE 6.14** InGaAs/InP avalanche photodiodes. (*a*) A mesa structure APD; long-wavelength photons are absorbed by the InGaAs layer, where the photogenerated carriers are multiplied in the high-field region in the InP $p^+n$ junction. (*b*) A planar construction of the InGaAsP APD. (*From T.Shirai et al., Electronics Letters, vol. 7, 1981, p. 826, with permission.*)

***Responsivity and Gain.*** The responsivity of an APD at unity gain has the same dependences as a PIN photodetector biased at a low voltage. At higher reverse biases, multiplication occurs and the photocurrent multiplication (gain) increases. Figure 6.15 shows the gain as a function of voltage of an $n^+ p\pi p^+$ APD. At low reverse bias, the $p$ region is only partially depleted and the device exhibits no gain. As the voltage increases, the electric field at the $p$ region increases, multiplication occurs rapidly, and the device gain rises sharply as shown. As the voltage increases further, the depletion region punches through the $p$ region into the lightly doped $\pi$ region. At this point, the device operates at high speed without a diffusion tail. Further increase in voltage is not followed by an increase in the electric field at the same rate as before because a significant amount of voltage is dropped across the $\pi$ region and results in a change in the slope of the multiplication vs. bias-voltage curve. A functional form of the multiplication curve can be expressed as[28]

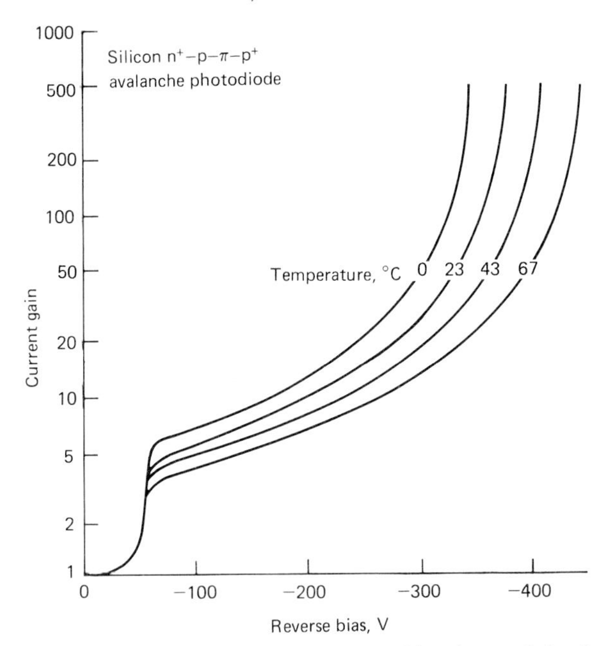


**FIGURE 6.15** Current gain vs. reverse-bias characteristic of a silicon $n^+p\pi p^+$ APD at different ambient temperatures measured with illumination at $\lambda = 0.825\ \mu\text{m}$. *(From H. Melchior, A. R. Hartman, D. P. Schinke, and T. E. Seidel, Bell System Technical Journal, vol. 57, no. 6, July–August 1978, p. 1791, with permission.)*

$$M = \frac{1}{1 - (V/V_{BR})^n} \tag{6.12}$$

where $V_{BR}$ is the reverse breakdown voltage and $n$ is an empirically determined exponent which depends on the device structure, detector material, and illumination condition. Usually, $n$ is less than 1.

The photocurrent gain depends strongly on temperature and decreases for an increase in temperature. Temperature behaviors of the ionization ratios $\alpha$ and $\beta$ and the device structure have also been considered. Temperature effects in punch-through APDs have been studied, with excellent agreement between theory and experiment.[29] Approximate expressions for temperature dependence are as follows:

$$V_{BR} = V_{B0} + a(T - T_0) \tag{6.13}$$

and

$$n = n_0 + b(T - T_0) \tag{6.14}$$

where $a$ and $b$ are positive constants readily obtained from the experiment.

***Reverse Leakage Current (Dark Current).*** The dark current increases exponentially with reverse bias voltage as the voltage approaches avalanche breakdown. This excessive dark current is related to the tunneling current under reverse bias.

In silicon APDs, the tunneling current is smaller ($10^{-11}$ A) because of the large bandgap of silicon. For germanium and InGaAs, because of their small bandgap, a large leakage current is observed. To reduce this tunneling current, as shown in Fig. 6.14, a structure that separates the light absorption region from the multiplication region can be used, where the absorption occurs in the narrow-bandgap material and multiplication takes place in the high-field region in the large-bandgap materials. For the InP/InGaAsP materials, with proper doping of InP and InGaAs (or InGaAsP) layers, the electric field can be sufficiently high to produce avalanche gain in the InP layer and still be low in the ternary (or quaternary) layer to avoid tunneling. In this way, leakage currents can be kept in the range of tens to hundreds of nanoamperes with a gain between 10 and 60.

***Capacitance.*** The structure of an APD is basically that of a reverse-biased photodiode. As in PIN photodiodes, the junction capacitance is governed by the area and thickness of the depletion region. Since the device is biased at the punch-through point, the junction capacitance is at the minimum value. For III-V materials, the area of APDs is reduced for another reason: the local breakdown due to the microplasma effect occurs more frequently in III-V materials, possibly because of defects propagated from the substrate materials. The junction capacitance $C_j$ can be approximated by a one-sided abrupt junction as

$$C_j = \frac{\varepsilon A}{W} = \varepsilon A \frac{q N_B}{2(V + \phi_B)} \tag{6.15}$$

where $\varepsilon$ = dielectric constant
$N_B$ = doping density of the depletion region
$\phi_B$ = built-in voltage

Since $C_j$ is proportional to $N_B$, a low doping density is required to reduce the capacitance.

***Speed.*** The speed of APDs depends on the same factors as those for PIN photodiodes, namely the depletion layer transit time, the *RC* time constant, and the diffusion time in the undepleted region. In addition, it depends also on the avalanche buildup time.

To design a high-speed APD, the area of the diode should be made as small as possible to reduce the depletion region capacitance and the series resistance should be reduced by using thinner layers and better ohmic contacts. Since, as in a photodiode, both transit time and capacitance are related to the depletion width, a tradeoff is needed to optimize speed and quantum efficiency.

The avalanche buildup time results from the fact that the multiplication process is not instantaneous. The buildup time $t_a$ depends on the number of collision processes and is proportional to the multiplication factor $M$:

$$t_a = \tau \langle M \rangle \tag{6.16}$$

where $\tau$ is the intrinsic response time. This time constant depends strongly on the ratio of the electron and hole impact ionization coefficients. If there is a large difference in these coefficients, the time constant will be smaller.

APDs typically exhibit a slightly asymmetric pulse response shape with a relatively fast rise time. The fast rise time is a result of electron collection, which occurs first. The remaining response characteristic is determined by the transit time of holes moving at a slower saturated velocity. The −3-dB bandwidth, normalized to the average carrier transit time, has been calculated with the hole-to-electron ionization ratio as a parameter, as shown in Fig. 6.16. The dashed curve ($M_0 = \alpha/\beta$) is for the low-frequency gain. When $M_0 > \alpha/\beta$ (the region below the dashed curve), the curves are straight lines, indicating a constant gain-bandwidth product and a dependence of multiplication on frequency as given by[30]

$$M(w) = \frac{M_0}{(1 + w^2 M_0^2 \tau_1^2)^{1/2}} \tag{6.17}$$

where $\tau_1$ is the effective transit time through the avalanche region:

$$\tau_1 = N\left(\frac{\alpha}{\beta}\right)\tau_2 \tag{6.18}$$

where $\tau_2$ is the actual transit time and $N$ is a number whose value varies slowly from ⅓ to 2 as $\beta/\alpha$ varies from 1 to 0.001.

At high frequency, the signal is distorted by the APD. A gain-bandwidth product larger than 200 GHz can be achieved easily in silicon APDs. Measurements[31] of the avalanche buildup time as a function of the width of the avalanche region and the illumination wavelength show that $\tau_1 \approx 5 \times 10^{-13}$, which relates to the

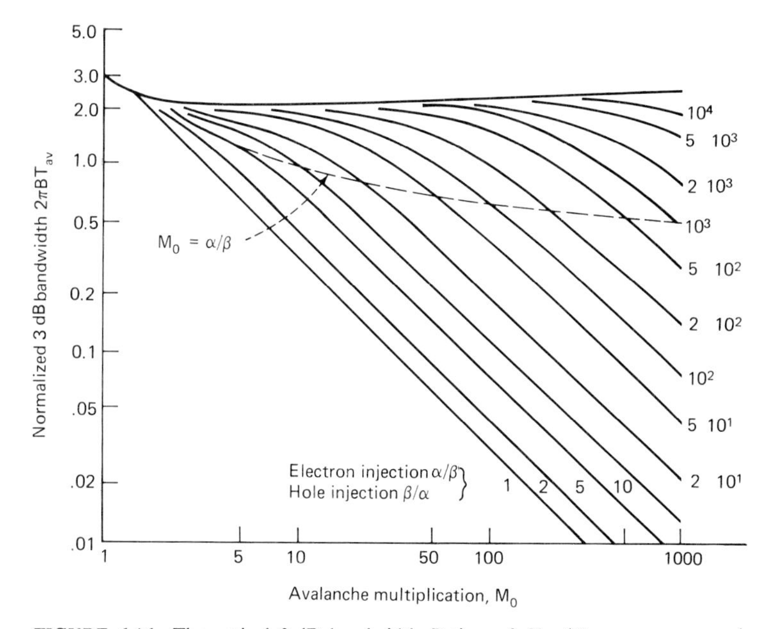


**FIGURE 6.16** Theoretical 3-dB bandwidth $B$ times $2pT_{av}$ ($T_{av}$ = average carrier transit time) of an avalanche photodiode plotted as a function of the low-frequency multiplication factor $M_0$ for various values of $\alpha/\beta$ for electron injection (or $\beta/\alpha$ for hole injection). Above the dashed curve ($M_0 = \alpha/\beta$) bandwidth is nearly independent of gain; below the dashed curve a constant gain-bandwidth product applies. *(After R. B. Emmons, J. Appl. Phys., vol. 38, 1967, p.3705.)*

total avalanche region buildup time, $t = \tau_1 M$. Since the gain-bandwidth product is so large, most practical systems are limited by $RC$ time constants and carrier transit time.

In long-wavelength cases, APDs are made with Ge or, for higher performance needs, with heterojunction InGaAs/InGaAsP/InP. Holes are trapped at the heterojunction between the InP and the InGaAsP, and the response speed is limited by the slow release of the trapped holes, which form a diffusion tail. A graded transition region has been shown to remove the long tail caused by the trapping effect.[31]

***Linearity.*** APDs show excellent linearity of multiplied photocurrent at very low-incident light levels. Several factors, however, can cause gain saturation at high-incident light levels. They are:

- Reduction of the electric field in the multiplication region (because of the external load resistor)
- Reduction of the electric field (because of space-charge effects in through-carriers drifting in the depletion region)
- Junction heating (which decreases the gain)

### 6.3.3 Photoconductors

Photoconductive effects, in which the radiation changes the electrical conductivity of the material upon which it is incident, have been known for many years. There are two basic types of photoconduction—extrinsic and intrinsic. In the intrinsic case, the photoconduction is produced by absorption of light to create a band-to-band transition across the bandgap, where the absorption coefficient is very large because of the large number of available electron states associated with the conduction and valence bands. Here the absorption coefficient $a$ is on the order of $10^4$ cm$^{-1}$ for photons near the bandgap energy. In the case of extrinsic photoconduction, the photons are absorbed at the impurity levels and, consequently, free electrons are created in the $n$-type semiconductor and free holes are created in the $p$-type semiconductor. Extrinsic photoconduction is characterized by a low absorption coefficient $a$ because of the small number of available impurity levels. Figure 6.17 shows a simple geometric model of a photoconductor. In nearly all cases, the change in conductivity is measured by means of electrodes attached to the sample. The transverse geometry is usually used, wherein the direction of the incident radiation is perpendicular to the direction in which the change of current is measured. The photosignal is detected either as a change in voltage across a load resistor in series with the photoconductor, or as a change in current through the sample. Frequently the signal is detected as a change in voltage across a load resistor matched to the dark resistance of the detector, although it may not be optimized for signal-to-noise ratio under all conditions.[32]

Traditional photoconductors are usually slow in response (in the microsecond range) and are not suitable for fiber-optic communication. With the advancement of microfabrication technology, however, photoconductive switches of various configurations have been fabricated in different materials giving responses on a picosecond time scale,[33] and which are therefore suitable for high-bit-rate, fiber-optic communication. Figure 6.18 shows such a photoconductor mounted on a microstrip line for high-speed applications.

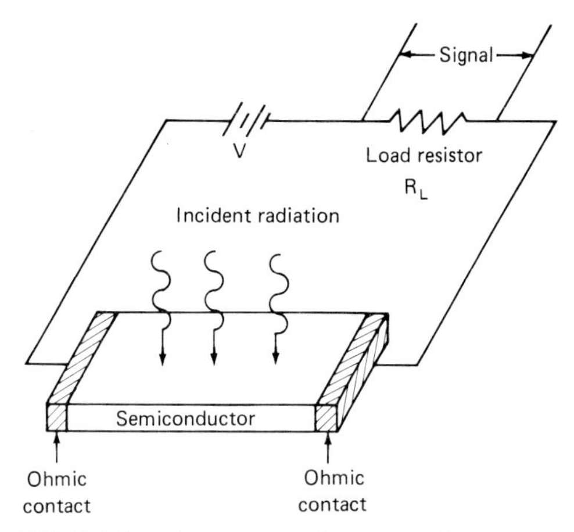


**FIGURE 6.17** Transverse photoconduction geometry and circuit.

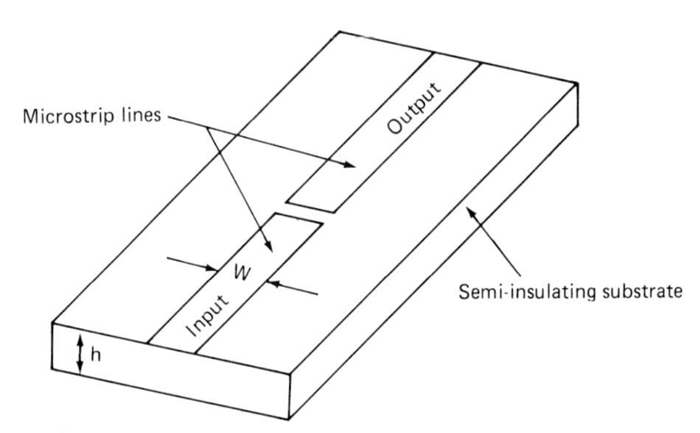


**FIGURE 6.18** A high-speed photoconductor fabricated in a microstrip-line configuration. [*After K. K. Li et al., in C. H. Lee (ed.), Picosecond Optoelectronic Devices, Academic, New York, 1984.*]

***Responsivity and Gain.*** If a steady optical signal with power $P_0$ and wavelength $\lambda$ is incident on the photoconductor, the optical generation rate of the electron-hole pairs as a function of distance $x$ into the sample, as shown in Fig. 6.19*b*, is given by

$$g(x)\,dx = \frac{(1 - r)P_0}{h\nu}\,\alpha e^{-\alpha x}\,dx \tag{6.19}$$

where $\alpha$ is the absorption coefficient at the energy $h\nu$ and $r$ is the surface reflectivity. If the sample has a thickness $D$ as shown in Fig. 6.19*a*, part of the incident photons will not be absorbed by the photoconductor and will be trans-

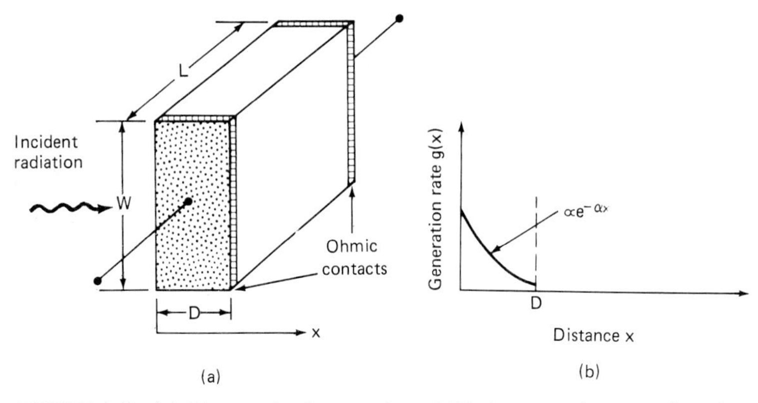


**FIGURE 6.19** (*a*) Photoconductive sample and (*b*) the generation rate of carriers throughout the sample thickness.

mitted through. The transmitted photons contribute to a reduction of the quantum efficiency $\eta$, which is then given by

$$\eta = (1 - r)(1 - e^{-\alpha D}) \tag{6.20}$$

For an intrinsic photoconductive detector, the change in the average steady-state electron and hole concentrations can be written as

$$\begin{aligned} \Delta n = \Delta p = \frac{G\tau_c}{WLD} &= \frac{1}{WLD}\int_0^D g(x)\tau_c\,dx \\ &= \frac{P_0}{WLDh\nu}(1 - r)(1 - e^{-\alpha D})\tau_c \end{aligned} \tag{6.21}$$

where $G$ is the electron and hole generation rate, $\tau_c$ is the excess carrier lifetime, and $W$, $L$, and $D$ are, respectively, the width, length, and thickness of the sample. Then the change in conductivity, $\Delta\sigma$, of the sample caused by the optical excitation of $\Delta n$ and $\Delta p$ is

$$\begin{aligned} \Delta\sigma &= q(\Delta n\mu_n + \Delta p\mu_p) \\ &= \frac{qG\tau_c}{WDL}(\mu_n + \mu_p) \end{aligned} \tag{6.22}$$

where $q$ is the electron charge and $\mu_n$ and $\mu_p$ are electron and hole mobilities, respectively.

For the circuit shown in Fig. 6.17 and the device dimensions shown in Fig. 6.19*a*, the photocurrent $i_p$ is given by

$$i_p = \frac{qGt_cV}{L^2}(\mu_n + \mu_p) \tag{6.23}$$

If $\tau_t$ is the transit time that the electrons and holes take to go across the sample, the photoconductive gain $M$ is given by[34]

$$M = \frac{\tau_c}{\tau_t} \tag{6.24}$$

***Dark Current and Off Resistance.*** The dark resistance of a photoconductor is determined by the resistivity of the material and the physical dimensions of the device. For the device shown in Fig. 6.19*a* the dark resistance is given by

$$R_D = \frac{L}{\sigma W D} \tag{6.25}$$

where $\sigma$ is the conductivity of the material.

The effect of the dark resistance on noise equivalent power has been determined[35] to be inversely proportional to $R_D$. Calculations[36] of photoconductor receiver sensitivity show that the sensitivity improves as $R_D$ increases. The limit to achieving a large $R_D$ in practical systems is the tradeoff between $R_D$ and speed.

***Capacitance.*** For traditional applications of photoconductors, the capacitance is usually not a limitation as the response speed is limited by the minority carrier lifetime (which is in the microsecond regime). As the frequency goes up, the effect of the stray capacitance becomes significant. For high-speed operation, the structure of the photodetector has to be modified as shown in Fig. 6.18. The modified structure consists of a microstrip line fabricated on a piece of semi-insulating semiconductor. A gap is made on the microstrip line, as shown. This gap separates the output from the input and forms the photoconductive junction of the device. Lower capacitance also improves the sensitivity of the optical receiver.[37]

***Speed.*** Since the photoconductivity effect depends on the photogenerated carriers, the response time of the photoconductor correspondingly depends mainly on the lifetime of the photogenerated carriers. The recombination mechanisms that determine the lifetime can include band-to-band recombination, recombination through midgap centers, and surface recombinations. In very high speed applications, deep levels such as those of chromium (Cr) in GaAs and iron (Fe) in InP are responsible for bringing the carrier lifetime from the microsecond range to the nanosecond range. Picosecond time response can be achieved by using the low-capacitance configuration, Fig. 6.18, with surface recombination.

## 6.4 TECHNOLOGY TRENDS

Recent efforts in photodetector research and development emphasize high responsivity and gain and low-noise photodetection, as well as monolithic integration of the receiver. Two important trends are (1) bandgap engineering based on the characteristics of heterojunctions, multiple quantum wells, and superlattices and (2) photodetector-amplifier integration.

### 6.4.1 Bandgap Engineering

Advances in material processing technology, such as molecular-beam epitaxy and metalorganic chemical vapor deposition, have led to new approaches to tailor the material structures, to grow ultrathin layers, and to control precisely the doping profiles. All of these approaches have facilitated improvements in photodetection. Due to the maturity of crystal growth technology, most of these approaches have been tested with GaAs-based materials. Consequently, photodetection based on these techniques has been restricted to the 0.8-μm and the far infrared (greater than 3 μm) regions. Much of the emphasis has been on improving photodetection devices based on avalanche multiplication.[38] As noted in Sec. 6.3.2, in order to reduce the noise generated in the avalanche multiplication process, the ratio of the ionization coefficients $\alpha$ and $\beta$ must be enhanced. This is difficult to achieve in most bulk III-V semiconductor materials and in Ge, since $\alpha$ and $\beta$ are nearly equal in these materials.

The so-called "graded-bandgap avalanche photodiode" is based on an approach designed to improve the $\alpha/\beta$ ratio. Inside a graded-bandgap region, the conduction band can be graded to a different extent than the valence band. This arises from the different band discontinuities in conduction and valence bands when one material is epitaxially grown on another.[38] As shown in Fig. 6.20, if the

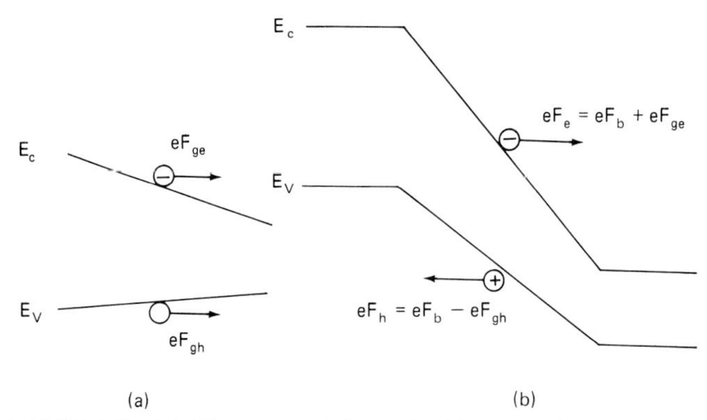


**FIGURE 6.20** (*a*) Effect of quasi-electric field in a graded-gap material. (*b*) Combined effect of applied and quasi-electric field. (*After F. Capasso, IEEE Trans. Electron. Dev., vol. ED-29, © 1982 IEEE.*)

conduction band has a larger grading than the valence band, electrons can be accelerated to a higher energy than holes under the same applied electric field (from external bias). Consequently, the observed $\alpha/\beta$ ratio can be increased, provided that the accelerating electrons travel across the different bandgap regions without suffering from phonon collision and reach the threshold energy for impact ionization. In practice, an ionization rate ratio $\alpha/\beta$ less than 10 can be obtained for a *pn* junction with a 0.4-μm, graded-bandgap $Al_xGa_{1-x}As$ region, in which $x$ ranges from 0 to 0.45.

Another scheme which utilizes the bandgap discontinuities between two adjacent layers to enhance the $\alpha/\beta$ ratio is the avalanche photodiode incorporating a superlattice region.[39] Figure 6.21 illustrates a structure that consists of alternating layers of large-bandgap materials (barrier layers) and small-bandgap materials (well layers), with unequal conduction- and valence-band discontinuities. In this structure, the electrons and holes experience a different drop in potential energy as they travel from the barrier layer to the well layer. Provided that the heterojunction is abrupt enough that the transition region is much shorter than the mean free path for phonon scattering, the potential energies gained by the

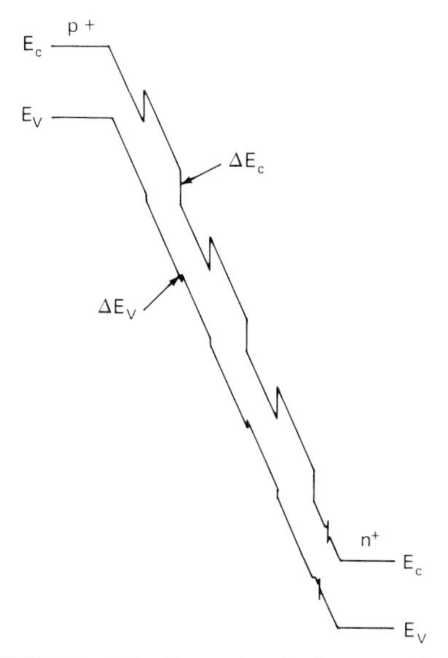


**FIGURE 6.21** Energy-band diagram of a superlattice APD with unequal conduction-band and valence-band discontinuities. (*After F. Capasso, Appl. Phys. Lett., vol. 40, 1982.*)

carriers can be added to those gained through the electric field. Since $\Delta E_c$ is observed to be larger than $\Delta E_v$ for GaAs/AlGaAs heterojunctions, and GaAs has a lower threshold energy for impact ionization than AlGaAs (because of the dependence of threshold energy on bandgap in the band-to-band impact ionization), electrons are more likely to initiate impact ionization in GaAs. For APDs, which are similar to the PIN photodiode in structure, except that the intrinsic region consists of 50 alternating layers of GaAs/$Al_{0.45}Ga_{0.55}As$ superlattice, an $\alpha/\beta$ ratio of 8 can be achieved.[39]

A more novel concept is to use a multistage graded-bandgap structure to achieve solid-state photomultiplication.[40] In the staircase solid-state photomultiplier, the electron ionization energy is entirely provided by conduction-band steps. As shown in Fig. 6.22, each stage in this photomultiplier structure is linearly graded in composition from a low bandgap to a high bandgap, with an abrupt transition back to the narrow-bandgap material. The energy step should be equal to or greater than the electron ionization energy. Since the conduction-band discontinuity in GaAs/AlGaAs materials is observed to be larger than the valence-band discontinuity, the corresponding grading and step for the valence band are considerably smaller.

### 6.4.2 Photodetector-Amplifier Integration

As mentioned previously, for receiver applications in high-speed, fiber-optic links, recovery of the low-power level, high-speed electrical signals generated in the photodetector has been a major challenge in receiver design.

Besides ensuring high speed and low-noise amplification, it is also necessary

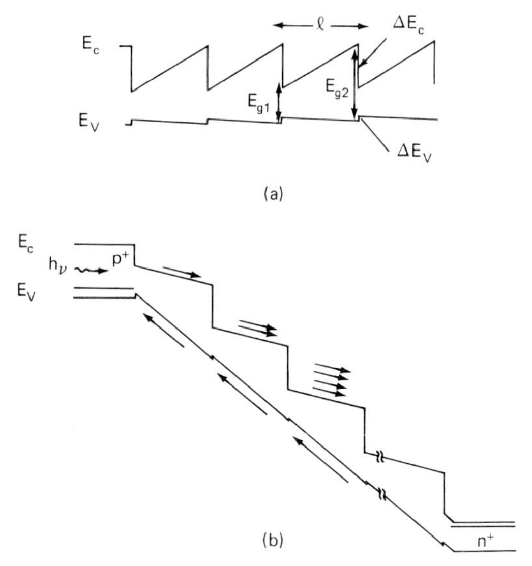


**FIGURE 6.22** Energy-band diagram of (*a*) unbiased graded-bandgap multilayer staircase photomultiplier and (*b*) biased photomultiplier. (*After F. Capasso, IEEE Trans. Electron. Dev., vol. ED-30, © 1983 IEEE.*)

to minimize the noise coupled inductively from nearby circuits, the preamplifier, and the photodetector. The monolithic integration of the photodetector with its accompanying amplifying circuits on the same substrate can reduce the extent of these problems primarily by reducing parasitic reactances.

Advances in InGaAsP/InP-based FETs, in contrast to advances in PIN photodiodes and APDs, have been slow. Schottky barriers made on InGaAsP/InP materials have a low and unstable barrier height, which has impeded the development of MESFETs in this material system. Other amplifiers such as InP metal-insulator-semiconductor FETs (MISFETs) and junction FETs (JFETs) and InGaAs JFETs have been proposed for integration with either PIN photodiodes or APDs.[41] In these structures, however, the inferior performance of the integrated amplifier version, compared to that of a hybrid GaAs MESFET construction, diminishes the claimed advantages of monolithic integration over hybrid integration, where low parasitic capacitance (largely only that of the package) has been achieved. Further improvements in InP-based amplifier technology, as well as in monolithic designs, are needed.

## REFERENCES

1. S. D. Personick, "Receiver Design for Digital Fiber Optic Communication Systems," pts. I and II, *Bell System Technical Journal,* vol. 52, no. 6, July–August 1973, pp. 843–886.

2. J. E. Goell, "An Optical Repeater with High-Impedance Input Amplifier," *Bell System Technical Journal,* vol. 53, no. 4, April 1974, pp. 629–643.

3. P. K. Runge, "An Experimental 50-Mb/s Fiber-Optic PCM Repeater," *IEEE Transactions on Communication,* vol. COMM-24, 1976, pp. 413–418.
4. J. L. Hullett and T. V. Muoi, "A Modified Receiver for Digital Optical-Fiber Transmission Systems," *IEEE Transactions on Communication,* vol. COMM-23, 1975, pp. 1518–1521.
5. J. L. Hullett and T. V. Muoi, "A Modified Receiver for Digital Optical-Fiber Transmission Systems," *IEEE Transactions on Communication,* vol. COMM-24, 1976, pp. 1180–1185.
6. K. Ogawa and E. L. Chinnock, "GaAs FET Transimpedance Front-End Design for a Wide-Band Optical Receiver," *Electronics Letters,* vol. 15, 1979, p. 650.
7. H. L. Van Trees, *Detection Estimation and Modulation,* vol. 1, Wiley, New York, 1968.
8. H. Hamaguchi, M. Makiuchi, T. Kumai, and O. Wada, "GaAs Optoelectronic Integrated Receiver with High-Output Fast-Response Characteristics," *IEEE Transactions on Electron Devices Letters,* vol. EDL-8, no. 1, 1987, pp. 39–41.
9. R. J. McIntyre, "Multiplication Noise in Uniform Avalanche Diodes," *IEEE Transactions on Electron Devices,* vol. ED-13, 1966, pp. 164–168.
10. R. D. Baertsch, "Noise and Ionization Rate Measurements in Silicon Photodiodes," *IEEE Transactions on Electron Devices,* vol. ED-13, 1966, p. 987.
11. R. D. Baertsch, "Noise and Multiplication Measurements in the InSb Avalanche Photodiodes," *Journal of Applied Physics,* vol. 38, 1967, pp. 4267–4274.
12. H. Melchior and W. T. Lynch, "Signal and Noise Response of High-Speed Germanium Avalanche Photodiodes," *IEEE Transactions on Electron Devices,* vol. ED-13, 1966, pp. 829–838.
13. R. J. McIntyre, "The Distribution of Gains in Uniformly Multiplying Avalanche Photodiodes: Theory," *IEEE Transactions on Electron Devices,* vol. ED-19, 1972, pp. 703–713.
14. S. D. Personick, "New Results on Avalanche Multiplication Statistics with Applications to Optical Detection," *Bell System Technical Journal,* vol. 50, no. 1, January 1971, pp. 167–189.
15. S. D. Personick, "Statistics of a General Class of Avalanche Detectors with Applications to Optical Communication," *Bell System Technical Journal,* vol. 50, no. 10, December 1971, pp. 3075–3095.
16. J. E. Mazo and J. Salz, "On Optical Data Communication via Direct Detection of Light Pulses," *Bell System Technical Journal,* vol. 55, no. 3, March 1976, pp. 347–369.
17. A. Van Der Ziel, "Noise in Solid State Devices and Lasers," *Proceedings of the IEEE,* vol. 55, 1970, pp. 1178–1206.
18. A. W. Smith and J. A. Armstrong, "Intensity Fluctuation and Correlations in a GaAs Laser," *Physical Review Letters,* vol. 16, 1965, pp. 5–6.
19. A. K. Laughton and Y. Kanabar, "Mode-Partition Noise in Gain-Guided Lasers at 850 nm and Its Impact on Analog Fiber-Optic Transmission Systems," paper TU03, *Digest of Technical Papers, Conference on Optical-Fiber Communication,* San Diego, February 11–13, 1985.
20. G. Grosskopf et al., "Laser Mode-Partition Noise in Optical Wideband Transmission Links," *Electronics Letters,* vol. 18, no. 12, 1982, pp. 493–494.
21. K. Petermann and G. Arnold, "Noise and Distortion Characteristics of Semiconductor Lasers in Optical-Fiber Communication Systems," *IEEE Journal of Quantum Electronics,* vol. QE-18, no. 4, 1982, pp. 543–555.
22. I. I. Ikushima and M. Maeda, "Self-Coupled Phenomena of Semiconductor Lasers Caused by an Optical Fiber," *IEEE Journal of Quantum Electronics,* vol. QE-14, 1978, pp. 331–332.
23. O. Hirota and Y. Suematsu, "Noise Properties of Injection Lasers due to Reflected Waves," *IEEE Journal of Quantum Electronics,* vol. QE-15, 1979, pp. 142–149.

24. M. A. Washington, R. E. Nahory, and E. D. Beebe, "High Efficiency InGaAsP/InP Photodetectors with Selective Wavelength Response between 0.9 and 1.7 μm," *Applied Physics Letters,* vol. 33, 1978, p. 854.

25. S. R. Forrest, "Performance of InGaAsP Photodiodes with Dark Current Limited by Diffusion, Generation Recombination, and Tunneling," *IEEE Journal of Quantum Electronics,* vol. QE-17, 1981, p. 217.

26. G. E. Stillman and C. M. Wolfe, "Avalanche Photodiodes," in P. K. Willardson and A. C. Beers (eds.), *Semiconductors and Semimetals,* vol. 12, Academic Press, New York, 1977, p. 291.

27. N. Susa, H. Nakagome, H. Ando, and H. Kanbe, "Characteristics in InGaAs/InP Avalanche Photodiodes with Separated Absorption and Multiplication Regions," *IEEE Journal of Quantum Electronics,* vol. QE-17, 1981, p. 243.

28. S. M. Miller, "Avalanche Breakdown in Germanium," *Physical Review,* vol. 99, 1955, p. 1234.

29. J. Conradi, "Temperature Effects in Silicon Avalanche Diodes," *Solid-State Electronics,* vol. 17, 1974, p. 99.

30. T. Kaneda, H. Takahashi, H. Matsumoto, and T. Yamaoka, "Avalanche Build-up Time of Silicon Reach-Through Photodiodes," *Journal of Applied Physics,* vol. 47, 1976, p. 4960.

31. J. C. Campbell, A. G. Dentai, W. S. Holden, and B. L. Kasper, "High Performance Avalanche Photodiode with Separate Absorption Grading and Multiplication Region," *Electronics Letters,* vol. 19, 1983, p. 818.

32. C. M. Penchina, "Note on Maximizing Signals from Photoconductive Detectors," *Infrared Physics,* vol. 15, 1975, p. 9.

33. C. H. Lee (ed.), *Picosecond Optoelectronic Devices,* Academic Press, Orlando, 1984, p. 31.

34. S. M. Sze, *Physics of Semiconductor Devices*, 2d ed., Wiley, New York, 1981, pp. 745–746.

35. R. H. Kingston, *Detection of Optical and Infrared Radiations*, Springer-Verlag, New York, 1978, pp. 62–64.

36. S. R. Forrest, "The Sensitivity of Photoconductor Receivers for Long-Wavelength Optical Communication," *IEEE Journal of Lightwave Technology,* vol. LT-3, no. 2, 1985, pp. 347–360.

37. H. L. VanTrees, *Detection Estimation and Modulation,* vol. I, Wiley, New York, 1968, pp. 36–45.

38. F. Capasso, "Physics of Avalanche Photodiodes," in W. T. Tsang (ed.), *Semi-Conductors and Semimetals,* vol. 22, pt. D, chap. 1, Academic Press, New York, 1985.

39. F. Capasso, W. T. Tsang, A. L. Hutchinson, and G. F. Williams, "Enhancement of Electron Impact Ionization in a Superlattice: A New Avalanche Photodiode with a Large Ionization Ratio," *Applied Physics Letters,* vol. 40, 1982, p. 38.

40. F. Capasso, W. T. Tsang, and G. F. Williams, "Staircase Solid-State Photomultipliers and Avalanche Photodiodes with Enhanced Ionization Rates Ratio," *IEEE Transactions on Electron Devices,* vol. ED-30, 1983, p. 381.

41. K. Ohnaka et al., "A Planar InGaAs PIN/JFET Fiber-Optic Detector," *IEEE Journal of Quantum Electronics,* vol. QE-21, 1985, pp. 1236–1240.

# CHAPTER 7
# MODULATION

**J. C. Daly**
**A. Fascia**
*Department of Electrical Engineering*
*University of Rhode Island*

## 7.1 INTRODUCTION

Modulation, in the case of fiber optics, is the process by which information is impressed on an optical carrier to be transmitted by the fiber. Demodulation is the process by which information is extracted from the carrier at the receiver end of the fiber.

With some notable exceptions, most fiber-optic communication systems use digital modulation, as represented by the typical digital fiber-optic communication system shown in Fig. 7.1. The information to be transmitted in this system is shown as a series of 1s and 0s. A coder organizes the 1s and 0s, and a modulator acts on this data by producing a current that turns a light-emitting diode (LED) or laser on and off. The resulting pulses of light containing the information are transmitted over the fiber. At the receiver, a detector converts the pulses of light into pulses of current. A demodulator extracts the information from the electrical pulses.

A variety of waveshapes can be used to represent binary data. The mapping of binary data to a waveshape is called *coding*. Impressing the information representing the code onto the optical energy is called *modulation*. Transmission of the

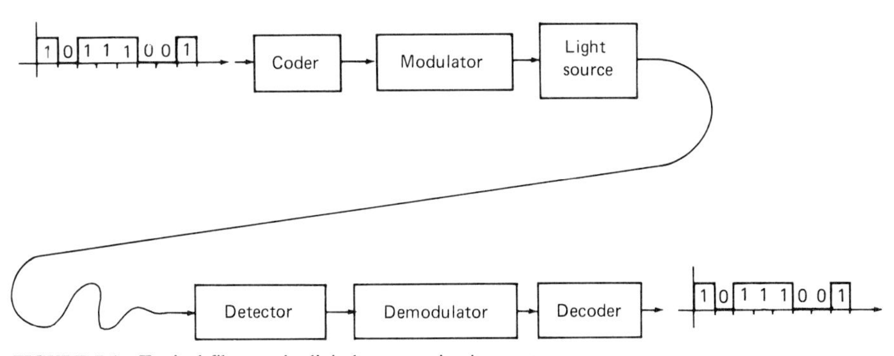


**FIGURE 7.1** Typical fiber-optic digital communication system.

information requires a modulator, an optical source, a fiber transmission path, a detector, and a demodulator, as shown in Fig. 7.1.

The fiber, the detector, and the demodulator introduce some degree of noise and distortion. This limits the distance and rate at which information can be transmitted. Waveform coding and modulation techniques can be chosen to minimize such negative effects.

## *7.2 DIGITAL SIGNALS AND CODES*

To understand digital signals and codes, it is useful to know the definitions of commonly used terms,[1] such as digital signal, signal element, unipolar, polar, data rate, and modulation rate. A *digital signal* is a series of discrete voltage pulses. Each individual pulse is a *signal element.* Binary data are transmitted by encoding each data bit into signal elements. All elements may not be positive voltage pulses. If both positive and negative voltage pulses are present the signal is *bipolar.* If only one polarity of voltage pulse is present the signal is *unipolar.* The *data rate* is the data transmission rate in bits per second. The amount of time it takes the transmitter to transmit one bit is the *bit duration.* The bit duration is $1/R$, where $R$ is the data rate. The *modulation rate* is the number of signal elements per second. It is expressed in bauds.

*Encoding* is the mapping of data bits to signal elements. Some of the common encoding schemes are shown in Fig. 7.2. These include:

- Non-return-to-zero (NRZ)
- Return-to-zero (RZ)
- Biphase
- Miller

Different codes have different properties and different advantages in different situations. Important properties include spectrum, synchronization capability, noise immunity, error detection capability, and complexity.

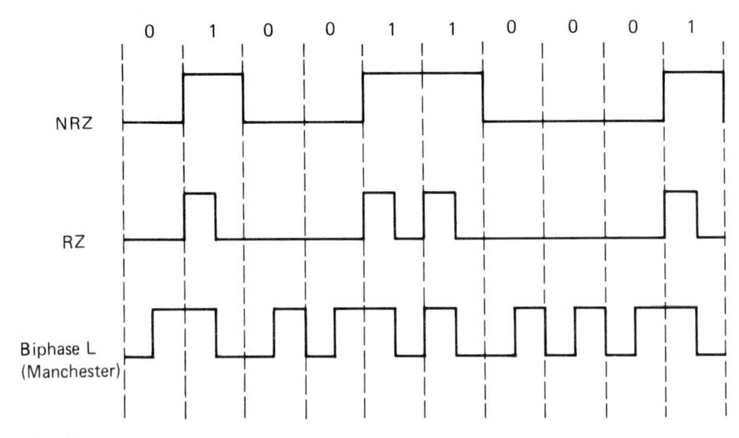


**FIGURE 7.2** Common encoding schemes.

Signal spectrum is important because it determines the bandwidth required. Also, if the spectrum has a DC component, direct connection between system devices is required. If no DC component is present, simpler AC coupling is used. If a code with a DC component is used in a system with AC coupling, a drift in the levels results for certain signals, such as a string of 1s.

Synchronization is important for a receiver. The receiver must decide if the signal element is a 1 or a 0. Each signal element occupies a definite time slot. A receiver must be synchronized with the incoming signal if it is to know when a signal element is occurring. Many digital communication systems employ a separate clock line. However, this requires an extra optical channel. It is usually avoided in fiber-optic systems by extracting the clock from the signal. Some codes are designed for easy extraction of timing information and clock signals.

Systems operating in a noisy environment or requiring low error rates use a code with error detection or correction capability.

### 7.2.1 Non-Return-to-Zero (NRZ)

The most common representation of binary data is in the form of a non-return-to-zero code, in which 1s and 0s are represented by voltage levels that are constant during the bit interval. Most computers use non-return-to-zero–level (NRZ-L) with a 1 represented by a high voltage level and a 0 represented by a low voltage level.

Some non-return-to-zero codes, such as non-return-to-zero–mark (NRZ-M) and non-return-to-zero–space (NRZ-S) are known as *differential codes.* In a differential code, the presence of a 1 or a 0 is determined with respect to the value of the voltage level during the previous time interval. If the code is NRZ-M and the present voltage level is different than the previous level, then a 1 is being transmitted. If there is no transition at the beginning of the interval then a 0 is being transmitted. For NRZ-S the inverse is true. A transition at the beginning of the interval represents a 0 and no transition represents a 1. These differential codes have the advantage that they do not depend on the polarity of the signal. If the polarity of the code is reversed by passing it through an inverting amplifier or by reversing the leads of a twisted pair, the interpretation of the code is unchanged. By comparison, if the polarity of an NRZ code is reversed, all 1s are interpreted as 0s and all 0s as 1s.

NRZ codes are convenient to use and make efficient use of bandwidth. The spectrum of NRZ codes is discussed in detail in Sec. 7.4, "Filtering to Reduce Intersymbol Interference." It is interesting to note the data requiring the most bandwidth are a string of alternating 1s and 0s. The NRZ code for this is a square wave with a period equal to two bit intervals. NRZ codes have the disadvantage of containing a DC component and they lack the facility for easy clock extraction. A long string of 1s or 0s represents a constant voltage and is difficult to detect because of voltage drifts and the absence of timing information in the signal.

### 7.2.2 Return-to-Zero

Return-to-zero codes differ from NRZ codes in that only one-half the bit duration interval is used for data. The voltage is zero during the second half of the bit duration. If a string of 1s is being sent, the code representation is a series of pulses each having a duration equal to one-half the bit rate. That is, a string of 1s is a square wave with a frequency equal to the bit rate. The modulation rate is defined as the number of signal elements per second. Since there are two signal elements (data and a space) per period, the maximum modulation rate for RZ codes is twice the bit rate. For NRZ codes, by comparison, the maximum modulation rate is equal to the bit rate. RZ codes have the same problems with the DC levels and synchronization. That is, the DC level depends on the number of 1s in the code. A long string of 0s represents a constant low voltage and provides no timing information.

### 7.2.3 Biphase

Biphase codes have at least one transition per bit interval. Biphase codes include, biphase-L (Manchester), biphase-S (S = space), and biphase-M (M = mark). Biphase codes are attractive in fiber-optic systems. They overcome the DC component and synchronization problems of NRZ and RZ codes. Biphase codes do this at the cost of greater bandwidth, but for many fiber-optic applications, bandwidth is available to trade off for ease of synchronization and elimination of DC drift. The maximum number of transitions per bit interval is two. Thus, the maximum modulation rate is twice the bit rate.

Manchester code has a transition in the middle of the bit interval. During the first half of the bit interval, the voltage level is high for a 1 and low for a 0. Therefore, a transition from high to low in the middle of the bit interval represents a 1 and a transition from low to high represents a 0.

Biphase mark and biphase space always have a transition at the beginning of the bit interval. A 1 or a 0 is represented by the presence or absence of a transition in the middle of the bit interval. For biphase mark, a transition represents a 1. For biphase space the absence of a transition represents a 1.

Differential Manchester always has a transition in the middle of the bit interval. A 1 is represented by no transition at the beginning of the bit interval. A 0 is represented by a transition at the beginning of the bit interval. If a 0 is transmitted, the voltage is the same as the previous bit interval. There are two transitions if the bit is 0 and one transition if the bit is 1.

Biphase codes have advantages of easy synchronization, no DC component, and some inherent facility for error detection. There is a predictable transition in each of the biphase codes on which the receiver can synchronize. Biphase codes are known as self-clocking codes. The absence of an expected transition can be used to detect errors. The absence of a DC component in the code allows AC coupling and eliminates errors caused by DC level drift in electronic circuits.

Although NRZ codes are most common, Manchester codes are also widely used in fiber-optic systems. Manchester code has been specified for the Institute of Electrical and Electronics Engineers (IEEE) Standard 802.3 for baseband coaxial cable using carrier sense, multiple access with collision detection (CSMA/cd).[2] It has also been used in the U.S. military standard MIL-STD-1553B for a twisted-pair bus system for noisy environments.[3] Differential Manchester has been specified for the IEEE Standard 802.5 token ring.

### 7.2.4 Miller

Miller coding incorporates at least one transition—but no more than two transitions—for every two bits. This gives it self-clocking properties but does not increase the modulation rate above the bit rate (the maximum rate for NRZ). A 1 is

**TABLE 7.1** Definition of Digital Signal Encoding Formats

Nonreturn-to-zero–level (NRZ-L)
  1 = high level
  0 = low level

Nonreturn-to-zero–mark (NRZ-M)
  1 = transition at beginning of interval
  0 = no transition

Nonreturn-to-zero–space (NRZ-S)
  1 = no transition
  0 = transition at beginning of interval

Return-to-zero (RZ)
  1 = pulse in first half of bit interval
  0 = no pulse

Biphase-level (Manchester)
  1 = transition from high to low in middle of interval
  0 = transition from low to high in middle of interval

Biphase-mark
  Always a transition at beginning of interval
  1 = transition in middle of interval
  0 = no transition in middle of interval

Biphase-space
  Always a transition at beginning of interval
  1 = no transition in middle of interval
  0 = transition in middle of interval

Differential Manchester
  Always a transition in middle of interval
  1 = no transition at beginning of interval
  0 = transition at beginning of interval

Miller (delay modulation)
  1 = transition in middle of interval
  0 = no transition if followed by 1
  Transition at end of interval if followed by 0

Bipolar
  1 = pulse in first half of bit interval, alternating polarity from pulse to pulse
  0 = no pulse

*Source:* From W. Stallings,[1] © 1984 IEEE.

represented by a transition in the middle of the bit interval. A 0 is represented by no transition at the middle of the bit interval. If the next bit is a 0, a transition is added to the end of the bit interval representing a 0. This always ensures transitions, even if, for example, the data are a string of 0s.

Table 7.1 provides a summary description of the various encoding schemes.

## *7.3 MODULATION TECHNIQUES*

Modulation is the process of impressing information on the optical signal. The amplitude, phase, or frequency of an optical waveform can be changed to represent the information to be transmitted. Digital modulation is the switching (keying) of one or more of the optical waveform parameters (amplitude, phase, or frequency) between discrete values. Analog modulation is the continuous variation of one or more of the optical waveform parameters.

An ideal signal, represented by $S(t)$, is a sinusoid of frequency $\omega$, amplitude $A$, and phase $\theta$:

$$S(t) = A \cos (\omega t + \theta) \tag{7.1}$$

Actual optical signals are more complex. An LED or laser diode output is the sum of components at series of frequencies determined by the resonances of the diode structure. The single sinusoidal frequency represented by Eq. (7.1) is required for coherent modulation schemes as discussed in Sec. 7.8 below. Coherent modulation can use the frequency, phase, or amplitude of the signal to carry the information. The development of single-mode diode lasers has allowed coherent systems to be developed. Initially, many fiber-optic communication systems were based on amplitude modulation of optical signals made up of many discrete sinusoidal terms. Amplitude modulation does not require a single-frequency sinusoid and allows lasers and LEDs with complex spectra to be used. Although Eq. (7.1) represents an idealized signal, it provides a basis for understanding the properties of modulation.

Optical frequencies are high. For example, an optical wavelength of 1.15 μm represents a frequency of $2.6 \times 10^{14}$ Hz [260 terahertz (THz)]. This extremely high carrier frequency permits high modulating frequencies and therefore high data rates.

### 7.3.1 Digital Modulation

Digital information, encoded according to one of the coding schemes such as the non-return-to-zero code or the Manchester code, is impressed on an optical signal by a modulation technique. For digital codes, modulation is the switching (keying) of an optical waveform parameter between levels. This impresses the code, and therefore the information, on the light.

***Amplitude-Shift Keying.*** Switching the amplitude of the optical signal between discrete levels is referred to as *amplitude-shift keying* (ASK). Usually there are two levels. One level is very low and the other is high. Different optical amplitude levels represent different power levels. Amplitude modulation is modulation of the source output power.

An important property of any modulation scheme is its spectral characteristic. This determines the bandwidth required for transmission. Some effects of frequency distortions, such as band limiting, can be seen from spectral characteristics. The spectrum of an amplitude-modulated signal can be obtained using the shifting theorem from Fourier analysis. The Fourier transform of a signal multiplied by an exponential function is the Fourier transform of the original signal shifted in frequency. That is, if the function $f(t)$ has a Fourier transform $F(\omega)$, then the Fourier transform of $e^{j\omega_a t}f(t)$ is $F(\omega - \omega_a)$, where $j$ is $\sqrt{-1}$, $t$ is time, $\omega_a$ is a constant frequency, and $\omega$ is the signal frequency in radians per second. Consider the following signal with a time-varying amplitude:

$$S(t) = f(t) \cos (\omega_c t) \tag{7.2}$$

where $\omega_c$ is the carrier frequency in radians per second. If cos $(\omega_c t)$ is written in exponential form and the shifting theorem from Fourier analysis is used, the Fourier transform for $S(t)$ can be written as

$$S(\omega) = \frac{F(\omega - \omega_c) + F(\omega + \omega_c)}{2} \tag{7.3}$$

where $F(\omega)$ is the Fourier transform of $f(t)$, the modulating function. Note that, since $f(t)$ and $S(t)$ are real functions of time, the magnitudes of $F(\omega)$ and $S(\omega)$ are even functions of $\omega$ and contain components at both positive and negative values of $\omega$. Amplitude modulation of cos $(\omega_c t)$ by the function $f(t)$ results in the spectrum of $f(t)$ being shifted in frequency by $\omega_c$, the carrier frequency. Since both positive and negative frequencies are shifted, the bandwidth required to transmit the amplitude-modulated carrier $S(t)$ is twice the bandwidth of the modulating signal $f(t)$.

For on-off shift keying, the amplitude of the optical signal is switched between 0 and some fixed level. A fixed value of unity for the on state can be used to illustrate the spectral properties of some on-off shift-keyed signals. Signals differ, depending on the bit pattern. The spectrum also depends on the bit pattern. Some information is obtained by assuming that the bit pattern results in a square wave signal amplitude. That is, the signal $S(t)$ is off for one bit period, then on for the next, resulting in a periodic square wave. Component amplitudes for the Fourier series representation of a symmetric square wave are

$$C_n = \frac{T}{2}\,\frac{\sin(\omega_n T/4)}{\omega_n T/4} = \begin{cases} \dfrac{T}{n\pi}(-1)^{(n-1)/2} & \text{for odd } n \\ 0 & \text{for even } n \end{cases} \tag{7.4}$$

where $\omega_n = 2\pi n/T$ and $T$ is the period of the square wave. The square wave is $f(t)$ in Eq. (7.2), and therefore becomes

$$f(t) = \sum_{n=0}^{\infty} C_n e^{-j\omega_n t} \tag{7.5}$$

The Fourier components of the signal $S(\omega)$ are the components of the envelope shifted in frequency as described by Eq. (7.3). That is,

$$S(\omega) = \sum_{n=0}^{\infty} \left\{ \frac{T}{2}\,\frac{\sin[(\omega_n - \omega_c)T/4]}{(\omega_n - \omega_c)T/4} + \frac{T}{2}\,\frac{\sin(\omega_n + \omega_c T/4]}{(\omega_n + \omega_c)T/4} \right\} \tag{7.6}$$

Figure 7.3 shows the positive half of this spectrum, i.e., the second term in Eq. (7.6).

***Frequency-Shift Keying.*** A frequency-shift-keyed (FSK) signal represents each discrete level of a code by a specific frequency. A binary code with two levels is represented by two discrete frequencies. Usually one frequency represents a logic 1 and the other a logic 0. For a bit stream consisting of alternating 1s and 0s, the FSK signal is a sinusoid, switching back and forth between two frequencies. It can be considered to be the sum of two ASK signals. For example, a 1 is sent by transmitting the signal frequency $f_1$ and a 0 by transmitting the signal fre-

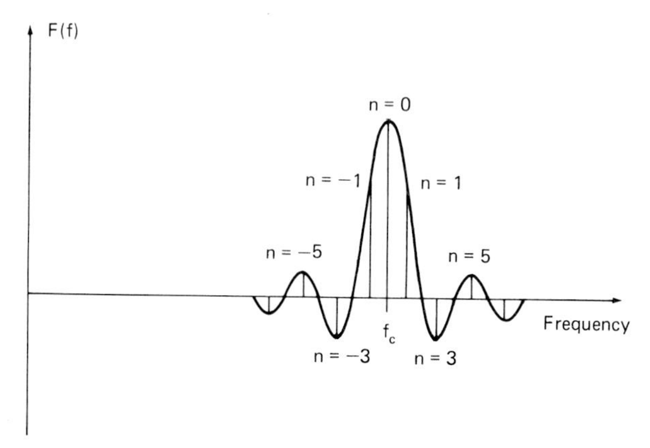


FIGURE 7.3 The frequency spectrum of an on-off shift-keyed signal representing alternating 1s and 0s.

quency $f_2$. The frequency spectrum of an FSK signal carrying a bit stream of alternating 1s and 0s is shown in Fig. 7.4. It is simply the sum of the spectrum of two ASK signals.

In practice in noncoherent optical systems, an FSK signal is formed and used to amplitude-modulate an optical carrier rather than direct FSK-modulate the optical carrier.

### 7.3.2 Amplitude Modulation

Direct amplitude modulation (AM) of the output power of a light-emitting diode or laser diode is achieved by varying the current through the device. Usually the modulating signal is a digital waveform, but it may also be a continuously varying analog signal. In the following, the properties of a pure sinusoidal signal with an amplitude that is varied in response to a modulating function are examined. The

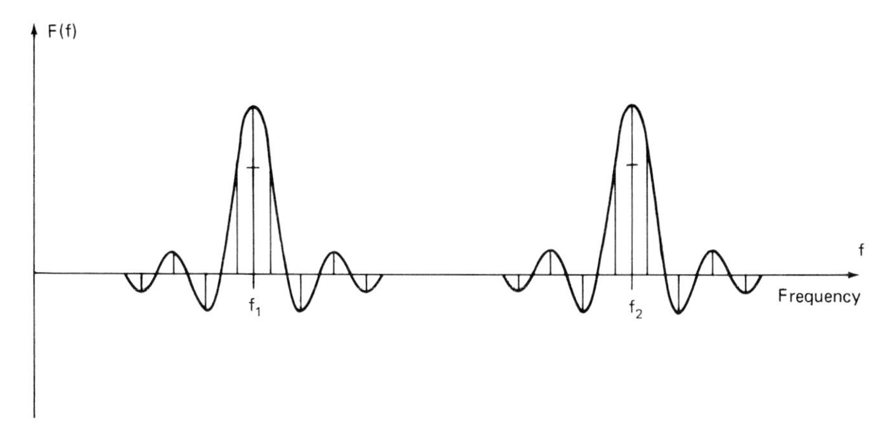


FIGURE 7.4 Frequency spectrum of an ASK signal consisting of alternating 1s and 0s.

modulating function is referred to as the *baseband* signal. The baseband signal can extend over a wide band, as it does for the case of an NRZ code with a bit rate in the gigabit-per-second range.

*Double-sideband* (actually an abbreviation of *double-sideband suppressed carrier*) modulation is a form of amplitude modulation that results when the amplitude of the carrier is proportional to the modulating signal $f(t)$:

$$S(t) = f(t) \cos(\omega_c t) \tag{7.7}$$

where $\omega_c$ is the carrier frequency in radians per second. From Eq. (7.2), the Fourier transform for $S(t)$ can be written as

$$S(\omega) = \frac{F(\omega - \omega_c) + F(\omega + \omega_c)}{2} \tag{7.8}$$

where $F(\omega)$ is the Fourier transform of $f(t)$, the modulating function. The modulation process has shifted $F(\omega)$ to the vicinity of $\omega_c$. Note that, since $S(t)$ is a real function of time, the magnitude of $S(\omega)$ is an even function of $\omega$ and contains components at both positive and negative values of $\omega$. The spectrum $S(\omega)$ is divided into upper and lower sidebands. The upper sideband is that portion of the spectrum lying in the region $|\omega| > \omega_c$. The portion of the spectrum lying in the region $|\omega| < \omega_c$ is called the lower sideband.

A double-sideband signal can be demodulated by multiplying $S(t)$ by $\cos(\omega_c t)$. This results in $y(t) = f(t) \cos^2(\omega_c t)$, or $y(t) = f(t)[1 + \cos(2\omega_c t)]/2$. Since the term multiplied by $\cos(2\omega_c t)$ is located in a frequency region centered about $2\omega_c$, it can be filtered out, leaving the desired function $f(t)$. This process is called synchronous detection. It requires a locally generated reference signal, $\cos(\omega_c t)$, at the receiver. A phase shift of $\theta$ between this signal and the original carrier reduces the output of the demodulator by a factor of $\cos\theta$.

The classic form of amplitude modulation is *double-sideband transmitted-carrier* modulation. Transmission of the carrier simplifies demodulation. The carrier is retained when a constant is added to the modulating function $f(t)$:

$$S(t) = K[1 + mf(t)] \cos(\omega_c t) \tag{7.9}$$

where $K$ = constant dependent on the power level
$m$ = modulation index
$\omega_c$ = carrier frequency, rad/s

It is assumed that $f(t)$ has a peak magnitude equal to 1 and $m$ is a positive number less than or equal to 1. This assures that the term $1 + mf(t)$ is never negative. The modulating function $F(t)$ and the amplitude modulated signal $S(t)$ are shown in Fig. 7.5a and b, respectively.

The frequency spectrum of $s(t)$ contains the carrier frequency $\omega_c$ and the upper and lower sidebands due to the product $Kmf(t) \cos(\omega_c t)$. An example of the spectrum of this classic AM is shown in Fig. 7.5c.

A *single-sideband* (SSB) signal contains only the upper or lower sideband. Since the sidebands contain redundant information, only one is required to recreate the signal at the receiver. Transmitting only one sideband cuts the bandwidth requirement in half. There are also power economies in SSB. In classic AM, with a carrier, upper sideband, and lower sideband, the carrier contains a minimum of half the transmitted power. A single-sideband wave may be obtained

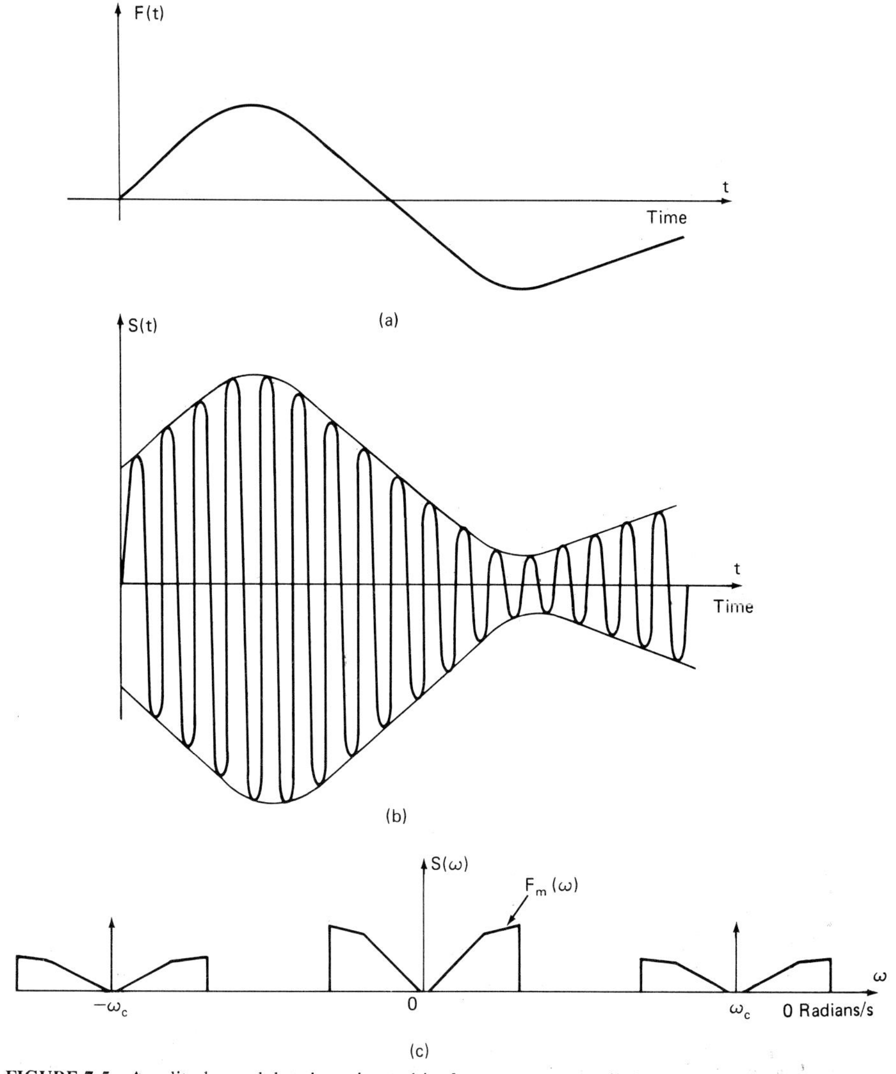


**FIGURE 7.5** Amplitude-modulated carrier and its frequency spectrum: (*a*) Baseband modulating signal; (*b*) amplitude-modulated carrier; (*c*) frequency spectrum of the amplitude-modulated carrier.

by passing a double-sideband signal through a filter that passes only one sideband.

Single-sideband systems have the drawback that they can not transmit a baseband spectrum down to zero frequency. The sharp filter characteristic for filtering out the undesired sideband also filters out the lower frequencies of the baseband signal. Some digital signals require DC to be retained. The filter also introduces phase distortion that can be troublesome in some systems.

*Vestigial-sideband* (VSB) modulation is an amplitude-modulation scheme that overcomes these difficulties by using a filter that has odd symmetry about $\omega_c$ and passes the low-frequency tail of the undesired sideband. When demodulated, the energy in the undesired sideband fills in the energy missing from the desired sideband. There is no loss of information as long as the filter characteristic has odd symmetry about $\omega_c$. The presence of a small amount of energy in the undesired sideband increases the bandwidth required for transmission over that required for SSB. The power requirements for VSB are the same as for SSB.

### 7.3.3 Frequency and Phase Modulation

Direct frequency modulation (FM) of an optical wave where the *optical* frequency or phase is varied in response to a baseband signal, requires the techniques of coherent optical communications discussed in Sec. 7.8. Noncoherent systems also employ phase and frequency modulation. For these systems, a phase- or frequency-modulated signal is used to amplitude-modulate the incoherent optical wave.

Phase and frequency modulation are closely related subsets of angle modulation. The relationship between phase and frequency modulation is easily understood by recalling the mechanical analogy of a rotating wheel. The angular velocity $\omega$ has units of radians per second and is the time derivative of the angular displacement $\theta$. Frequency is analogous to angular velocity and phase is analogous to angular displacement. For a sinusoidal signal the instantaneous frequency $\omega$ is the derivative of the phase $\theta$. If the frequency is constant, the phase increases linearly with time. That is, if the constant frequency is $\omega_c$, the phase is $\omega_c t$ plus a constant of integration.

If, in response to a baseband signal such as $f(t)$ in Fig. 7.6*a*, the frequency of a sinusoidal signal is made to vary about its constant carrier value, the phase will also vary. Such a frequency-modulated sinusoid is shown in Fig. 7.6*b*. A common method of generating FM signals is to phase-modulate a carrier with the integral of the baseband signal. The phase of the sinusoid is $\omega_c t$ plus the integral of the

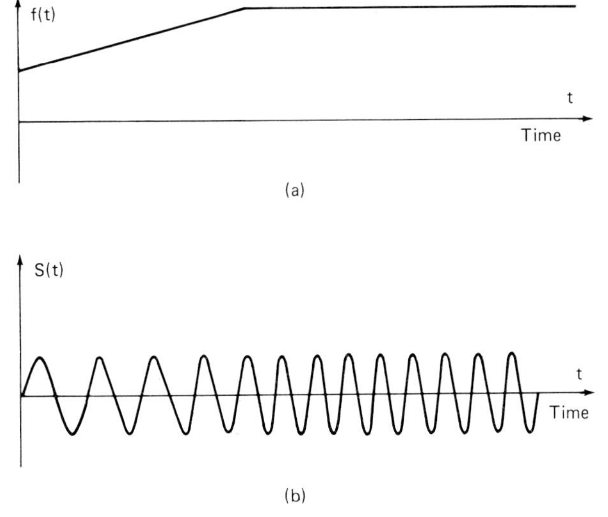


**FIGURE 7.6** (*a*) Baseband modulating signal. (*b*) Frequency-modulated signal.

baseband signal. Since the frequency is the derivative of the phase, the frequency is the constant $\omega_c$ plus the baseband signal.

A frequency-modulated signal is represented by

$$S(t) = A \cos \{2\pi [f_c + \Delta f \cdot f(t)]t\} \tag{7.10}$$

where $A$ is the amplitude, $f_c$ is the carrier frequency, $\Delta f$ is the frequency deviation, and $f(t)$ is the baseband signal. The absolute value of $f(t)$ is less than 1. The maximum instantaneous frequency deviation of the frequency from the carrier frequency $f_c$ is $\Delta f$. The actual bandwidth occupied by the FM signal depends on the amplitude and bandwidth of the baseband signal as well as the frequency deviation $\Delta f$. There is no precise expression for the bandwidth of an FM signal, but a widely used rule of thumb, called *Carson's rule,* states that the bandwidth required for an FM signal is $2 \cdot (\Delta f + f_m)$, where $f_m$ = maximum bandwidth of the baseband signal.

***FM Distortion Immunity.*** Nonlinear elements such as LEDs, lasers, and detectors distort signals. FM has an inherent immunity to distortion introduced by nonlinearities. Consider an FM signal:

$$S(t) = A \cos \theta \tag{7.11}$$

where $\theta = \omega_c t + 2\pi\Delta \cdot f(t)t$. The input, shown in Fig. 7.7*a,* to the nonlinear device is $S(t)$ and the output of the device is $Y(t)$. Since $S(t)$ is a periodic function of $\theta$, $Y(t)$ will also be periodic in $\theta$, and therefore may be represented by a Fourier series with $\theta$ as independent variable:

$$Y(t) = \sum_{n=0}^{\infty} B_n \cos (n\theta + \phi_n) \tag{7.12}$$

where $B_n$ and $\phi_n$, the component amplitudes and phases, are constants determined by the nonlinearity. For example, if the nonlinearity is produced by the hard limiter shown in Fig. 7.7*b,* $Y(t)$ is a square wave as shown in Fig. 7.7*c*. Then $B_n$ and $\phi_n$ are the Fourier amplitude and phase constants for a square wave periodic in $\theta$. The term in the summation corresponding to $n = 1$ is an undistorted FM signal that differs from the input only by an amplitude constant and phase shift. Distortion immunity exists if higher-order components ($n > 1$) can be filtered out, leaving only the component containing the undistorted signal as shown in Fig. 7.7*d* and *e*. This occurs if other components in the summation do not overlap with the $n = 1$ component. Since each term in the summation can be considered an FM signal, Carson's rule can be used to predict component bandwidths. There will be no overlap of the $n = 1$ component by the distortion-generated terms if

$$f_c > 3(\Delta f + f_m) \tag{7.13}$$

where $f_c$ = carrier frequency, Hz
$\Delta f$ = peak frequency deviation, Hz
$f_m$ = maximum frequency of the modulating signal, Hz

***Tone-Modulated FM.*** The frequency spectrum of an FM or phase-modulated signal is complex. This is illustrated by the simple example of frequency modulation

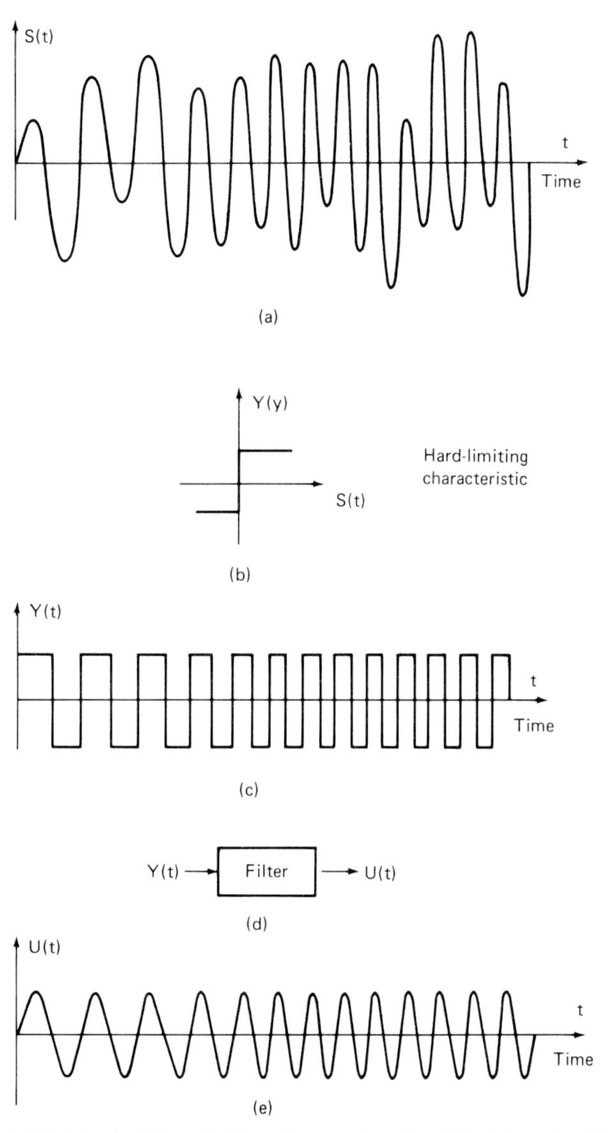


**FIGURE 7.7** Hard limiting characteristic for FM. Note that if $f_c > 3(\Delta f + f_m)$, the FM signal is undistorted. (*a*) Amplitude-distorted FM signal; (*b* limiter characteristic; (*c*) square wave $f_m$ out of limiter; (*d*) low-pass filter; (*e*) undistorted FM signal.

by a sinusoid. When the modulation function is a tone, $f(t) = \cos(\omega_m t)$, the frequency-modulated signal is

$$S(t) = A \cos\left[\omega_c t - \left(\frac{\Delta f}{f_m}\right) \sin(\omega_m t)\right] \tag{7.14}$$

where $\omega_c$ = carrier frequency, rad/s
$\Delta f$ = peak frequency deviation, Hz
$f_m$ = frequency of the modulation tone, Hz
$\omega_m$ = $2\pi f_m$, rad/s

This may be rewritten by using Bessel's trigonometric identities:[4]

$$S(t) = \sum_{n=-\infty}^{\infty} J_n(\beta) \cos(\omega_c t - n\omega_m t) \tag{7.15}$$

where $\beta = \Delta f/f_m$ and $J_n(\beta)$ is the $n$th-order Bessel function of the first kind. Since $J_{-n}(\beta) = (-1)^n J_n(\beta)$, the first few terms of $S(t)$ are

$$S(t) = J_0(\beta)\cos(\omega_c t) + J_1(\beta)[\cos(\omega_c t - \omega_m t)] - J_1(\beta)[\cos(\omega_c t + \omega_m t)] + J_2(\beta)[\cos(\omega_c t - 2\omega_m t)] + J_2(\beta)[\cos(\omega_c t + 2\omega_m t)] + \cdots \tag{7.16}$$

This shows that a single modulating tone produces all sets of sidebands displaced from the carrier by all possible multiples of the modulating frequency. It indicates that the bandwidth requirements of FM may be large. When $\beta$ is small, $J_n(\beta)$ is negligible for $n > 1$. For this case, only the first set of sidebands is significant. This is called narrowband FM. For large $\beta$, $J_n(\beta)$ is negligible if $n > \beta$. That is, the number of sidebands is limited. Since $\beta = \Delta f/f_m$, the number of pairs of sidebands with significant energy is $\Delta f/f_m$. Therefore, the bandwidth required for transmission of this tone example is $2\beta f_m = 2\,\Delta f$, when $\beta$ is large. Consider the quasi-stationary case. For this case, $\Delta f$ is large and $f_m$ is small. The frequency varies slowly between $f_c - \Delta f$ and $f_c + \Delta f$. The bandwidth required is $2\,\Delta f$.

When a sinusoidal carrier is phase-modulated by a sum of $N$ sinusoidal tones, the spectrum is more complex than the case of modulation by a single tone.[5] When a sum of tones modulates the phase,

$$S(t) = \cos\left[\omega_c t + \sum_{r=1}^{N} m_r \cos(\omega_r t)\right] \tag{7.17}$$

where $m_r$ are constants. When Eq. (7.17) is expanded,[5]

$$S(t) = \sum_{n_1=-\infty}^{\infty} \cdots \sum_{n_N=-\infty}^{\infty} \prod_{r=1}^{N} J_{nr}(m_r) \cos\left(\omega_c t + \sum_{r=1}^{N} n_r \omega_r t + \sum_{r=1}^{N} \frac{n_r \pi}{2}\right) \tag{7.18}$$

The symbol $\Pi$ denotes that all $N$ of the $J_{nr}$ coefficients are multiplied together. Sidebands will be displaced from the carrier at all possible multiples of the modulating frequencies and at all possible sums and differences of the modulating frequencies.

## 7.4 FILTERING TO REDUCE INTERSYMBOL INTERFERENCE

In fiber-optic communication systems where data are sent as pulses of light, very short-duration pulses are possible because of the extremely large fiber band-

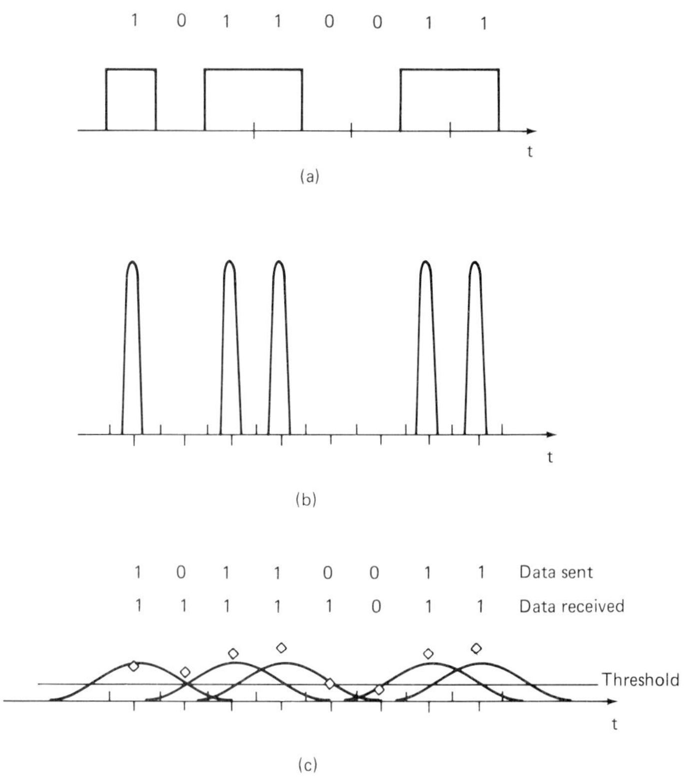


**FIGURE 7.8** A signal experiencing intersymbol interference due to pulse spread. (*a*) Data sent; (*b*) transmitted optical pulses; (*c*) received optical pulses showing spreading. Diamonds indicate pulse sums at sampling instants.

width. There is a reciprocal relationship between bandwidth and duration of pulses. When the bandwidth of a pulse is limited, it will spread out and occupy more time. Intersymbol interference occurs when one pulse spreads over into the time slot allocated to another. Figure 7.8 shows a signal experiencing intersymbol interference due to pulse spreading. At the receiver the signal is sampled at a specific time. The receiver decides if a pulse has or has not occurred on the basis of the received energy at the sampling instant. An error can occur because of intersymbol interference if, for example, a zero was sent but other pulses had spread to the sampling instant with sufficient energy to cause the receiver to indicate a pulse had been sent.

When the spectrum of the incoming signal has specific properties, intersymbol interference can be reduced. Specifically, pulses have nulls at the sampling instants of other pulses. The physically unrealizable but theoretically interesting ideal low-pass characteristic is a frequency spectrum that represents a time function with zero intersymbol interference. Consider the frequency spectrum shown in Fig. 7.9*a*:

$$\begin{aligned} H(\omega) &= 1 \qquad \text{for } \omega < \omega_c \\ H(\omega) &= 0 \qquad \text{otherwise} \end{aligned} \tag{7.19}$$

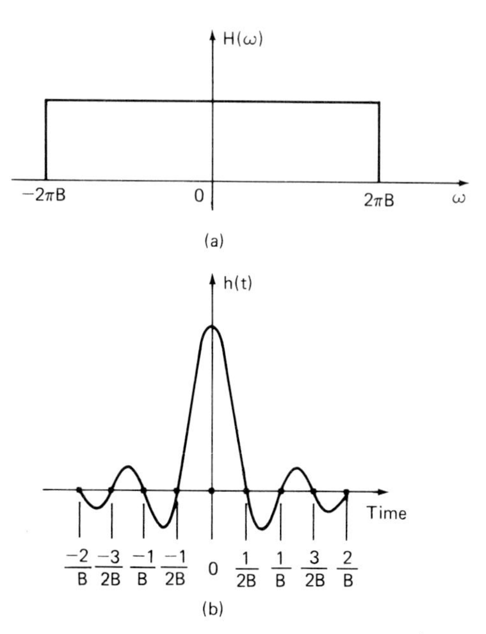


**FIGURE 7.9** (*a*) Ideal low-pass frequency spectrum; (*b*) corresponding time function.

where $\omega_c = 2\pi B$ and $B$ = bandwidth, Hz. The inverse Fourier transform of $H(\omega)$, $h(t)$, is shown in Fig. 7.9*b*. Note that $h(t)$ has nulls uniformly distributed along the time axis, separated by $1/2B$ s. If other pulses occur and are sampled at these times, there would be no intersymbol interference. If each pulse represents one bit of information, this corresponds to a bit rate of $2B$ b/s. This is not physically realizable because of two drawbacks of the ideal low-pass spectrum:

1. An ideal low-pass filter with sharp ("brick-wall") cutoff is unattainable.
2. Signal energy at nulls is sensitive to clock jitter.

Clock timing jitter is a displacement of the incoming signal with respect to the receiver clock. When it occurs, the sampling instants at the receiver do not correspond to the best time for the received pulses. Jitter causes the sampling instant to be offset from the nulls of the other pulses. The tails of these pulses decrease as $1/t$, where $t$ is the time from the occurrence of the pulse. The sum of the tails can grow without limit if they are summed at times other than their nulls. All systems possess some clock jitter. It can not be totally eliminated.

### 7.4.1 Raised Sine and Cosine Spectra

The ideal low-pass spectrum can theoretically be used to transmit $2B$ pulses per second, where $B$ is the bandwidth in hertz. Raised sine and cosine spectra overcome some of the difficulties of the ideal low-pass spectrum while retaining re-

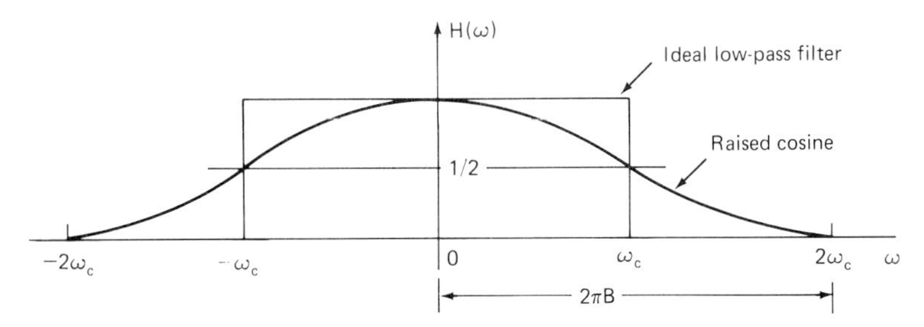


**FIGURE 7.10** Raised cosine spectrum.

duced intersymbol interference.[6] This is done at the expense of bandwidth requirements, however.

Consider the raised sine spectrum $H(\omega)$, shown in Fig. 7.10:

$$H(\omega) = \begin{cases} 1 & \text{for } |\omega| < \omega_c - \omega_x \\ \dfrac{1}{2}\left\{1 - \sin\left[\dfrac{(\pi/2)(\omega - \omega_c)}{\omega_x}\right]\right\} & \text{for } |\omega - \omega_c| < \omega_x \\ 0 & \text{for } |\omega| > \omega_c + \omega_x \end{cases} \tag{7.20}$$

$H(\omega)$ has a bandwidth of $B(1 + \omega_x/\omega_c)$ Hz. In Fig. 7.10, $\omega_x = \omega_c$. $2\omega_x$ is the frequency band used to transition $H(\omega)$ from 1 to 0. If $\omega_x$ is zero, $H(\omega)$ represents the ideal low-pass spectrum. In Fig. 7.10, $H(\omega)$ is the raised cosine spectrum, since $\sin [(\pi/2)(\omega/\omega_c - 1)] = \cos [(\pi/2)(\omega/\omega_c)]$.

The pulse represented by the spectrum $H(\omega)$ given in Eq. (7.20) is

$$h(t) = \frac{(\omega_c/\pi)[\sin(\omega_c t)/\omega_c t] \cos(\omega_c t)}{1 - (2\omega_x t/\pi)^2} \tag{7.21}$$

By definition, $t = 0$ is the time at which the pulse occurs. Note that the term $\sin(\omega_c t)$ causes pulses to have nulls every $1/2B$ s, where $\omega_c = 2\pi B$. When pulses occur every $1/2B$ s, there will be no intersymbol interference if there is no clock jitter. With clock jitter, intersymbol interference is reduced because pulse tails drop more rapidly when $\omega_x$ increases. This improvement over the ideal low-pass spectrum has been achieved at the expense of increased bandwidth requirements. The bandwidth required is $B(1 + \omega_x/\omega_c)$ Hz, where $T$, the time between pulses, is $1/2B$. Practical receiver filters can be implemented with a bandwidth that is 20 to 30 percent wider than the ideal low-pass filter.[7]

## 7.5 DEVICE CHARACTERISTICS

Optical sources, detectors, and the optical fiber itself have characteristics that influence modulation and detection systems. In this section, physical characteristics of the devices that make up a fiber-optic communication system are reviewed for the purpose of illustrating their effect on modulation and detection methods.

### 7.5.1 Modulation of Light-Emitting Diodes

Light is produced in light-emitting diodes by recombination of electrons and holes at the *pn* junction. These electrons and holes enter the junction as components of a forward current through the device. Recombination is the movement of an electron from a conduction state into a vacant valence state of an atom. This vacant state is called a hole. The energy difference between the two states is given off as a photon. In gallium arsenide (GaAs), the transition is termed *direct* because the momentum of the electron in the conduction state is the same as the momentum of the electron in the valence state. In indium gallium arsenide phosphide (InGaAsP), the transition is called *indirect* because electrons are first trapped at isoelectronic centers, where they combine with holes. Isoelectronic centers are formed by adding impurities to the semiconductor material. The electron-hole pair is then annihilated, generating a photon whose energy equals the bandgap energy minus that present at the isoelectronic center.

The peak wavelength $\lambda$ emitted by a direct-transition LED is related to the bandgap energy $E_g$ of its semiconductor material by

$$\lambda = \frac{hc}{E_g} \tag{7.22}$$

where $h$ = Planck's constant ($6.62 \times 10^{-34}$ J · s) and $c$ = speed of light ($3 \times 10^8$ m/s). The peak wavelength emitted by indirect-transition LEDs varies with the level of impurities present. The wavelengths of several LED types are shown in Table 7.2.

A fundamental limitation on modulation rate of an LED is the carrier lifetime $\tau_{lf}$. *Carrier lifetime* is the average length of time a carrier remains in the active region before recombining and producing a photon. At very high frequencies, carriers move in and out of the active region without remaining long enough to recombine and produce a photon. The *modulation bandwidth,* also called the *LED cutoff frequency,* is the frequency in hertz at which the optical output power $P_{opt}(\omega)$ decreases 1.5 dB.[8] The optical output power of a typical LED at high forward current bias is

**TABLE 7.2** Listing of LEDs, Indicating Peak Emission Wavelengths and Transition Characteristics

| LED type | Peak wavelength, nm | Transition |
|---|---|---|
| Ge | 1880 | Indirect |
| Si | 1140 | Indirect |
| GaAs | 910 | Direct |
| GaP | 560 | Indirect |
| $GaAs_{.60}P_{.40}$ | 650 | Direct |

$$P_{opt}(\omega) = P_{opt}(0)[1 + (\omega\tau_{lf})^2]^{-1/2} \tag{7.23}$$

The modulation bandwidth $\Delta\omega_{3dB}$ is therefore

$$\Delta\omega_{3dB} = \frac{1}{\tau_{lf}} \tag{7.24}$$

The 3-dB subscript appears in the modulation bandwidth expression because the 1.5-dB drop in optical output power translates into a 3-dB drop in detected electric power at the receiver. Hence, the modulation bandwidth is also the frequency at which the detected electric power equals $\frac{1}{2}P_e(0)$, or

$$P_e(\Delta\omega_{3dB}) = \frac{1}{2} P_e(0) \tag{7.25}$$

In addition to carrier lifetime, modulation speed is influenced by how fast charge can be moved in and out of the active junction region. An equivalent capacitance is used to model the process of moving charge (electrons and holes) into the junction to set up conditions for recombination. Resistance-capitance ($RC$) time constants of the diode and drive circuit limit the current pulse response of an LED. The LED capacitance has two components, the space-charge capacitance of the entire junction area and the diffusion capacitance of the active area where recombination takes place. Current into the junction passes mainly through the active area. For a nonlinear diode, as current increases, resistance decreases. At current levels normally encountered, current is crowded into the small active region and response time is limited by carrier lifetime. In lasers, carrier lifetime is reduced further by stimulated emission. This is an additional mechanism for recombination that reduces carrier lifetime and increases speed. Recombination in LEDs is spontaneous and slower than in lasers. LED carrier lifetime can be reduced by increased doping or reducing the active area thickness. Both reduce the quantum efficiency and therefore the optical output power. This results in a tradeoff between power output and modulation bandwidth. Modulation rates of 100 Mb/s are common for LEDs. Some LEDs can be modulated up to about 200 MHz. Figure 7.11 is a plot showing power and bandwidth limitations of several types of LEDs.[9]

An LED DC bias current can be used to decrease resistance and reduce the effect of diffusion capacitance. The space-charge capacitance of the entire junction is reduced by etching away the junction surrounding the active area. The equivalent circuit shown in Fig. 7.12 predicts the pulse response of the surface-emitting LED shown in Fig. 7.13.[10] If the diode capacitance $C_d$ is known, a simple series $RC$ circuit may be used to simulate the diode response. The diode resistance $R_d$ can be calculated from

$$R_d = \frac{\tau_{lf}}{C_d} \tag{7.26}$$

where $\tau_{lf}$ is the carrier lifetime.

LEDs radiate incoherent light over a broad emission spectrum. Unlike laser diodes, they are operated in the spontaneous emission region. LED data sheets from manufacturers normally include information such as the peak wavelength $\lambda_p$, a graph of wavelength vs. relative intensity, the diode capac-

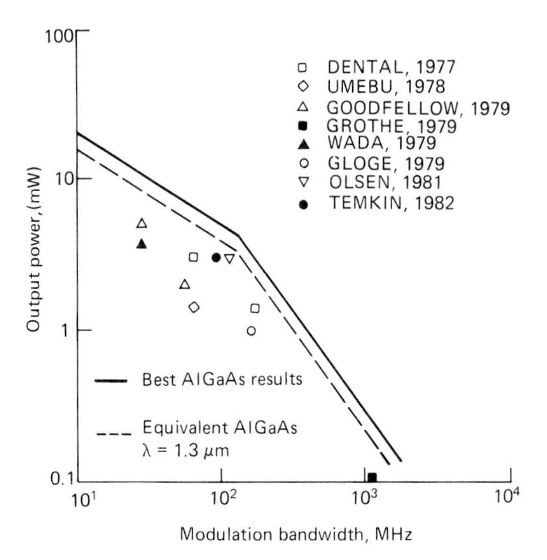


**FIGURE 7.11** Power and bandwidth limitations of an LED. *(From Saul,[9] © 1983 IEEE.)*

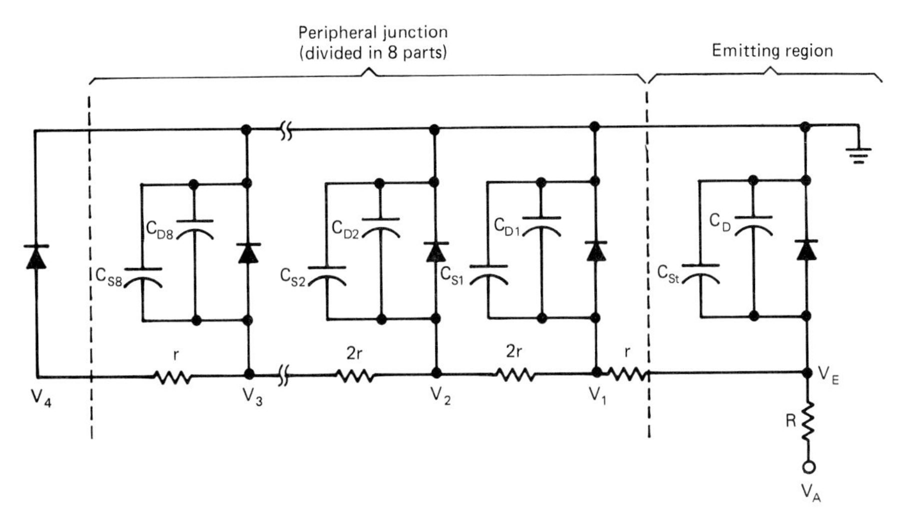


**FIGURE 7.12** Equivalent circuit used to predict the performance of the LED shown in Fig. 7.13. *(From Saul,[9] © 1983 IEEE.)*

itance measured at a certain frequency between the anode and cathode with a bias current applied, the relative optical power response vs. frequency, the relative radiant intensity vs. the angle subtending the diode axis, the relative optical output power vs. forward current, and the temperature dependence of optical output power.[11]

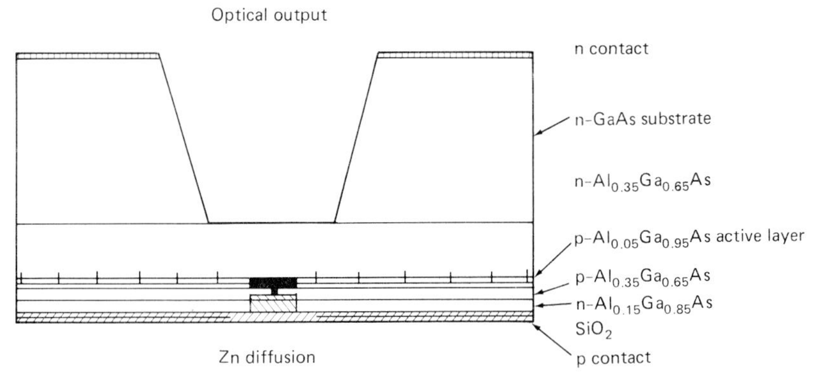


**FIGURE 7.13** Structure of a surface-emitting LED. *(From Hino and Iwanoto,[10] © 1979 IEEE.)*

### 7.5.2 Modulation of Laser Diodes

Diode lasers utilize optical feedback to achieve stimulated recombination of electron-hole pairs in the active region. Optical feedback is usually obtained with reflecting surfaces. A portion of the light produced by the diode is reflected back through the active area to produce additional stimulated electron-hole recombination. A photon produced by stimulated emission has the characteristics of the stimulating photon. Its frequency and direction of propagation are the same as the original photon. Important laser characteristics such as narrow linewidth and output beam divergence are determined by the structure that recirculates photons back through the active region. This structure is called a *cavity.* A simple cavity is made by cleaving the laser crystal to form two mirrors. The mirrors form an optical resonator, with the resonant frequency determined by the optical path length between mirrors. The resonance wavelength is a multiple of the round-trip optical cavity length. Photons bounce back and forth between the mirrors, making multiple passes through the active region, where they may induce stimulated recombination and additional photons. In addition to the longitudinal confinement of the light by the mirrors, lateral confinement is also desirable. Lateral confinement increases optical flux density and eliminates higher-order transverse mode patterns. Lateral confinement is achieved with a heterojunction structure. Aluminum doping is used to increase the index of refraction and decrease the bandgap. Increased index of refraction acts to confine the light in the region of aluminum doping, where the bandgap is also reduced. Carriers concentrate in the reduced bandgap region. Carrier confinement reduces the size of the active region and increases current density. This reduces the total current required for laser operation.

Above threshold, laser output power is a linear function of drive current. Early lasers without lateral confinement exhibited a kink in the output vs. drive current characteristic, because of shifts in transverse mode power patterns as a function of drive current.

The pulse response of a diode laser is influenced by the interaction of the optical energy in the cavity and charge stored in the active region. When a pulse of current is applied to the diode, there is a delay of 1 to 2 ns before light is emitted.[8] During this time, charge is building up in the junction. For laser action the gain experienced by light making a round trip in the resonator must be greater than the losses. Gain depends on a sufficient number of carriers being in place for stimulated recombination. When the number of carriers reaches a threshold, laser action begins. In practice, this delay time is eliminated by biasing the laser above threshold.

There are two energy-storage mechanisms in a laser that interact to produce overshoot and ringing in the output. These mechanisms act together like an oscillating pendulum with kinetic and potential stored energy. Energy is stored as photons in the optical cavity and as carriers in the active region. As a pendulum oscillates, energy shifts between kinetic and potential. In a laser, oscillatory output results from shifts between energy stored in carriers and energy stored as photons in the optical cavity. When the laser is turning on and the threshold has been reached, there are fewer photons and an excess of carriers. As laser operation proceeds, additional photons are produced which deplete the carriers through stimulated recombination. This proceeds until there are an excess of photons and a corresponding carrier deficiency. When fewer carriers are present, stimulated recombination decreases. Therefore, the number of photons is again reduced. This oscillation of energy between carriers and photons has a characteristic frequency in the gigahertz range that depends on a number of factors including the drive current level. A higher current produces a higher resonance frequency. The step response with characteristic pulsations is shown in Fig. 7.14.[8] The small-signal AC response contains the same resonance. This resonance limits the modulation bandwidth of lasers. It also results in increased noise at the resonance frequency.

The presence of carriers in the active region changes the index of refraction. This changes the optical length of the cavity and therefore the optical resonant frequency. This wavelength dependence on drive current is called *chirping.* It results in pulse amplitude distortion in dispersive transmission systems.[12] It is useful in that it allows the drive current to frequency- or phase-modulate the optical signal.

Many types of laser diodes exist.[13] Illustrated in Fig. 7.15 are the typical optical spectra and linewidths of several important lasers to be discussed. A more comprehensive treatment of lasers (and LEDs) is found in Chap. 5.

A gain-guided laser consists of an active region sandwiched between a *pn*

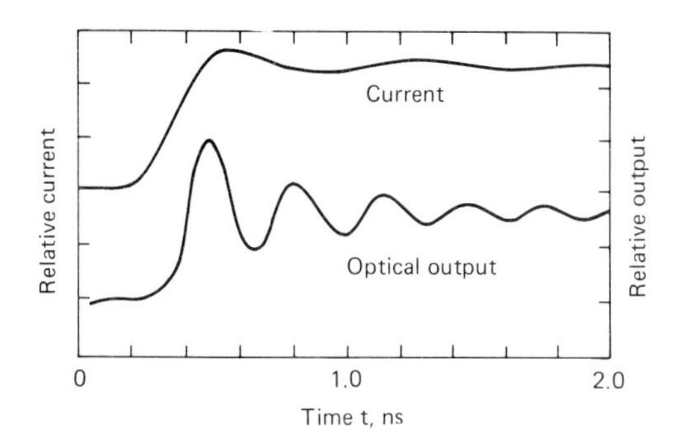


**FIGURE 7.14** Typical step response of a laser diode. *(Reproduced with the permission of the publisher, Howard W. Sams & Co., Indianapolis, Optical-Fiber Transmission, edited by E. E. Basch, © 1986.)*

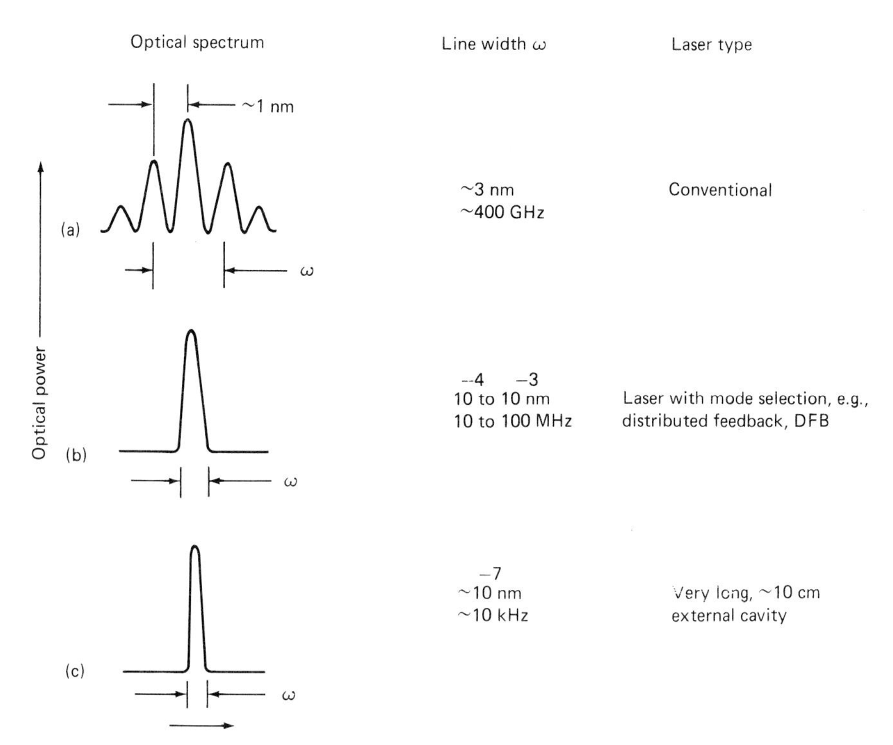


**FIGURE 7.15** Typical optical spectra and linewidths of various types of lasers. *(From I. W. Stanley,[14] © 1985 IEEE.)*

junction, as illustrated in Fig. 7.16. Parallel sides $M_1$ and $M_2$ are cleaved to form an optical cavity of length $L$. When current is injected into the active region, population inversion occurs. Because of the population inversion, optical emission also takes place in this region. The spatial distribution of photons emitted is dictated by the current density distribution.

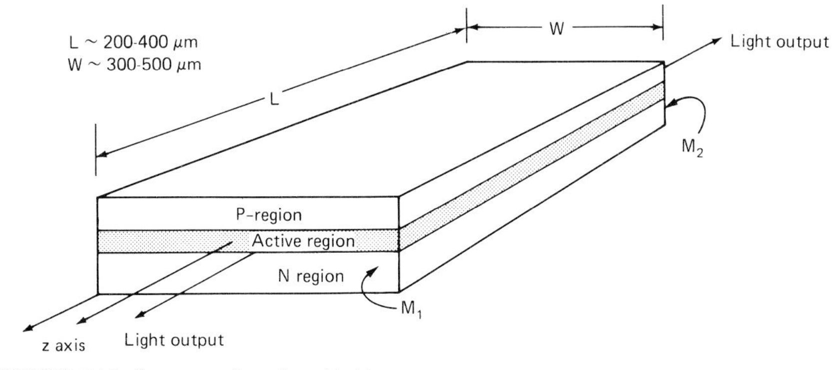


**FIGURE 7.16** Structure of a gain-guided laser.

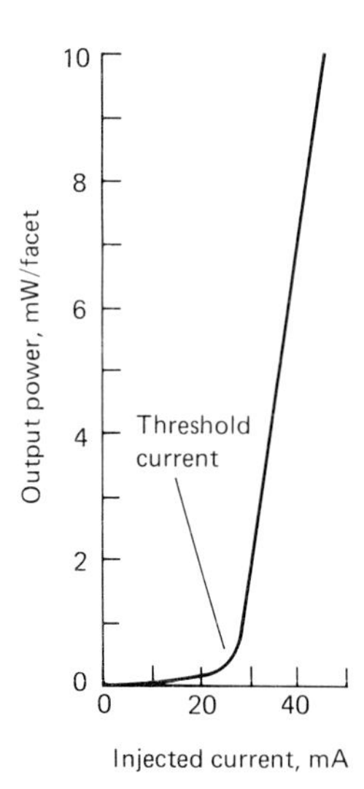


**FIGURE 7.17** Typical output power as a function of injected current for a gain-guided laser. (*From N. K. Dutta.*[15] *Reproduced with permission of the publisher, Howard Sams & Co., Indianapolis, Optical-Fiber Transmission, edited by E. E. Basch, © 1986.*)

The relationship between the $z$-axis output power and injection current $I_i$ of a typical gain-guided laser is shown in Fig. 7.17. Below the threshold current level $I_{th}$, the laser undergoes spontaneous emission of photons. Optical output power is incoherent in this region. Above $I_{th}$ in the stimulated-emission region, photonemission is nearly coherent. Most often lasers are biased above threshold, and are modulated such that current $I_i$ does not fall below threshold. As the optical spectrum of the conventional gain-guided laser shown in Fig. 7.15*a* indicates, the device supports many longitudinal modes. Modulation of any semiconductor below $I_{th}$ causes spectral broadening due to a degradation in the coherence of emitted photons.[14] However, the penalty for not turning the laser off completely is that the circuit suffers from a finite extinction ratio. The extinction ratio $E_x$ is defined as the ratio of the high-level optical signal to the low-level optical signal:

$$E_x = \frac{P_{Hopt}}{P_{Lopt}} \tag{7.27}$$

At low frequencies, setting the bias current of the laser below threshold is sometimes desirable. For example, consider the circuit shown in Fig. 7.18*a*. As Fig. 7.18*b* indicates, modulating the laser below threshold ensures that the optical output power drops to zero. The extinction ratio of a circuit in which $P_{Lopt} = 0$ is $E_x = \infty$.

Index-guided lasers produce narrower optical spectra than gain-guided lasers because of their stripe geometry. A regrown layer acts to confine the output beam laterally. This eliminates high-order transverse modes and generates a narrow linewidth in the optical output power.

A third type of laser is the distributed-feedback laser. In distributed-feedback lasers, a grating is etched along the cavity length, providing frequency-selective feedback by reflection. Similarly, the distributed-Bragg-reflector laser is formed by etching the grating near the ends of the cavity. Figure 7.19*a* and *b* illustrates distributed-feedback and distributed-Bragg-reflector lasers.[15] The typical laser linewidth of a distributed-feedback laser is shown in Fig. 7.15*b*.

Another frequency-selective laser is the cleaved coupled cavity, or $C^3$, type. $C^3$ lasers amplify the two coincident longitudinal modes of two coupled cavities to generate a narrow optical spectrum. An example of a $C^3$ laser is shown in Fig. 7.20.[15]

In coherent optical communication systems, extremely narrow laser linewidths are necessary.[12] By coupling an external cavity to a laser diode, fur-

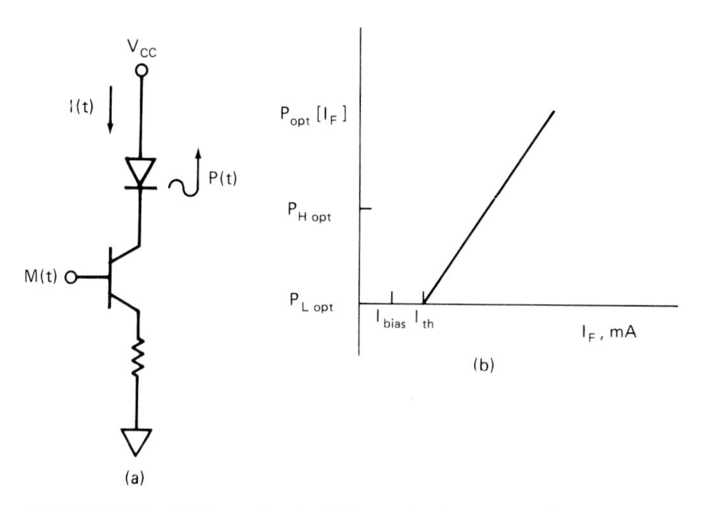


**FIGURE 7.18** (*a*) Laser circuit; (*b*) laser circuit response diagram.

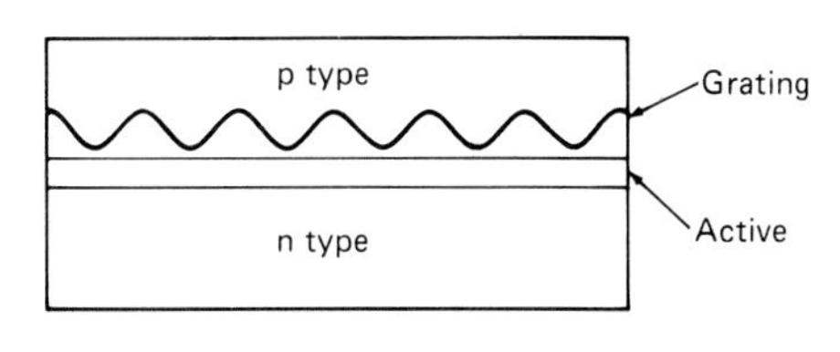


(a) DFB laser

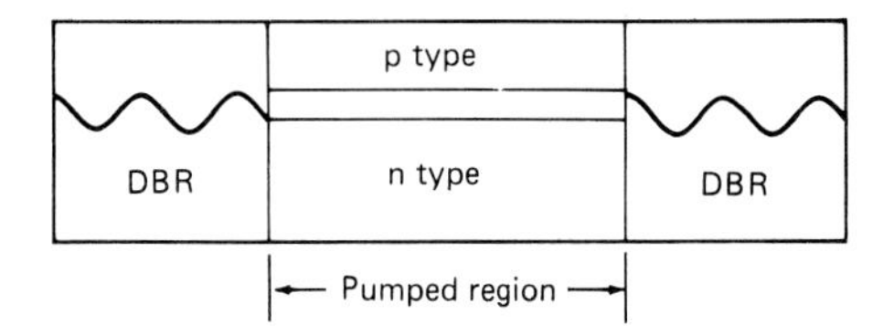


(b) DBR laser

**FIGURE 7.19** Structure of (*a*) distributed-feedback laser and (*b*) Bragg-reflector laser. [*Reproduced with the permission of Howard W. Sams & Co., from N. K. Dutta,*[15] *in E. E. Basch, (ed.), Optical Fiber Transmission,* © *1986.*]

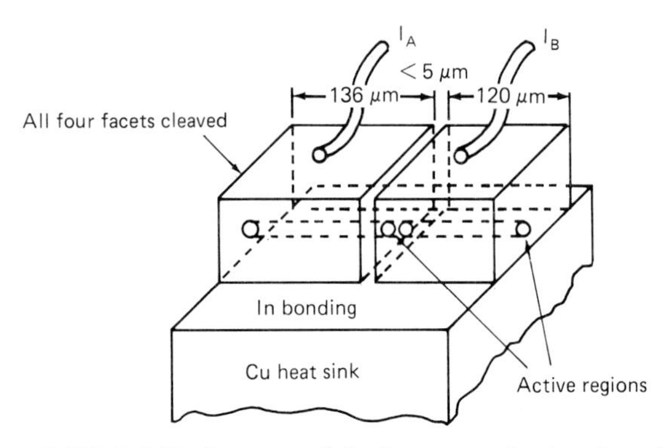


**FIGURE 7.20** Structure of the frequency-selective cleaved coupled cavity laser. [*Reproduced with the permission of Howard W. Sams & Co., from N. K. Dutta,* [15] *in E. E. Basch (ed.), Optical Fiber Transmission,* © *1986.*]

ther reduction in laser linewidth can be obtained. An example of an external cavity laser is shown in Fig. 7.21.[14] Figure 7.15 shows a typical external cavity laser linewidth.

Bit rate, modulation scheme, bit error rate (BER), signal-to-noise ratio (SNR), etc. all contribute to establishing the performance requirements of components in an optical communication system.[16] Sometimes requirements are lenient enough that a laser source is not needed. In those cases, light-emitting diodes normally suffice.

### 7.5.3 External Optical Modulators

The two major types of external optical modulators which are commercially available are acousto-optical and electro-optical modulators. Increased interest in high-speed data transmission within the last few years has stimulated the development of a large variety of these modulators.[17–20] Therefore, only essential parameters generic to these modulators will be presented.

***Acousto-Optical Modulators.*** Acousto-optical modulators operate on the principle of Bragg diffraction. One type of acousto-optical modulator consists of a piezoelectric transducer bonded to the end of a glass bar.[21] See Fig. 7.22. The transducer generates acoustic waves which change the material's index of refraction, causing diffraction and hence modulation.

Manufacturers often provide graphs (Fig. 7.23)[21] of the normalized optical output intensity response vs. normalized modulation bandwidth. The modulation bandwidth $f_0$ is the frequency at which the optical output intensity drops 1.5 dB (3 dB electrical). It may be computed from the rise time $t_r$ by the formula

$$f_0 = \frac{1}{t_r} \tag{7.28}$$

where $t_r = 0.65d/v$. The parameter $v$ = acoustic velocity (approximately $4 \times 10^6$ mm/s) and $d$ = optical beamwidth at the $1/e^2$ intensity points. The rise time for an

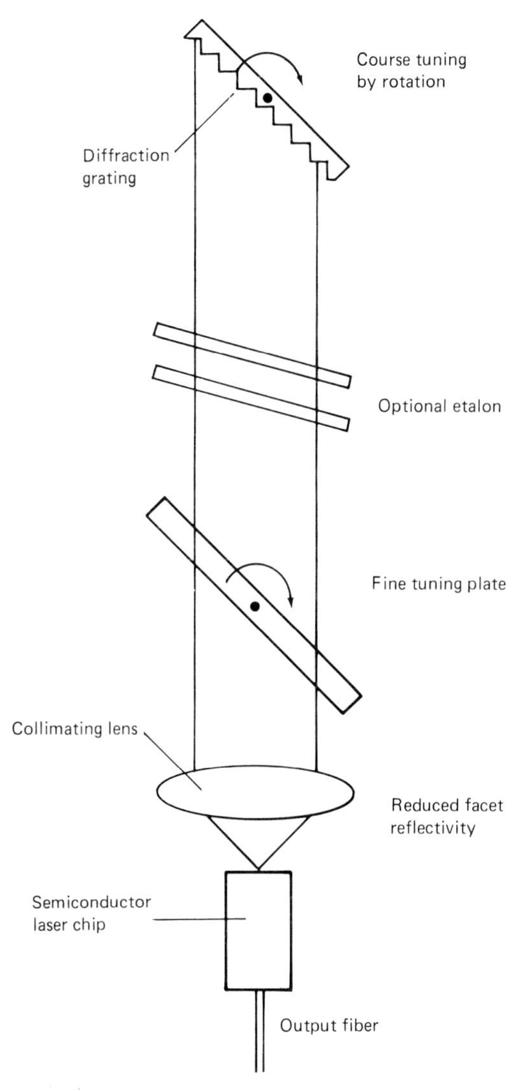


**FIGURE 7.21** External cavity laser. *(From I. W. Stanley,[14] © 1985 IEEE.)*

acousto-optical modulator with a modulation bandwidth $f_0$ = 20 MHz is therefore $t_r$ = 50 ns.

Acousto-optical modulators with bandwidths between 100 kHz and 50 MHz are commercially available. Although they can handle higher laser power than electro-optical modulators, they also have greater insertion loss, and require higher drive power per unit bandwidth. Presently, acousto-optical modulators capable of being modulated at speeds above 1 Gb/s have not been developed. These modulators are most useful in megabit-rate optical transmission systems.

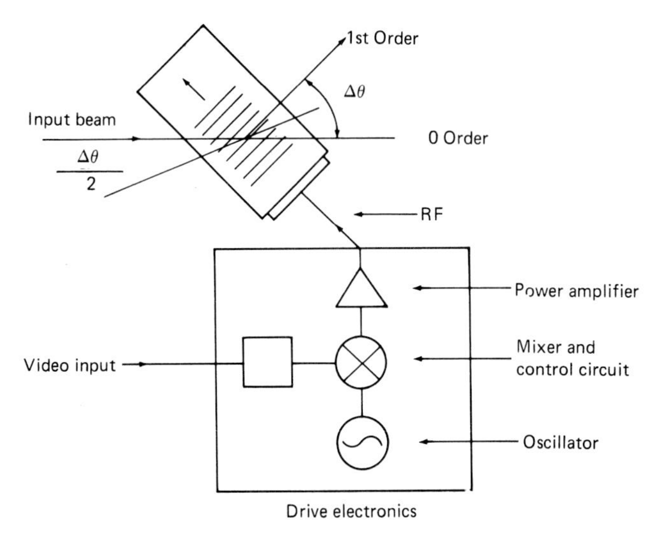


**FIGURE 7.22** Acousto-optic modulator. *(Reproduced with permission of Anderson Laboratories,[22] Bloomfield, Conn.)*

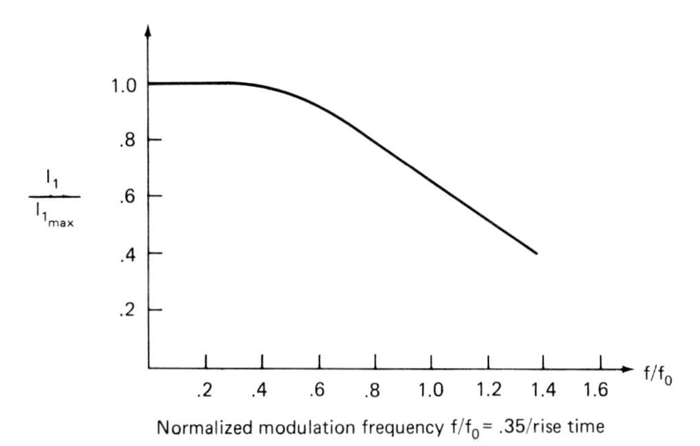


**FIGURE 7.23** An example of the response of an acousto-optic modulator as a function of modulation frequency. *(Reproduced with permission of Anderson Laboratories,[22] Bloomfield, Conn. )*

***Optical-Guided-Wave Modulators.*** Electro-optic modulators utilize the electro-optic Pockel's effect. This effect is a result of the linear relationship that exists between the applied voltage across the electrodes attached to a crystal and the induced phase shift in the light propagating through the crystal.[8,18,19] Shown in Fig. 7.24 is a diagram of a titanium–lithium niobate (Ti:$LiNbO_3$) optical-guided-wave phase modulator. Shown in Fig. 7-25 is a diagram of a Ti:$LiNbO_3$ optical-guided-wave Mach-Zehnder interferometer intensity modulator.[22] Tables 7.3 and 7.4 display some of the typical electrical and optical characteristics of each optical-guided-wave modulator type.

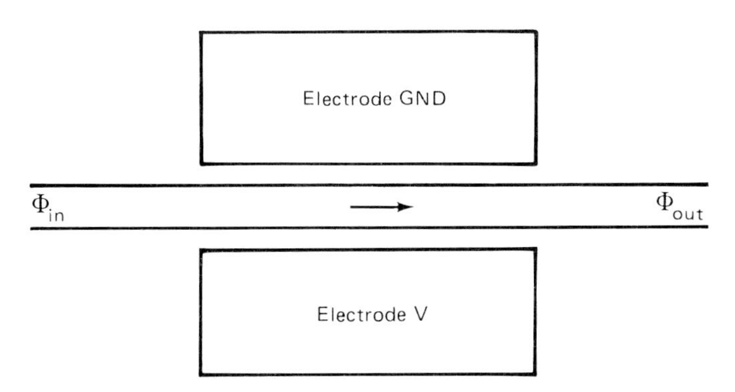


**FIGURE 7.24** Optical-guided-wave phase modulator. *(Reproduced with permission of Crystal Technology, Palo Alto, CA.)*

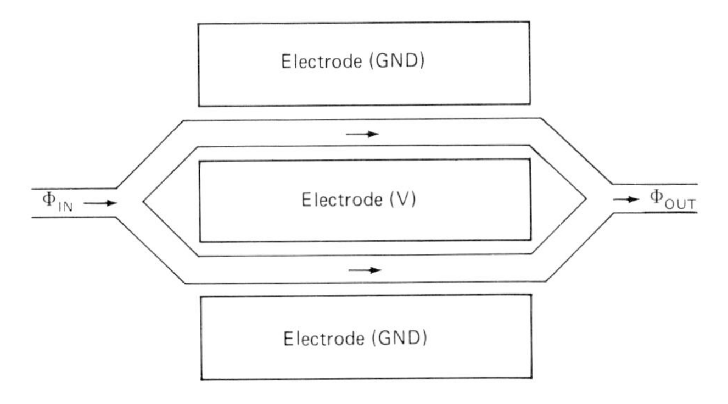


**FIGURE 7.25** Optical-guided-wave Mach-Zehnder interferometer intensity modulator.

**TABLE 7.3** Typical Electrical and Optical Characteristics of an Optical-Guided-Wave Phase Modulator

| Characteristic | Value |
|---|---|
| Bandwidth (small signal) | DC to 3 GHz |
| Drive voltage (for a 180° shift) | 10 V (typical) |
| Wavelength | 850 nm, 1300 nm, 1550 nm |
| Termination impedance | 50 Ω |

**TABLE 7.4** Typical Electrical and Optical Characteristics of an Optical-Guided-Wave Mach-Zehnder Interferometer Intensity Modulator

| Characteristic | Value |
|---|---|
| Bandwidth (small signal) | DC to 3 GHz |
| Drive voltage | 7 V (typical) |
| Extinction ratio | 15 dB (minimum) |
| Wavelength | 850 nm, 1300 nm, 1550 nm |
| Termination impedance | 50 Ω |

Now consider the phase modulators. An optical waveguide channel is formed by diffusing a titanium strip between the device electrodes. An applied voltage at the ungrounded electrode introduces a change in the refractive index of the waveguide. Variation of the applied voltage alters the velocity of light passing through the waveguide, and hence phase modulation occurs.

In the intensity modulators, an optical input signal is split between parallel waveguides. A phase shift proportional to the applied voltage is introduced in one of the waveguides. Before reaching the output, the signals are summed together where they undergo destructive interference. Thus, the magnitude of the output signal is regulated solely by the applied voltage.

*Optical Response of Optical-Guided-Wave Modulators.* For the optical-guided-wave modulators shown in Figs. 7.24*a* and 7.25*a*, the output light intensity $I$ as a function of electrode voltage $V_e$ can be expressed as[23]

$$I = I_0 \cos^2 \left( \frac{\pi V_e}{2V_\pi} + \frac{\phi}{2} \right) \tag{7.29}$$

where $V_\pi$ = voltage required to change $I$ from a maximum to a minimum value
$\phi$ = static phase shift

The equivalent lumped-element model[23] is shown in Fig. 7.26. A plot of optical output power vs. frequency of a typical electro-optical modulator is shown in Fig. 7.27.

Although the lumped-element model accurately predicts the output light intensity response with respect to frequency, the input impedance of the model does not actually represent the measured input impedance of the device. The magnitude and phase of the input impedance for a typical optical-guided-wave phase modulator is shown in Fig. 7.28*a* and *b*.[24] It is equivalent to the input impedance of an ordinary bandstop filter. This is expected, since the configuration of the electrodes is the same as that of a low-pass microstrip filter at microwave frequencies. The impedance of the optical-guided-wave Mach-Zehnder interferometer is similar to that of the optical-guided-wave phase modulator. Its configuration is the same as that of a bandstop microstrip filter.[25]

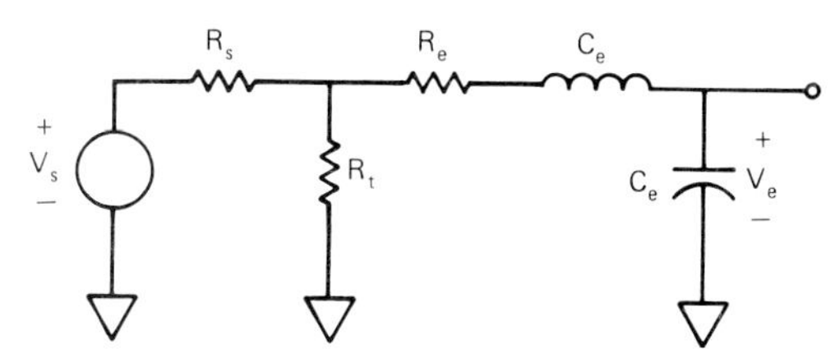


**FIGURE 7.26** Lumped-element equivalent circuit model representing an optical-guided-wave modulator's optical response. $V_s$ = source voltage, $R_s$ = source resistance, $R_e$ = electrode resistance, $R_t$ = terminating resistor shunting generator to ground, $L$ = inductance of the electrodes and bonding wires, and $C$ = capacitance of electrodes.

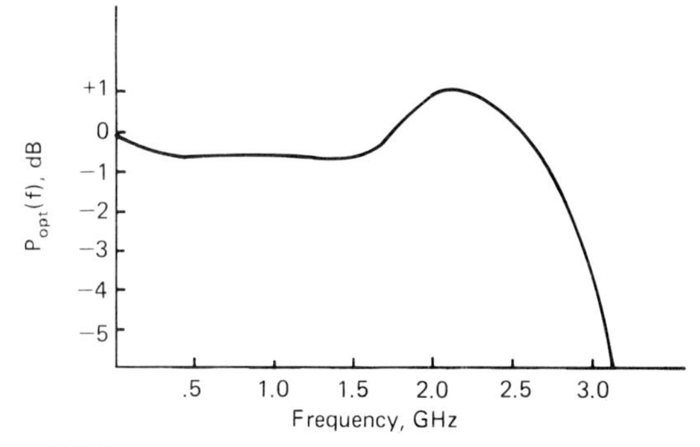


**FIGURE 7.27** Typical plot of optical output power $P_{opt}(f)$ vs. frequency of an optical-guided-wave modulator. *(Reproduced with permission of Crystal Technology, Palo Alto, Calif.)*

***Comparison of Acousto-Optical and Optical-Guided-Wave Electro-Optical Modulators.*** The optical and electrical characteristics of acousto-optical modulators are not as salient as those of electro-optical modulators. Although useful for megabit-per-second fiber-optic systems, the narrow bandwidth of acousto-optical modulators precludes them from being incorporated into gigabit-per-second, coherent fiber-optic links. Conversely, electro-optic modulators have wide bandwidths, and have already been implemented in developmental gigabit-per-second fiber-optic systems.[26] Optical-guided-wave modulators typically have a modulation bandwidth of several gigahertz. Also, optical-guided-wave modulators have lower optical absorption and scattering loss than acousto-optical modulators, and they require less electric power per unit bandwidth.[8] However, optical-guided-wave modulators do suffer from insertion loss and undergo chirping.[27]

### 7.5.4 Detectors

Photodetectors are optical transducers which change incident light energy into electrical signals. The two most commonly used photodetectors in high-

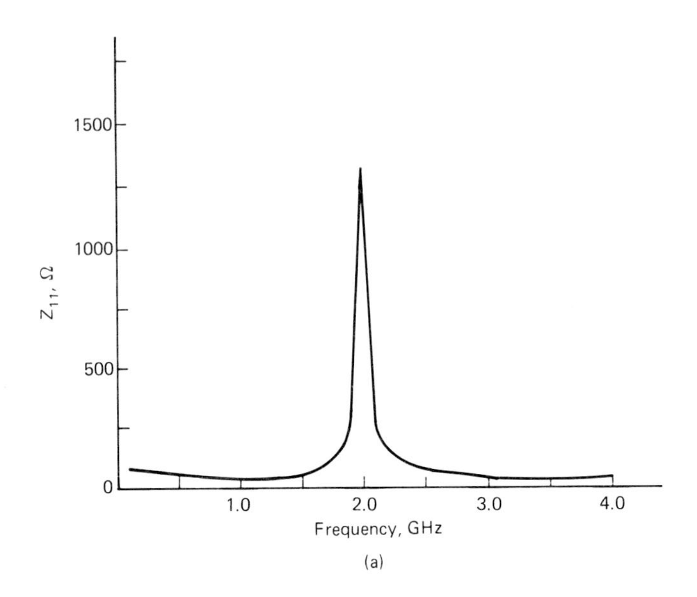


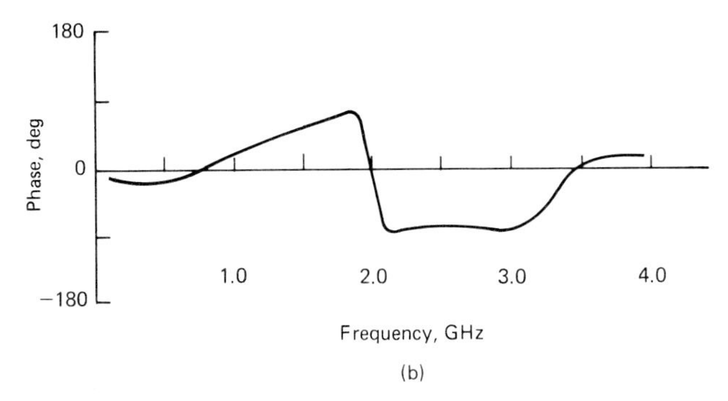


**FIGURE 7.28** Plot of typical input impedance $Z_{11}$ characteristics of an optical-guided-wave modulator. (*a*) Magnitude vs. frequency; (*b*) phase vs. frequency.

speed data transmission are avalanche photodiodes (APDs) and PIN photodiodes.[11,28,29] A few important properties of PIN photodiodes and APDs will now be discussed.

*Avalanche Photodiodes.* The induced photocurrent $i_p(t)$ in an APD is directly proportional to the absorbed optical power $P(t)$. APDs use the mechanism of impact ionization to further amplify the photocurrent. However, the avalanche gain mechanism not only amplifies the photocurrent induced by the optical field, but also those components of noise current that are spontaneously generated. APD noise current will be addressed shortly.

The constant of proportionality relating optical power and photocurrent is actually the product of two constants, $R$ and $M$. The constant $R$ is called the *responsivity*. It has units of microamperes per microwatt. The other constant, $M$, is called the *avalanche multiplication factor* or *avalanche gain*. The expression for induced photocurrent is

$$i_p(t) = RMP(t) \tag{7.30}$$

The responsivity of the photodiode is a measure of the transducer's optoelectrical effectiveness. It is a function of the electronic charge ($e = 1.6 \times 10^{-19}$ C), the wavelength $\lambda_p$ of illuminating light, the photodiode's quantum efficiency $\eta$, Planck's constant ($h = 6.62 \times 10^{-34}$ J · s), and the speed of light ($c = 3 \times 10^8$ m/s) and is expressed as

$$R = \frac{e\lambda\eta}{hc} \tag{7.31}$$

The principal noise generators in APDs are the bulk dark current, the surface dark current, and the shot noise current (also called the quantum noise current). The surface dark current $i_{\text{sd}}$ is a leakage current determined by the surface area of the APD, surface aberrations of the APD, dirt on the surface of the APD, and APD bias voltage. The bulk dark current $i_{\text{bd}}$ is the thermal noise current produced by the APD. The shot noise current $i_q$ is a noise current originating from the photodetection process. This process follows the same statistics as a Poisson probability process. The bulk dark current and the quantum noise current are the components of noise current that are spontaneously generated. The complete noise response of APDs is characterized by Eqs. (7.32) through (7.35):

$$i_n^2 = i_{\text{bd}}^2 + i_{\text{sd}}^2 + i_q^2 \tag{7.32}$$

where the parameter $i_n^2$ is the mean-square noise current generated by the APD;

$$i_{\text{sd}}^2 = 2ei_{\text{sd}}\,\Delta f \tag{7.33}$$

where the parameter $i_{sd}^2$ is the mean-square value of surface dark current;

$$i_{\text{bd}}^2 = 2ei_{\text{bd}}M^2F\,\Delta f \tag{7.34}$$

where the parameter $i_{bd}^2$ is the mean-square value of bulk dark current;

$$i_q^2 = 2ei_pM^2F\,\Delta f \tag{7.35}$$

where the parameter $i_q^2$ is the mean-square value of quantum noise current. Other terms in the preceding equations are defined below:

$F$ = avalanche excess noise factor

$P_r$ = expected value of received optical power

$i_p$ = expected value of photocurrent

$\Delta f$ = noise equivalent bandwidth

The term $\Delta f$ will be discussed in Sec. 7.6. APDs are usually operated in reverse bias. Bias voltages range between 100 and 300 V.[8,28,29]

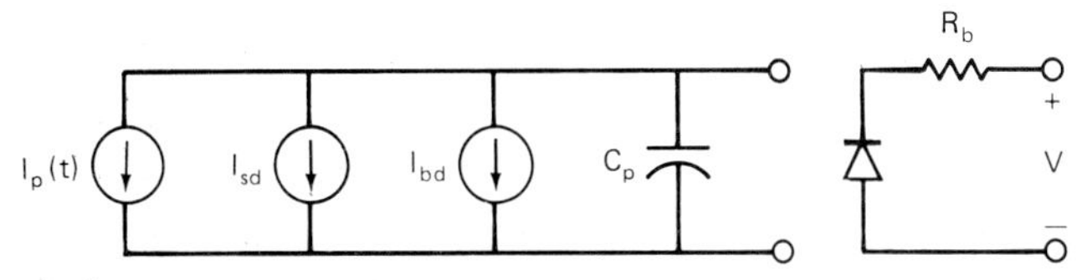


**FIGURE 7.29** Equivalent circuit for PIN photodiode.

*PIN Diodes.* PIN diodes have a lightly doped $n$ region separating the $p$- and $n$-doped regions. When a reverse bias is applied to the diode, the depletion region extends through the lightly doped region, also called the *intrinsic* region. The intrinsic region increases the volume of silicon interacting with optical energy to produce photocurrent. Electron-hole pairs generated by optical absorption are swept out of the depletion region by the bias field and contribute to the photocurrent. The responsivity of PIN diodes is a function of the same parameters as for APDs, except that there is no avalanche gain. The induced photocurrent equation is the same as that for APDs, i.e., Eq. (7.30), except that $M = 1$.

The noise current $i_n$ generated by PIN diodes also differs from that of APDs because it does not undergo the avalanche gain process. Equations (7.32) to (7.35) hold true for PIN diodes except that $M = 1$ and $F = 1$.

The lumped-element equivalent circuit of a PIN photodiode is shown in Fig.7.29. The shunt capacitance $C_p$ depends on the diode's physical dimensions and the dielectric constant of the semiconductor. $C_p$ is usually specified in the manufacturer's data sheets. The diode-biasing resistor is $R_b$. From $C_p$ and $R_b$, the diode cutoff frequency $f_c$, rise time $t_r$, and fall time $t_f$ can be approximated by the following formulas:

$$f_c = \frac{1}{2\pi R_b C_p} \tag{7.36}$$

and

$$t_f = t_r = \frac{2.2}{2\pi f_c} \tag{7.37}$$

The diode cutoff frequency is important in determining a circuit's signal-to-noise ratio and bit error rate. Signal-to-noise ratio and bit error rate will be discussed in later sections.

Photodiodes are normally operated in reverse bias. Bias voltages for PIN diodes range from about 5 to 20 V.[8,28,29]

## 7.6 NOISE AND INTERFERENCE

### 7.6.1 Probability Terms

The expected value of a periodic function $f(t)$ with a period $T$ is $E[f(t)]$. It is expressed by

$$E[f(t)] = \frac{1}{T}\int_{-T/2}^{T/2} f(t)\, dt = \mu_f \tag{7.38}$$

where $\mu_f$ = constant.

The variance $\sigma^2$ of the same periodic function $f(t)$ can be expressed by

$$\sigma^2 = \frac{1}{T}\int_{-T/2}^{T/2} [f(t) - \mu_f]^2\, dt \tag{7.39a}$$

or

$$\sigma^2 = \frac{1}{T}\int_{-T/2}^{T/2} [f(t)]^2\, dt - \mu_f^2 \tag{7.39b}$$

The variance parameter $\sigma^2$ is also called the *mean-square value.* When it appears as $\sigma$, it is called the *root-mean-square* (rms) *value* or the *standard deviation.*

The gaussian probability density function $g(x)$ is expressed by

$$g(x) = \frac{1}{\sqrt{2\pi\sigma_g^2}}\, e^{-(x-\mu_g)^2/2\sigma_g^2} \tag{7.40a}$$

The expected value of $g(x)$, $E(g(x))$, is

$$E(g(x)) = \mu_g \tag{7.40b}$$

The variance of $g(x)$ equals $\sigma_g^2$, as it appears in Eq. (7.40). The area enclosed by the curve of $g(x)$ is such that

$$\int_{-\infty}^{\infty} g(x)\, dx = 1 \tag{7.41}$$

See Fig. 7.30 for an illustration of $g(x)$.[30]

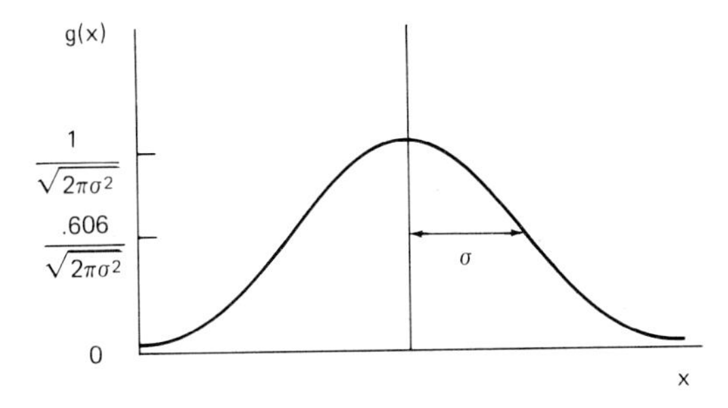


**FIGURE 7.30** Gaussian probability density function.

### 7.6.2 Noise Terms

The three principal types of noise are thermal noise, low-frequency noise, and shot noise. Shot noise of APDs and PIN photodiodes was discussed in Sec. 7.5.4. Information on low-frequency noise may be found in Ref. 31. This section will address only the subject of thermal noise.

***Noise Equivalent Bandwidth.*** The power gain of a system, $G(f)$, describes the system's power response in the frequency domain. Most systems have a power gain which is continuous over a range of frequencies from DC to an upper cutoff frequency $f_c$, like the one shown in Fig. 7.31.

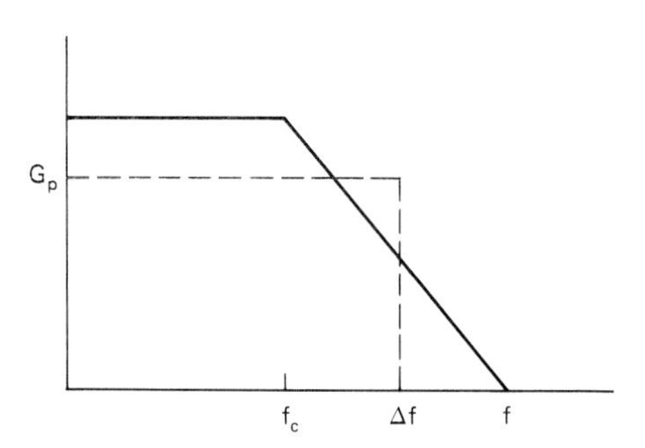


**FIGURE 7.31** Typical power gain curve.

The noise equivalent bandwidth of a system, $\Delta f$, is equal to the area enclosed by the power gain of the system divided by the peak power gain $G_p$. In equation form, this becomes

$$\Delta f = \frac{1}{G_p} \int_0^{\infty} G(f)^2 \, df \tag{7.42}$$

Since the power gain of a system is directly proportional to the transfer function squared, Eq. (7.42) can also be written[28,31]

$$\Delta f = \frac{1}{|H(p)|} \int_{-0}^{\infty} |H(f)|^2 \, df \tag{7.43}$$

***Thermal Noise Power.*** The thermal noise power of a source is the amount of noise power that the source can deliver to a resistive load equal to its own source resistance. The thermal noise power $P_{\text{th}}$ of a conductor is expressed by

$$P_{\text{th}} = KT \, \Delta f \tag{7.44}$$

where $K$ = Boltzmann's constant ($1.38 \times 10^{-23}$ W · s/K)
$T$ = absolute temperature of the conductor, K
$\Delta f$ = noise equivalent bandwidth, Hz

For a load resistor $R$, the thermal noise power can be expressed as a thermal noise voltage $V_{th}$ or thermal noise current $I_{th}$ by substituting into Eq. (7.44):

$$V_{th}^2 = 4KTR\,\Delta f \tag{7.45}$$

and

$$I_{th}^2 = \left(\frac{4KT}{R}\right)\Delta f \tag{7.46}$$

Thermal noise is also known as *Johnson noise.* The thermal noise voltage $V_{th}$ is equivalent to the root-mean-square noise voltage. Likewise, the thermal noise current is equivalent to the rms noise current.[31]

*Spectral Power Density.* A noise signal with a power density $N(f)$ may have one discrete frequency component, several discrete frequency components, one continuum of frequencies, or several continua of frequencies. The spectral power density of a noise signal, $S_n(f)$, is a function which describes the power of that signal per unit bandwidth; i.e.,

$$S_n(f) = \frac{N(f)}{\Delta f} \tag{7.47}$$

The spectral power density of white noise, $S_{nw}(f)$, is shown in Fig. 7.32.

When a noise signal with a spectral power density of $S_{ni}(f)$ is passed through a linear system with transfer function $H(f)$, the output noise spectral power density $S_{no}(f)$ is expressed by

$$S_{no}(f) = |H(f)|^2 S_{ni}(f) \tag{7.48}$$

and

$$S_{no}(f) = \frac{N(f)}{\Delta f} \tag{7.49}$$

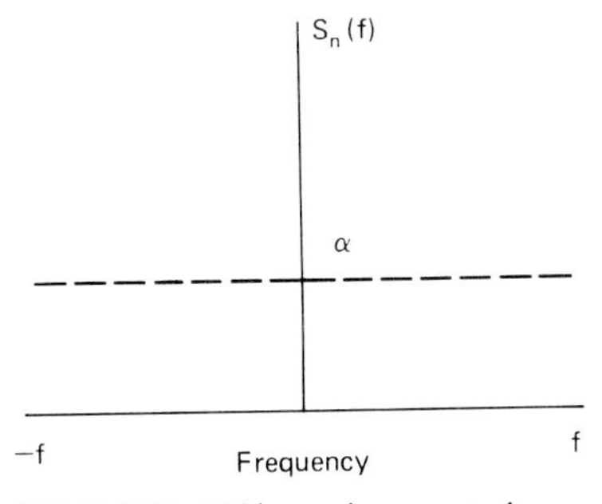


FIGURE 7.32 White noise spectral power density.

The total output noise power $P_{no}$ is then[28,31]

$$P_{no} = \int_{-\infty}^{\infty} S_{no}(f)\, df \tag{7.50}$$

***Amplifier Noise.*** The noise equivalent circuit for an amplifier is shown in Fig. 7.33. $I_a^2$ and $V_a^2$ are the mean-square noise current and noise voltage, respectively. The noise voltage and noise current of commercial amplifiers are normally specified in manufacturers' data sheets.[31]

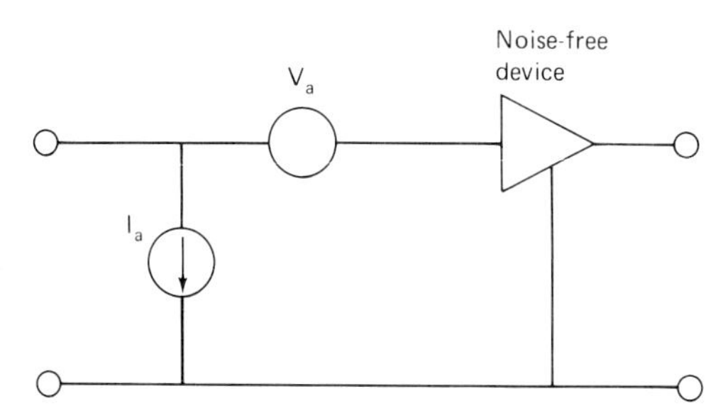


**FIGURE 7.33** Amplifier noise equivalent circuit.

### 7.6.3 Modal Noise

Modal noise is introduced by the propagation characteristics of the fiber. Changes in the phase of different modes propagating through multimode fiber is one mechanism which brings about modal noise. These modes are the transverse propagation modes in the fiber. They cause the speckle interference pattern in the spot of light emitted from a multimode fiber. Speckle patterns occur when the time delay between fiber modes is less than the coherence time of the optical source. For modal noise to be present, mechanisms which alter the speckle pattern must be present. Splices, connectors, microbends, fiber core discontinuities, and fiber-detector coupling are all speckle-dependent loss mechanisms. In other words, the speckle pattern depends upon these loss mechanisms. The loss mechanisms cause propagation mode changes which result in a variation of the interference pattern through the fiber. Detectors with nonuniform responsivity can cause modal noise. Wavelength shifts in the laser source, the phenomenon known as *chirping,* contribute to modal noise.

The easiest means of eliminating modal noise is to use a source with zero coherence time. Using incoherent signal sources, such as LEDs, at the transmitter can eliminate speckle patterns, and therefore modal noise.[21,28,29]

### 7.6.4 Mode Partition Noise

Mode partition noise is dependent on the propagation characteristics of the fiber. It is a result of fluctuations in the distribution of energy among the laser oscillating modes. These are the longitudinal modes of the laser and are separated in frequency by $L_c/2c$, where $L_c$ is the cavity length and $c$ is the velocity of light.

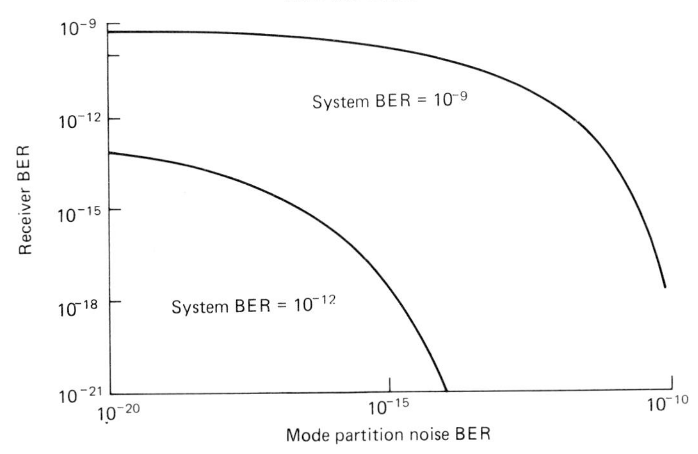


**FIGURE 7.34** The effect of mode partition noise on bit error rate. *(From E. E. Basch, R. F. Kearns, and T. G. Brown,[32] © 1985 IEEE.)*

Although the total source power remains almost constant, power contained in individual modes fluctuates randomly. Chromatic dispersion in the fiber, together with shifts in the laser mode distribution, creates noise in the received signal and degrades signal-to-noise ratio and bit error rate. See Fig. 7.34 for a graph of mode partition noise vs. BER.[31]

### 7.6.5 Chirping

Chirping is the change in the laser optical output frequency due to a shift in the resonance frequency of the optical cavity caused by a change in laser injection current. Therefore, direct modulation of the injection current can dynamically broaden the laser linewidth. This linewidth broadening occurs near the leading and trailing edges of transmitted pulses, the times in which the injection current changes. Therefore, along the length of the pulses, dispersion becomes unevenly distributed, resulting in distortion and possible intersymbol interference.[15]

### 7.6.6 Intersymbol Interference

Dispersion in optical fibers fluctuates with changing environmental conditions, such as temperature and stress. Dispersion causes pulse broadening. Certain bit periods can be elongated by fiber dispersion so much that they extend into adjacent bit periods. This phenomenon, called intersymbol interference, sometimes causes incorrect bits to be registered at the decision circuit in the receiver. Another source of intersymbol interference is jitter. Ideally, a transmitted bit stream ought to have a constant bit period. In real systems, the bit periods vary. This condition is called jitter.[13,15]

### 7.6.7 Nonlinear Distortion

Lasers, APDs, PIN diodes, amplifiers, transistors, etc., suffer, to some extent, from nonlinear distortion mechanisms called *harmonic distortion* and *intermodu-*

*lation distortion.* A device produces harmonic distortion when it responds to an input sine wave of frequency $\omega_1$ by generating output signals with frequencies of $\omega_1$, $2\omega_1$, $3\omega_1$, etc. Intermodulation distortion occurs when the device responds to a pair of input signals with frequencies $\omega_1$ and $\omega_2$ by generating output signals with frequencies $\omega_2 - \omega_1$, $\omega_2 + \omega_1$, $\omega_2 - 2\omega_1$, $\omega_2 + 2\omega_1$, etc., i.e., with frequencies equal to the sum and difference of the input signals and their harmonics.[28]

## 7.7 ASK, PSK, AND FSK DEVICE METHODS

In this section, important aspects of amplitude-shift keying (ASK), phase-shift keying (PSK), and frequency-shift keying (FSK) will be discussed as they apply to fiber-optic transmission. The relative merits of each technique[8] will now be summarized.

### 7.7.1 Amplitude-Shift Keying

On-off keying is the simplest intensity-modulation scheme available. Directly driving the laser with a binary bit pattern simply turns the optical power source on and off. For direct detection using a PIN or APD photodetector, this approach is adequate for low-frequency transmission. However, in coherent optical transmission systems, the change in injection current causes laser chirp and introduces excessive noise at the receiver.

An alternative method to generating an ASK signal is to use an electro-optical modulator. By applying the modulating bit pattern to the electrodes of this device, wideband performance with good extinction ratio can be achieved. Although optical insertion loss reduces the SNR at the receiver, the elimination of laser chirp makes this method viable for coherent detection systems.[8]

### 7.7.2 Phase-Shift Keying

One method for phase-shift-keying an optical signal is to use an electro-optical crystal to introduce the phase shift. This is simply accomplished because the optical output phase of an optical-guided-wave modulator is directly proportional to the applied electrode voltage.

Phase-shift-keying receiver circuitry is more complex than either ASK or FSK circuitry; however, PSK offers the best theoretical minimum receiver sensitivity of all coherent modulation methods.[8]

### 7.7.3 Frequency-Shift Keying

Direct modulation of a laser is one technique used to frequency-shift-key an optical carrier. By varying the laser injection current, mode resonance frequencies can be changed. Thus, the optical frequency spectrum can be shifted between two frequencies to produce an FSK signal. Another method uses an external electro-optical phase modulator. If a 1 bit is generated by applying a sawtooth voltage waveform with slope $a$ to the device electrodes over time interval $T$, and

**TABLE 7.5** Optical Signal Phase $\theta(t)$ Corresponding to Logic States 1 and 0 in an FSK Process

| $\theta(t)$ | Data bit |
|---|---|
| $\alpha t$ | 1 |
| 0 | 0 |

a 0 bit is generated by applying a constant zero-voltage waveform to the device electrodes over a time interval $T$, then the optical signal phase $\theta(t)$ varies as shown in Table 7.5.

Since frequency is the time rate of change of phase, i.e.,

$$\omega(t) = \frac{d\theta}{dt} \tag{7.51}$$

then a digital signal frequency keyed as shown in Table 7.6 would be produced.[8]

**TABLE 7.6** Digital Signal Frequencies $\omega(t)$ Corresponding to Logic States 1 and 0 in an FSK Process

| $\omega(t)$ | Data bit |
|---|---|
| $\alpha$ | 1 |
| 0 | 0 |

## 7.8 DIRECT AND COHERENT TRANSMISSION SYSTEMS

Fiber-optic transmission systems can be divided into two major types: coherent and direct. A coherent system offers several advantages over a direct system. As shown in Fig. 7.35, a coherent system can operate with lower receiver sensitivity.[12] Repeater stages having lower receiver sensitivity may be spaced at larger separation distances for long-haul transmission applications. In addition, coherent systems allow more efficient use of fiber bandwidth. Many broadband channels may be transmitted by heterodyning baseband signals with subcarriers as high as the microwave frequency range. The optical carrier is then modulated by the microwave subcarrier. In comparison, a direct system implementing wavelength-division multiplexing (WDM) can accommodate only a few broadband channels. Despite the technological advantages of coherent systems, direct systems are typically more economically feasible.

### 7.8.1 Direct-Detection Fiber-Optic Communication Systems

In most direct-detection systems, a laser diode is amplitude-modulated and the emitted optical power is coupled into a multimode fiber. The multimode fiber acts

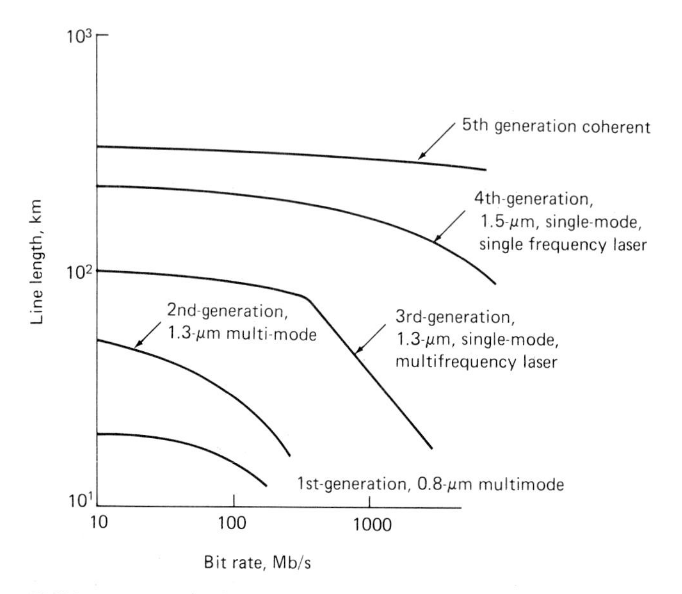


**FIGURE 7.35** A comparison of various transmission systems. *(From E. E. Basch and T. G. Brown,[12] © 1985 IEEE.)*

as a lightguide which connects the source with the receiver. The receiver implements either a PIN or an APD photodiode to convert the optical signal back into an electrical signal.

Consider the following example. The optical output power of a laser diode is proportional to the the driving current $I$. A laser diode begins lasing at a threshold current level $I_{th}$. The bias current $I_{bias}$ may be set slightly above or below $I_{th}$, depending upon the system requirements. Recall Fig. 7.18*b*, which shows the optical output power of a lase, diode which is amplitude-modulated by a pulse-code modulated (PCM) signal, $m(t)$. The instantaneous optical output signal $P(t)$ can be expressed mathematically as

$$P(t) = P_0 + P_m(t) \cos [\omega_{op}(t) + \phi] \tag{7.52}$$

where $P_m(t)$ = modulation envelope
$\omega_{op}$ = optical carrier frequency (about $10^{14}$ Hz)
$P_0$ = DC optical offset due to biasing in the spontaneous-emission region

This optical signal from the laser is then coupled into a fiber. At the opposite end of the fiber, $P(t)$ illuminates the PIN photodetector in the receiver circuitry.

For systems with low reflections and negligible dispersion, this optical signal is proportional to the transmitted signal $P(t)$; i.e.,

$$P_{rec}(t) = \beta P(t) \tag{7.53}$$

where $\beta$ = constant of proportionality.

The current response of the PIN photodetector, $I_p(t)$, is proportional to the modulation envelope of the transmitted power $P_m(t)$:

$$I_p(t) = \alpha R P_m(t) \tag{7.54}$$

where $R$ = PIN diode responsivity and $\alpha$ = constant which accounts for optical power losses in the link.

The receiver amplifies this current, filters out the DC component as well as extraneous noise, and recovers the original PCM signal, $m(t)$. Thus, the output signal of the receiver, $S_o(t)$, is a regeneration of the original modulating signal:

$$S_o(t) = m(t) \tag{7.55}$$

### 7.8.2 Wavelength-Division Multiplexing

Wavelength-division multiplexing can be applied to direct systems. In a typical WDM system, information from parallel electronic circuits is transmitted by modulating several laser sources of different wavelength. The laser light is then coupled into one fiber which guides it to its destination. At the destination, wavelength-discriminating photodetectors receive the light from the fiber and retrieve the original information, once again operating in parallel.

One disadvantage of WDM systems is their inherent waste of available fiber bandwidth. Consider the following example. The typical optical linewidths of two semiconductor lasers are shown in Fig. 7.15. The optical frequency spread $\Delta f$ is directly related to the laser linewidth $\Delta\lambda$ by the relationship

$$\Delta f = \frac{-c\,\Delta\lambda}{\lambda^2} \tag{7.56}$$

The linewidth of the conventional semiconductor laser is $\Delta\lambda$ = 1 to 5 nm. The corresponding frequency spread, calculated from Eq. (7.56), is $\Delta f$ = 133 to 655 GHz.

Depicted in Fig. 7.36 is a direct-detection fiber-optic link employing WDM. If each WDM channel transmits 500 Mb/s PCM data (approximately equal to 1 GHz of data bandwidth), each would require between 399 GHz ($\Delta\lambda$ = 1 nm) and 1965 GHz ($\Delta\lambda$ = 5 nm) of optical frequency spectrum. In comparison, a WDM scheme employing coherent techniques could transmit an equivalent amount of data and require approximately 3 GHz of optical frequency spectrum, i.e. three channels spaced 1 GHz apart.[14]

### 7.8.3 Coherent Fiber-Optic Communication Systems

In heterodyne systems, the data signal to be transmitted, $m(t)$, is first modulated by a subcarrier at an intermediate frequency $\omega_{if}$. That signal, $m_{if}(t)$, defined as

$$m_{if}(t) = m(t)\cos(\omega_{if}t + \phi_1) \tag{7.57}$$

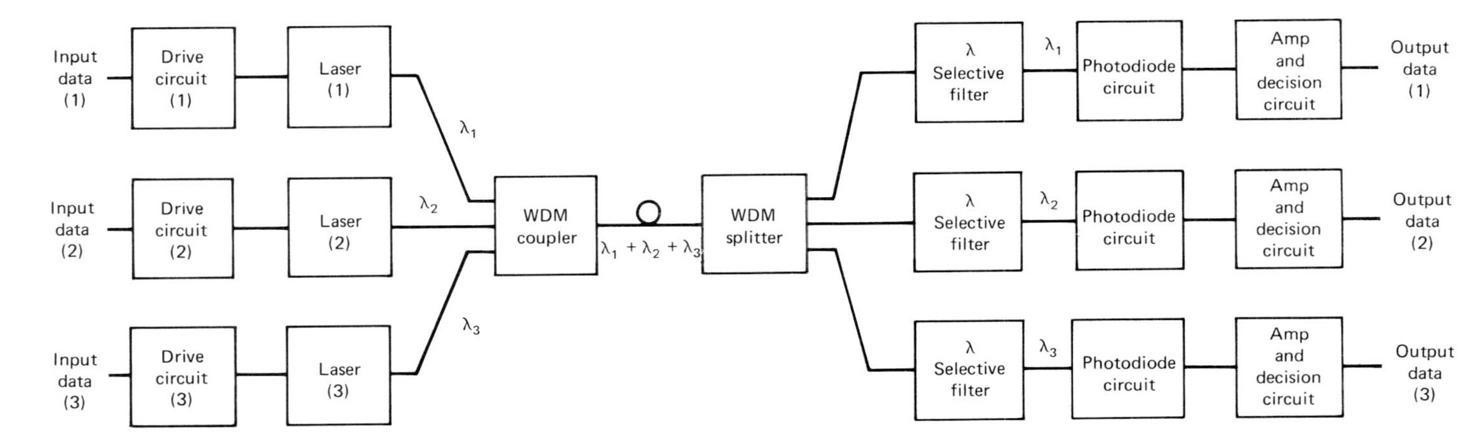


**FIGURE 7.36** Block diagram of wavelength-division multiplexing system.

modulates the optical source. The optical power transmitted $P(t)$ then becomes

$$P(t) = m_{\text{if}} \cos(\omega_{\text{op}} t + \phi_1) \tag{7.58}$$

Homodyne systems do not modulate the data signal with a subcarrier. Therefore $\omega_{\text{if}} = 0$, and the input data signal is modulated only at the optical frequency $\omega_{\text{op}}$. At the receiver, the optical local-oscillator frequency equals the optical carrier frequency.

The optical source may take different forms. For the system diagramed in Fig. 7.37, the optical source is either a laser diode or an external electro-optical modulator. The external modulator receives a continuous-wave optical signal from the laser diode. The modulator electrodes are stimulated with the modulation signal generated by other electronics.

For a coherent system, the laser source must have a very narrow linewidth and be extremely stable in frequency. Typically, linewidths must be less than 1 MHz. Using the relationship

$$\Delta\lambda = \frac{\lambda^2 \, \Delta f}{c} \tag{7.59}$$

one can calculate the linewidth for $\Delta f = 10^6$ Hz to be

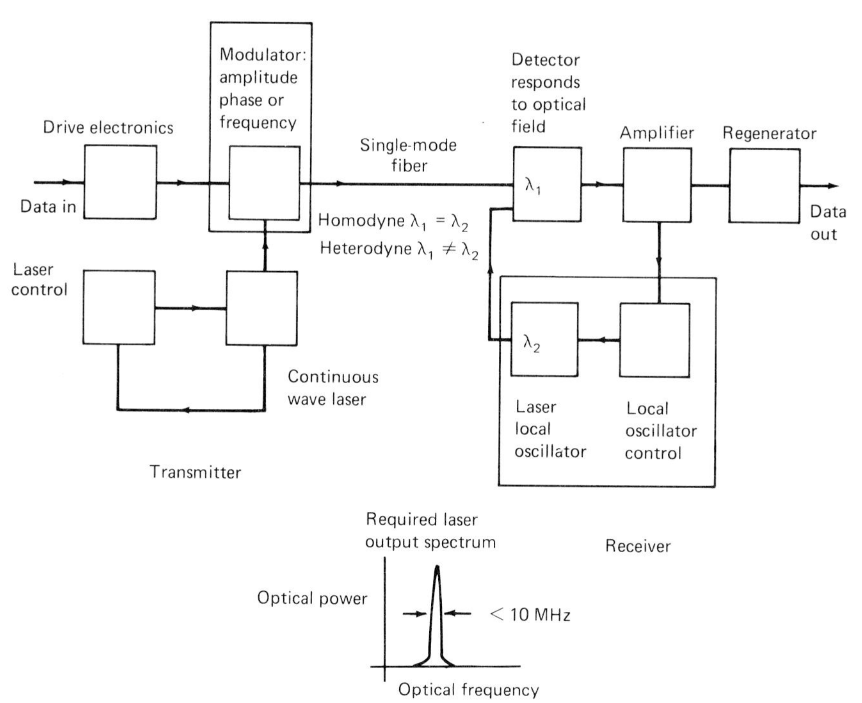


FIGURE 7.37 Heterodyne and homodyne coherent detection. *(From I. W. Stanley,[14] © 1985 IEEE.)*

$$\Delta\lambda = \frac{(1.3 \times 10^{-6})^2(10^6)}{3 \times 10^8} \approx 10^{-5} \text{ nm}$$

Among the three lasers shown in Fig. 7.15, only the laser diode regulated by the external cavity meets this stringent frequency span requirement.

In both heterodyne and homodyne systems, a local oscillator signal from a second laser source of wavelength $\lambda_2$ is combined at the receiver with the transmitted signal $\lambda_1$. The induced photocurrent $I_p(t)$ through a PIN detector illuminated by optical power $P_{rec}(t)$ was given previously in Eq. (7.54). The instantaneous power collected by the photodetector is a function of the optical electric field $E(t)$ illuminating the PIN surface and is given by

$$P_{rec}(t) = \frac{E^2(t)}{Z} \tag{7.60}$$

where $Z$ is a constant of proportionality. Hence, the instantaneous current can be defined also by

$$I_p(t) = \frac{RE^2(t)}{Z} \tag{7.61}$$

If the data signal $m(t)$ is assumed to be a sinusoid of frequency $\omega_d$,

$$m(t) = \cos(\omega_d t) \tag{7.62}$$

and the optical signal frequency is $\omega_s$, then the transmitted optical electric field can be expressed in the notation of Eqs. (7.57), (7.58), and (7.61) as

$$E_s(t) = E_{so} \cos(\omega_s t) \cos(\omega_{if} t) \cos(\omega_d t) \tag{7.63}$$

where $E_{so}$ is the peak electric field intensity of the signal. To simplify this expression, we rewrite Eq. (7.63) as

$$E_s(t) = E_{sm} \cos(\omega_s t) \tag{7.64}$$

where

$$E_{sm} = E_{so} \cos(\omega_{if} t) \cos(\omega_d t) \tag{7.65}$$

The local-oscillator electric field $E_l(t)$ is

$$E_l(t) = E_{lo} \cos(\omega_l t) \tag{7.66}$$

where $E_{lo}$ is the peak electric field intensity of the local oscillator and $\omega_l$ is the frequency of the local oscillator. The total electric field on the PIN surface, $E_t(t)$, is

$$E_t(t) = E_{sm} \cos(\omega_s t)\, E_{lo} \cos(\omega_l t) \tag{7.67}$$

The new expression for the photocurrent is then

$$I_p(t) = \left(\frac{R}{Z}\right)[E_s(t) + E_l(t)]^2 \tag{7.68}$$

By expanding Eq. (7.68) and applying the following trigonometric equalities:

$$\cos^2(\omega t) = \left(\frac{1}{2}\right)[1 + \cos(2\omega t)] \tag{7.69}$$

and

$$\cos(\omega_1 t)\cos(\omega_2 t) = \left(\frac{1}{2}\right)\{\cos[(\omega_1 - \omega_2)t] + \cos[(\omega_1 + \omega_2)t]\} \tag{7.70}$$

a new expression for $I_p(t)$ can be obtained:

$$I_p(t) = \left(\frac{R}{Z}\right)\left\{\left(\frac{1}{2}\right)(E_{\rm sm}^2 + E_{\rm lo}^2) + \left(\frac{1}{2}\right)E_{\rm sm}^2\cos(2\omega_s t) + \left(\frac{1}{2}\right)E_{\rm lo}^2\cos(2\omega_l t) + E_{\rm sm}E_{\rm lo}\cos[(\omega_s + \omega_l)t] + E_{\rm sm}E_{\rm lo}\cos[(\omega_s - \omega_l)t]\right\} \tag{7.71}$$

Since PIN diode gain-bandwidth products extend only into the gigahertz frequency range, PIN diodes cannot respond to the $2\omega_s$, $2\omega_s$, and $\omega_s + \omega_l$ frequency components. Also, the DC electrical response of the PIN diode is usually filtered out. The resulting signal current $I_s(t)$, from Eq. (7.71) becomes

$$I_s(t) = \left(\frac{R}{Z}\right)E_{\rm sm}E_{\rm lo}\cos[(\omega_s - \omega_l)t] \tag{7.72}$$

Now, setting $\omega_s = \omega_l$, one arrives at

$$I_s(t) = \left(\frac{R}{Z}\right)E_{\rm so}E_{\rm lo}\cos(\omega_{\rm if}t)\cos(\omega_d t) \tag{7.73}$$

In heterodyne systems, where $\lambda_1 \neq \lambda_2$, circuitry following the PIN receiver must modulate the data signal from $\omega_{\rm if}$ down to baseband.

In homodyne systems, $\lambda_1 = \lambda_2$, and (as stated earlier) $\omega_{\rm if} = 0$. Since the recovered signal is already at baseband, one can write

$$I_p(t) = \left(\frac{R}{Z}\right)E_{\rm so}E_{\rm lo}\cos(\omega_d t) \tag{7.74}$$

Both Eqs. (7.73) and (7.74) indicate that, for heterodyne and homodyne systems, the PIN photocurrent varies linearly with the amplitude of the transmitted electric field $E_{so}$ and the local-oscillator electric field $E_{lo}$. Thus, improvement in receiver sensitivity can be achieved by increasing the local-oscillator power.[12–14,32–34]

## 7.9 NOISE AND ERROR MECHANISMS

### 7.9.1 Signal-to-Noise Ratio in Direct-Detection Optical Systems

The noise equivalent circuit for a typical AC fiber-optic receiver circuit using a photodiode is shown in Fig. 7.38. For that circuit, $I_t^2$ is the total mean-square noise current; i.e.,

$$I_t^2 = I_n^2 + I_{th}^2 + I_a^2 \tag{7.75}$$

where $I_n$ is the shot noise from the photodetector, $I_{th}$ is the thermal noise from resistor $R_s$, and $I_a$ is the amplifier noise current. The voltage source $V_a$ is the amplifier noise voltage. The mean-square noise voltage generated at the amplifier input is found by the solving the following integral:

$$V_n^2 = \int_0^{\Delta f} (1 + 4\pi^2 f^2 R_s^2 C_P^2) V_a^2 \, df + R_s^2 I_t^2 \tag{7.76}$$

The signal-to-noise ratio with an APD detector at the input to the amplifier stage can then be computed by

$$\text{SNR} = \frac{R_s^2 I_s^2}{1 + \frac{4}{3}\pi^2 (\Delta f)^2 C_p^2 R_s^2 V_a^2 + R_s^2 I_t^2} \tag{7.77}$$

The photocurrent $I_p^2$ is the mean-square value of the photocurrent received by the photodetector:

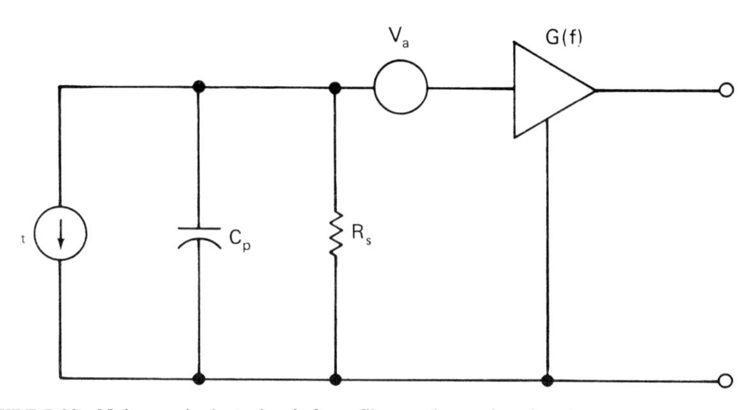


**FIGURE 7.38** Noise equivalent circuit for a fiber-optic receiver input.

$$I_p^2 = \frac{1}{T}\int_{-T/2}^{T/2} i_p^2(t)\,dt \tag{7.78}$$

For an amplitude-modulated signal $s(t)$:

$$I_p(t) = MRP_0(1 + \gamma s(t)) \tag{7.79}$$

where $P_0$ = peak optical power received and $\gamma$ = modulation index of the signal. When the signal $s(t) = \cos(\omega t + \phi)$, then the signal current $i_s(t)$ is

$$i_s(t) = MRP_0\gamma s(t) \tag{7.80}$$

The mean-square value of signal current $I_s^2$ is

$$I_s^2 = \frac{M^2R^2\gamma^2P_0^2}{2} \tag{7.81}$$

When a PIN detector is used, Eq. (7.75) remains the same. However, in evaluating $I_t^2$ of Eq. (7.73) by first determining $I_n^2$ using Eqs. (7.31) through (7.34), one must set $M = 1$ and $F = 1$.[12–14,34–36]

### 7.9.2 Signal-to-Noise Ratio in Coherent Systems

The signal-to-noise ratio in Eq. (7.77) is the same in both heterodyne and homodyne systems, as in the direct-detection case. The only difference is that the terms $I_s^2$ in Eq. (7.77) and $I_q^2$ in Eq. (7.32) are different because $i_p(t)$ takes a different form. From Eqs. (7.38) and (7.71), the expected value of photocurrent induced in a heterodyne detector, $I_p$, is

$$I_p = \left(\frac{R}{2Z}\right)(E_{so}^2 + E_{lo}^2) \tag{7.82}$$

or

$$I_p = \left(\frac{R}{2}\right)(P_{so} + P_{lo}) \tag{7.83}$$

where $P_{lo}$ is the peak local-oscillator power and $P_{so}$ is the peak source power. Usually this expression is simplified further by assuming that $P_{lo} \gg P_{so}$. Then it is rewritten in terms of the rms local-oscillator power $P_{lo\text{-rms}}$ by substituting

$$P_{lo\text{-rms}} = \frac{P_{lo}}{2} \tag{7.84}$$

and thus we get

$$i_p = RP_{lo\text{-rms}} \tag{7.85}$$

The signal current $i_s(t)$ was shown earlier to be

$$i_s(t) = \left(\frac{R}{Z}\right) E_{so} E_{lo} \cos(\omega_{if} t) \cos(\omega_d t) \tag{7.86}$$

In the homodyne case, $\omega_{if} = 0$. The mean-square value of the recovered signal current $I_s^2$ is simply the coefficient preceding $\cos(\omega_d t)$:

$$I_s^2 = \left(\frac{R}{Z}\right) E_{so}^2 E_{lo}^2 = P_{so} P_{lo} R^2 \tag{7.87}$$

For a heterodyne detection scheme, we must first evaluate the integral

$$I = \frac{1}{T} \int_{-T_D/2}^{T_D/2} \cos^2(\omega_{if} t)\, dt \tag{7.88}$$

By expanding this, using Eq. (7.67), and integrating, $I_s^2$ is computed as

$$I_s^2 = \left(\frac{1}{2}\right) P_{so} P_{lo} R^2 \tag{7.89}$$

Calculation of SNR for other modulation schemes may be found in Ref. 36.

## 7.10 RECEIVER SENSITIVITY

### 7.10.1 Bit Error Rate of Digital Receivers

In order to calculate the probability of a decision circuit registering a 1 when a 0 has been transmitted, $P_{e01}$, or a 0 when a 1 has been transmitted, $P_{e10}$, the normal procedure is to assume that the input to the decision circuit has a gaussian distribution (see Fig. 7.39). The total probability of error $P_{eT}$, for a signal with an equal amount of 1s and 0s and decision threshold voltage $v_{th}$, is

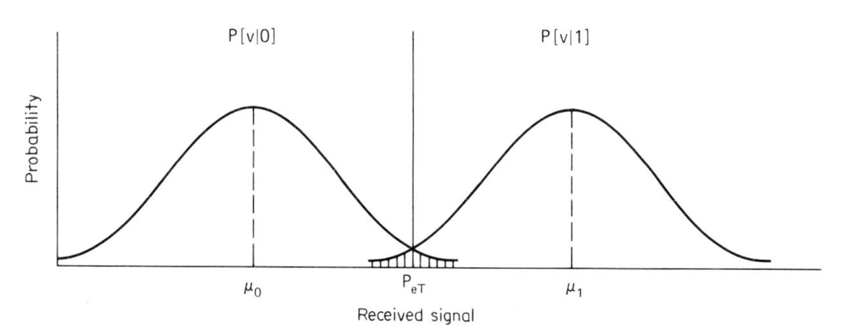


FIGURE 7.39 Receiver decision circuit probability distribution.

$$P_{eT} = P_{e01} + P_{e10} \tag{7.90}$$

where

$$P_{e01} = \left(\frac{1}{2}\right) \int_{-\infty}^{v_{th}} P[v|1]\, dv \tag{7.91}$$

is the probability that the input comparator voltage was less than $v_{th}$ when a 1 was transmitted and

$$P_{e10} = \left(\frac{1}{2}\right) \int_{v_{th}}^{\infty} P[v|0]\, dv \tag{7.92}$$

is the probability that the comparator voltage was greater than $v_{th}$ when a 0 was transmitted. In addition, we assume that an equal number of 1s and 0s were transmitted, which means ½ is the probability of a 1 being transmitted; ½ is also the probability that a 0 was transmitted. In Eq. (7.92), $P_{e10}$ equals the product of the probability a 1 has been transmitted and the probability a 0 is detected, given that a 1 has been transmitted.

Consider the term $P_{e10}$, where we assume the probability density function is gaussian. In reality, the distribution of $P_{e01}$ and $P_{e10}$ are poissonian, but approximating them with gaussian distributions introduces only a small error. Therefore

$$P_{e01} = \left(\frac{1}{2}\right) \int_{-\infty}^{v_{th}} \frac{1}{\sqrt{2p\sigma_1^2}} e^{-(v-\mu_1)^2/2\sigma_1^2} \tag{7.93}$$

The expected value of signal voltage at the comparator for a 1 is $v_1$; i.e., $v_1 = \mu_1$. The expected value of signal voltage for a 0 is $v_0$; i.e., $v_0 = \mu_0$. Thus, in order to find the probability of error for a receiver, one needs only to calculate $\mu_1$ and its corresponding $\sigma_1^2$, in which

$$\sigma_1^2 = v_{\mathrm{nrms1}}^2 \tag{7.94}$$

Here $v_{\mathrm{nrms1}}$ is the rms noise voltage when a 1 is transmitted. In addition, one may calculate $\mu_0$ and its corresponding $\sigma_0^2$, in which

$$\sigma_0^2 = v_{\mathrm{nrms0}}^2 \tag{7.95}$$

In this case $v_{\mathrm{nrms0}}$ is the rms noise voltage when a 0 is transmitted. The probability of error, $P_e$, is also called the *bit error rate.* It can be expressed as a function of $Q$, where

$$Q = \frac{v_{th} - v_0}{\sigma_0} = \frac{v_1 - v_{th}}{\sigma_1} \tag{7.96}$$

Thus

$$\text{BER} = P_e(Q) = \left(\frac{1}{2}\right) \text{erfc}\left(\frac{Q}{\sqrt{2}}\right) \tag{7.97}$$

in which erfc $(x)$ is the complementary error function. Manipulating Eq. (7.96), one can calculate the required difference voltage, $v_1 - v_0$, for a given BER, using

$$v_1 - v_0 = Q(\sigma_1 + \sigma_0) \tag{7.98}$$

One phenomenon not explained in this section is the dependence of BER on the spectral width of the local oscillator. See Fig. 7.40 for a graph of the BER vs. spectral width of a differential PSK system. Derivation of this effect is somewhat lengthy. More detailed explanations on BER vs. spectral width may be found in Refs. 28 and 29.

### 7.10.2 Receiver Sensitivity of Direct vs. Coherent Systems

The major advantage of using homodyne and heterodyne methods vs. the direct detection method, as mentioned earlier, is that circuits employing these techniques can achieve better receiver sensitivity. Table 7.7 compares the BER formulas and theoretical limits for several different modulation schemes.[12] Table 7.8 provides values for erfc $(x)$. When the quantum efficiency $\eta$, the frequency $\nu$, and the data rate $B$ are held constant, the formulas in Table 7.7 indicate that the BER of each modulation scheme decreases with increasing optical power at the receiver, $P_s$. When $\eta$, $\nu$, and $P_s$ are held constant, increasing the data rate $B$ also increases the probability of making bit errors at the receiver. Moreover, the most outstanding facet of the calculation shown in Fig. 7.40 is the difference in theo-

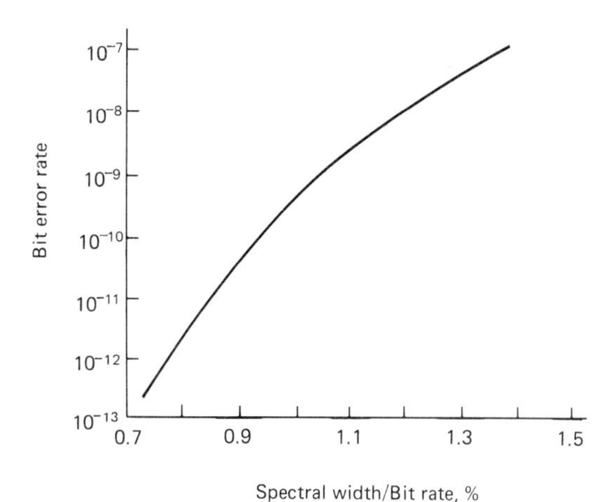


**FIGURE 7.40** The dependence of bit error rate on laser spectral width in differential phase-shift-keying modulation systems. *(From E. E. Basch and T. G. Brown,[12] © 1985 IEEE.)*

**TABLE 7.7** Bit Error Rate for Several Modulation-Demodulation Schemes

| System type | Modulation-demodulation technique | BER formula | Number of photons for $\eta = 1$ and BER $= 10^{-9}$ |
|---|---|---|---|
| Coherent | ASK heterodyne | $0.5 \text{ erfc } \sqrt{\eta P_s/4h\nu B}$ | 72 |
| Coherent | ASK homodyne | $0.5 \text{ erfc } \sqrt{\eta P_s/2h\nu B}$ | 36 |
| Coherent | FSK heterodyne | $0.5 \text{ erfc } \sqrt{\eta P_s/2h\nu B}$ | 36 |
| Coherent | PSK heterodyne | $0.5 \text{ erfc } \sqrt{\eta P_s/h\nu B}$ | 18 |
| Coherent | PSK homodyne | $0.5 \text{ erfc } \sqrt{2\eta P_s/h\nu B}$ | 9 |
| Coherent | Direct detection | $0.5 \exp(-\eta P_s/h\nu B)$ | 21 |
| Direct | Typical receiver | | 400–4000 |

*Source:* From E. E. Basch and T. G. Brown.[12] © 1985

**TABLE 7.8** Complementary Error Function Table

$$\text{erfc}(x) \equiv \frac{2}{\sqrt{\pi}} \int_x^{\infty} e^{-u^2} du$$

| $x$ | erfc $(x)$ | $x$ | erfc $(x)$ |
|---|---|---|---|
| 0.0 | 1.000 | 2.8 | $7.9 \times 10^{-5}$ |
| 0.2 | 0.777 | 3.0 | $2.3 \times 10^{-5}$ |
| 0.4 | 0.572 | 3.2 | $6.3 \times 10^{-6}$ |
| 0.6 | 0.396 | 3.4 | $1.6 \times 10^{-6}$ |
| 0.8 | 0.258 | 3.6 | $3.7 \times 10^{-7}$ |
| 1.0 | 0.157 | 3.8 | $8.0 \times 10^{-8}$ |
| 1.2 | $8.97 \times 10^{-2}$ | 4.0 | $1.6 \times 10^{-8}$ |
| 1.4 | $4.87 \times 10^{-2}$ | 4.2 | $2.9 \times 10^{-9}$ |
| 1.6 | $2.37 \times 10^{-2}$ | 4.4 | $5.0 \times 10^{-10}$ |
| 1.8 | $1.09 \times 10^{-2}$ | 4.6 | $7.9 \times 10^{-11}$ |
| 2.0 | $7.21 \times 10^{-3}$ | 4.8 | $1.2 \times 10^{-11}$ |
| 2.2 | $1.86 \times 10^{-3}$ | 5.0 | $1.6 \times 10^{-12}$ |
| 2.4 | $6.9 \times 10^{-3}$ | 4.32 | $1.0 \times 10^{-9}$ |
| 2.6 | $2.4 \times 10^{-4}$ | | |

retical receiver sensitivity for BER $= 10^{-9}$. Although coherent circuitry is more complex than direct-detection circuitry, the benefits gained in improved receiver sensitivity make coherent technology very attractive[12,14] for those applications requiring higher performance.

## REFERENCES

1. W. Stallings, "Digital Signaling Techniques," *IEEE Communications Magazine,* vol. 22, no. 12, December 1984, pp. 21–25.
2. W. Stallings, *Local Networks: An Introduction,* Macmillan, New York, 1984.

3. D. Mandelkern, "Rugged Local Network Follows Military Aircraft Standard," *Electronics,* April 7, 1982.
4. M. Abramowitz and I. A. Stegun, *Handbook of Mathematical Functions with Formulas, Graphs, and Mathematical Tables,* Applied Mathematics Series 55, National Bureau of Standards, December 1972, p. 361.
5. *Transmission Systems for Communications,* 5th ed., Bell Telephone Laboratories, Holmdel, N.J., 1982.
6. M. Schwartz, *Information Transmission Modulation and Noise,* 3d ed., McGraw-Hill, New York, 1980, p. 183.
7. E. E. Basch and H. A. Carnes, "Digital Optical Communication Systems," in James C. Daly (ed.), *Fiber Optics,* CRC Press, Boca Raton, Fla., 1984, p. 174.
8. S. Stone, "Modulation of Optical Sources," chap. 10 in E. E. Basch (ed.), *Optical-Fiber Transmission,* Sams, Indianapolis, 1987.
9. R. H. Saul, "Recent Advances in the Performance and Reliability of InGaAsP LEDs for Lightwave Communication Systems," *IEEE Transactions on Electron Devices,* vol. ED-30, no. 4, April 1983, pp. 285–295.
10. I. Hino and K. Iwanoto, "LED Pulse Response Analysis Considering the Distributed CR Constant in the Peripheral Junction," *IEEE Transactions on Electron Devices,* vol. ED-26, no. 8, August 1979, pp. 1238–1242.
11. S. Gage, M. Hodapp, D. Evans, and H. Sorensen, *Optoelectronics Applications Manual,* McGraw-Hill, New York, 1981.
12. E. E. Basch and T. G. Brown, "Introduction to Coherent Optical Fiber Transmission," *IEEE Communications Magazine,* vol. 23, no. 5, May 1985, pp. 23–30.
13. J. Gowar, *Optical Communication Systems,* Prentice-Hall International, London, 1984, pp. 342–343.
14. I. W. Stanley, "A Tutorial Review of Techniques for Coherent Optical Fiber Transmission Systems," *IEEE Communications Magazine,* vol. 23, no. 8, August 1985, pp. 37–53.
15. N. K. Dutta, "Optical Sources for Lightwave System Applications," chap. 9 in E. E. Basch (ed.), *Optical-Fiber Transmission,* Sams, Indianapolis, 1987.
16. P. S. Henry, "Lightwave Primer," *IEEE Journal of Quantum Electronics,* vol. QE-21, December 1985, pp. 1862–1879.
17. R. C. Goodfellow, B. T. Debney, G. T. Rees, and J. Buus, "Optoelectronic Components for Multigigabit Systems," *IEEE Journal of Lightwave Technology,* vol. LT-3, no. 6, December 1985, pp. 1170–1179.
18. R. C. Alferness, "Optical Guided-Wave Devices," *Science,* vol. 234, November 1986, pp. 825–829.
19. R. C. Alferness, "Waveguide Electrooptic Modulators," *IEEE Transactions on Microwave Theory and Techniques,* vol. MTT-30, no. 8, August 1982, pp. 1121-1137.
20. H. M. J. Otten, "Fiber-optic Communications," *Electronic Components and Applications,* vol. 3, no. 2, February 1981, pp. 87–100.
21. "Standard Acousto-Optic Laser Modulators and Drivers," data sheets, Anderson Laboratories, Bloomfield, Conn.
22. "Optical Guided Wave Devices," data sheets, Crystal Technology, Palo Alto, Calif.
23. R. A. Becker, "Broadband Guided Wave Electrooptic Modulators," *IEEE Journal of Quantum Electronics,* vol. QE-20, no. 7, July 1984, pp. 723–727.
24. A. Fascia, "A Wideband Medium Power Microwave Amplifier," Master's thesis, University of Rhode Island, Kingston, R.I., 1988.
25. G. Matthaei, L. Young, and E. Jones, *Microwave Filters, Impedance-Matching Networks, and Coupling Structures,* Artech House, Dedham, Mass., 1980.

26. S. K. Korotky et al., "4-Gbit/s Transmission Experiment over 117 km of Optical Fiber using Ti-$LiNbO_3$ External Modulator," *IEEE Journal of Lightwave Technology,* vol. LT-3, no. 5, October 1985, pp. 1027–1031.

27. F. Koyama and K. Iga, "Frequency Chirping in External Modulators," *IEEE Journal of Lightwave Technology,* vol. LT-6, no. 1, January 1988, pp. 87–93.

28. Gerd Keiser, *Optical Fiber Communications,* McGraw Hill, New York, 1983.

29. H. Kressel (ed.), *Semiconductor Devices for Optical Communication,* Topics in Applied Physics, vol. 39, Springer-Verlag, New York, 1987.

30. J. E. Freund and R. F. Walpole, *Mathematical Statistics,* Prentice-Hall, Englewood Cliffs, N.J., 1980.

31. C. D. Motchenbacher and F. C. Fitchen, *Low Noise Electronic Design,* Wiley, New York, 1973.

32. E. E. Basch, R. F. Kearns, and T. G. Brown, "The Influence of Mode Partition Fluctuations in Nearly Single Longitudinal Mode Lasers on Receiver Sensitivity," *IEEE Journal of Lightwave Technology,* vol. LT-4, no. 5, May 1986, pp. 516–519.

33. T. Okoshi, "Heterodyne and Coherent Optical Fiber Communications: Recent Progress," *IEEE Transactions on Microwave Theory and Techniques,* vol. MTT-30, no. 8, August 1982, pp. 1138–1149.

34. T. Okoshi, "Ultimate Performance of Coherent Optical Fiber Communications," *IEEE Journal of Lightwave Technology,* vol. LT-4, no. 10, October 1986, pp. 1556–1562.

35. Y. Yamamoto and T. Kimura, "Coherent Fiber Transmission Systems," *IEEE Journal of Quantum Electronics,* vol. QE-17, no. 6, June 1981, pp. 919–934.

36. E. Basch and T. Brown, "Introduction to Coherent Fiber-Optic Communication," in E. E. Basch (ed.), *Optical-Fiber Transmission,* Sams, Indianapolis, 1987.

# CHAPTER 8
# OPTICAL FIBER SENSORS

**G. D. Pitt**
*Renishaw Transducer Systems Ltd. and Queen Mary College, University of London*

## 8.1 INTRODUCTION AND BACKGROUND

### 8.1.1 Perspective

The primary application of optical fibers will remain in the many areas of telecommunications. The advantages of optical fibers in terms of wide bandwidth, lightness, and reduced volume, for short- and long-haul communications applications, are overriding. However, benefits are also to be found in the use of optical fibers for sensing, control, and instrumentation.[1–13] In these applications, several of the apparent technical problems associated with optical fibers for telecommunications can be turned to advantage. It has been demonstrated, for example, that optical fibers can (inadvertently) be highly sensitive to their environment. At the simplest level, the effects of microbending and the resulting transmission loss, which can be measured, illustrate how the fiber response can provide a means of detecting pressure or strain.

The field of sensors, however, is extremely large and competitive in the types of technologies that can be employed. Some criteria for the selection of an optical-fiber sensor technology over other techniques for an engineering design are listed:

- The design should meet the user's sensitivity, accuracy, and repeatability requirements.
- Cross-perturbing effects should be minimized so that the measurement specification can be met.
- The environmental operating conditions should not cause problems for optical fibers and their packaging.
- The fiber-related components (e.g., couplers) should be compatible with the host system in terms of standardization and availability, as well as in terms of output signals for subsequent processing.
- Mounting and packaging specifics should be defined and within requirements.
- Cost should be competitive for the required performance.

The terms *sensor* and *transducer* are often interchanged. For the purpose of this chapter, the sensor is an element which changes in some way because of alterations in the environment, e.g., chemical change or dimensional change. A transducer includes the sensor, but converts the change into a recognizable signal for measurement purposes.

It is appropriate to consider the sensor in the context of a transducer system, having output signals which can directly and easily interface to the system. Semiconductor sensors, for example, can be constructed with on-chip signal processing, as in the case of piezoresistive silicon diaphragms. The complete device is therefore capable of both sensing and processing the signal, which can then be transmitted over some distance on wire. Similarly, it is important to ensure from the outset that a fiber-optic sensor system should be capable of easy interfacing with a fiber-optic transmission line.

*Optical-Fiber Sensor Advantages.* Specific applications for optical fiber sensors may be inferred from the list of advantages in Table 8.1.

Before attempting to exploit fiber-optic sensors, the designer is cautioned to keep the technical advantages listed in Table 8.1 in a practical perspective. Some cautionary notes are provided:

1. The fiber-optic approach should show clear technical (and, ideally, cost) advantages over available and competing technologies. Historical examples exist wherein a technical enthusiasm for a given fiber-optic sensor project proved costly, particularly when alternative technologies also improved in parallel and resulted in better, robust sensors at reduced cost.
2. Experience has shown that user acceptance of any new sensor technology can be slow to develop. Taking electrical safety as an issue, much effort has been applied to conventional sensors, over the years, to realize devices that are now considered safe and can be approved to accepted international standards. Such devices are also standardized in terms of connections and input and output ratings, allowing easy replacement when necessary. This degree of standardization for optical-fiber components has yet to be be achieved. In spite of the many safety attributes of optical fibers in hazardous (e.g., explosive) environments, full acceptance has not yet been achieved. Many sensor users will remain conservative and may still prefer to use established and approved technology.
3. Laboratory developments, which have often indicated high potential for a given optical-fiber sensor, do not necessarily provide a guarantee that a fully ruggedized system is available for use in the harsh world outside the laboratory. Much engineering work is often necessary to properly package fiber-optic sensors.

Particular emphasis has been placed on the aforementioned issues in this chapter.

*Optical-Fiber Sensor Disadvantages.* Some technical disadvantages of optical fibers for sensors are listed in Table 8.2. Here the fundamentals of sensor system design are again emphasized. The dimensional aspects of the problem (i.e., one can be dealing with 50-μm fiber cores, or much less, for single-mode fiber) means that cost-effective precision engineering and design are required to ensure correct alignment and location of the fiber within the sensor enclosure. Materials selection for packaging can be crucial, since the fiber will also respond to all forms of induced strain, e.g., thermal. Ideally the design of the packaging should also enhance the response to the variable to be measured, and reduce the cross-

**TABLE 8.1** Advantages of Optical-Fiber Sensors

| | |
|---|---|
| Safety | Nonmetallic construction precludes conduction of hazardous voltages; no spark risk. (But fiber breakage at high optical power levels can produce explosion; see Sec. 8.7.) |
| Small cable size and weight | Useful in aerospace applications (e.g., fly-by-light), offshore platforms, and extremely remote applications where dispensing heavy cabling is costly and difficult (low-loss fiber needed). |
| Electromagnetic interference (EMI) immunity | Can be collocated with electrical power cables and in other high electrical fields (e.g., near transformers) without crosstalk. |
| Radio-frequency-passive | No RF emissions; covert in RF context. |
| Low thermal and inertial mass | Useful for fast-response (less than 1 μs) temperature sensing and for integrated measurement along a length; applicable to accelerometry; but can be affected by cross-thermal effects. |
| Small sensor size | Can be inserted into small and awkward volumes (as in medical applications) or used in difficult inspection situations. |
| Selective surface sensitivity | Total internal reflection and surface plasmon effects can be used to detect chemical species (but reversibility can be a problem). |
| Geometric versatility | Can be wound or shaped into various configurations, as for hydrophone arrays, magnetic gradient sensing, and fiber gyro coils. |
| Radiation sensitivity | Non-radiation-hardened fibers can be applied to dosimetry. |
| Power transfer | Efficiency of silica fibers can support optical powering of remote sensors or actuators; can eliminate need for local electrical power supply at sensor site. |
| Multiplexing | Optical transmission affords additional degree of freedom for multiplexing (i.e., wavelength-division multiplexing). |

perturbing effect of other environmental changes. This analysis, however, can also be said to be applicable to the packaging of other sensor devices, e.g., integrated silicon pressure sensors.

Apart from careful mechanical design, several electronic schemes can be used to filter out extraneous variations, given some initial determination of likely fre-

**TABLE 8.2** Some Disadvantages of Optical-Fiber Sensor Systems

| | |
|---|---|
| Fragility | Careful attention to packaging is often needed to ensure robustness. |
| Small scale of optical components | The small dimensions of optical fibers can lead to device manipulation and alignment problems, requiring special techniques and facilities during assembly and field repairs. |
| Sensitivity to multiple environmental parameters | Thermal and acoustic/vibrational interference can be a problem with high-sensitivity devices. Special packaging and signal processing may be necessary. |
| Limited optical bandwidth | Spectroscopic applications are limited by availability of infrared-transmitting (wavelength > 3 μm) fibers. Special handling is anticipated for emerging infrared fibers. |
| Cost | Sources and other components are often in specialty categories, and priced accordingly. Many available fiber components and techniques derive from telecommunications requirements and are therefore not optimum for all sensors. |

quency effects, e.g., low-frequency thermal variations. Optical-fiber sensors may offer further such advantages when integrated-optic and advanced optical processing techniques become more widely available.

### 8.1.2 Sensor Classification

Various methods of classifying optical-fiber sensor systems have been proposed. For convenience, in this chapter they are divided into three broad groups. These are listed, with examples, in Table 8.3. Figure 8.1 provides a highly schematic realization of the three groups.

***Extrinsic Sensors.*** *Extrinsic optical-fiber sensors* are those wherein optical fibers are used to transmit radiation to and from the point or region to be sensed. The radiation is released from the transmitter fiber and modulated externally by some induced or environmental change. Such a change may be caused by beam interruption by a pressure release switch, back-reflection from a moving reflector (as a function of angular, axial, or lateral displacement), absorption by a gas (spectroscopy), passage through a material (e.g., a semiconductor) having a temperature-sensitive absorption edge, or perturbing effects from environmentally sensitive birefringent elements. Most of the early and more commercially successful fiber-optic sensors have been of this type, where the fiber has been used largely as a transmission medium.

In several applications, the fiber can also have certain advantages as small,

**TABLE 8.3** Optical-Fiber Sensor Mechanisms and Corresponding Application Examples

| | Extrinsic |
|---|---|
| Light interruption/reflection | On-off sensors, microswitches, frequency-out sensors, displacement sensors |
| Spectroscopy | Absorption, light scatter, fluorescence and scintillation sensors, laser velocimetry |
| Birefringence | Photoelastic effect, interposed birefringent elements (e.g., Pockels effect), current and voltage measurement |
| Distributed effects | Optical time-domain reflectometry, fiber break sensors |
| | Evanescent |
| Refractive index | Temperature, level sensors |
| Microbending | Temperature, level, pressure sensors |
| Surface absorption/plasmons | Chemical species sensors, color change sensors |
| | Intrinsic |
| Phase measurement | Gyroscope, hydrophone, magnetometer |
| Polarimetry | Birefringent properties of fibers for acoustic, magnetic field, temperature measurement |
| Microbending | Pressure, strain, displacement, acoustic sensors |
| Distributed effects | OTDR, intrusion sensors, temperature sensors, Raman backscatter |

remote, well-defined, single or multiple light sources having a well-defined geometry or arrangement. Examples are in fiber sources for surface inspection and vibrometry using reflective techniques and for light scatter detection with improved angular resolution (by comparison with typical larger-area silicon detector arrays).

***Evanescent Sensors.*** *Evanescent sensor devices* rely on the measurable loss of guidance from an optical waveguide as a means of detecting external changes. Examples are level detection (when the effective refractive index of surrounding material around an optical fiber or exposed optical element changes—for example, at a liquid-to-air interface), detection of specific gravity, and detection of chemical reactions, chemical species, and pH.

Microbending effects on the fiber can also promote the loss of light from the optical waveguide and provide a means of measuring strain. The sensitivity depends on the critical angle for internal reflection within the deformed fiber. The transfer of radiation from core to cladding of the fiber within the total waveguide structure by microbending, however, means that such devices could also be classified as intrinsic sensors, since the light is still retained within the overall fiber structure.

***Intrinsic Sensors.*** In some cases, the fiber itself can act as the responsive element. Such devices can range from highly sensitive interferometric systems,

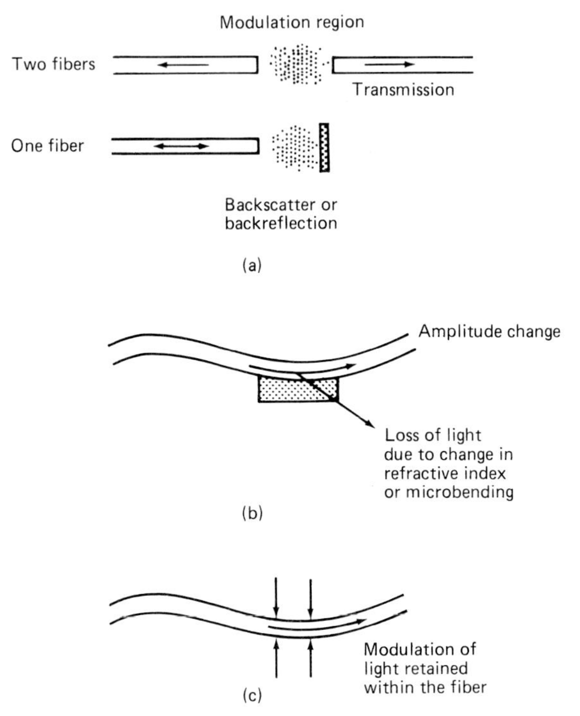


**FIGURE 8.1** Schematic representation of the three broad groups of optical-fiber sensors. (*a*) Extrinsic sensors; (*b*) evanescent sensors; (*c*) intrinsic sensors.

where a coherent source is used and phase measurements are made, to less sensitive systems where the polarization state in the fiber is disturbed.

Otherwise, loss mechanisms within the fiber can provide sufficient backscatter signal for detection, as in the case of optical time-domain reflectometry (OTDR). Figure 8.2 provides a schematic series of diagrams of different fiber-optic interferometric configurations. The further effects of modal transfer between core and cladding (which can also be classed as intrinsic sensing mechanisms) and the different modulating effects on the optical-fiber geometry for sensing have become major subjects (both theoretical and empirical) in their own right.

Technologies for applying specialized coatings to a fiber to enhance its sensitivity to external effects have also been developed. Examples are aluminum coatings for temperature sensors, metallic glass or nickel coatings for magnetic sensors, and special gas- or chemical-species-absorbing materials. The strain applied to the fiber by these coatings can be detected.

The methods of mounting or winding optical fibers for intrinsic sensors are crucial to their design. Intrinsic fiber-optic sensors have the advantage that radiation is not lost from the fiber. There are, therefore, possibilities for measurement systems such that a single continuous loop of fiber can be interrogated remotely to detect changes at specific or distributed points on the loop. It is theoretically possible to multiplex a single fiber in this way and thereby avoid

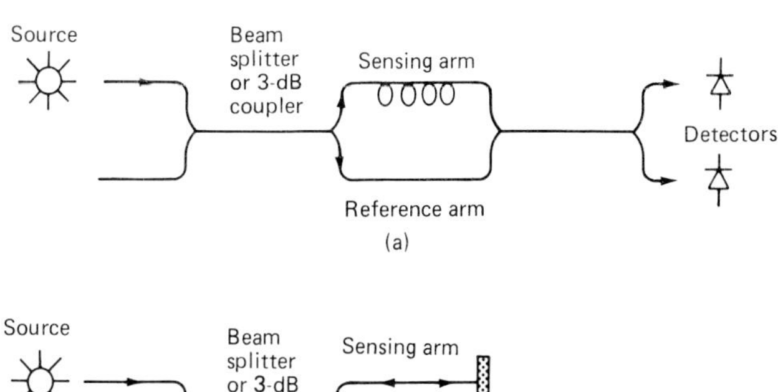


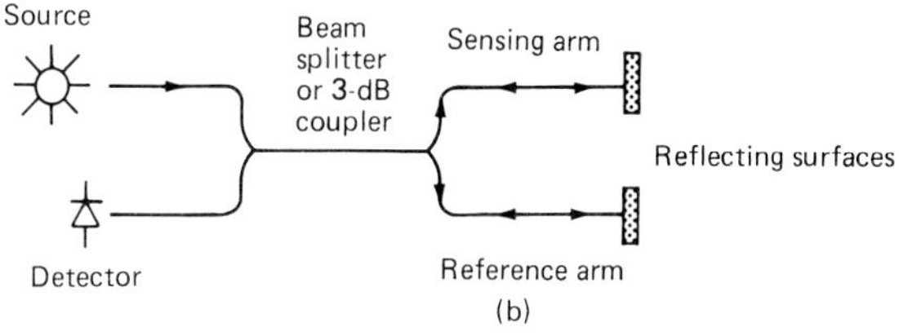


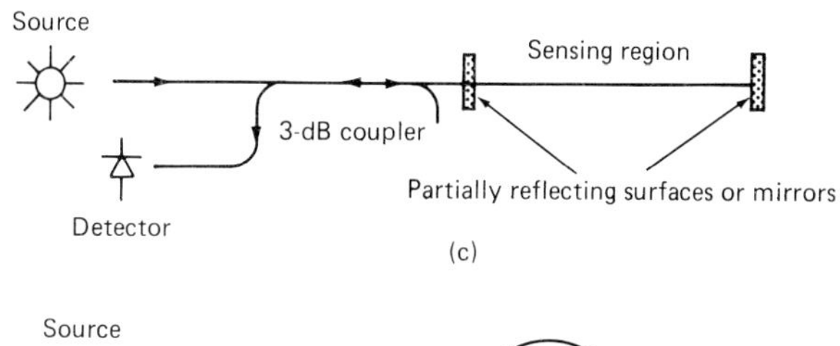


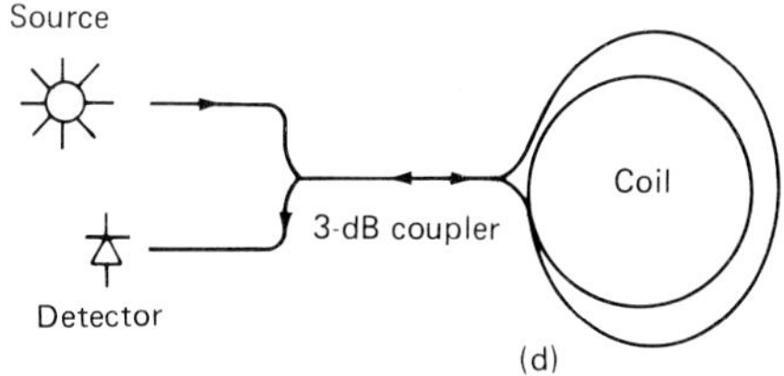


**FIGURE 8.2** Schematic diagrams of various fiber-optic interferometric sensor configurations. (*a*) Mach-Zehnder; (b) Michelson; (*c*) Fabry-Perot; (*d*) Sagnac.

numerous, expensive couplers. Currently, however, such multiplexing possibilities are rather limited. The state of the art for sensor loops (e.g., hydrophone arrays) tends to require the use of expensive telecommunications or other highly specialized couplers.

Several other intrinsic optical-fiber sensors have been proposed wherein it is necessary that the (high or low) birefringent properties of the fiber must be carefully controlled over the fiber length, as in the case of electric field sensors. This implies further requirements for special fiber preparation, couplers, and cabling. A description of some of the developments of specialized optical-fiber components, promoted by the increased awareness of optical-fiber sensors, is given in Sec. 8.5.

### 8.1.3 Systems Engineering Approach

As described earlier, the schematic fiber sensor configurations shown in Figs. 8.1 and 8.2 must be packaged and engineered. A full design engineering and systems approach is required for any optical-fiber sensor development. It is significant that the encroachment of optical fibers into control applications, as distinct from telecommunications, has been cautious. This may be illustrated by examining briefly the case of sensor and control systems as applied to offshore platforms.

A typical offshore hard-wired electrical system is shown in Fig. 8.3. A central control unit is linked to different local control centers, which may also have local regulated power supplies. Each center can have subcenters, and multiplexing at the local center is carried out for several sensors, e.g., gas alarm, smoke, fire, temperature. To save on installation costs, each local center is modularized, with fabrication and testing being carried out onshore; the final system is assembled offshore. Shipboard installations are often carried out in a similar fashion. Offshore, it would be reasonable to assume that optical fibers would offer a great advantage in terms of safety. In fact, the initial impetus has come from the possibility of laying fibers for communications next to power cables, to preclude interference and to reduce cable installation costs, as well as to obtain a substantial weight reduction (in comparison to laying additional electrical cables). For offshore platforms, optical fibers are being used for inter-control-room links, while for ships, sensing systems are currently being designed and tested (e.g., for naval fire and flash detection, damage limitation, and control applications). On board ships, since increasing amounts of machinery are being concentrated together, new installations are again more concerned with the advantages of optical-fiber systems in eliminating electromagnetic interference than in increasing safety. For process control applications, where cost is still more important, the incorporation of optical-fiber sensors in multiplexed systems is likely to proceed at a conservative pace. A cautious fiber sensor development trend is noted in industries requiring measurement and control.

Some sensor systems, such as the fiber-optic gyro, could be classified as

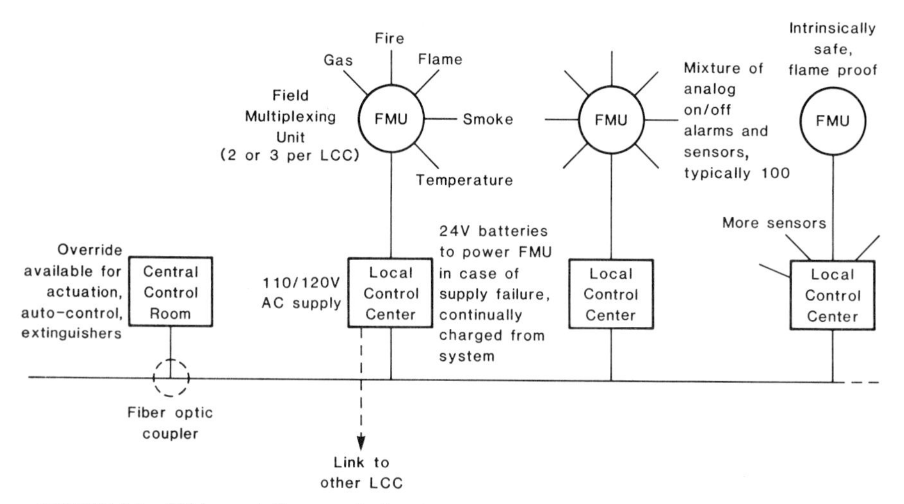


FIGURE 8.3 Offshore-platform control system.

stand-alone devices and can compete in niche applications without necessarily being incorporated in a multiplexed or control loop scheme. Nevertheless, optical-fiber sensors should still be considered as systems calling for careful mechanical design, materials selection, optical loss budgeting, electro-optic interfacing, and signal processing.

### 8.1.4 Scope of Chapter

In this chapter, an engineering and systems approach will be taken to describe the use of optical fibers for sensor systems. Thus, while some space will be devoted to a description of the basic sensing mechanisms, emphasis will be placed on those sensor examples which have progressed beyond the laboratory. The broad classification of Sec. 8.1.2 will be used as a basis for describing the various fiber-optic sensors.

An important aspect of sensor development is the ability to use optical power to both actuate and sense. This has led, for example, to novel methods of optically activating the sensor and then measuring changes in oscillation of the device to provide "frequency-out" information. The control loop has been closed using purely optical techniques. Some careful design engineering is necessary to ensure that the light levels through the system are optimized.

The final section of this chapter will enumerate some of the challenges for optical sensor development, as determined for various applications such as medicine, oil exploration, and defense.

The number of papers in the field of fiber-optic sensors has grown rapidly since the mid-1970s, when serious activity in this area began. Numerous useful reviews have been completed and those wishing to design optical-fiber sensor systems are advised to consult these (see "References" at the end of the chapter) prior to embarking on a search of the vast research and patent literature on this subject.

## *8.2 INTENSITY-MODULATED SENSORS (EXTRINSIC)*

The engineering aspects of several extrinsic sensors are discussed in the following, with reference to the schematic examples shown in Fig. 8.1. In several cases, difficulties have subsequently emerged with these sensors when some form of multiplexing has been required, largely because of the low-amplitude signal levels that can be returned to the detector from the sensing region.

When light is emitted from one fiber into another, as in standard couplers for telecommunications applications, it is typical to estimate a loss of the order of 1.0 dB for system light-budget calculations. When this emitted radiation is then modulated by interaction with an added sensor element or elements before returning to the same or another fiber, the loss can be very high (e.g., 50 to 60 dB.) This loss, when factored with the relatively high cost (for industrial applications) of low-loss, fiber-optic couplers, means that the engineering possibilities for incorporating the typical extrinsic sensor into a multiplexed system must be carefully examined. The added complexity and associated cost of referencing the intensities for extrinsic intensity-modulated sensors must also be taken into account.

## 8.2.1 Interruption Effects

Interruption-effect sensors can be grouped into two general categories: microswitches and back-reflection/displacement sensors.

Microswitches are used, for example, in the gas industry to provide alarms when there is a sudden pressure rise or release in operating machines or pipelines. Devices can be placed strategically at points where the release or control actuation should occur. Applied gas pressure can force a spring-loaded plunger between opposed optical fibers, or between the fiber and an adjoining reflective surface. Several commercial microswitches on this theme have been offered, but have found difficulty in overcoming user conservatism in replacing the traditional, protected electrical microswitches approved for use in hazardous atmospheres. Typical optical-fiber devices of this class have been constructed using design features of both the electrical microswitch and optical-fiber connector technologies.

Figure 8.4 shows three schematic examples of optical microswitches. In order to reduce crosstalk to acceptable levels and, particularly, to provide ease of alignment at low assembly cost, mechanical design criteria can be surprisingly complex. Large-volume applications, where there are requirements for large numbers of devices, are needed to justify the investment in accurate plastic molding and manufacturing techniques. Figure 8.5 shows a commercially available optical microswitch device.

Variations of the shutter approach can generate variable attenuation as a measure of shutter displacement, rather than a simple on-off indication. Examples of shutter variants[1] include:

- Moving wedges
- Graded scales
- Moving gratings

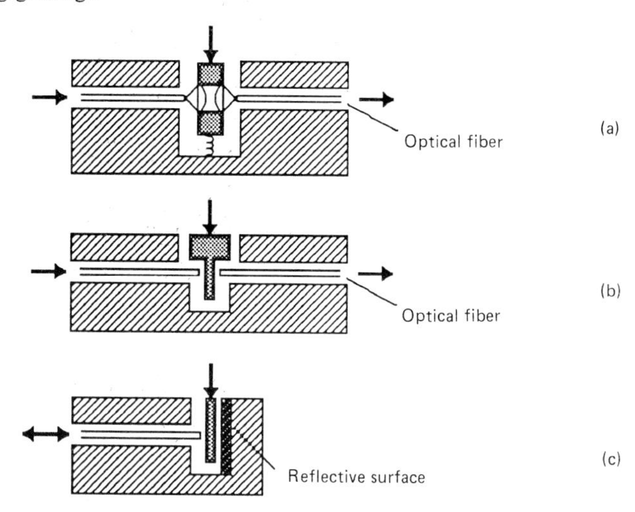


**FIGURE 8.4** Schematic diagrams of typical optical microswitches. (*a*) Expanded-beam connector type; (*b*) shutter type; (*c*) reflective variant of shutter type (one or two fibers can be used).

FIGURE 8.5 Fiber-optic pushbutton switch. (*a*) Rear view; (*b*) front view. (*Courtesy of EOTec Corp.*)

Several levels of sensitivity can be obtained, since various electronic interpolation methods can be applied. The grating sensor, attached to a movable diaphragm, has achieved sensitivity levels close to those required for underwater hydrophone applications. Such devices can become more complex when referencing of intensity variations through the system is required. This is further discussed in Sec. 8.2.3.

Representative of the back-reflection/displacement-type sensor is the Fotonic[16] sensor illustrated in Fig. 8.6. This device measures the variation in intensity of the back-scattered radiation from a movable surface near the fiber end. Different geometrical configurations of dual or multiple fibers can also be used. In the design shown, fiber bundles with send and receive fibers alternately juxtaposed can be used. It is seen in Fig. 8.6*b* that there could be ambiguities in the displacement reading over an extended range if only one sensor is used.

Since the mechanical design requires that temperature variations and other intensity-modulating effects in the system should be compensated, referencing can be assisted by using two detector fibers in parallel with an illuminating fiber, where the distance between the detector fibers is known (as in Fig. 8.7). Thus the

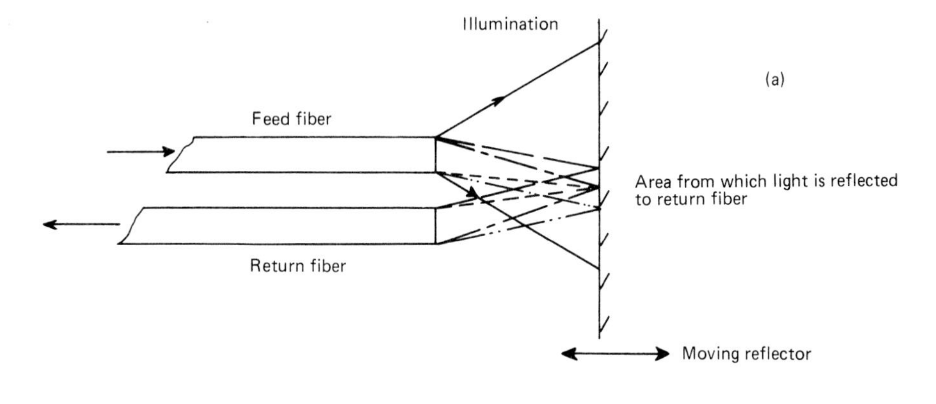


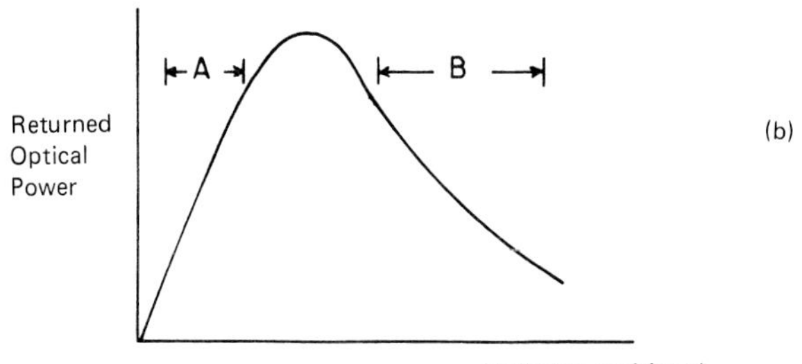


**FIGURE 8.6** The Fotonic fiber-optic displacement sensor. (*a*) Principle of operation; (*b*) optical signal as a function of reflector position. (*Courtesy of Mechanical Technology Inc.*).

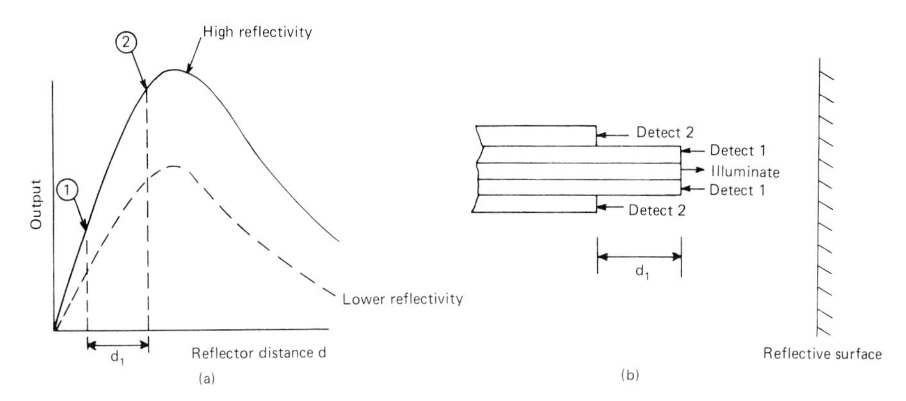


**FIGURE 8.7** Referencing in backscatter/reflection sensor using two detector fibers. (*a*) Detector output curves; (*b*) "two-fiber" geometry. (*Courtesy of Mechanical Technology, Inc.*)

availability of output curves for both fibers provides a means of compensation. This assumes that no preferential attenuation of one sensor relative to the other occurs during measurement. With suitable electronics, the ability to update and compensate the system using periodic calibration checks can be achieved. Such devices can be used for surface proximity measurement (as in machine tools), and adapters can be fitted to the end of the fiber for different displacement ranges.

Several other variations on the theme of the Fotonic reflective sensor have been made commercially available. One of these, a temperature sensor, is illustrated in Fig. 8.8.

The reflective technique has also been adapted to measure the rotational rate in a flowmeter. In some respects, this requirement is less demanding, since on-off signals are to be obtained, rather than analog displacement measurements. In one implementation[3] of a fiber-optic flowmeter, two fibers are connected to a standard make-and-break diaphragm relay having a soft-iron pole piece fixed to the relay body. The diaphragm is moved backward and forward as the poles of a series of coils rotate within the flowmeter housing. The resulting optical output is a sawtooth wave, the frequency of which gives the rotation rate. As with the adaptation of all microswitch technologies, the fiber alignment is crucial in optimiz-

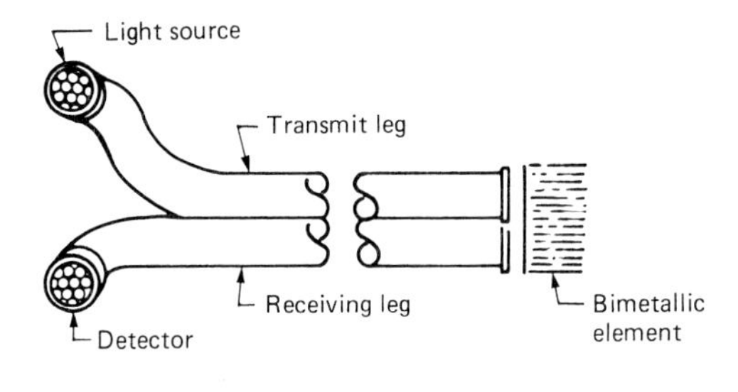


(a)

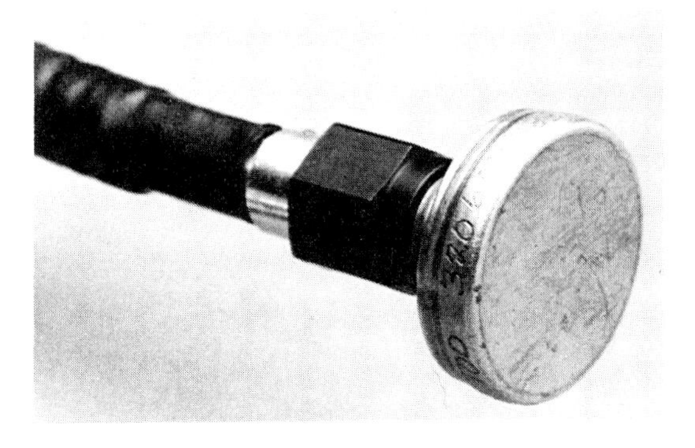

(b)

**FIGURE 8.8** Fiber-optic temperature sensor with bimetallic element. (*a*) Schematic diagram; (*b*) pictorial representation. (*Courtesy of EOTec Corp.*)

ing signal levels and minimizing crosstalk. Light-budget calculations and tests have confirmed that operation with silica multimode fiber is possible over 500 m from the source. Other related methods for turbine flow measurement, using spindle encoders or interruption of a traversing beam by the turbine blades within a flow path, have been demonstrated in commercially available devices.

### 8.2.2 Light Scatter Effects

The incorporation of fiber optics into laser doppler velocimeter (LDV) systems has had considerable impact.[14] Some of the advantages of fiber-optic-based velocimetry are

1. Miniaturization (applicable to measurement of blood flow and to wind tunnel studies, where small probe size is critical)
2. Flexibility (applicable to situations in which the source can remain in a vibrationally isolated region, while the probes extend to the point of measurement, as in ship's propeller turbulence studies)
3. Safety (because of elimination of metallic conductors—see also Table 8.1)

Note that the availability of polarization-preserving fibers to guide radiation is important in some applications to reduce polarization fading (i.e., to improve fringe contrast) caused by thermal and vibrational effects on the fiber leads. This can be important in industrial environments, but comes at the expense of tighter tolerances on the stability of source launching and coupling. A more detailed description of the use of polarization-preserving fibers for sensing applications is provided in Sec. 8.5.

Figure 8.9 shows a schematic diagram of a laser doppler anemometer which uses polarization-preserving fibers.[15] The input radiation, consisting of the two orthogonal polarizations from a single longitudinal-mode coherent source, e.g., a helium-neon (He-Ne) laser at 633 nm, is fed into a single polarization-preserving fiber after one of the polarized beams has first been frequency-shifted by a Bragg cell. The orthogonally polarized beams are separated again in the miniaturized probe head and are focused to define the typical LDV interference fringes within the measurement volume. The backscatter from particles passing through the fringes of known separation in the fluid provides a means of measuring velocity. The frequency shift feature allows the direction of particle travel to be determined. The returned radiation is carried by a step-index multimode fiber, the diameter of which is selected to be compatible with the image of the measuring volume. Alternatively, two polarization-preserving fibers can be used from the source to the sensing head, eliminating the need for most of the polarization and beam-splitting optics. When cabled with the return fiber, the response to thermal and vibration variations is common, but if a single fiber is used, this common-mode feature is enhanced.

The optical design and specialized components required for fiber-optic LDV means that such systems are relatively expensive. In this particular applications area, however, the advantages of fiber are very clear and it is likely that the use of such devices will grow. Optical-loss budget calculations and analysis of system noise factors for fiber LDV systems can be found in the literature.[15]

Forward-scattered radiation has been successfully applied to the measurement of the oil content in the operational discharge of ballast water from supertankers.[3] This technique, in which optical fibers transmit radiation to and from tanker

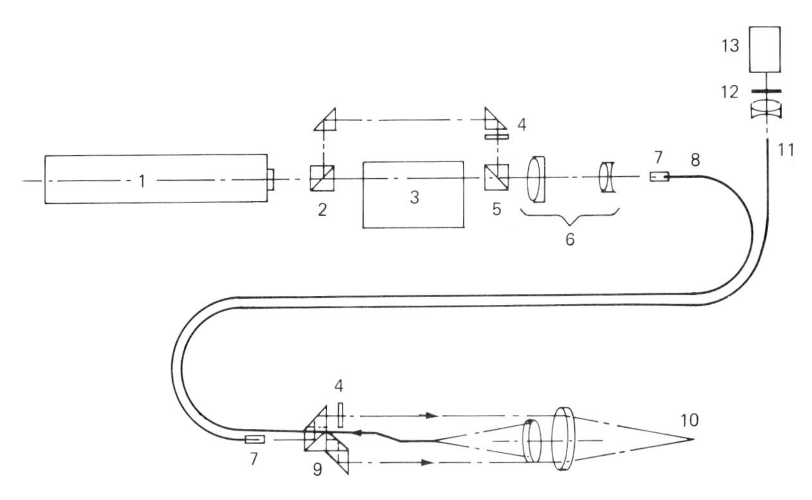


**FIGURE 8.9** Layout of laser doppler anemometer system. (1) Polarized laser; (2) beam splitter (50/50); (3) Bragg cell (40 MHz); (4) 90° polarization rotator; (5) polarization beam splitter; (6) ×4 beam compressor; (7) GRIN-rod lens; (8) single-mode polarization-preserving fiber; (9) polarization beam splitter with beam separation; (10) measuring volume; (11) receiving multimode fiber; (12) interference filter; (13) photomultiplier. (*Courtesy of Dantec Electronics Limited.*)

pump rooms, is schematically illustrated in Fig. 8.10. Over 1000 of these units have been installed to ensure compliance with the International Maritime Organization pollution control requirements, as well as the safety legislation of the international classification societies. The units utilize radiation from a double-heterostructure laser diode at approximately 830 nm coupled to a 10-m-long fiber-

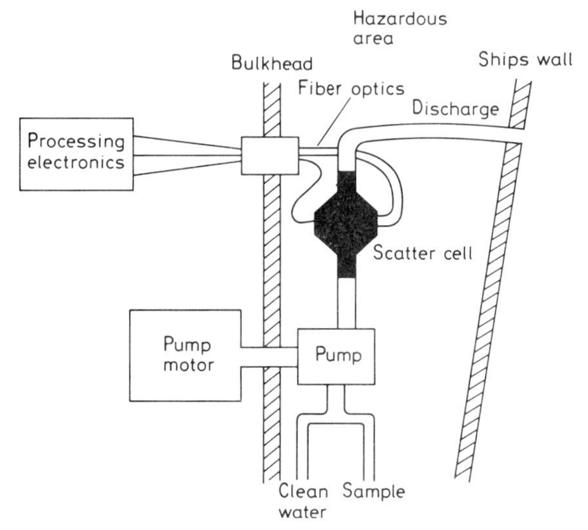


**FIGURE 8.10** Schematic diagram of the oil-in-water scatter cell.

optic bundle, which is passed through a fireproof bulkhead. At the measurement cell, more optical fibers, situated at different angles relative to the incident beam, detect the scattered radiation from the oil droplets in the process stream. Some discrimination can be made between oil and other suspended solids as a function of particle size and refractive index by suitably intelligent signal processing of the detected signals. Extensions of this principle, using multiple sources and fiber-optic detector arrays, have been proposed for cross-correlation flow measurement and color and particle shape analysis.

### 8.2.3 Spectroscopic Effects

Spectroscopic sensors have been developed for remote measurement of gases. Examples include sensors to detect methane ($CH_4$) leaks on oil platforms or in coal mines.[16] Unfortunately, the relevant strong absorption bands for most noxious gases lie in the infrared beyond the normal transmission range of silica optical fibers. There are, however, several sharp gas-absorption-band harmonics in the near-infrared range of 1.3 to 1.7 μm, which can be exploited. One example is the 1.33-μm absorption line for methane ($CH_4$). Such wavelengths conveniently coincide with those of available narrowband quaternary lasers [e.g., indium gallium arsenide phosphide (InGaAsP)] and of minimum absorption for silica fibers. InGaAsP lasers can be tuned by temperature control to the correct absorption line with the help of a reference cell.[3] Referencing can be carried out by thermal modulation of the source, such that the source wavelength can be moved on and off the absorption line peak. Frequency-stabilization techniques for semiconductor lasers have been developed for interferometric sensors (see Secs. 8.4 and 8.5), and adequate thermal stability, coupled with wavelength shift control, has been demonstrated. Provided that care is taken in the design to eliminate those problems inherent in single-mode fibers and narrow-line lasers (i.e., mode hopping and modal noise), remote gas detection can be demonstrated over fibers that are tens of kilometers in length.

Depending on the strength of the absorption line to be measured and on the gas concentration limits to be detected, the fiber can be taken to a remote White transmission cell and the absorption information collected by a multimode fiber. Alternatively, a reflective cell can be used. A single-fiber cable, transmitting in single-mode and receiving in multimode, can be located with its remote fiber terminus at the focus of a 10-mm focal-length concave mirror. The loss through such a detection system with correct alignment has been demonstrated to be as low as 3 dB. Given the relatively low losses for such a design, it is possible to accommodate it in a multiplexed system. On cost grounds, for the detection of methane leakage in mines or offshore, the solid-state Pellistor gas-sensing device has a distinct advantage. The optical-fiber gas sensor is highly specific in detection and it provides a means of eliminating the characteristic poisoning effects from other gaseous species, which can lead such solid-state gas sensors to give false alarms (as from nearby welding fumes).

The development of infrared-transmitting fluoride or chalcogenide glasses is at an early stage, and the glasses will require careful packaging before extended lengths can be used in such difficult environments. For combustion-control applications, however, it is possible that short lengths could be used, provided that the fiber and packaging can withstand—or be protected from—high temperatures. There is little doubt that their application to sensors will be considered.

In the discussion of shutter devices (Sec. 8.2.1), reference was made to an analog output representing displacements that cause the obscuration of an optical beam (by means of a grating or a graded wedge). Compensation for drift due to source intensity variations or losses at couplers (before or after the connection), however, has remained a problem. Several solutions have been proposed, including the use of a dual-wavelength input to the fiber. The shutter can then take the form of a filter (interference or absorption) which alters one wavelength but not the other. Such systems rely on the stability of the filters, couplers, and detectors with respect to wavelength transmission properties over the operating temperature range and the projected lifetime of the system. It is the amplitude of the signals at the defined wavelengths which is eventually measured and, inevitably, there will be associated errors, which must be quantified.[17]

When wavelength-division multiplexing is to be employed in a local-area network, there are two options: selection of a range of narrow-line sources or use of a wideband source [e.g., a light-emitting diode (LED)] in conjunction with expensive optical-grating multiplexing and demultiplexing components. In either case, there are difficulties in injecting the radiation into the fiber and in analyzing the radiation output from the fiber. It is possible that the multiplexing components (e.g., zone plates, prisms, and gratings) can themselves be used to sense position, with the corresponding wavelength shifts being a direct measure of position.[18] This requires detection optics specifically designed to detect small wavelength changes, and their wavelength sensitivity should not drift with time.

A suitable semiconductor element can be inserted between the ends of opposed optical fibers or between a fiber and end reflector to provide a temperature sensor.[19] The absorption edge of the semiconductor should be compatible with the output spectrum of the source and the detector wavelength sensitivity. The shift of the absorption edge can then be measured. Wavelength shifts in the photoluminescence peak of a sensor crystal with temperature can also be detected. Such devices operating over the 0 to 200°C temperature range have been constructed with an absolute accuracy of ±1°C and resolution of 0.1°C. Figure 8.11 illustrates the construction of the Takaoka Electric Manufacturing Co. temperature probe, which is based on this technique.

A fluoroptic thermometer manufactured by Luxtron Corp. consists of a fiber link leading to a rare-earth phosphor, gadolinium europium oxysulfide [$(Gd_{0.99}Eu_{0.01})O_2S$], at the fiber tip, which is excited by an ultraviolet (uv) line near 300 nm. Sharply defined emission peaks occur near 500 to 600 nm. The intensities of the a line at 630 nm and the c line at 540 nm vary in a known fashion. The curve for the ratio of the intensities, as measured by a pair of silicon diode detectors, allows measurements to be taken with a resolution of 0.1°C, from −50 to +300°C. A reference temperature is needed for electronic compensation of drift effects, and some time averaging is required to achieve this level of accuracy. Since uv radiation is used to provide the excitation, and absorption levels in silica fiber are high at such wavelengths, the length of the fiber-optic lead to the sensor is necessarily limited. Other variations on this technique and time decay of different fluorescence phenomena have been proposed.[13]

One of the original targets of the Takaoka photoluminescence temperature sensor was to meet the specialized requirement for measuring transformer oil temperatures without reducing the sensor performance by interference effects. The design is a novel marriage of semiconductor and fiber-optic technologies. Such devices could well be the precursors of other such hybrid designs (see also Sec. 8.2.5). A range of devices has been produced with package diameters ranging from 0.55 to 1.2 mm, to cover different applications. The packaging is designed to be either oil-resistant or chemically inert, or for general-purpose use. A

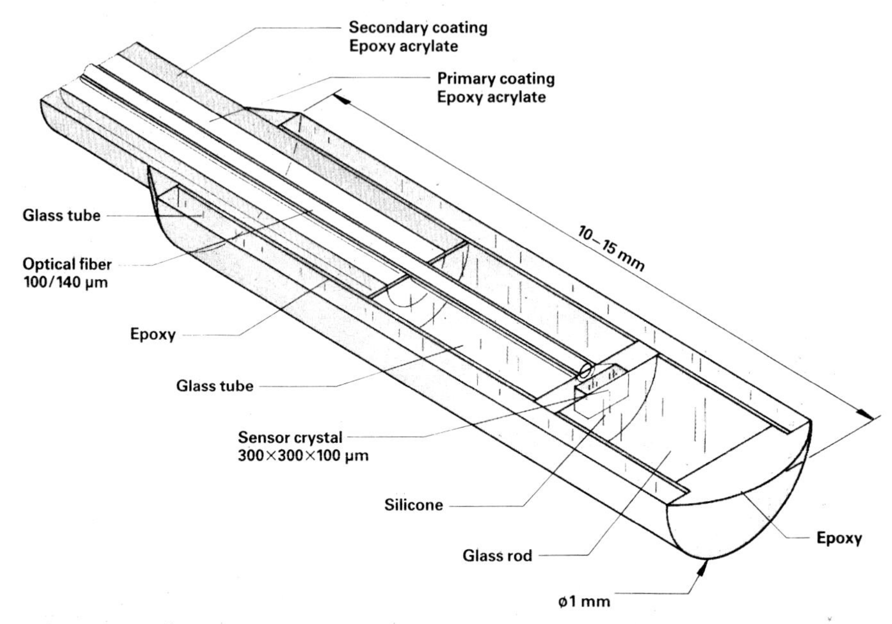


**FIGURE 8.11** Fiber-optic temperature sensor with semiconductor sensing element. *(Courtesy of Takaoka Electric Manufacturing Co., Ltd.)*

change in wavelength of the photoluminescence peaks is measured, putting constraints on the wavelength stability of the special detectors and processing circuitry. Figure 8.12*a* illustrates the type of spectrum analysis that is carried out, and Fig. 8.12*b* is a schematic diagram of the total system. The photoluminescence spectra were recorded for sensor temperatures between 0 and 200°C.

Several fiber-optic blackbody radiation detectors (i.e., pyrometers) have been developed principally for measurement of high temperatures in turbines and nuclear plants. The Land Infrared Ltd. fiber-optic pyrometer (Fig. 8.13) uses a sapphire lens to focus the radiation onto an air-purged (for cooling) fiber-optic bundle. Various options are possible, e.g., using a lens with a length of flexible fiber-optic bundle. A further materials technology development, based on original work by the U.S. National Bureau of Standards, is coating the fiber tip with a metallic emitter for contact temperature sensing. The sensor can be constructed from a sapphire rod (approximately 10 to 20 cm in length) with noble metal coatings to provide the blackbody cavity source. Measurements from 500 to 2000°C are possible, with an accuracy of ±0.005 percent. Measurement of engine gas dynamics in real time (to 10 kHz) has been carried out with such a sensor within an internal combustion engine.

For fiber-optic pyrometric measurements, no fiber-optic source is required, since the radiation being measured is provided by the target object itself. In this case, the peak of the blackbody radiation shifts as a function of temperature; at higher temperatures, the peak shifts to a lower wavelength. It is possible to measure and analyze the total spectrum, but there are alternative techniques whereby it is possible to analyze the outputs at two wavelengths selected by narrowband filters. The advantages in using fiber optics for remote pyrometry are

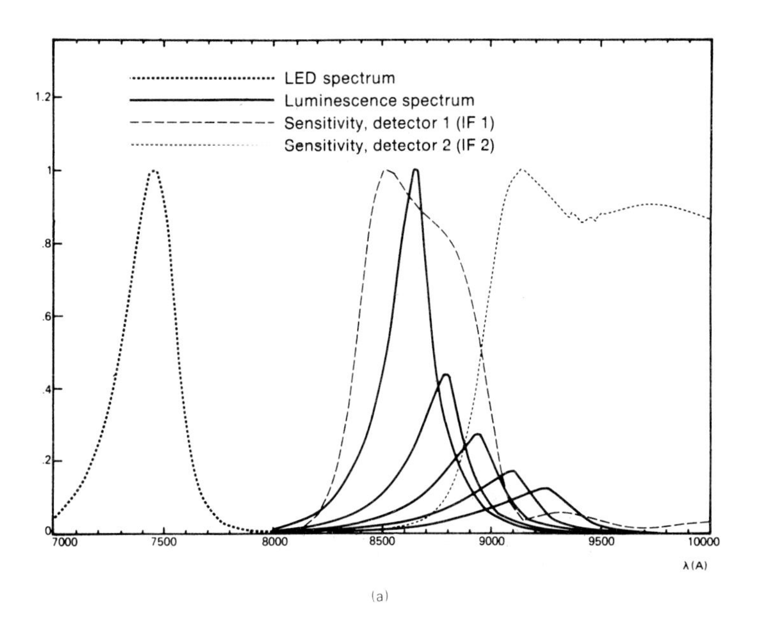


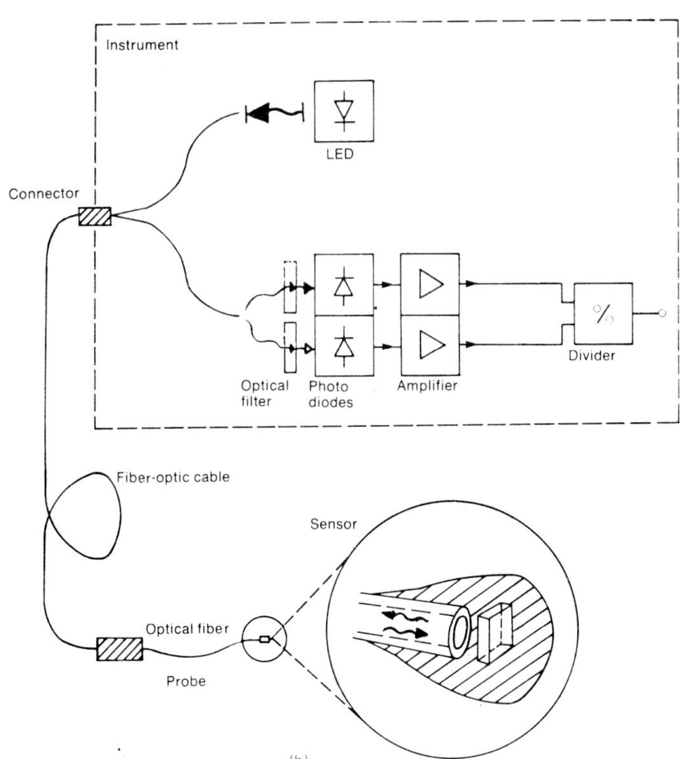


**FIGURE 8.12** Takaoka fiber-optic temperature probes. (*a*) Spectral curves relating to the LED, sensor crystal, and detectors; (*b*) schematic diagram of the associated electronics and fiber. (*Courtesy of Takaoka Electric Manufacturing Co., Ltd.*)

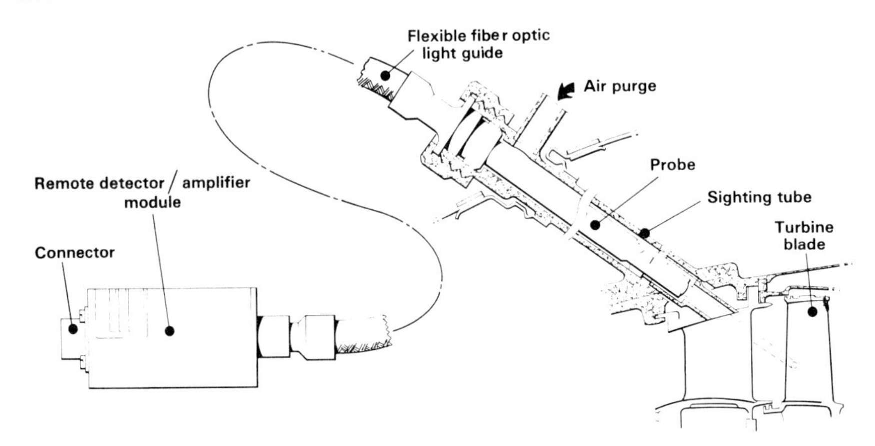


**FIGURE 8.13** Blackbody radiation sensor for remotely measuring turbine blade temperatures. Design drawing shows focusing optics and air purge system to maintain fiber and lens coupling at reduced temperature. (*Courtesy of Land Infrared Ltd.*)

1. The ability to see around corners
2. The ability to measure the temperature of small targets
3. The removal of the detector to a safer and lower temperature zone (which also provides for lower noise)

The ranges over which such devices can operate are partially limited by the near-infrared (ir) transmission characteristics of the silica fiber used. The advent of a suitably robust ir-transmitting fiber will further enhance this technology. Over a typical operating range of 600 to 1100°C, the quoted uncertainty and repeatability depend upon the relative part of the range and the algorithms employed. Higher temperature ranges can be covered, particularly for furnace monitoring. Other design parameters which can affect accuracy are surface emissivity of the material to be measured, the coating, and the efficiency of the probe contact at the sensing point.

Liquid-crystal optical switches which operate within specific temperature ranges are also possible.[20] Such devices can switch from transparency to opacity to visible wavelengths within 0.5°C windows, albeit with relatively slow time constants (e.g., 10 ms). Costs of specially designed, reliable liquid-crystal mixtures and packages, however, can be high. Also, the reliability and stability (i.e., resistance to aging) of liquid-crystal devices are still to be proved in sensing applications.

A range of chemical sensors, responding to refractive index changes in surrounding or interposed liquids, is likely to prove effective for medical applications, e.g., measurement of pH. Several devices could also be classed as evanescent sensors and are described in Sec. 8.3. A summary of typical sensor devices using coated sensors or interposed elements is given in Table 8.4.

### 8.2.4 Encoders

Several designs of optical encoder are based on optical fibers either in transmission or in reflection mode. NASA developed a transmission encoder consisting of a rotating disk with holes around the periphery, interrogated by an array of input

**TABLE 8.4** Summary of Typical Optical-Fiber Sensor Devices Using Coated Sensors or Interposed Elements

| Type | Applications | Range | Comments |
|---|---|---|---|
| Semiconductor (adsorption edge) | Temperature | −50 to 150°C | ± 1°C accuracy typical |
| Photoluminescence | Temperature | 0 to 200°C | 0.1°C resolution; ± 1°C accuracy |
| Fluoroptic (luminescence) | Temperature | −50 to 300°C | 0.1°C resolution, but some averaging required |
| Blackbody | Temperature | 300 to 2000°C | Accuracy and resolution depend on the sensor and the range of operation |
| Liquid crystal | Temperature | −50 to 200°C | ± 0.5°C accuracy; operates as a switch |
| Photoelastic | Acoustic/pressure | 0 to 2 MPa (pressure) | Materials are available with 0.25 MPa/fringe/m of path length to 10 Pa/fringe/m. With glass, the minimum pressure is about 1.4 Pa and the calculated dynamic range is 123 dB |

and output fibers, to give angular or linear displacement.[21] Similar devices have been taken as the basis for an analysis of multiplexing schemes.[22] Several reflective or transmissive on-off sensors or encoders can be multiplexed by separate fibers of differing lengths (Fig. 8.14). The timing and fiber lengths are easily calculated, but the insertion losses and total system need careful assessment. Multiplexing from a single fiber is described in Sec. 8.6 (and illustrated there, in Fig. 8.40). Figure 8.15 is a photograph of a 14-bit encoder developed by GEC Re-

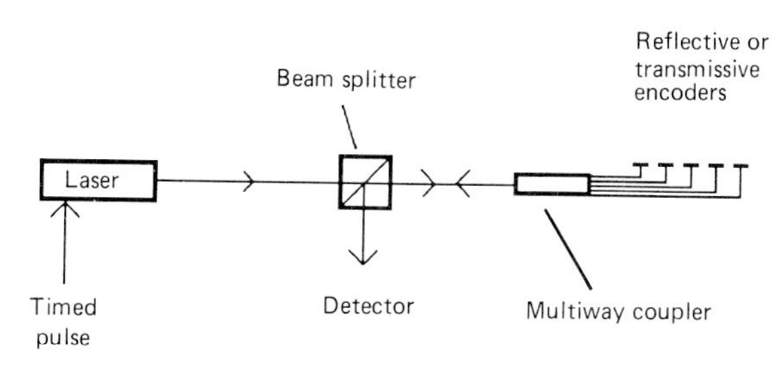


FIGURE 8.14 Schematic of reflective encoder system using fiber optics.

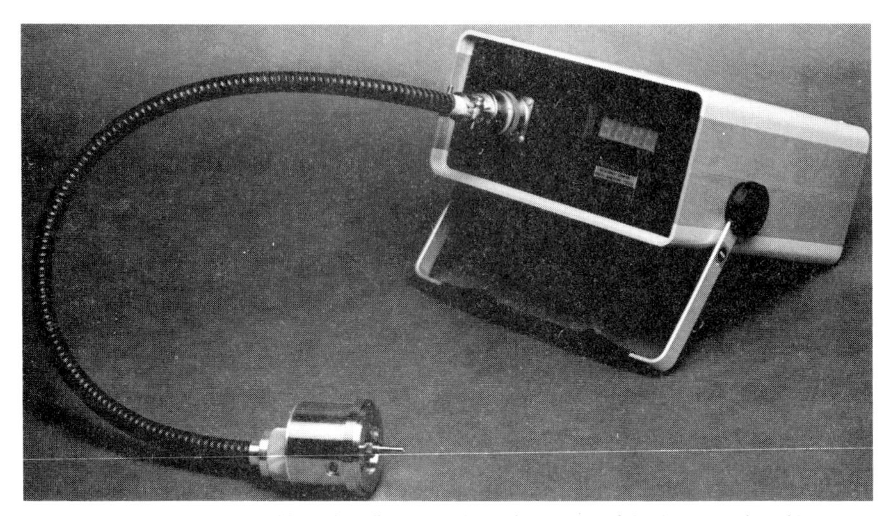

**FIGURE 8.15** Fourteen-bit optical-fiber encoder. *(Courtesy of GEC Research Ltd.)*

search Ltd. Figure 8.16 illustrates a rotating-drum dial encoder for 360° measurement using a reflective coded strip. A single 6-dB discrimination requirement for reliable switching of the electronics for returns from reflective and nonreflective portions was achieved by suitable materials selection, e.g., electroforming the pattern in polished (or bright) nickel and black nickel on a phosphor bronze base. Reflectivity of the black nickel areas can be reduced by etching prior to electroforming. Given the advances being made in avionic "fly-by-light" systems, there will be an increased requirement for such devices, with emphasis on higher angular resolution, possibly in combination with wavelength-multiplexing techniques. Materials selection, both for reflective coding (so that only one fiber cable to the encoder is needed) and for reduced weight combined with ruggedness will be extremely important in new designs.

### 8.2.5 Frequency-out Sensors

From the previous descriptions of extrinsic sensors, it is clear that problems exist with the added complexity and cost of referencing amplitude- and wavelength-dependent systems. These problems are exacerbated when requirements for multiplexing are added and when the system must cope with time and temperature variations of coupler and components. One solution may emerge from the development of transducers directly transmitting frequency-out information, as a function of the strain applied. Following the trend illustrated in the previous sections, the juxtaposition of semiconductor and fiber-optic technologies provides a series of attractive options for reducing the size of sensors and for taking advantage of the high-precision micromachining techniques used in both technologies. Careful mechanical design and packaging of such devices is required to minimize optical losses at the sensing head. With the new semiconductor etching techniques becoming available, alignment of the fibers within appropriately designed micron-sized grooves can be achieved. Table 8.5 summarizes various options for frequency-out sensors.

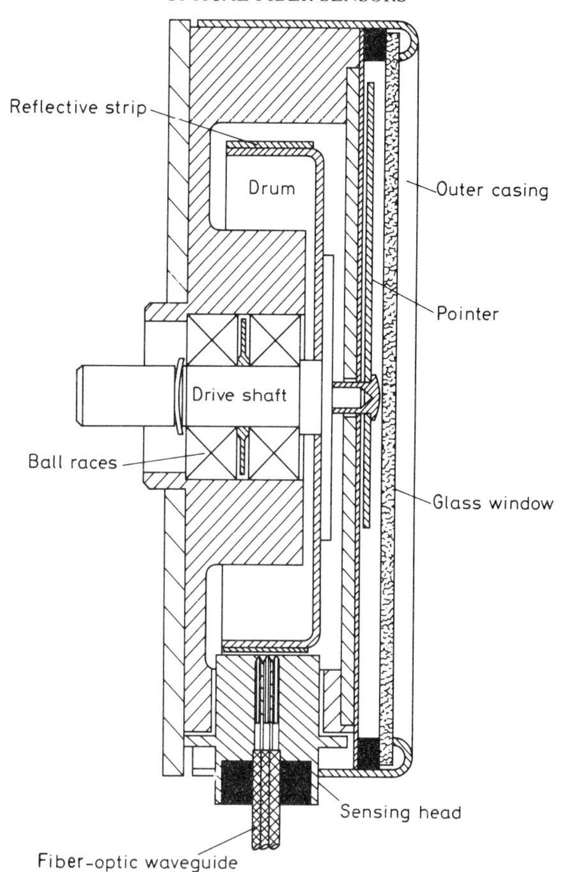


**FIGURE 8.16** Reflective drum dial encoder, with gray-coded strip encoded by a single, 10-fiber cable. *(Courtesy of STC Technology Ltd.)*

The principal types of frequency-out sensors are

1. *Vibrating Wire:* A sensing technique that does not require expensive investment in semiconductor technology and uses available components is to detect the movement of a tensioned, resonantly vibrating wire. Radiation transmitted from a light-emitting diode via an optical fiber and photocell excites the coil of a soft-iron magnet. A wire situated between the magnet poles is made to oscillate at the driven frequency. The movement of the wire is detected by two additional optical fibers. Although three fibers are now used, no doubt single-fiber solutions will emerge. It has been shown that optical powers required to excite the wire into motion are low and therefore meet the safety requirements for numerous process control applications. The frequency range is relatively low (up to 2.5 kHz) and is dependent on the mechanical properties of the diaphragm and wire.[23] Likely applications include remote pressure and differential pressure measurement in hazardous process control areas.

**TABLE 8.5** Types of Frequency-Out Point Sensors

| Type | System | Performance | Comments |
|---|---|---|---|
| Vibrating wire | Three fibers to sensor with matching transformer. LED provides return signal. Taut wire oscillated between poles of an energized magnet. | Electric power for intensity; 50 pW. Reported superlinear to 2.4 kHz for tension to 0.2 N. Relatively low $Q$; high insertion loss. | Practical: uses available components and systems. |
| Quartz | Two fibers to sensor with return signal. Modulated radiation (25 mW input to fiber) adjusted to resonance via phase-lock loop after detection. | Range 1 kg; resolution 0.5 g; $Q$ of 2700 in air. Insertion loss $\approx$ 57 dB. Linear output. | Batch processing of quartz cantilevers possible. |
| Micromachined silicon | Can use two fibers. Single fiber possible. Frequency output is related to strain. Can be attached to diaphragm. Also used as an accelerometer for vibration sensing. | 20 $\mu$W for direct actuation and interferometric sensing of vibration. Higher powers could simplify detection technique. High $Q$ ($\approx$ 2000) is possible. Linearity depends on resonant frequency selected. Mechanical resonance at 1 kHz demonstrated. | Useful for totally passive multiplexing with mixture of FDM, TDM, WDM. Batch processing of etched grooves in silicon can facilitate fiber alignment. Totally optical/mechanical/digital operation. |
| Micromachined GaAs | Two fibers for send and receive; for vibration sensing. | Mechanical resonance at 1 kHz demonstrated. | Good temperature stability. Useful for multiplexing (as with silicon above). Batch processing. Totally optical/mechanical operation. |

*Source:* Adapted from Ref. 3.

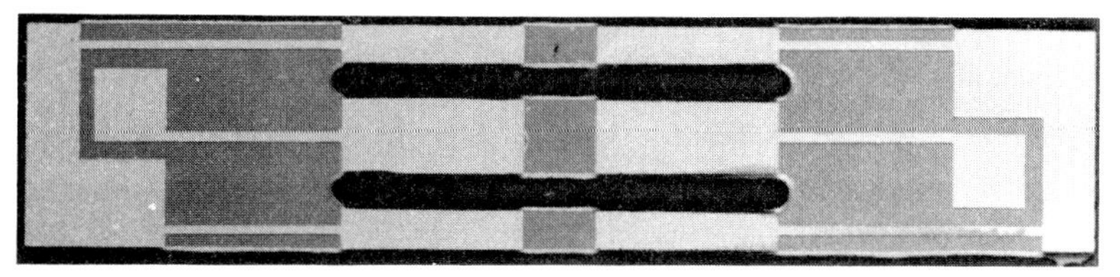

FIGURE 8.17 Photograph of quartz cantilever beam. The resonance frequency varies with applied stress. (*Courtesy of GEC Research Ltd.*)

2. *Vibrating Quartz:* Quartz cantilever beams (Fig. 8.17) can be optically excited into resonance. The frequency varies linearly with axial stress loading. Such devices have been demonstrated and full optical-loss budget calculations have been carried out. Insertion loss for the system illustrated in Fig. 8.18 is

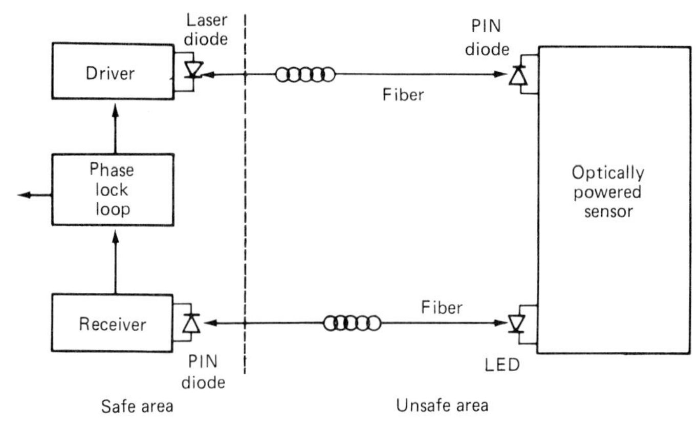


FIGURE 8.18 System diagram for optical actuation and detection of quartz vibrating sensors. (*Courtesy of GEC Research Ltd.*)

typically 57 dB. Adequate mean power levels of typically 2.5 mW can be supplied by solid-state sources through 200-μm core fiber to the optically powered electronics. The pulsed light is received by a photodiode connected to the primary coil of a transformer. The beams of the quartz crystal are then forced into oscillation at the applied resonant frequency, which is returned to the transmitter via a lower (e.g., 3 nW) optical signal along the return optical fiber. Under load, the resonant frequency changes and a phase-locked loop at the base is used to alter the frequency of the transmitter. This error signal is used as a measure of the force applied. Devices made by GEC Research Ltd. have a measurement range of 1 kg and a resolution of 0.5 g. Resonant frequencies tend to be near 15 kHz and the devices can have a $Q$ of about 2700 in air. With this system, as described, electronics are still required at the sensor. However, the possibility that such devices can also be driven optically, as discussed below for micromachined silicon, should not be ignored.

3. *Semiconductor Devices:* Selectively etched gallium aluminum arsenide (GaAlAs) devices are available for temperature sensors (see Sec. 8.2.3). Etched semiconductor and silica beam devices operating in reflective mode

have also been developed for vibration sensors.[24] Such devices may dispense with local powering at the sensing head to provide initiation of the vibration; i.e., the sensor is totally passive. The Takaoka device in Fig. 8.19 has been designed for monitoring the acceleration of mechanical components in high-voltage environments. The sensor chip is excited by a near-ir LED and returns two optical signals of differing wavelengths along the same fiber. One is the measurement signal and the other is the reference provided by photoluminescence from a piece of neodymium-doped glass situated at the sensor head. The acceleration frequency range of the device is 5 to 1000 Hz ($\pm 10$ percent), with a root-mean-square (rms) resolution of 0.1 $ms^{-2}$. The dimensions of the package at the end of the fiber are 17 × 14 × 9.5 mm. It measures primarily in one orientation, so careful mounting is necessary. The device is sensitive to temperature changes of 0.5 percent per °C or less.

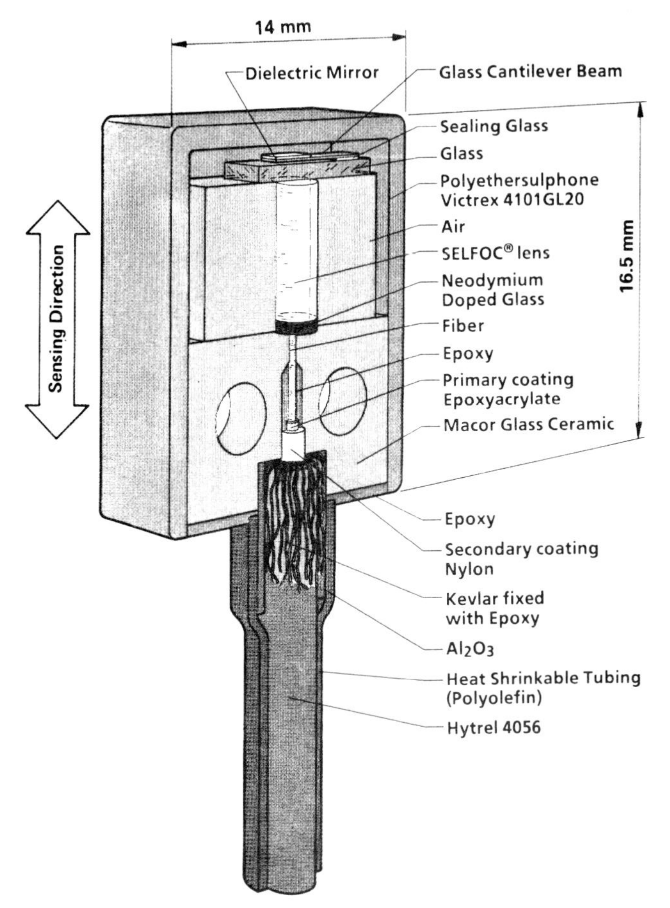


**FIGURE 8.19** Diagram of a vibration sensor using a vibrating cantilevered beam and wavelength referencing techniques. (*Courtesy of Takaoka Electric Manufacturing Co., Ltd.*)

Optically actuated micromachined silicon devices (i.e., no local photodetector circuit) have been demonstrated with special thermally absorbing surfaces on miniature (10-μm-long) silicon or silica bridges.[24,25] Thermal strains can be induced sufficiently to produce 10-nm displacements with optical powers of less than 20 μW. However, at these displacements, costly interferometric detection techniques are required, and higher powers are needed to provide sufficient deflection for simple amplitude variations to be detected. For single-fiber operation, there is an inherent conflict in that the surface of the deflecting element should be designed to be simultaneously reflective and also absorbing (to provide the thermal strain). The mechanical excitation of these silicon structures can be explained by photoacoustic effects. The efficiency of the mechanical design and photoacoustic coupling, as well as the contributions of thermal and electronic strain effects, are complex subjects. Further effort is needed before satisfactory devices can be practically realized, and the potential aging and failure mechanisms will have to be identified.

Figure 8.20 shows a magneto-optically actuated scanning device. A single multimode fiber transmits radiation to a polished, micromachined silicon paddle (1 mm × 1 mm area), suspended by 10- to 20-μm tensioned cantilevers.[3] Smaller reflectors and structures are possible. Light is reflected from the etched surface of the paddle onto a photoelectric cell coupled with a suitable capacitor (for delay). This can provide sufficient electric power (2 μW) to energize a small coil magnet, which attracts a small magnetic strip attached to the back of the paddle. The mirror is then deflected so that the radiation misses the photodetector, and the coil is no longer energized. As a result, the paddle is forced into oscillation. Given the excellent mechanical properties of silicon, little signal degradation is observed, at low-frequency operation, with time. Mechanical shock, however, can have a deleterious effect, so careful package design is necessary. Scanning

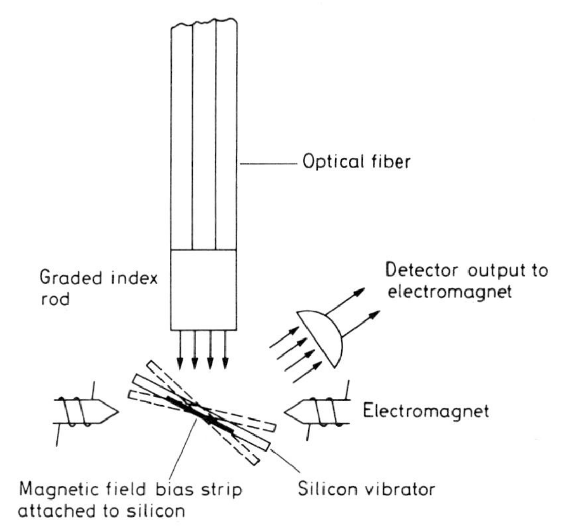


**FIGURE 8.20** Magneto-optical actuation of a micromachined silicon vibrating element with a remote photoelectric cell and a small permanent magnet attached to the silicon. *(From Ref. 3.)*

applications are possible over large angles (e.g., 270°) without failure, and at reproducible scanning rates (adjusted by suitable delay circuitry at the photodetector). Angular rotational scans beyond 360° have not met with success. As with all material sensors operating in a vibrational mode, it is vitally important that a full failure-mode analysis be carried out.

### 8.2.6 Birefringent Modulation

Birefringent changes in optical fibers, particularly those present in single-mode interferometric devices, can provide added unwanted cross-perturbing effects in any measurement scheme. For the more sensitive applications, the extent of the problem has led to the design of special polarization-preserving fibers, where the birefringence difference in two orthogonal directions is maintained within the fiber at a high value. However, the insertion of birefringent elements between fibers can also be used for sensing purposes. Birefringent sensor techniques are conveniently divided into those using the photoelastic effect and those using electro-optic effects.

The photoelastic effect is used routinely by mechanical and civil engineers for assessing stresses in model structural systems. Transparent plastic models can be constructed and the transmission of light through the model can be observed with crossed polarizers under different structural loading conditions. Stress-induced birefringence generates fringes, the obscuration effects of which can be analyzed. A fiber-optic photoelastic system[26] designed for hydrophone applications is illustrated schematically in Fig. 8.21. Stress is applied to the birefringent element via a diaphragm. The induced birefringence causes the two orthogonally polarized components of the incident beam to travel through the material, shaped to enhance the birefringence effect at the measurement point, at different velocities. With the analyzing optics, the fringe intensities can be measured to provide a measure of the stress. Unfortunately, the more sensitive materials, such as polyurethane, suffer from temperature sensitivity and creep problems. Stiffer ma-

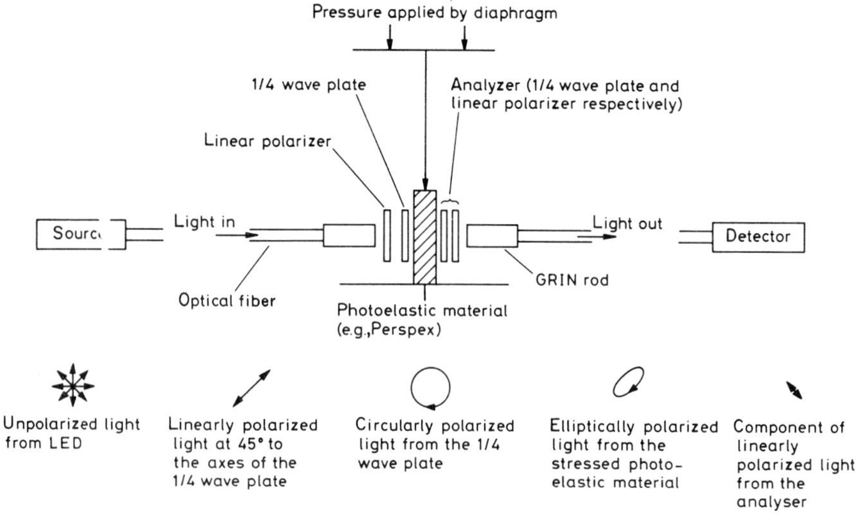


FIGURE 8.21 Schematic diagram of photoelastic sensor. (*From Ref. 3.*)

terials, e.g., polymethyl methacrylate, have been demonstrated to provide sufficient fringe resolution to a pressure of 2 MPa. Mechanical design to maximize stress transference in hydrophones, and still maintain sufficient sensitivity at depth, has proved to be important. With careful mechanical design and differential electronic processing techniques, good sensitivities have been obtained. Data on this kind of device are compared with those for interferometric hydrophone devices in Sec. 8.4. Photoelastic devices have also been suggested for use as accelerometers.[27]

Electro-optic elements may also be interposed between optical fibers for measuring current and voltage in power systems.[6,28] Optical designs are similar to those for the photoelastic effect; i.e., they use crossed polarizers and include a quarter-wave plate to produce a $\pi/2$ retardation and enhance linearity of response. Detailed mechanical design is necessary to maximize outputs from such devices when they are attached to the bus bar. Measurements have been carried out on gas-insulated systems having suitable fiber-optic penetrations. A range of Pockels and Faraday electro-optic materials is available. A bismuth germanium oxide ($Bi_{12}GeO_{20}$) has been preferred over lithium niobate ($LiNb0_3$) for voltage (Pockels) measurement because of its relatively low temperature coefficient, i.e., less than $\pm 0.5$ percent from $-25$ to $+85°C$. For the Faraday effect, flint glass has been selected by several workers because of its low temperature coefficient.

The responses of different Pockels and Faraday effect materials are collected in Table 8.6 with their relevant figures of merit and equations. The Faraday effect, as observed within the fiber itself, has also been used as a means of measuring magnetic field (see Sec. 8.4.2). It is found that both the Kerr and Faraday effect within single-mode fibers provide sources of noise, or unwanted bias, in such highly sensitive phase measurement interferometric systems as the fiber-optic gyroscope (Sec. 8.4.1).

## 8.3 EVANESCENT SENSORS

Evanescent sensors are those in which guided light is lost to the external environment, which is being sensed, from a stripped or specially coated silica fiber core. There are other "guiding" transfer effects, such as from core to cladding and from core to core in step-index and multicore fibers, respectively. Since light is not lost from the total fiber structure, these latter effects are classed as intrinsic sensor effects, and they will be described in the following section.

Plastic-clad silica fiber can be stripped of its polymer coating so that the bared core can be exposed to external environments. When the bared core is dipped into a water stream, for example, the light is retained within the fiber, since the refractive index of water allows it to act as an effective cladding. However, when oil in the water attaches itself to the bared fiber, there is an attenuation of the signal in the fiber.[30] This effect has been shown to provide a measure of the oil content in water, and was at one time considered a viable candidate for pollution-control instrumentation. It was shelved because of doubts about the robustness of a bared fiber in a fast-flowing and dirty process stream. It was found that, with time, the surface quality of the sensor deteriorated; i.e., surface microcracks developed. Nevertheless, this type of approach is still valid for several kinds of sensor, provided that any one of the following conditions prevails:

**TABLE 8.6** Polarization Modulation Materials and Effects

| Pockels materials—voltage sensors | | | |
|---|---|---|---|
| Material | Electro-optic coefficient, $10^{-1}$ cm/V | Wavelength, μm | Refractive index |
| ZnSe | 2.0 | 0.5–0.6 | $n_0 = 2.66$ |
| CdTe | 6.8 | 1.06 | $n_0 = 2.60$ |
| $Bi_{12}GeO_{20}$ | 3.22 | 0.666 | $n_0 = 2.54$ |
| $Bi_{12}SiO_{20}$ | 4.35 | 0.87 | $n_0 = 2.45$ |
| $LiNbO_3$ | 3–30 | 0.633 | $n_0 = 2.286$<br>$n_e = 2.20$ |
| $LiTaO_3$ | 2–30 | 0.633 | $n_0 = 2.176$<br>$n_e = 2.180$ |

Faraday materials—current and magnetic field sensors

$$\phi = V_r HL$$

where
$\phi$ = Faraday rotation angle
$V_r$ = Verdet constant
$H$ = AC magnetic field produced by the current in the line
$L$ = optical path length through the crystal

| Material dependence | Verdet constant | Wavelength, μm | Temperature variation, % (−20 to +80°C) |
|---|---|---|---|
| Flint glass | 0.04 | 0.85 | $< \pm 0.5$ |
| $As_4S_3$ | 0.10 | 0.90 | $< \pm 1$ |
| ZnSe | 0.21 | 0.82 | $\approx \pm 1$ |
| $Bi_{12}GeO_{20}$ | 0.188 | 0.85 | $\approx \pm 1.5$ |
| FR-5 glass | 0.11 | 0.85 | $\approx \pm 1.5$ |
| YIG* | 9.0 | 1.3 | $\approx \pm 8$ |
| $(Tb_xY_{x-1})IG$ | 15.6 | 1.15 | $\approx \pm 1.5$ |

*Yttrium iron garnet.
*Source:* After Ref. 29.

1. The medium to be investigated remains clean.
2. Methods of surface cleaning can be incorporated into the design.
3. The device can be made so cheaply as to be disposable after a limited number of sensing operations.

Specialized coatings can be applied such that, for example, the refractive index varies with temperature or in the presence of various chemically adsorbed or absorbed species.

Evanescent sensors lend themselves to some potentially simple multiplexing schemes, when combined with OTDR techniques. Several examples of evanescent sensors are listed in Table 8.7. Two examples—level sensing and leak detection—are briefly discussed here.

Several level-sensing designs have been advanced for optical-fiber sensors using "transparent" optical sensing tips (e.g., corner cubes) at the end of the fiber. Several companies, including Hughes Aircraft and Honeywell in the United States and Delta Controls Ltd. in the United Kingdom, market such sensors. The

**TABLE 8.7** Examples of Evanescent Sensors

| Sensor type | Description | Application |
|---|---|---|
| Liquid level | Uses corner cubes or shaped fiber ends to enhance extinction ratios between the fiber end and the liquid to be detected. Provides on-off switching devices, although analog outputs of two phase concentrations in flow have been attempted. | Remote level sensing of clean, white petroleum products. Battery level, specific gravity. |
| Liquid absorption temperature | Immersion in thermochromic (e.g., cobalt salt) solutions or dye indicators. | Medical |
| pH | Refractive index change with pH. | Medical |
| Temperature | Refractive index change of polymer coating produces loss of guidance at low temperatures; interrogation by optical time-domain reflectometry. | Leakage detection in refrigeration plant. |

Note: Many of these devices, with their small package size and inert nature, can be used for *in vivo* medical measurement and drug process control. The potential of such devices for cheap, disposable medical sensors is discussed in Sec. 8.8.

shaping and selection of the optimum refractive index and materials for the sensing tip are important for

- Maximizing the detector output difference between total internal reflection and loss of guiding (i.e., ensuring that the refractive indexes of the sensing tip and the liquid to be sensed are optimized)
- Eliminating unwanted foreign material or extraneous coatings on the sensor end

Other factors are the cost of couplers and referencing the system. For example, fiber-optic sensor systems have been proposed for level monitoring and splash control for hazardous road-tanker cargoes. However, it has proved difficult to develop a sufficiently cheap and robust coupler to cope with, for example, repeated use and adverse treatment when a road tanker is connected to and disconnected from its cab. Figure 8.22 shows different variations on this fiber guidance theme for liquid level measurement. By shaping the fiber end, or by providing the optimum radius of curvature for the fiber, the extinction ratio can be maximized.

Changes in refractive index of the surrounding materials can also then be measured. This can result in an analog output refractometer (as distinct from the on-off level switch) with which, for example, hydrocarbons have been detected.[31] The analog hydrocarbon detector uses optical fibers with special oleophilic coatings. The fiber refractometer technique has also been applied to the measurement of specific gravity of storage batteries.[32]

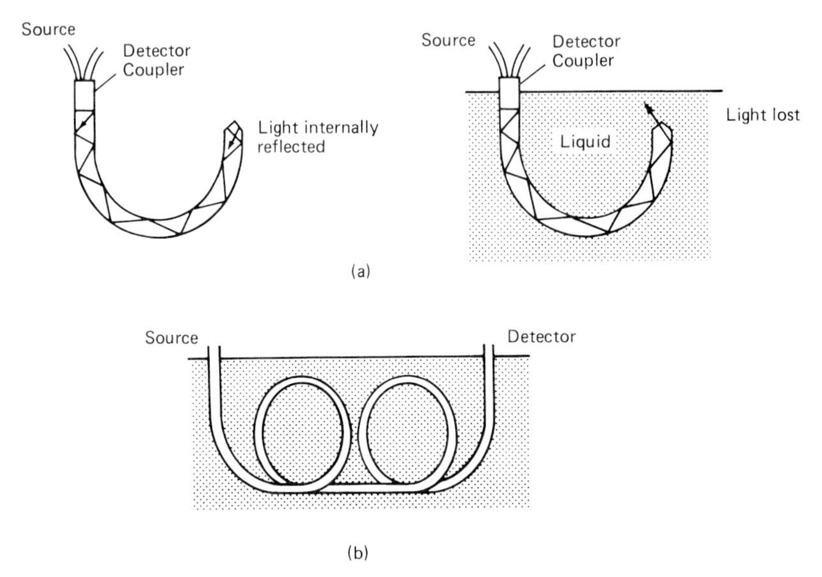


**FIGURE 8.22** Level sensing and refractometer measurement using shaped optical waveguides. (*a*) Shaped level sensor with inverted tip (to inhibit droplet retention); (*b*) coiled sensor, which can also be used for refractive index measurement.

In the leak detection example, evanescent light loss from specially clad fibers is used to detect liquid nitrogen leaks in an industrial environment. In this case, the refractive index of the polymer coating of a plastic-clad silica-fiber cable reaches that of the fiber core (at low temperature) and light guidance is lost. This provides a signal, by means of OTDR, that is equivalent to that of a fiber break. Distributed OTDR sensor systems are described in Sec. 8.4.4.

## 8.4 INTRINSIC SENSORS

### 8.4.1 Interferometric and Phase-Modulated Sensors

***Effects of Pressure and Strain.*** Most of the early evaluations of the sensitivities of interferometric sensors were carried out using a Mach-Zehnder interferometer configuration. As a research tool, the Mach-Zehnder interferometric system has proved useful, since the environmentally sensitive arm of the interferometer can be compared with an undisturbed reference arm, and it is possible to separate the contributions due to various environmental factors, e.g., strain and temperature. Also, it has been useful for investigating the effects of different fiber coatings for enhancing or nulling a fiber's sensitivity relative to a range of variables. Much of this original work was carried out in the United States by the Naval Research Laboratory in development of highly sensitive underwater hydrophones and arrays.[1]

Figure 8.23 shows a more detailed schematic diagram of a fiber-optic Mach-Zehnder interferometer. A stable, single-longitudinal-mode laser source, with low phase noise, is used. The sensing arm is exposed to the environment, and the

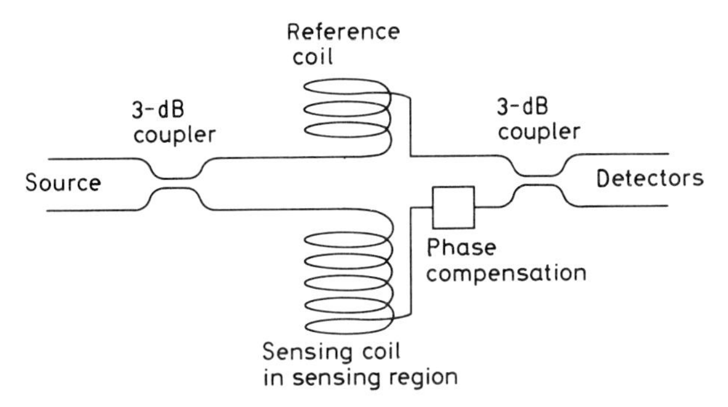


FIGURE 8.23 Schematic diagram of an all-fiber Mach-Zehnder interferometric sensor with phase compensation in the reference arm. (*From Ref. 3.*)

reference arm of approximately equal length is maintained in a controlled environment. Phase compensation (or modulation) in the reference arm is used to provide the necessary adjustment so that both arms are 90° out of phase (i.e., to achieve the quadrature condition, which maximizes sensitivity). Most of the subsequent fiber-optic sensor development on Mach-Zehnder devices since 1980 has been analysis and reduction of noise sources throughout the system. This has led to the development of improved semiconductor laser stabilization techniques (to be described in Sec. 8.5) and the emergence of all-fiber couplers, as distinct from the noisy bulk optical-beam splitters used in early tests.

The different single-mode fiber strain parameters and typical phase-shift magnitudes are gathered in Table 8.8. It is seen that the longitudinal dimensional strain has the greatest effect on the phase, while thermal effects are also high. The refractive index changes originate from hydrostatic pressure and birefringence changes within the fiber core. The contribution of the wavelength dispersion term ($\partial\beta/\partial D$, where $D$ is the fiber core diameter) to the phase shift is negligible.[33]

The strain effect on the fiber can be enhanced or reduced for measurement of a given environmental parameter by the application of appropriate coatings to the cladding of the fiber (Table 8.9). Some adjustment can also be made by careful mechanical design of the winding of the sensing and reference arms and the relative position and orientation of both in response to variables, e.g., temperature gradients. Thus, in the hydrophone it is possible to maintain the device's dynamic range and reduce hydrostatic pressure effects by careful packaging and mounting of both arms in the remote environment. As described below, different magnetostrictive coatings can also be applied to magnetic sensors, to enhance the strain effects in the presence of an applied magnetic field. However, this latter area is extremely difficult to model because of nonlinear magnetization phenomena.[41] Further practical data are needed to increase the confidence level that long lengths (i.e., kilometers) of sensitized fiber can be consistently produced at low cost.

Alternative technologies have been proposed to enhance the phase shifts occurring within the optical-fiber core. These can include doping of the fiber itself, or providing composite polymer, or adding polymer or metal coatings (Table 8.9). For the design engineer there must be reservations to such an approach, given the likely costs and technical risks involved in special fiber preparation. Special-

**TABLE 8.8** Phase Modulation Calculations Based on Strain Applied to Silica Fibers

| Strain | Formula | Phase change at 0.633 μm, rad |
|---|---|---|
| Axial | $\Delta\phi = \beta \left\{1 - \frac{n^2}{2}[\rho_{12} - \gamma(\rho_{11} + \rho_{12})]\right\}_3$ | 11.4 |
| 2-D radial pressure (fiber axially unconstrained) | $\Delta\phi = \beta \left\{\frac{2\gamma}{1-\gamma} + \frac{n^2}{2}\left[(\rho_{11} + \rho_{12}) - \frac{2\gamma\rho_{12}}{1-\mu}\right]\right\} \frac{1-2\mu}{E} \Delta P$ | 11.7 |
| 2-D radial pressure (fiber axially constrained) | $\Delta\phi = \beta \left[-\frac{n^2}{2}(\rho_{11} + \rho_{12})\right] \frac{(1-\mu)(1-2\mu)}{E} \Delta P$ | −6.4 |
| 3-D hydrostatic pressure | $\Delta\phi = \beta \left[-\frac{n^2}{2}(\rho_{11} + \rho_{12})\right] \frac{1-2\mu}{E} \Delta P$ | 3.9 |
| Thermal | $\Delta\phi = \beta \left(\frac{1}{n}\frac{dn}{dt} + \alpha\right) \Delta T$ | 9.1 |

Definitions

| Parameter | Symbol | Value at 0.633 μm |
|---|---|---|
| Optical-fiber length | $L$ | 1 m |
| Propagation constant, $2\pi\bar{n}/\lambda$ | $\beta$ | |
| Young's modulus | $E$ | $7.29 \times 10^{16}$ N/m$^2$ |
| Poisson's ratio | $\mu$ | 0.17 |
| Applied pressure | $\Delta P$ | $10^5$ N/m$^2$ |
| Applied strain | $\epsilon_3$ | $10^{-6}$ |
| Photoelastic constant | $\rho_{11}$ | 0.21 |
| Photoelastic constant | $\rho_{12}$ | 0.27 |
| Effective refractive index | $\bar{n}$ | 1.458 |
| Temperature change | $\Delta T$ | 1°C |
| Thermal expansion coefficient | $\alpha$ | $0.45 \times 10^{-6}$/°C |
| Temperature dependence of $\bar{n}$ | $d\bar{n}/dT$ | $8.5 \times 10^{-6}$/°C |

**TABLE 8.9** Examples of Sensor Coatings Applied to Optical Fibers*

| Coating | Sensor applications |
|---|---|
| Polymers of varying thickness[34] | Acoustic |
| Metallic coatings[35] | Acoustic, magnetic, temperature, acoustic |
| Nickel,[36] metallic glass (magnetostrictive)[37] | Magnetic |
| PVDF (radially poled)[38] | Electric field |
| Aluminum[39] | Temperature |
| Palladium[40] | Gas |

* Other examples are possible; plastic-clad silica fiber, for example, can be stripped and sensitized to different or specific chemical species over limited lengths (e.g., for disposable medical applications).

ized optical fibers with good birefringence control, e.g., high birefringence with a deliberately induced strain within the fiber itself (for polarization preservation along a length) or very low birefringence for electrical current measurements, are described in Sec. 8.4.2.

Figure 8.3 outlines some optical-fiber interferometers. All require single-mode optical fiber and costly components and stabilization techniques. Hence, such devices may be appropriate only for specific, non-cost-sensitive applications. It is appropriate, therefore, that the rest of the description in this section should be "application-driven," and specific uses of interferometric devices for acoustic, magnetic, and gyroscopic purposes will be described. The acoustic application is used first to describe, in greater detail, the design of Mach-Zehnder and Fabry-Perot interferometers (including partially reflecting devices).

*Acoustic Sensors.* Most applications for acoustic fiber-optic sensors are aimed at underwater hydrophones. Their advantages can be summarized as follows:

- *Lightweight:* They can potentially be rapidly and easily dispensed from small vessels.
- *Passivity:* They do not inject energy into their environment and can be configured to be acoustically transparent.
- *Geometrical versatility:* Special windings (e.g., gaussian) can allow for direct beam forming or extended windings can reduce noise in turbulent flow (e.g., in a towed acoustic hydrophone array).
- *High sensitivity:* Potentially, long lengths of sensitized fiber can be produced.

*Mach-Zehnder.* The design of a Mach-Zehnder interferometer has been briefly described, and this geometry was used for the first demonstration[42] of acoustic sensors. Measurement limits are set by the ability to detect $10^{-5}$ to $10^{-6}$ rad of phase difference between the interferometer arms. This corresponds to an optical path length change of $10^{-11}$ m. A comparison plot of the acoustic sensitivities for different fiber-optic sensors is given in Fig. 8.24. Based on theoretical estimates for different lengths of fiber, it is seen that noise equivalent to sea state zero (see caption of Fig. 8.24 for explanation) is theoretically attainable with a sensor having the Mach-Zehnder configuration. However, because of difficulties in compensating for laser noise (to be described later), it has proved necessary on occasion to accurately match the lengths of interferometer arms.[43]

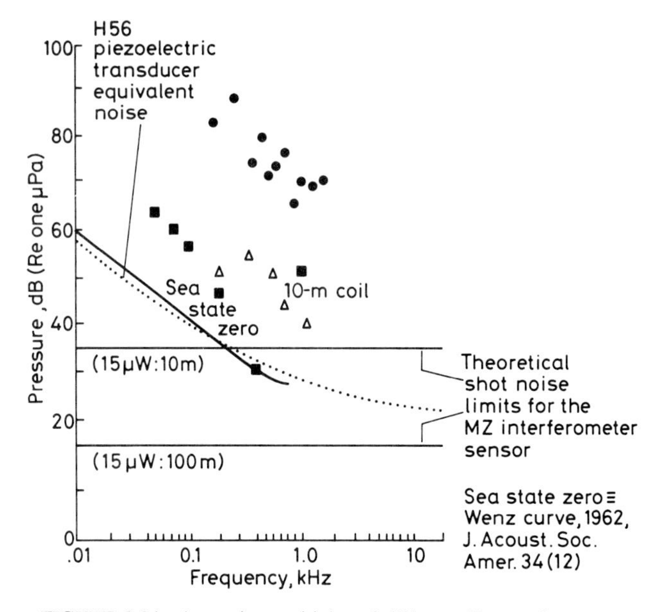


**FIGURE 8.24** Acoustic sensitivity of different fiber-optic sensors relative to sea state zero, where • represents a microbend sensor, ■ represents a photoelastic sensor, and △ represents a 10-m coil Mach-Zehnder sensor. (There are numerous values given for sea state zero in the sensor literature. We have taken the absolute minimum data; very often this cannot be measured because of the high ambient, such as that in harbors. Sea state zero is defined by the Wenz curve; see *Journal of the Acoustical Society of America,* vol. 34, no. 12, 1962, pp. 1936–1956.) *(From Ref. 3.)*

There are problems in determining the phase difference with the cosine-squared fringe pattern obtained from the Mach-Zehnder output. A detection scheme is necessary to provide linearity and to maximize the sensitivity. Figure 8.25 illustrates the shape of typical intensity fringes obtained from the output fibers at the second coupler of the Mach-Zehnder interferometer. As the path length—and hence phase difference—changes under the influence of the acoustic field, so the intensity changes. It is clear that, for maximum sensitivity, as described earlier, it is necessary to balance the interferometer at quadrature. A further problem occurs when the fringe contrast fades simultaneously for both arms of the interferometer. Fading is further discussed below.

Methods of detection can fall into two broad categories:

*Active*—where electrically driven modulating components are acceptable in the design of the interferometer for inclusion at the sensing head. Lengths of fiber wound around piezoelectric devices in different configurations can give phase modulation, phase compensation, and frequency modulation. The high sensitivity and geometrical flexibility of fiber optics for magnetic sensors and gyroscopes are such that the incorporation of such components is often acceptable. However, for the hydrophone application, the passivity remains important, and active components at the sensor head are not considered an attractive option.

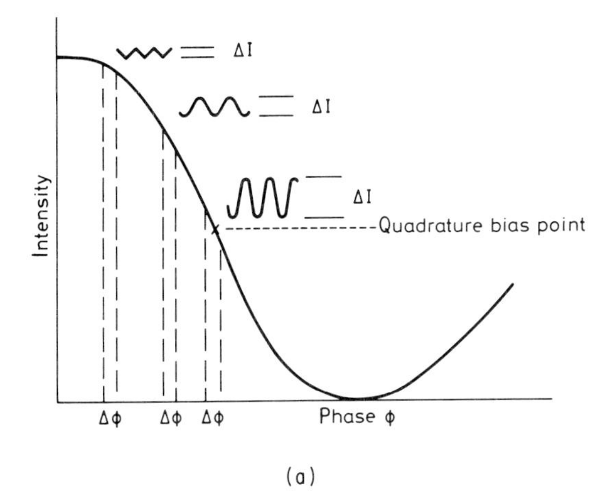


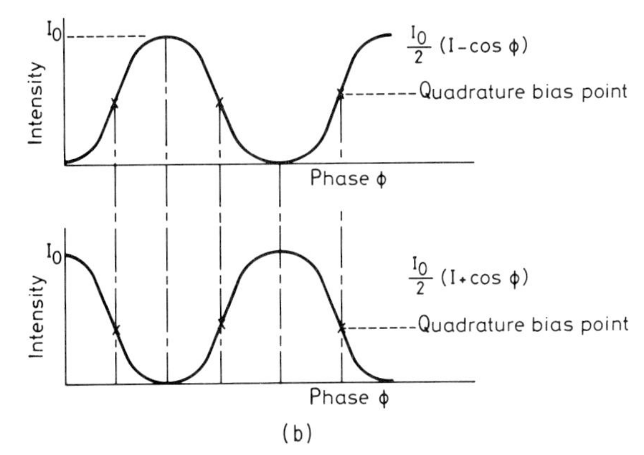


**FIGURE 8.25** Illustration of the requirement to balance the interferometer at quadrature to obtain maximum sensitivity. (*a*) Intensity modulation is largest when the interferometer is set at quadrature. (*b*) Both signals are in antiphase; both channels fade simultaneously. $\Delta\phi$ is the modulation in $\phi$ due to external signal. (*From Ref. 3.*)

*Passive*—where optical demodulation techniques are used. Special coupler designs can be used (e.g., 3 × 3 couplers) whose specific output phase relationships can be analyzed to provide the necessary information for detection.[44] These and other specialized optical devices for fiber-optic sensors are described in Sec. 8.5.

For interference of the beams, the two output optical signals should have identical polarization states. A deterioration in this situation will cause fading (i.e., reduction in the contrast ratio of the fringes). For acoustic sensors to be used for sensing noise underwater, the state of polarization can vary in the fiber with strain, pressure, and temperature as a result of birefringence changes. If the polarization states drift sufficiently in both arms, they could become orthogonal and total signal fading will occur. High-birefringence fiber (i.e., polarization-preserving fiber) can be used to retain the polarization state.

While the Mach-Zehnder interferometer has provided a very useful test bed in the laboratory for determining sensitivities of acoustic sensing devices, relatively few practical design data are available for systems which have been tested for the underwater hydrophone application. The Naval Research Laboratory originally designed a Mach-Zehnder hydrophone in which the single-mode laser source and the sensing and reference fibers were placed in the hydrophone at the end of an extended electrical cable. Successful results with high sensitivities were claimed.[1] Other laboratory water tank tests have been carried out with devices where only the sensing coil was placed in the acoustic field and the reference coil and electrical modulation were situated in a protected environment. While this configuration is useful, in that the modulation is more accessible and can be readily controlled, there are problems in noisy down-lead sensitivity of the fiber-optic cable to the sensing arm.

Alternatively, both reference and sensing arms can be situated in the hydrophone capsule and passive optical detection techniques used. In this case the cable to the sensor carries only optical signals. Careful design of the packaging is required to maximize the sensitivity of the sensing coil relative to the reference arm.

Figure 8.26 shows the scheme of the remote passive hydrophone made by STC Technology Ltd., and Fig. 8.27 shows the mechanical configuration within the hydrophone capsule. Essentially, the two-compartment structure containing both reference and sensing mandrels allows for compensation of hydrostatic pressure effects. This type of development again illustrates the importance of mechanical design in the enhancement of signals and the nulling of extraneous influences in optical-fiber sensors. For Mach-Zehnder hydrophones, several specialized optical components are necessary. When the system is extended into

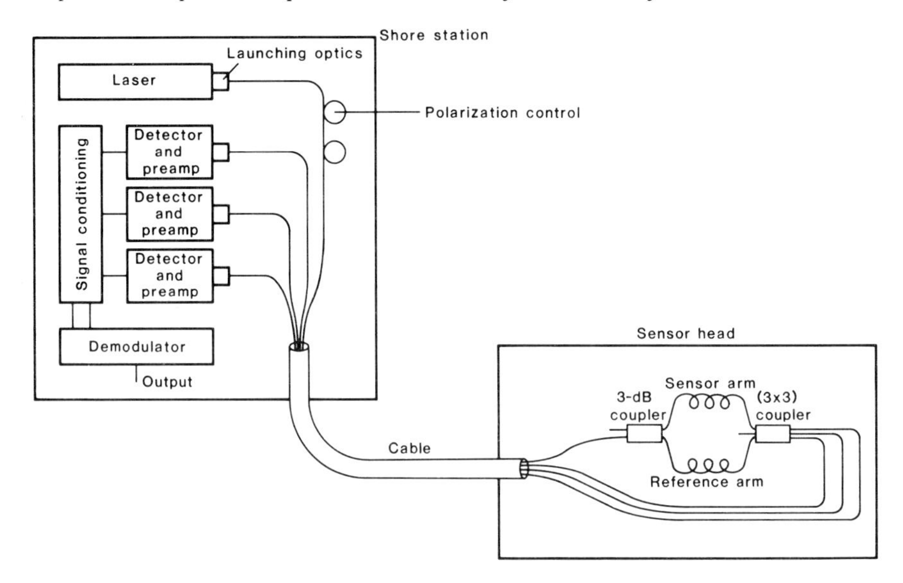


**FIGURE 8.26** Schematic layout of the STC remote Mach-Zehnder hydrophone. *(Courtesy of STC Technology Ltd. and Admiralty Research Establishment, U.K.)*

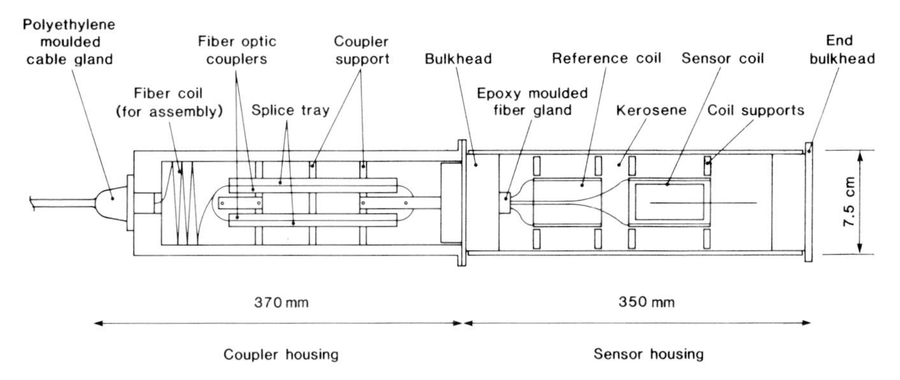


**FIGURE 8.27** Mechanical design of the STC hydrophone. (*Courtesy of STC Technology Ltd. and Admiralty Research Establishment, U.K.*)

many acoustic sensors and arrays, the number of components (e.g., couplers) multiplies. Several proposals for the design of Mach-Zehnder arrays have been advanced in conceptual form, and these are briefly described in Sec. 8.6.

*Fabry-Perot.* Variations on the Fabry-Perot interferometer configuration have been advanced for acoustic sensors, since the system offers some potential for simplified multiplexing. A reduction in the number of optical beam-splitting or recombining devices is possible. Experimental all-fiber Fabry-Perot sensors have been demonstrated, where the fiber ends themselves have been made partially reflecting by means of a silver coating.[45] Other, earlier designs have used partially reflecting mirrors external to the fiber. A Fabry-Perot fiber sensor is a multipass device with several reflected traverses backward and forward along the fiber. With high-reflectivity fiber ends there is a sharpening of the Fabry-Perot peaks, but there is also a commensurate increase in sensitivity to other interfering sources, e.g., thermal effects. External Fabry-Perot cavity point sensors, with micromachined silicon reflectors linked to the optical fiber,[46] are also of interest, although these should perhaps be classed as extrinsic sensors.

A variation on the Fabry-Perot technique, using a pulsed reflectometric heterodyning interferometer for passive underwater hydrophone arrays, has been demonstrated by Plessey Naval Systems Ltd.[47] The system is illustrated schematically in Fig. 8.28. A coherent laser injects radiation through a Bragg cell, such that pairs of pulses at different frequencies ($f_1$ and $f_2$) are passed down-lead to the sensor array. The sensor array consists of a series of fiber sensors connected by semireflecting splices at each fiber end. The length of the fiber sensor is such that the combination of the transit time to and fro along the sensor equals the pulse separation. The different frequencies are therefore superposed at the detector and a heterodyne difference frequency can be obtained. When subjected to an acoustic field, the output signal, suitably linearized, is encoded as a phase modulation on the difference frequency. This can then be demodulated. The hydrophones therefore could be said to use a combination of Fabry-Perot and time-domain multiplexing techniques. Down-lead sensitivity to environmental perturbations can be electronically filtered out. Much of the engineering design has been concentrated on

- Demodulation and noise-rejection electronics (to reduce the effect of extraneous environmental effects)

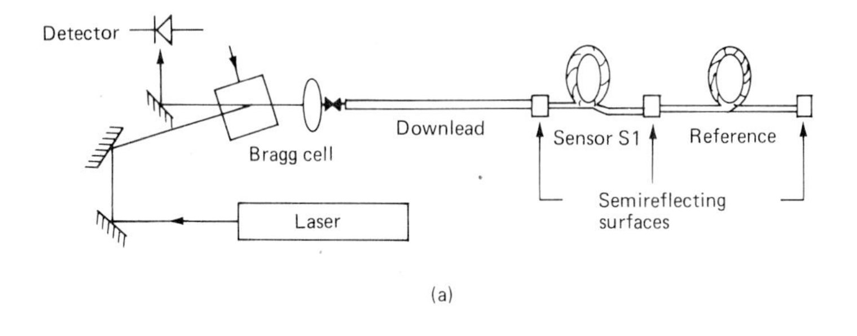


(a)

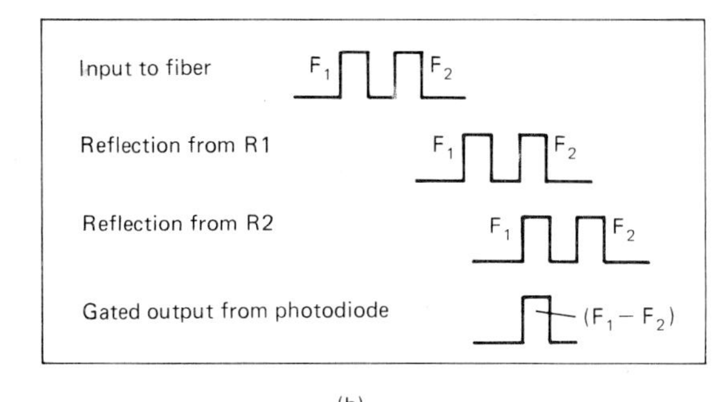


(b)

**FIGURE 8.28** Reflective splice hydrophone. (*a*) Optical system; (*b*) pulse timing. (*Courtesy of Plessey Naval Systems Ltd.*)

- Mechanical mounting of the reflective fiber coils with suitable encapsulating materials (to provide a flat frequency response and to maximize sensitivity)

Results quoted for a single hydrophone show an average sensitivity of −172 dB relative to 1 rad/(μPa · m). Theoretical calculations have shown that sensitivities below sea state zero are possible for multielement arrays in the frequency range below 1 kHz.

Table 8.10 compares fiber-optic hydrophones based on the Mach-Zehnder and reflective-splice approaches with other available hydrophone technologies. Reference should be made to Fig. 8.24 for a summary of the acoustic sensitivities of different fiber-optic sensors relative to sea state zero.

*Fiber-Optic Gyroscope.* The following discussion of fiber-optic gyroscopes is divided into three areas: operation, comparative advantages, and noise sources.

*Operation.* In the Sagnac interferometer (Fig. 8.29), the input radiation is split so that it traverses the fiber coil in opposite directions. Ideally, the clockwise (CW) and counterclockwise (CCW) beams should travel along exactly the same paths (reciprocity condition) before the interference fringes are detected. Rotation of the coil about its axis creates a difference in the propagation times of the two beams, giving an optical phase difference $\phi$ and interference fringes which can be detected. The phase difference is

$$\phi = \frac{8\pi NA\Omega}{\lambda_0 c} = \frac{4\pi RL\Omega}{\lambda_0 c} \tag{8.1}$$

**TABLE 8.10** Simplified Design Comparisons for Fiber-Optic and Piezoelectric Hydrophones

| Characteristic | Fiber-optic (Mach-Zehnder) | Fiber-optic (reflective splice) | PZT | PVDF | Ceramic composite* | Piezoelectric cable† |
|---|---|---|---|---|---|---|
| Sensitivity at 1 Hz | Below sea state zero for 100 m of fiber. Sea state 1 demonstrated. | Sea state 1 demonstrated | Below sea state zero | Below sea state zero | Same as PZT | Not quoted at 1 Hz; data only for above 30 Hz. |
| Minimum detectable pressure | Depends on length of fiber | −172 dB re 1 rad/(μPa · m) | 30–57 dB re 1 μPa typical (10–1000 Hz) | 40–50 dB re 1 μPa | Same as PZT | Not quoted |
| Frequency response | Reduction in sensitivity at low frequencies because of circuitry | Sensitivities below sea state zero possible below 1 kHz | Nonresonant designs give flat response; nonlinearities at very low frequencies. | Same as PZT (preamp limited) | Same as PZT | Probably same as PZT and flat beyond 10 Hz |
| Geometry | Versatile, i.e., coiled or extended. *S/N* is not limited by geometry. Difficult to multiplex in arrays. | Same as Mach-Zehnder. Potential for multiplexing with a reduced number of fibers and couplers. | Usually cylindrical, with variable size as required. *S/N* is dependent on geometry. | Flexible, can be molded to surface shapes. Can be made into arrays with contact processing on one sheet. *S/N* same as PZT | Flexible, but over limited areas. Requires "hinges." | Flexible; can be wound. Other extended tube types proposed can have less flexibility but still provide an extended sensor. |
| Relative dielectric constant | Not applicable | Not applicable | 1300–2000 | 10–15 | 40 (typical) | 30 |

*Composites of PZT and more-compressible compounds (e.g., epoxy or polymer).

†Cable with piezoelectric composite between central conductor and outer conducting sheath (e.g., NGK in Japan). AT&T Bell Laboratories has developed a flexible casing with a composite of piezoelectric particles and polyurethane polymer.

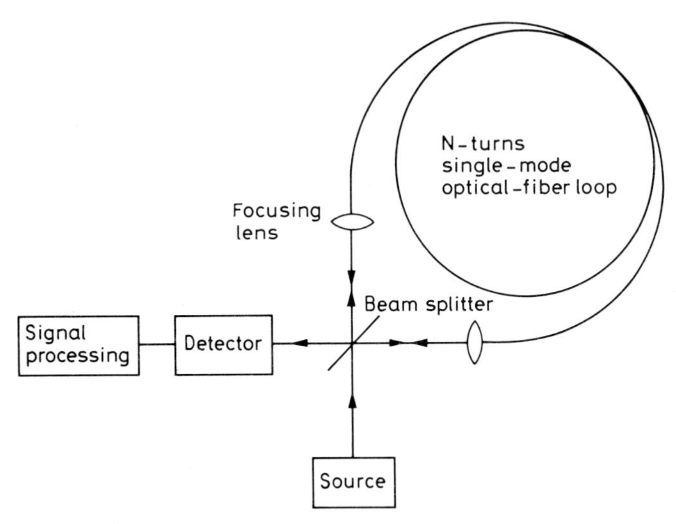


**FIGURE 8.29** Schematic diagram of an early optical-fiber Sagnac interferometer using bulk optics. (*From Ref. 3.*)

where $A$ = coil area
$R$ = loop radius
$N$ = number of fiber turns
$L$ = loop length
$\Omega$ = angular rotation rate
$\lambda_0$ = vacuum wavelength of the radiation
$c$ = speed of light

A phase shift is obtained since the light, propagating in two opposite directions, CW and CCW, around the loop, travels greater and lesser distances, respectively, as a result of rotation. The phase shift is effectively doubled when the beams are recombined, giving a path difference of $R\Omega\tau$, where $\tau$ is the transit time in the loop, and this difference is doubled to $2R\Omega\tau$. A phase difference of $1 \times 10^{-6}$ rad is obtained for a 1-km fiber length, 0.1-m-radius coil, and a rotation rate of 0.1°/h, when $\lambda_0$ is 0.85 μm.

*Comparisons and Justification for Development.* Competing devices are the ring laser gyro and mechanical gyro, both of which are based on relatively mature technologies. New types of mechanical gyroscope are still being investigated and improved. Numerous varieties exist, from cheap and less accurate devices to extremely stable precision-engineered models. The mechanical gyro is likely to continue to dictate the market price for some time. A comparison summary is provided in Table 8.11.

Many of the advantages put forward to justify development of the fiber-optic gyro can also be attributed to the ring laser gyro. The latter requires a stabilized, low-noise, long-coherence-length optical source (i.e., a gas laser). However, these sources can have limited operating and shelf lives. In practice, the ring laser gyro has further problems in that the counterpropagating beams can couple together in frequency to produce a frequency lock-in effect, leaving the ring laser gyro with a nonlinear scale factor. To reduce this effect (thought to be caused by backscatter from the ring laser gyro mirrors), much development has concen-

**TABLE 8.11** Comparison of Fiber-Optic Gyroscopes with Mechanical Gyroscopes

| Fiber-optic | Mechanical |
|---|---|
| Potentially inexpensive | Can be inexpensive |
| Gravity-insensitive (with correct coil mounting) | Gravity-sensitive |
| Temperature-sensitive (can be reduced by Shupe windings, etc.) | Temperature-sensitive |
| Low power consumption (more or less independent of rotation rate) | Power consumption proportional to rotation rate |
| Can measure and withstand high rotation rates | Maximum measurable rotation is, in general, dictated by maximum temperature rise permitted to maintain closed-loop operation |
| $<1$ ms switch-on time (dictated by electronics only) | $>1$ ms switch-on time (dictated by how long it takes to achieve wheel spin speed or mechanical resonance) |
| Can withstand high acceleration or vibration and still measure rotation rates | In general, mechanical gyros require stops to withstand high acceleration or vibration and hence cannot measure rotation rates during such periods |

trated on the accurate separation, alignment, and production of very high quality, low-scatter mirrors.

The advantages of short warm-up time and long shelf life therefore make the fiber-optic gyro particularly attractive for several missile applications, since it can potentially be manufactured for low cost, with a reduced space requirement. The ability to make it cheaply and in large numbers has yet to be demonstrated, since the expensive and complex fiber micromanipulating techniques used for prototypes will not be allowed in manufacture.

The rationale for the use of fiber-optic gyros in low-drift, highly stable navigation applications, when compared with ring laser gyros, is more difficult to justify. Nevertheless, theoretical calculations show that adequate drift rates might be possible and it is therefore worth analyzing (1) the various factors which can affect fiber-optic gyro sensitivity and (2) the origins of the various noise sources within the system. Many variables must be controlled. However, since 1976, when the first fiber-optic gyro was demonstrated, the degree of stability has steadily improved, as advanced signal processing techniques and specialized optical components have been incorporated. Much effort has been expended on the packaging and system integration of devices, and this has helped to reduce further the several extraneous environmental effects.

Passive optical-fiber ring resonators[48] are still at an early research stage, so only a brief mention is made here. High-performance optical couplers are needed in a passive optical-fiber ring resonator in order to achieve ring laser gyro navigation performance. The passive optical-fiber ring resonator can also have low-rotation-rate lock-in problems, which may not present in the fiber-optic gyro. One advantage of the fiber-optic passive optical-fiber ring resonator is that only short lengths of fiber are necessary, while the fiber-optic gyro coil can require relatively long fiber and hence add to the expense.

*Noise Sources (Bias Errors) in Fiber-Optic Gyro Designs.* One group of noise sources is associated with the system reciprocity requirement, i.e., that the light beams should ideally traverse exactly the same paths in counterpropagating fashion and produce the same phase and intensity relationship on recombination. Environmental effects, provided that their time constants are much greater than the loop transit time—i.e., they are low in frequency—influence both beams equally as they traverse the loop and hence lead to zero phase shift. However, phenomena having comparable frequencies can affect the two beams in different ways, leading to drift. Other noise sources are related to the optical components used, e.g., light-emitting diodes or lasers, and the type of fiber. A list of the different sources of noise is given in Table 8.12 together with a brief description of the methods which have been employed to resolve these problems.

NONRECIPROCAL EFFECTS. Within a single-mode fiber, as a result of high optical intensities, there are changes in the propagation constants, termed the *Kerr effect.* Nonlinear optical effects result, and, because the two beams have different intensities, an effective phase shift occurs on recombination. The Kerr shift is difficult to separate from the Sagnac shift; hence it must be classed as a source of noise. The solution is to use a wideband source, i.e., a superluminescent diode.

**TABLE 8.12** Sagnac Interferometer Fiber-Optic Gyroscope: Some Problem Areas and Proposed Solutions*

| Noise effect | Solution |
|---|---|
| Nonreciprocal effects: | |
| Faraday rotation of plane of polarization | Effect reduced by magnetic shielding or use of polarization-preserving fiber. |
| Kerr effect—nonlinear optical effect can affect propagation constants and hence the counterpropagating beams | Use broadband superluminescent diode so that Kerr effects are averaged out over the wavelength range. |
| Polarization mode coupling in fibers produces fading of signals | Use low-coherence source (superluminescent diode) with depolarizer, or use polarization-preserving fiber. |
| Coherent Rayleigh backscatter within the fiber and at coupler splices and joints with integrated-optic (IO) chips | Use low-coherence source (superluminescent diode) or correct integrated-optic chips to minimize backscatter. |
| Thermal gradients and transients | Optimize coil windings to minimize the effect, and use good thermal packaging practices. Also, temperature sensors can be used to control the thermal stability. |
| Acoustic noise (e.g., bias error proportional to the difference in effective areas of the coil for the two counterpropagating beams) | Optimize mechanical design of the coil or use electronic filtering. |

*This is not a comprehensive list. There exists a series of complex interactions between the different solutions offered.

The sum of the different wavelength components traversing the fiber then provides an effective phase shift of zero.

*The Faraday effect* in a fiber-optic gyro coil under an applied magnetic field changes the plane of polarization and hence the velocity of the beams. The velocity increase or decrease depends on the orientation of the coil in the magnetic field gradient. Once again, this leads to a phase shift which can be confused with the rotational Sagnac phase shift. Solutions are to shield the coil within a box of highly magnetic material (e.g., mu metal) or alternatively to use polarization-preserving fiber so that relative changes to such a rigidly confined polarization state are greatly reduced. Depolarizers can also be used at both outputs of a single-mode fiber sensor coil.

MATERIALS AND COMPARISON EFFECTS. To reduce noise and bias from the nonreciprocal effects just described, wideband sources and polarization-preserving fibers can be used.

For reciprocity, the counterpropagating beams should have consistent and identical polarization states through the fiber. Since single-mode fibers support two orthogonally polarized states, there can be problems in mode coupling, caused by stress-induced birefringence or inconsistencies in the physical dimensions and cross-sectional shape of the core over the length of the fiber, leading to varying propagation velocities of the modes. In the presence of acoustic fields, and particularly for applied thermal strain effects, this mode-coupling effect can be enhanced. Therefore, the technique for winding the coil to minimize such errors is most important. The use of polarization-preserving fiber helps to solve some of these difficulties, but careful control of the fiber parameters during coil winding and packaging is still required, since it is difficult to fully eliminate mode-coupling effects. The polarization state can be defined by a polarization element at the input to the coil, and if a fiber or integrated-optic coupler is used, the polarization state stability and environmental response of such devices will also be important if the counterpropagating beams are to maintain their polarization states.

The effects of thermally induced nonreciprocity in optical-fiber coils have been identified.[49] If the length of the coil is sufficiently long, so that the transit time is of the order of microseconds, then the two beams can be differentially affected if the temperature has changed at points on the coil within this time scale. To minimize this effect, different methods of winding the coil have been tried. Apart from transient thermal variations, there are other systematic thermal variation effects, caused by standing thermal gradients across the coil, which will also provide erroneous bias effects. To isolate the coil from external thermal variations, several designs have been proposed, some of which are comparable to typical gas laser mode-stabilization schemes; i.e., the coil is well insulated and heated evenly to stabilize the temperature uniformly above ambient. Temperature sensors for thermal control have also been found to be useful.

A coherent optical source leads to added noise from Rayleigh backscatter and also enhances noise from back-reflections within the system. Various component interfaces are present within the fiber-optic gyroscope system, e.g., fiber splices, joints between fiber and integrated-optic chips (where the refractive indexes can differ), and fiber-coupler joints. For OTDR-based sensors (Sec. 8.4.4), coherent sources are required to obtain good quality signals from back-reflections, but in the case of the Sagnac interferometer, where there are two opposed beams, rapidly varying back-scattered signals can provide unwanted effective phase shifts at both outputs. Once again, the solution has been to use a wideband LED source.

Even a device of this type has a limited coherence length (less than 1 mm), but this is still sufficient coherence for the detection of fiber-optic gyroscope phase shifts.

The presence of modulation effects from back-reflecting surfaces also has practical mechanical design implications. As in the case of the Fabry-Perot interferometric sensor, techniques can be used whereby the fibers and integrated-optic chip can be appropriately angled, so that the reflected light is not returned to the source. The interfaces are also designed to minimize losses. Other techniques, involving special designs within the integrated-optic chips themselves, can also be used to reduce bias (spurious phase shifts) from back reflections.

In summary, the noise sources within the fiber-optic gyroscope tend to be caused by the source itself, stress birefringence, and nonreciprocal effects. Unwanted bias effects can be largely reduced by the judicious use of wideband sources or polarization-preserving fiber. For other interferometric optical sensors, the requirement for reciprocity is not necessary. There are common factors, however, in the use of all-fiber couplers and the use of source frequency stabilization techniques. Selection of the optimum operational wavelength can be important in terms of the theoretical shot noise detection limitations of the fiber-optic gyroscope system. Calculated quantum limits for gyroscopes as a function of wavelength are available, where the shot noise limits of different detectors are used. Most fiber-optic gyroscope systems to date have been operated at 0.85 μm, but with the drive in telecommunications to longer wavelengths, there will be a consequent influence on component availability at longer wavelengths. Some effort has therefore been applied toward designing FOGs to operate at 1.3 μm.

Several approaches to the detection of the Sagnac shift have been reported. As in other interferometric sensors, the phase of fringe patterns should be detected to levels of $10^{-6}$ rad, typically, and signal linearization is required because of the sinusoidal nature of the fringes. For air navigation applications, the dynamic range should extend over 7 orders of magnitude ($10^7$), and additional electrical and optical signal processing techniques are therefore necessary. Two solutions to this problem have involved phase modulation and frequency nulling. In both these cases one introduces an extra phase shift into the light path of the fiber-optic gyroscope to counteract or null the Sagnac effect. It is the size of this nulling signal which is then the measurement of the rotation rate. A comparison and brief description of these two methods, with their specialized counting techniques, are presented in Table 8.13. Figures 8.30 and 8.31 illustrate schematically the different types of fiber-optic gyroscope signal processing. Table 8.14 describes various fiber-optic gyroscope operational techniques. Figure 8.32 shows an integrated-optic chip realization of a fiber-optic gyro, and Fig. 8.33 shows, as an example, a photograph of the STC Technology Ltd. prototype fiber-optic gyroscope, which uses an integrated-optic chip.

The various design options are complex, but can be summarized briefly as follows:

- Fiber-optic gyroscope devices could be made cheaply if integrated-optic chips and the suitable interconnection technology are developed. Initial applications would be in missiles with lower sensitivity requirements.
- The optical components are available for "all-fiber" systems, which theoretically could reach sensitivities suitable for inertial navigational systems. Considerable technical advances are necessary, however, before such devices can seriously compete with laser ring gyros. Integrated-optic solutions for larger volume applications (e.g., missiles), with reduced dynamic range, are possible

**TABLE 8.13** Fiber-Optic Gyroscope: Some Optical System Architecture Options*

| Gyroscope | Technology | Comments |
|---|---|---|
| All-fiber gyroscope | All-fiber couplers, polarizers, phase and frequency modulators. Fiber-fiber splicing required. | PZT fiber modulators are limited to frequencies below 1 MHz. Coupler technology is now robust. May offer a route to high-sensitivity devices. Low photorefractive index. |
| Integrated-optic (IO) gyroscope with fiber coil | Wavequide in rectangular lithium niobate substrates. High-dynamic-range phase shifters, with large frequency bandwidth; mode filters available on one chip. | Devices are lossy. Coupling of optical fibers to the rectangular geometry of the chips requires further effort to ensure robust devices. Photorefractive index reasonably large and hence a source of bias errors. Potential integrated-optic chips with low photorefractive index are being sought. Offers the best route for cheap, robust devices, but integrated-optic chips remain costly. |
| Fiber coil | | |
| Polarization-preserving fiber or single-mode fiber plus polarizers and depolarizers | Described in text. | For design and different applications, the choice of polarization-preserving fiber or ordinary fiber will be based on cost, if mounting and locating of polarizers is correct. |

*A further major option is the choice of source, i.e., laser or relatively incoherent superluminescent diode. This is discussed in the text and in Table 8.12.

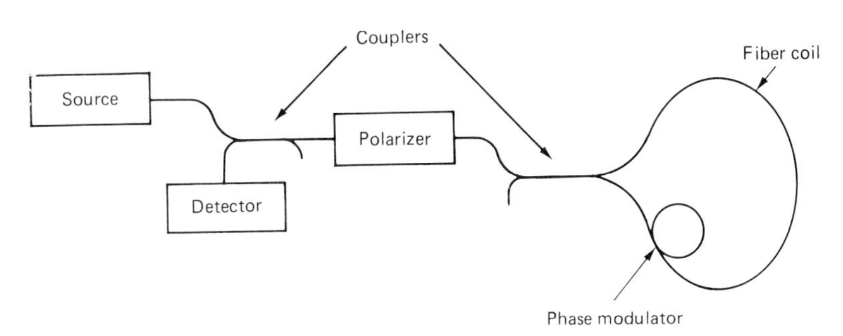


**FIGURE 8.30** Analog fiber-optic gyroscope.

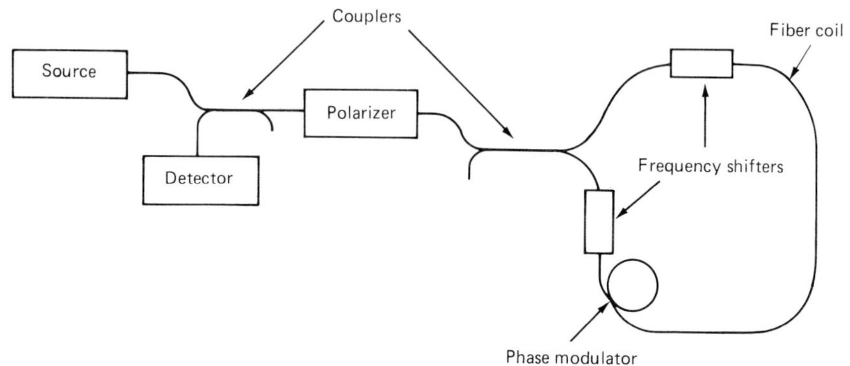


**FIGURE 8.31** Digital fiber-optic gyroscope.

once the design technologies for low-cost manufacture and robustness in operation are confirmed.

- Packaging and coil winding are crucial parts of the design. They could provide the limiting factors in determining the stability of the fiber-optic gyroscope, regardless of architecture.

*Magnetic Sensors.* The proposal[51] that single-mode optical fibers with magnetostrictive coatings could provide highly sensitive magnetic sensors or gradiometers was originally made in 1980. Initial calculations based on AC measurements and subsequent extrapolations based on extended fiber lengths indicated that measurement down to $10^{-6}$ T was feasible ($10^{-5}$ gauss = 1 gamma = $10^{-9}$ T). For the more numerous DC applications, however, measurement sensitivities are somewhat lower.

Fiber-optic magnetic sensors measure the magnetic induction in free space (the unit of measure is typically the tesla); it has become the convention to refer to this magnetic induction as the magnetic field. Fiber-optic measurement of AC fields provided advantageous results from the outset, since the low-frequency environmental variations could be filtered out. Measurements at DC, on the other hand, are more difficult. Extrapolations of sensitivity based on early measurement techniques have proven to be optimistic, although sensitivity has been improved to some extent by innovative geometrical fiber windings and specially shaped magnetostrictive strain elements.

In the presence of an applied magnetic field, a strain is induced in the magnetostrictive coating and is then coupled into the fiber core. When a coated fiber is used in one arm of a Mach-Zehnder interferometer, a phase shift can be detected because of the change in optical path length. Most reported data have centered on the sensitivity of relatively short lengths (e.g., 0.5 m) of coated optical fiber or on the effect of strain on fibers wound on magnetostrictive cylinders or mandrels (see also polarimetric sensors, Sec. 8.4.2). Coating technologies of this type are still at an early stage, although electroplating of nickel on a fiber has been achieved.[36]

Test results have also been obtained with sensors created by bonding optical fibers to amorphous metal strips (i.e., metallic glass, described below) which have high magnetostrictive coefficients. Experimental tests have been reported in which alternating magnetic fields were superimposed on a DC bias field. Sensi-

**TABLE 8.14** Fiber-Optic Gyroscope Operational Techniques

*These techniques are independent of fiber-optic gyroscope architecture*

| Operation | Description | Advantages | Disadvantages |
|---|---|---|---|
| Quadrature drive<br>$\Omega$, $-\Omega_{max}$, $+\Omega_{max}$, $\phi$ | Analog output followed and controlled about the quadrature point by nonreciprocal phase shifter in the fiber coil. | Simplest form of gyro design. $\Omega_{max}$ can be quite high, provided the coil is small enough. Cheapest concept, suitable for high-rotation-rate, low-dynamic-range applications. | Analog output. Low dynamic range (e.g., 3000). Scale factor correction ~1%. |
| Quadrature plus servo drive<br>Quad signal phase; $\tau$, transit time of coil<br>Voltage; Time; $\phi$; $\tau$; $V \propto 2\pi$<br>$\phi = \frac{8\pi NA\Omega}{\lambda_0 c}$ | Behaves like a rotational scale, giving a pulse out for a fixed angle of rotation. | Gyroscope output is digital. Some error signals can be used for self-calibration. Dependent only on coil area and not on coil length. | For perfect operation, requires infinitely fast reset with restart of phase ramps. Requires high linearities. |

**TABLE 8.14** Fiber-Optic Gyroscope Operational Techniques (*Continued*)

*These techniques are independent of fiber-optic gyroscope architecture*

| Operation | Description | Advantages | Disadvantages |
|---|---|---|---|
| Quadrature plus square wave<br>(a) Quadrature drive: $2\tau$<br>(b) Closed loop drive: $V \propto \Omega$ | Output voltage directly proportional to rotation rate. Similar to mechanical gyro in output. | Can gate out signal because of finite rise and fall time of drive signal. Mismatch in phase between two signals can also be gated out. | Useful signal available only in every other time period. Output is analog, therefore analog-to-digital converter required at the output. |
| Quadrature plus step phase ramp drive signal<br>$V \propto \Omega$; $V \propto 2\pi$ | Behaves like a rotational scale, giving a pulse out for a fixed angle of rotation. | Useful signal obtained in every time period. Can gate out signals because of finite rise and fall time of electronics. Error signals can easily be obtained to receive $2\pi$ modulation. Digital output. | Must generate the required waveform and monitor reset. |

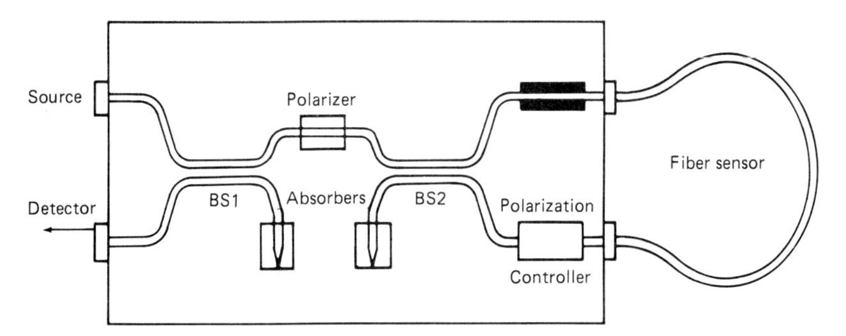


FIGURE 8.32 Integrated-optic chip realization[50] of a fiber-optic gyroscope. *(From Ref. 3.)*

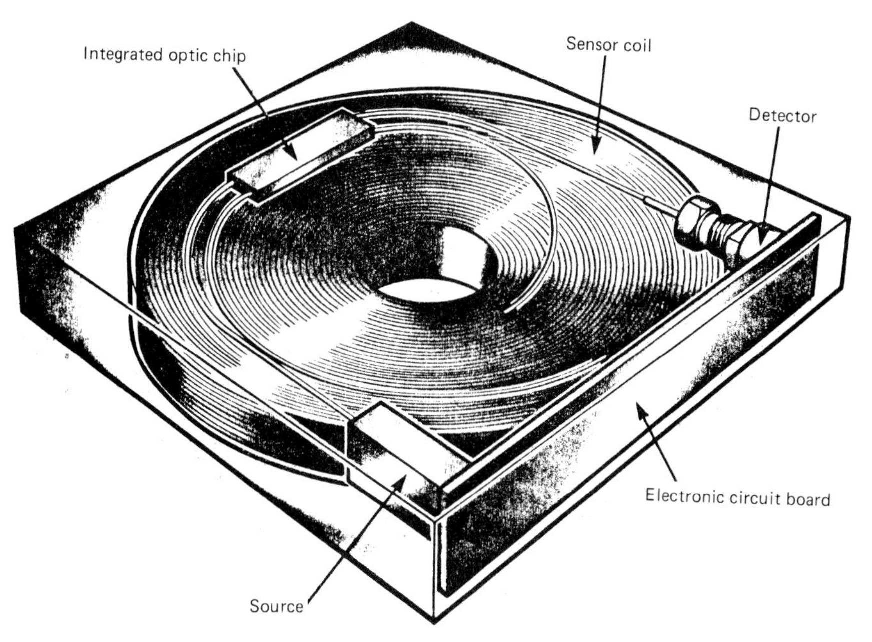


**FIGURE 8.33** A representative fiber-optic gyroscope prototype schematic. *(Courtesy of STC Technology Ltd.)*

tive phase detection techniques can then be readily employed. Work has been carried out to develop DC magnetic gradiometers using the advantageous quadratic magnetostrictive response of metallic glass to reduce the effects of unwanted environmental perturbations near 0.1 Hz.[44,45] This technique and two types of DC fiber-optic magnetic sensor are described below, after a brief theoretical discussion.

The relationship between magnetostriction $\lambda$ and a small magnetic field $B$ is parabolic (Fig. 8.34):

$$\lambda = KB^2 \tag{8.2}$$

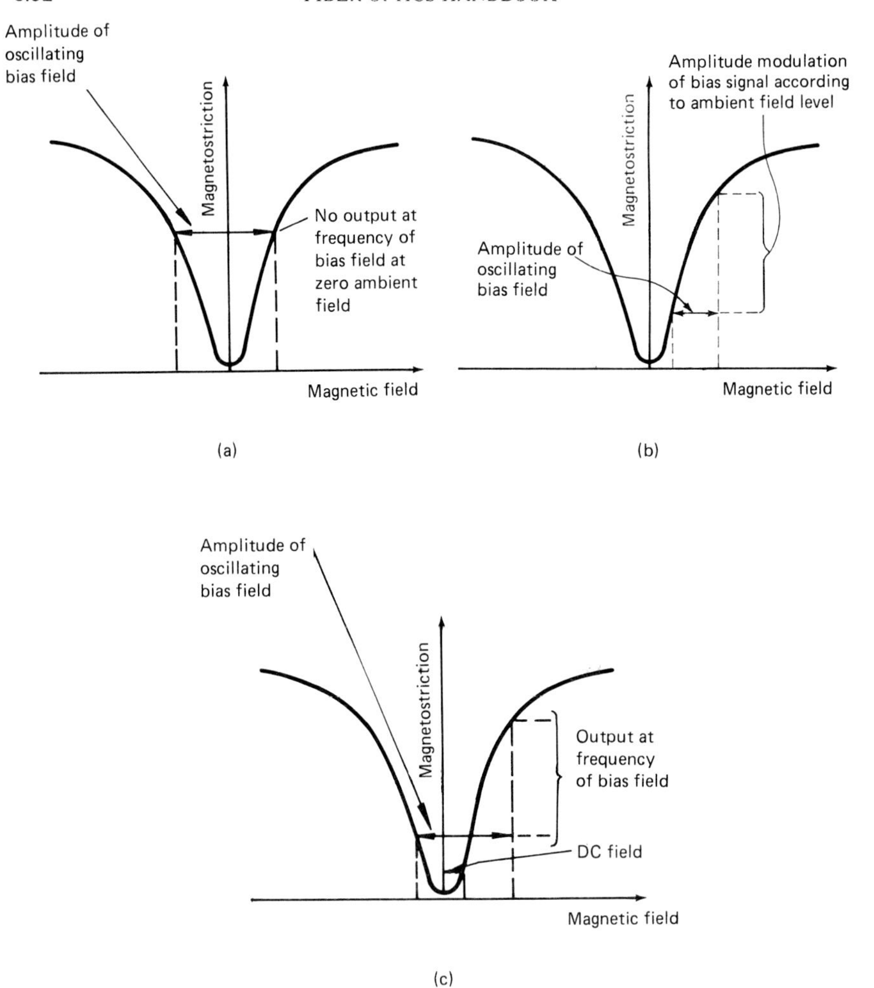


**FIGURE 8.34** Nonlinear relationship of magnetostrictive coefficient against magnetic field, indicating how a DC field can be measured with an oscillating bias field. (*a*) Bipolar bias field (zero field); (*b*) unidirectional bias field; (*c*) bipolar bias field (small DC field present). (*Adapted from Ref. 3.*)

Differentiating Eq. (8.2) gives

$$\frac{d\lambda}{dB} = 2KB \tag{8.3}$$

The magnetostrictive output with an applied alternating bias field is proportional to the field to be measured. Hence it is possible to also measure the sign of the field. A simple magnetostrictive response will provide only the magnitude, since $\lambda$ is proportional to $B^2$. With an applied bipolar bias field, there is no output at the frequency of the bias field at zero ambient field. Behavior is comparable to that of a fluxgate magnetometer in that the sensor gives a second harmonic output at zero field, and the fundamental appears when an ambient field is applied.[37,45] An alternative to a bipolar bias field is a unidirectional field. In this case, the nonlinear shape of the magnetostriction curve leads to amplitude modulation of the bias signal. The advantages and disadvantages of closed-loop vs. open-loop systems and other detection techniques have been discussed.[52,53]

For magnetic sensors, active PZT (phase shifting control) elements can be used with a Mach-Zehnder configuration. With a 400-Hz bias signal and a 10-cm length of fiber sensitized by attachment to a high-magnetostrictive-coefficient metallic glass, sensitivities of the order of $10^{-10}$ T · m for DC fields have been obtained.[37,45] The gradient magnetometer,[54] consisting of two coils separated by a known distance (e.g., 10 cm), offers possibilities for highly accurate gradient measurement and balancing of thermal expansion effects between coils. As with the gyroscope and ultrasonic sensor, the crucial design factors lie in the mounting and packaging of the coils and innovative winding geometries (see below).

Earlier linear extrapolations of the minimum field that can be measured, based on increasing the length of fiber, have proved to be incorrect. The sensitivity is limited by the demagnetizing field caused by the free poles generated at the ends of the coated fiber by the magnetization. Thus the sensitivity $S$ (radians/tesla) changes from the relationship

$$S \propto \frac{\lambda L}{B} \tag{8.4}$$

where $L$ is the length of sensitized fiber, to the more complex

$$S \propto \frac{\lambda}{B(1 + N\mu)} \tag{8.5}$$

where $N$ is the demagnetizing factor and $\mu$ is the relative permeability of the coating.

This effect has been experimentally verified and has further design implications for the geometry of the sensor. To reduce noise effects further, polarization-preserving fiber is required. A comparison of various magnetic field sensing techniques is given in Table 8.15.

*Other Interferometric Sensors.* At different times, examples of Michelson fiber-optic sensor configurations have been advanced.[7–10] The problems in signal processing are similar in several respects to those for Mach-Zehnder devices. One interesting example of the use of a Michelson configuration has been for the investigation of slow strain phenomena, occurring over hours or days.

Other applications for interferometric sensors are briefly analyzed. Such techniques can be expensive when compared with available sensing methods. Only all-fiber interferometric solutions are described.

*Temperature.* The calculated strain sensitivity of fiber to thermal variations is large. Fiber-based techniques offer the possibility of an integrated temperature measurement along a length. Thus, applications are possible for highly specific measurements requiring low thermal inertia (i.e., fast response times) in environments where conflicting vibration effects and strains do not cause insurmountable problems. Typically, for an optical fiber of 1-m length, theoretical sensitivities of the order of $10^{-8}$°C are possible,[1] which compares with $10^{-4}$°C for a thermocou-

**TABLE 8.15** Magnetic Sensors: Simplified Comparison

| Type | Minimum detectable field, T* | Comments |
|---|---|---|
| Fluxgate | $10^{-10}$–$10^{-11}$ | Well-accepted, can now be made very small, but with loss in sensitivity. 3-axis devices are available. High-frequency limitations. |
| Magnetoresistors | $10^{-6}$–$10^{-9}$ | Permalloy films on glass or silicon. Small and robust. Suitable for high-frequency applications. |
| Hall effect | $10^{-7}$–$10^{-9}$ | Relatively new GaAs devices with flux concentrators. Johnson noise limited. Small and robust. |
| Nuclear magnetic resonance and optically pumped vapor devices | $10^{-10}$–$10^{-12}$; $10^{-12}$ typical | Bulky, expensive. Used for undersea and airborne anomaly detection, mineral exploration. |
| Superconducting quantum interference device (SQUID) | $10^{-14}$–$10^{-15}$ (possibly optimistic) | Cryogenic cooling required. Bulky, expensive. Now starting to be used for geophysical exploration. |
| Optical fiber | $10^{-10}$/m at DC. Quoted AC sensitivities are $10^{-12}$–$10^{-13}$/m at 1 kHz. | No cooling required, geometrical versatility. Development of long lengths of magnetostrictively clad fiber required. Gradient magnetometers possible with suitable windings. |

*1 gamma = $10^{-5}$ gauss = $10^{-9}$ tesla.
*Source:* Adapted from Ref. 3.

ple. However, sensitivities of this order are probably not required in practice, and some sensitivity reduction by suitable fiber packaging can be an advantage. If ease of signal processing is an important consideration, it is possible that Fabry-Perot or polarimetric sensors would be preferred to Mach-Zehnder or Michelson systems.

*Acceleration.* A single-axis fiber-optic accelerometer consists of a mass suspended between two fibers. This acts as a simple harmonic oscillator, and when it is accelerated in a direction parallel to the fiber, a strain is induced proportional to acceleration. Accelerations have been measured to within submicrogravity accuracy with a frequency response up to 600 Hz, when a "mass-fiber" construction is placed in the arm of an interferometer. However, the interferometric system is complex, when compared with other small fiber-semiconductor accelerometers or even with miniature quartz and micromachined silicon accelerometers.

### 8.4.2 Polarimetric Sensors

The use of stress-induced birefringent effects has already been described in Sec. 8.2.6, with respect to Faraday rotational and photoelastic phenomena within crystals and materials acting as extrinsic sensors. However, the silica fiber itself is also subject to birefringent changes which give rise to polarization fading effects in interferometric devices. The use of this effect within the fiber has been proposed for several sensing applications. Polarization-preserving fiber (having a preferentially oriented high birefringence ratio) can be used, and the relative optical delays of the two orthogonal polarizations at the end of the sensing fiber can be measured by observing the polarization state. These sensors therefore require optics at either end of the fiber sensor to ensure that (1) the input polarization states are correct relative to the birefringence axes and (2) the amplitudes of the output polarization states can be measured. Because initial attempts at such sensors predated the availability of polarization-preserving fiber, some early testing[55] was conducted on strained fiber (e.g., fiber wrapped around a magnetostrictive nickel form for magnetic field sensing and around mandrels for hydrophone applications), in order to strain-induce the necessary orthogonal birefringence difference in the single-mode fiber, and make one or the other of the orthogonal modes preferentially lossy in the presence of strain. It was thought that such devices would provide advantages, since both the sensing and reference beams might effectively be separated into the two orthogonal polarizations within the fiber and—provided the sensor mechanical design was such that one of the orthogonal polarizations could be preferentially perturbed—the common cross-perturbations (e.g., pressure and temperature) could be canceled out. Unfortunately, as described previously, it is extremely difficult, even with polarization-preserving fiber, to fully restrict mode cross-coupling when the fiber is under stress. Attempts to develop compensated polarimetric sensors have met with some success. One such device uses a rotated 90° coupler, where the sensor is spliced into two equal lengths and then the orthogonal polarizations are realigned to form a balanced interferometer.[56]

While the polarimetric sensors described above rely on the use of polarization-preserving or high-birefringence (hi-bi) fiber, considerable effort has also been applied to the development and use of fiber with well-controlled low birefringence for current and voltage measurement.[28] Linearly polarized light is injected into a low-birefringence monomode fiber, which is coiled around an electrical power bus bar. Cladding modes are stripped out and the measured rotation of the polarization vector in the presence of the applied horizontal magnetic field is taken to be proportional to the current (i.e., proportional to the line integral of the magnetic field around the loop). Depolarization optics, consisting of a Wollaston prism beam splitter, are used to separate the two orthogonally polarized beams. Devices of this type have been tested in the field, where currents up to 14 kA have been detected, with nonlinearities less than 1 percent.

There remain problems in terms of cross-perturbing effects for polarimetric fiber sensors. The mechanical mounting of such devices requires great care, in order not to induce unwanted strains and not to promote cross-coupling of the polarization modes. Tight specifications on the control and quality of fiber birefringence over the 100-m lengths required for the more sensitive applications (e.g., hydrophones and magnetic detectors) imply a high cost of manufacture. For current and voltage measurement, the same constraints apply, and the argument in favor of using remote electro-optic crystals and polarizing optics with more standard fiber is powerful, particularly when costs and the degree of flexibility in local mechanical design for placement of the remote sensor are improved

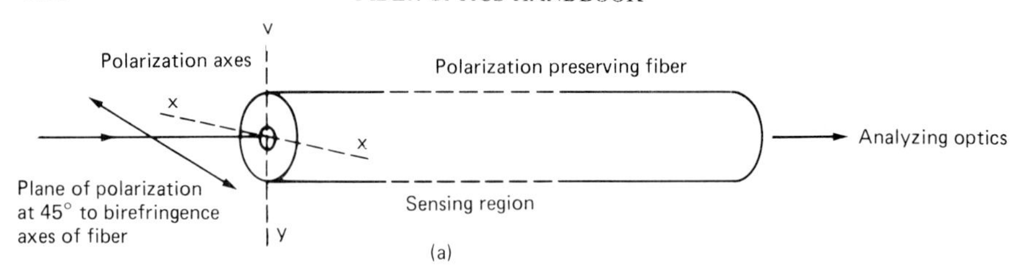


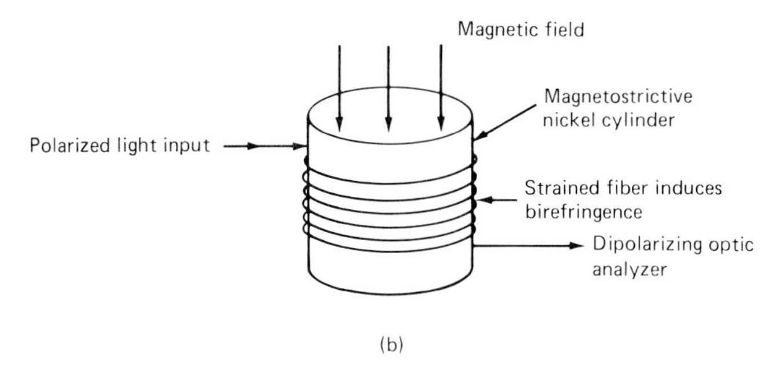


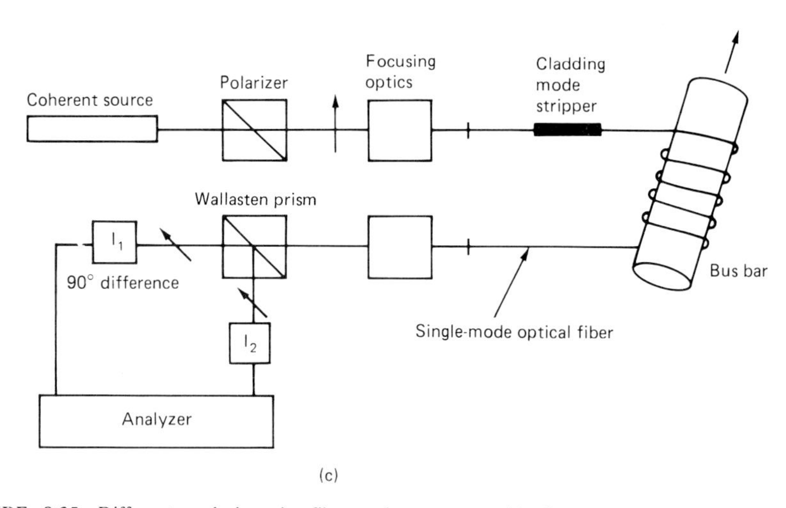


**FIGURE 8.35** Different polarimetric fiber-optic sensors. (*a*) Sensor using polarization-preserving fiber; (*b*) magnetic field sensor using strained/wound monomode fiber; (*c*) electric current sensor using low-birefringence fiber.

(despite the added requirement for couplers at the sensor). Figure 8.35 summarizes the range of intrinsic polarimetric sensors. Provided that the cost of polarizing optics (couplers, etc.) can be reduced, such devices may find a niche for very special sensing applications.

### 8.4.3 Microbend Sensors

Judging from the published literature, perhaps the most popular intrinsic fiber-optic sensor involving multimode fiber is the microbend sensor. Various proposals have been put forward, ranging from potentially very cheap devices for use with OTDR distributed systems (see also Sec. 8.4.4) to highly sensitive hydrophone systems.[1–13] Figure 8.36 illustrates the principle. A multimode fiber is laid between grooved or serrated plates. On compression, the fiber is forced into the grooves and the critical angle for total internal reflection within the core is exceeded. Light is lost to the cladding. Sensors can be constructed to measure the amount of light entering the cladding (dark field), or the loss of light from the core, or both.

The theory is not complex,[57] and calculations for different core and cladding diameters in both step-index and gradient-index multimode fiber have been carried out, giving good agreement with experimental data. The calculations can also take account of both modal transfer from the core to the cladding and the light transferred back. Comparable systems involving contained evanescent coupling of this type have also been proposed for single-mode fibers,[58] high-birefringence fibers,[59] and dual-core fiber[60] (with transfer between the cores within a joint cladding). The disadvantages of the last system are the special and costly nature of the fiber and the further associated cost of coupling optics.

Single-mode devices offer a means of sensing and possibly separating multivariable inputs to the fiber, i.e., using the added degree of freedom of wavelength. Inherent in such optical systems, however, is the cost of matching beat lengths and applied grating structures with input wavelengths and the high cost of the special optical alignment and wavelength multiplexing devices required in manufacture and maintenance. Eventually, perhaps, the control and preparation of such specialized fibers will permit the design of effective practical systems

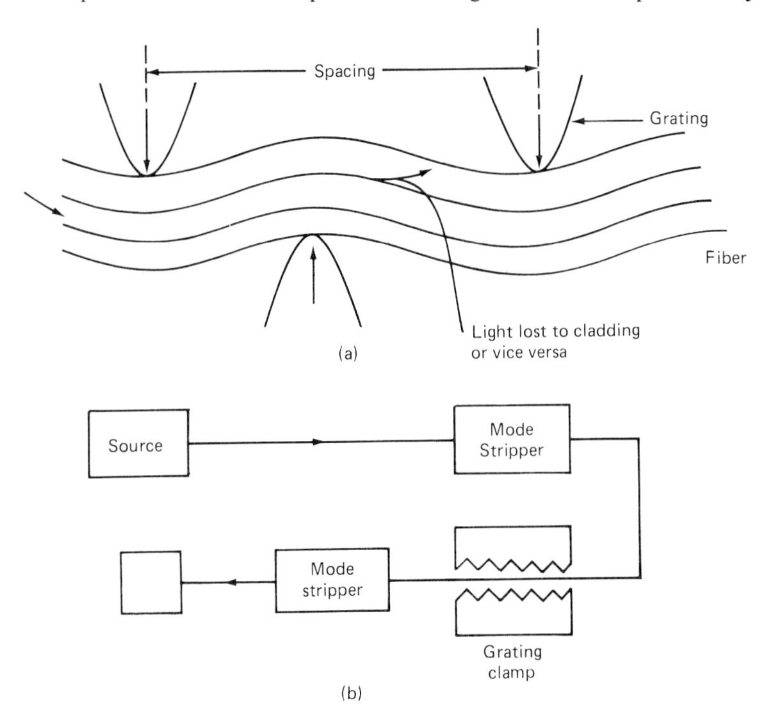


FIGURE 8.36 Principle of the microbend sensor. (*a*) Detail of loss mechanism; (*b*) system schematic.

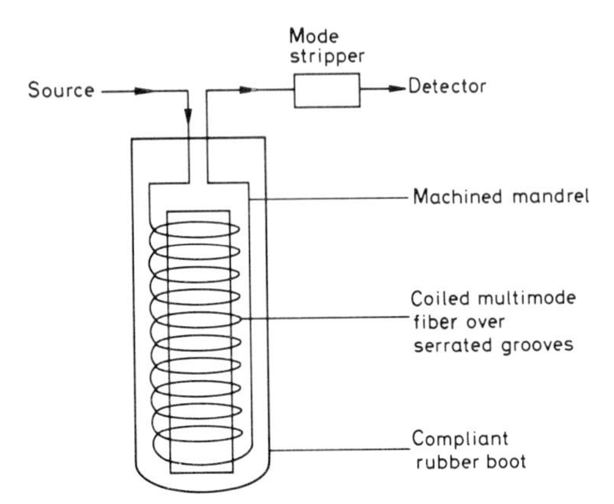


**FIGURE 8.37** The Naval Research Laboratory microbend hydrophone sensor. (*After Ref. 57.*)

with greater certainty, but much of this work remains at the research stage. The following discussion is therefore restricted to microbend sensors constructed from conventional multimode fiber, although the use of single-mode systems in conjunction with OTDR will be described in Sec. 8.4.4.

Figure 8.37 shows the basic Naval Research Laboratory multimode microbend sensor for hydrophone applications.[57] Results of tests on the microbend hydrophone are included in Fig. 8.24 for comparison with other proposed interferometric hydrophone devices. The fiber is coiled over a serrated mandrel having a specifically designed profile and wavelength, i.e., height and pitch. The outer surface is a compliant boot which transmits the acoustic pressure to the fiber. The mechanical design and mounting are critical, since such hydrophone units can be subject to a range of conflicting variables (particularly if towed), such as vibration, acceleration, hydrostatic pressure and temperature.

The microbend sensor essentially measures displacement, and strain is then transferred to the fiber. For example:

- The fiber can be attached to a moving bimetallic strip for temperature sensing.
- Pressure can be applied directly to the fiber for load sensing.
- For chemical sensing or leakage detection, the material to be detected provides a detectable exothermic reaction or is absorbed within a material which then expands.

For multiplexing, an array of microbend devices can be attached at different points along a fiber for different sensing applications. Remote interrogation by OTDR can then be carried out. Such measurements are relatively easy for digital on-off sensors. Resolution of analog displacements can be carried out, but compensation for drifts remains a problem.

### 8.4.4 Optical Time-Domain Reflectometry

Figure 8.38 illustrates schematically the standard technique used in telecommunications for the detection of breakage along lengths of optical fiber. An input

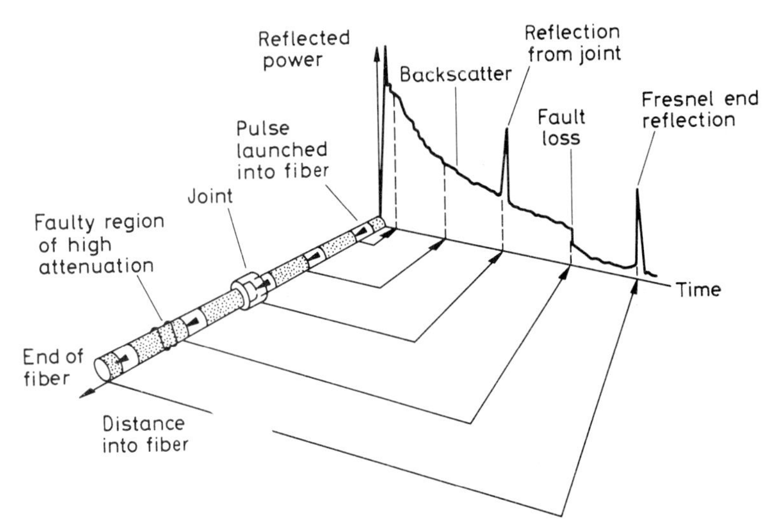


**FIGURE 8.38** The principle of optical time domain reflectometry for detecting fiber breakage. (*From Ref. 3.*)

pulse of radiation can be back-scattered from a defect, splice, or breakage, and, by analyzing the time delay, the position of the defect along the cable can be identified for maintenance. Discontinuities in the decay curve of back-scattered radiation provide the time-delay data. This technique has obvious applications:

- Breakage of a fiber in a security fence can provide an alarm, or interruption of a signal in a fiber attached to a large structure can indicate strain. British Maritime Technology Ltd. has produced a series of strain sensors for use on offshore structures. Much of the effort has been put into developing consistent methods of bonding the fiber to the structure (Fig. 8.39). Ideally, in such systems, there should be interrogation electronics at ends of the fiber, so that more than a single measurement is possible.
- Analog outputs can be obtained from microbend displacement sensors. Fibers can be coated with a strain-sensitive material, or gaps between the ends of fibers can be filled with materials of variable refractive index, or partially reflecting splices can generate signals that can be analyzed in the time domain. This list comprises a mixture of intrinsic, extrinsic, and evanescent sensors, but the common interrogation technique involves variations on OTDR. Several systems have been proposed in which up to about 100 sensors can be placed in one loop. For pipeline leakage and security applications, the sensors can be placed at specific points along the pipe. For a 1-km fiber length, sensor separations of 50 cm can be resolved.

The combination of wavelength multiplexing techniques with OTDR has yet to be properly investigated, although it has been proposed[61] that the phase information returned with the back-scattered radiation could be interpreted for sensing. Several attempts at using variations in higher-intensity nonlinear effects[62,63] (e.g., Stokes, anti-Stokes, and Raman scattering effects) have also been proposed (see also Sec. 8.7.1). Practical devices are likely to be expensive and suffer from fur-

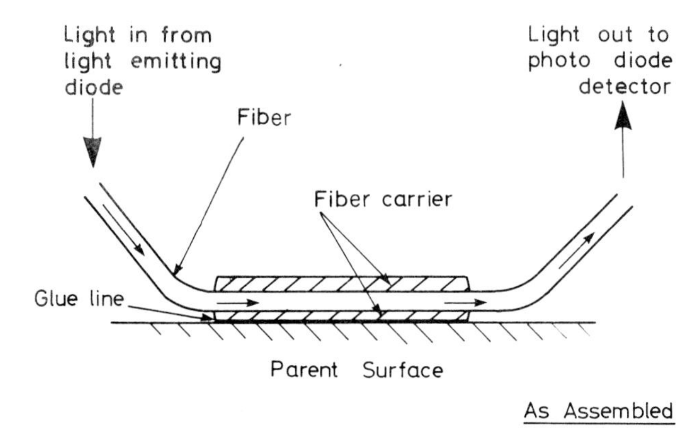


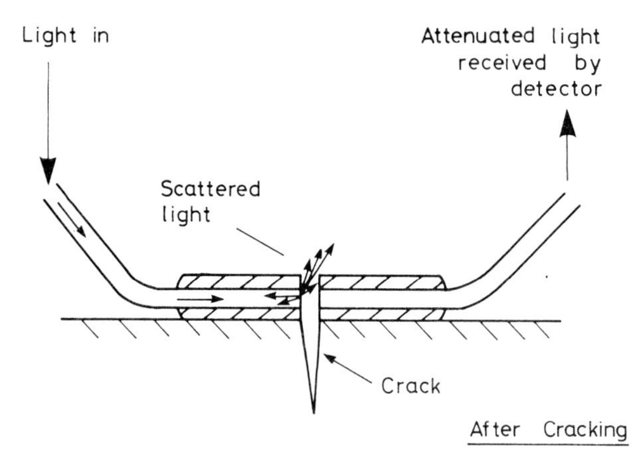


**FIGURE 8.39** Schematic of the sensing element interrogated by OTDR. (*a*) As assembled; (*b*) after cracking. (*Courtesy of British Maritime Technology Ltd.*)

ther perturbing effects, apart from added safety hazards in the use of high optical power levels. Nevertheless, this area can be viewed with interest for the future.

## 8.5 OPTICAL COMPONENTS FOR FIBER SENSOR SYSTEMS

The previous discussions have indicated that, to make several sensor systems viable, specialized optical-fiber components are required. It is not the intention here to describe the fundamental optical properties of such devices; much of this is covered in other chapters. In any optical-fiber sensor system development, apart from the basic sensing mechanism and mechanical design of the sensor, it is the cost, specification, and availability of the necessary optical components which provide the limiting factors. Several examples are briefly reviewed.

### 8.5.1 Optical Sources

Solid-state coherent sources for interferometric applications tend to be more robust and acceptable from a packaging viewpoint than gas lasers. Suitable technologies have been developed for launching light into fibers, based on the parallel telecommunications developments. The wavelength stability and linewidth (for coherence) of such devices are crucial, however, in maintaining the low noise levels required in highly sensitive interferometric systems. The use and advantages of wideband sources, e.g., superluminescent diodes, in fiber-optic gyroscopes have also been described earlier. The most important contributions to noise are

*Phase noise:* Gas lasers can have a long coherence length and a stable frequency with a narrow linewidth, since the emission involves transitions between discrete energy levels in the gas. They are therefore ideal for more general interferometry applications and are used widely for accurate calibration of coordinate-measuring machines. Attention has been focused, however, on solid-state lasers, particularly gallium aluminum arsenide (GaAlAs) diodes, since they are at present more reliable and compact. Emitted radiation comes from inter-energy-band transitions, however, and coupled with their very short cavity length, gives them wider linewidths and shorter coherence lengths. The frequency stability with temperature will vary with the energy bandgap movement and cavity length change, while for HeNe lasers the largest contribution to frequency change comes from minute changes in internal mirror spacing. The effect of such fluctuations in solid-state lasers can lead to noise effects at the detector of a Mach-Zehnder interferometer which are indistinguishable from the phase changes to be measured. Various techniques have been proposed to reduce such noise effects and frequency-stabilize laser diodes, including using passive optical feedback with external cavities to produce line narrowing, multiple external mirrors, an all-fiber external Fabry-Perot stabilizer, and reflections from the coherent backscatter of a coiled fiber. Temperature stability is also necessary for external cavities, and they should be appropriately mounted with the laser in a thermally controlled environment. Typical phase noise improvements of approximately 50 dB can be achieved when several of these techniques are combined.

*Amplitude or intensity noise:* Free-running solid-state lasers can have noise levels which will provide equivalent phase errors, because of variations in signal frequency, of greater than $10^{-6}$ radian at low frequencies (less than 1 kHz). This can be exacerbated if only a small percentage (less than 0.01 percent) of light is reflected back to the laser from splices and couplers in the interferometer. Hopping between longitudinal modes can occur. This can then cause line broadening and a decrease in effective coherence length, leading to an unwanted requirement to match the lengths of the arms in a Mach-Zehnder interferometer. To overcome unnecessary back-reflections, the design should incorporate angled fiber ends, so that light does not reenter the laser cavity.

For multimode systems and industrial applications, cheaper incoherent LEDs are generally preferred. For multiplexed systems, with numerous sensors attached to the bus, there can be a requirement for high-power sources, e.g., up to 200 mW. To some extent, this negates the safety aspects which are a feature of optical fibers (see Sec. 8.7.3) by posing the potential threat of eye damage or risk of explosion in hazardous spaces, should the fiber break. However, for special applications, e.g., fly-by-light, where the advantages of lightness, low bulk and immunity to electromagnetic interference are paramount, such considerations can be secondary.

The trend in telecommunications is toward longer-wavelength systems, leaving the long-term availability of shorter-wavelength sources questionable. The larger communications-driven demand will maintain the high cost of these devices, and it will often prove difficult to justify their use for sensing applications. The advent of the laser disk, with (almost) visible-wavelength GaAlAs lasers, and other advances to produce improved visible-wavelength devices, provides opportunities for use in other sensing systems, although their stability, linewidth, and operating characteristics must be carefully checked, should they be required for single-longitudinal-mode applications. There are uses for these and other visible optical sources coupled to plastic fibers. Over short distances, for automotive sensors and illuminators, there are several applications. Splicing and alignment of visible light within plastic fibers is attractive for applications requiring regular maintenance. In the future, pumped-fiber lasers and diode-pumped ring lasers will come into consideration for some applications.

### 8.5.2 Detectors

A discussion of detectors for optical-fiber communications systems has been provided in Chap. 6. For sensing systems, however, there are clear cost and reliability advantages in using silicon detectors rather than the longer-wavelength devices. The quantum limits of sensor systems, after calculating optical loss budgets, are governed by the shot noise limits of the detector over the system temperature range. The use of optical-fiber detector arrays for improved angular resolution in systems should always be considered.

### 8.5.3 Fibers and Cables

Sensing applications tend to be highly specific and require special designs of the probe head for variable ambient conditions. Thus high-temperature applications, such as combustion control and automotive commutator and injection monitoring, require special high-temperature polymer or metallic coatings on the exposed fiber. These can provide added perturbations in their own right, if care is not taken in the design of the fiber packaging. However, the removal of the source and detector from the high-temperature sensing region can have considerable benefit in signal-to-noise ratio. There are other cases, e.g., down-well sensing, where the application is made possible only by the availability of a cable that can withstand high temperatures, high static fatigue, and a corrosive environment. If they are to be used in hazardous environments, the flame-retardant quality or breakdown voltage of the coating is a significant factor.

The composition of the fiber can often be critical. Graded-index fiber, for example, can be more sensitive than step-index multimode fiber for microbend devices. Doping of optical fiber with rare-earth metals can enhance Faraday rotation and secondary emission effects. Special polymer fiber arrays are also used as scintillators for gamma and x-ray detection.[64] In polarization-preserving fiber, the birefringence of the fiber is made deliberately asymmetric by including regions of high stress or by making the core or cladding in an elliptical geometry. Despite the induced preservation of polarization in high-birefringence fibers, care in coiling and winding for sensors is still necessary, if cross-coupling of the orthogonal modes is to be prevented. Several basic references to research in this field are provided at the end of this chapter. Descriptions of special polarization-preserving fiber preparation, relevant to sensor applications, can be found in the literature.[65,66]

### 8.5.4 Couplers

Many systems have failed to reach the marketplace because of the high cost of couplers. Even a simple intensity-displacement sensor requires a Y coupler. When several devices are to be multiplexed, several couplers (with their associated losses) have to be incorporated. While multiway couplers are available, sensor markets, by their very nature, are still often insufficient to warrant increased manufacturing investment to reduce costs. At the top end of the range, in interferometric systems, there are the requirements for 2 × 2 (i.e., 3-dB) single-mode fiber couplers for branching the radiation along the sample and reference arms of a Mach-Zehnder interferometer. A significant decrease in noise level of early experimental sensor systems was achieved when fused single-mode couplers were first used. The 3 × 3 coupler was subsequently developed for passive demodulation schemes for hydrophones and magnetic sensors. The outputs of couplers can vary with wavelength and temperature, hence their packaging design is particularly important if they are to be used in a sensor or in wavelength-multiplexed systems.

### 8.5.5 Other Components

Other components are being developed for fiber sensor systems, e.g., frequency shifters (Bragg cells) and integrated-optic devices for local-area network (LAN) switches and modulators. However, robust coupling of fibers to integrated-optic chips remains a specialized and expensive art, rather than an accepted manufacturing practice. Several sensor designs have actually evolved from simple coupler technology (e.g., shutter devices).

## 8.6 SYSTEM MULTIPLEXING OF SENSORS

It was emphasized in the introduction to this chapter that fiber-optic sensors should ultimately fit into networks that are now electrical, such as the system shown in Fig. 8.3. In such a scheme, cables are led from local control centers to different sensors. Redundancy of sensors for highly safety conscious or critical applications is often included in the design, with hard wiring back to the central control unit for reliability. The example in Fig. 8.3, taken from the offshore environment, is a special case, but it illustrates the degree of conservatism in such designs and the reluctance to use new technology, even when there are clear

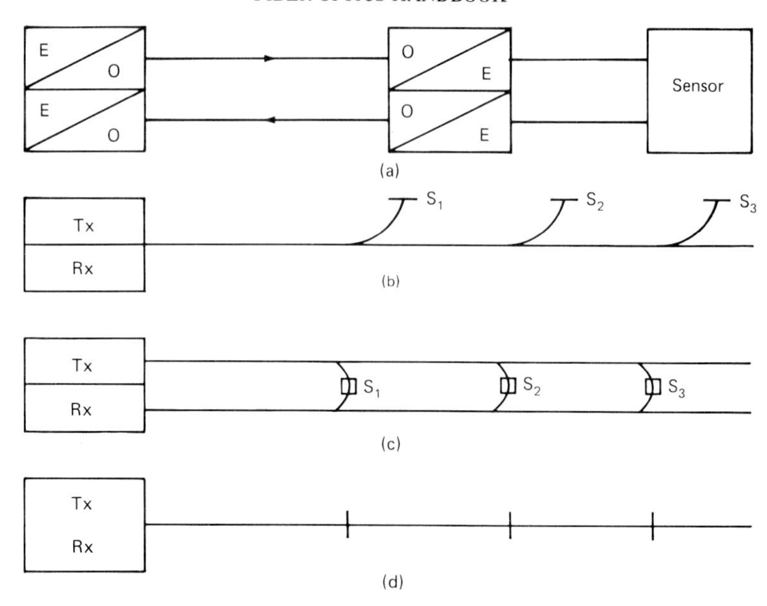


**FIGURE 8.40** Trends in optical-fiber sensor multiplexing. (*a*) Electrically operated sensors with fiber-optic transmission; (*b*) acoustic modulation from sensors to the fiber-optic bus; (*c*) transmission sensors coupled to the highway; (d) reflective splices interrogated in the time domain (i.e., via OTDR).

safety advantages. Only since the 1980s have fiber-optic cables been tentatively used for inter-control-center communications offshore, with occasional trial links to and from specific sensors. Little attempt had been made to try out more advanced multiplexing schemes with fiber optics.

Nevertheless, fiber optics are gradually encroaching into process control, leading to revisions in the design of the sensor head. Figure 8.40 shows the anticipated gradual trend for multiplexing sensors, starting with hybrid electrical and fiber-optics systems to fully optical fiber systems. Figure 8.40*a* shows simple fiber links to and from already commercially available sensors. Such a system has been made by ITT Barton/Fuji for process control applications, interfacing with differential pressure transmitters, pressure sensors, and flowmeters. A schematic diagram of this system is shown in Fig. 8.41. Figure 8.40*b* and *c* are schematic line diagrams of a proposed sensor encoder system. This consists of a series of multimode fiber links of varying length to reflective on-off sensors. Similarly, transmissive devices can be incorporated, provided there is a return path. Analysis is carried out in the time domain, where the speed of processing depends on the differential optical path length to the sensors. Figure 8.40*b* can also be construed as the line diagram for multiplexing remotely powered frequency-out transducers employing quartz, micromachined silicon, or other devices described in Sec. 8.2.5. With wholly optically excited frequency-out sensors, it is possible to visualize systems combining time-division multiplexing (TDM), frequency-division multiplexing (FDM), and wavelength-division multiplexing (WDM), where all the information is carried on one fiber bus. While the system theoretically should not require complex referencing techniques, there are a number of

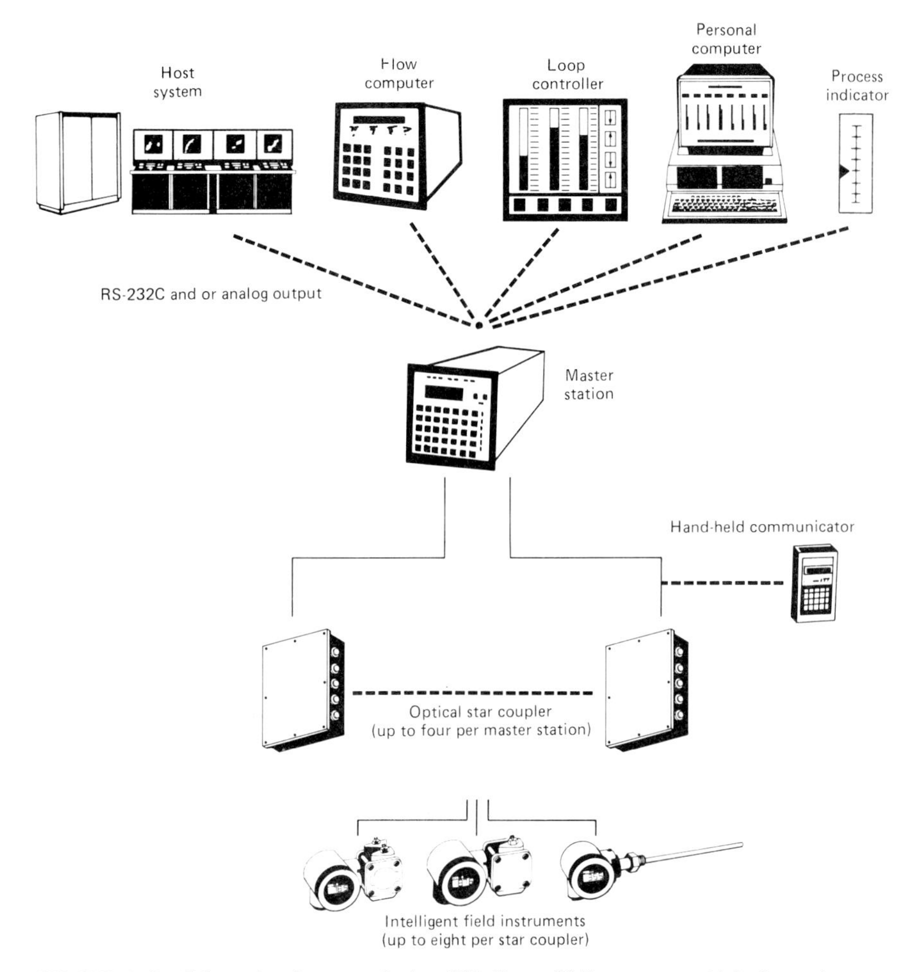


**FIGURE 8.41** Schematic diagram of the ITT Barton/Fuji sensor multiplexing scheme. (*Courtesy of ITT Barton/Fuji Ltd.*)

couplers and special filters to be used, which must first become available. Figure 8.40*d* shows how OTDR can be used on a single fiber. Other variations on OTDR techniques can be used in conjunction with strain sensors such as microbend devices or reflective splices. With mode strippers before each microbend sensor, it is possible to tap off the added light coupled into the cladding after the sensor. Difficulties arise over optical-loss budgets, since such systems require considerable precision of alignment for the taps and splices.

For intensity sensors, the problem of referencing in multimode systems provides complications. For reflective systems, this can be done by doubling the number of fibers and using a dummy sensor which is not directly subjected to the environmental stress to be measured. Alternatively, as discussed in Sec. 8.2, wavelength-sensitive filters can be used for referencing, with WDM techniques, although the stability of such devices should be verified over the operating range of the system.

Single-mode interferometric sensors pose particular problems for multiplexing (apart from the cost of couplers, etc.), since it is the phase which must usually be measured. Potential phase-based sensors analogous to those in radar have been suggested.[67,68] Frequency-modulated carrier wave techniques, similar to those used in FM ranging systems, have been proposed. Frequency-chirped lasers having a large (1-GHz) modulating bandwidth and a reasonably linear frequency-modulation characteristic can be used.[69] It is possible to address a series of parallel interferometers so that adequate coherence is present but beating between different interferometers is outside the coherent length of the source. However, this relies on obtaining lasers of sufficient stability and adequate coherence length relative to the loops to be multiplexed.

Further techniques have been based on the reflective splice technique previously discussed for hydrophone applications (Fig. 8.28). Care is necessary to obtain reproducible reflective splices that do not degrade with time. Experimental data on a two-moded fiber system have been obtained[70] using either the two orthogonally polarized modes in high-birefringence fiber, or the two sets of modes in a step-index fiber defined at a wavelength just before cutoff of third-order modes. FM CW laser operation is used. After allowing for noise effects, including thermal variations in the linearity of the frequency traverse and phase noise, a potential resolution of 5 cm should be possible for a single-sensor system. Most phase- or frequency-modulated systems are still conceptual, and further work is necessary. A summary of several multiplexing techniques is given in Table 8.16.

## *8.7 POWER TRANSMISSION, ACTUATION, AND SAFETY ASPECTS OF DESIGN*

In this section, the power transmission limitations in optical fibers are briefly discussed. These are relevant to sensing applications, since optical power has been used, either directly or through remote optically switched battery power, for the actuation of valves and other alarms or switch mechanisms. Nonlinear effects at high power levels in single-mode fiber have also been proposed for distributed sensors using variations of optical reflectometry.[62,63] For several multiplexing applications, however, even on local-area networks, for the system to have a viable number of nodes, higher-power sources are necessary. As a result, 500-mW CW (or greater) laser diode devices are attractive. These power levels demand commensurate safety considerations, particularly in the industrial environment.

### 8.7.1 Power Transmission

At high power levels, it has been demonstrated that 400 W CW can be passed along meter lengths of large-diameter multimode optical fiber for robotic welding applications, while at the lower end, optical power levels of less than 10 mW have proved adequate for remotely powering pneumatic valves. In multimode fibers,

**TABLE 8.16** Fiber-Optic Sensor Multiplexing: Some Examples of Multimode Techniques

| Multiplexing type | Limitations | Typical applications | Comments |
|---|---|---|---|
| Separate links | Separate fibers to and from sensor, with electrical multiplexing at source. Initial systems are similar to available electrical communications and sensor links. | Limited sensing applications, e.g., separate detectors with one control box for sequential switching and alarms. Local control centers with remote batteries. | Technique may not be cost-effective for very large numbers of sensors. Usually selected on basis of communications advantage. |
| Optical power to local photodetectors and frequency out transducers | Uses separate fibers to each sensor. Electrical multiplexing at source, in that the source frequency is electrically driven. | As above, with the possibility of eliminating batteries. | Under development. |
| Wavelength-division multiplexing | Wideband source/sources with multiway couplers. | Increased number of sensors with frequency-selective inputs to data bus | Size and robustness of wavelength splitters and detection systems require evaluation and development. Nodes are limited by linewidths of sources, wavelength selective couplers, fiber dispersion. Components can provide crosstalk limitations. Stability of filters and wavelength sensitivity of components (e.g., couplers) needs investigation. |

**TABLE 8.16** Fiber-Optic Sensor Multiplexing: Some Examples of Multimode Techniques (*Continued*)

| Multiplexing type | Limitations | Typical applications | Comments |
|---|---|---|---|
| Time-domain multipexing | Sequential optical sampling of channels. Each channel sampled at a rate of 2 × highest significant frequency. Techniques are possible for equally spaced and separated sensors. | Hydrophone arrays. | High-intensity sources often required. Increased numbers of couplers adds to loss budget. Transmission distance depends on overall loss budget. For optical fiber arrays, sampling rate must be increased. |
| Frequency-domain reflectometry | Chirped laser pulse transmitted. | Reflective transducers. Interferometric sensors. | FM processing has wideband capability, which improves resolution with respect to pulsed time-domain techniques. |
| Optical time domain | Combination of TDM and reflectometry. | Digital reflective/transmissive transducers. Can also use microbend sensors, e.g., with dark or bright field taps. | Reference signal required. Full evaluation necessary for analog output devices. |
| PZT modulation onto data bus | Attached PZT modulators to optical fiber. | Inputs to fiber bus without splicing. | Not wholly passive. |
| Frequency-division multiplexing | With point frequency-out transducers, combinations with TDM and WDM can be used. | Point sensors for displacement. | Under development. |

*Source:* Adapted from Ref. 3.

the onset of nonlinear Raman and stimulated Brillouin scattering occurs at much higher power levels than for single-mode fiber.

Proposals have been advanced for distributed temperature sensors using OTDR to measure the change in intensity of Rayleigh backscattering from liquid-filled fiber cores[62] and for those using also Raman scattering.[63] (See also Sec. 8.6.) Temperature values can be obtained by measuring the ratio of anti-Stokes and Stokes components of back-scattered radiation in multimode fiber. The ability to measure temperatures from 77 to 800°K with a spatial resolution of less than 3 m in sensor lengths of 1 km has been demonstrated. Argon ion and semiconductor laser sources were used to produce pulses of 5 W (at a repetition rate of 40 kHz) and 3 W (at 10 kHz), respectively. Considerable filtering and signal analysis were required.

For telecommunications, wavelength dispersion effects are important and hence narrow-line coherent sources are used. Stimulated Brillouin scattering can then limit power transmission to only a few milliwatts over long lengths (e.g., 15 km) on long-wavelength single-mode (9-μm core) fiber. If a system objective is to simply pass optical power, good performance can be obtained by using a range of wideband, multiple-wavelength sources.

Multimode fibers, in contrast to single-mode fibers, offer a high launch efficiency from a semiconductor diode laser, and the effective power density can be reduced by using larger-core fiber. With a pulsed neodymium-doped yttrium aluminum garnet (Nd:YAG) source, a 50-μm-core multimode fiber can carry in excess of 100 W over 1 km.[71] The implications for sensing systems are considerable, in that cooled solid-state lasers and multiple fibers, can be used to provide power level transmissons of several tens of milliwatts over tens of kilometers. Thus, the technology is available for installing several remotely powered sensing nodes at kilometer distances. The efficiency problems arise at launch and energy conversion.

### 8.7.2 Actuation

Figure 8.42 shows a schematic diagram from a GEC Research Ltd. demonstration of actuation with a fiber attached to a remote process control valve. (Earlier, an optically powered telephone system had been demonstrated by AT&T Bell Laboratories using a laser-fiber-photovoltaic cell arrangement to provide the necessary power to return the transmission via a second optical source.) Pulse-width-modulated LEDs are used to operate a servo pilot valve via a remote photovoltaic cell. With further remote pneumatic amplification, a full 3 to 15 lb/in$^2$ (20 to 100 kN/m$^2$) pneumatic signal can be made available for valve control.

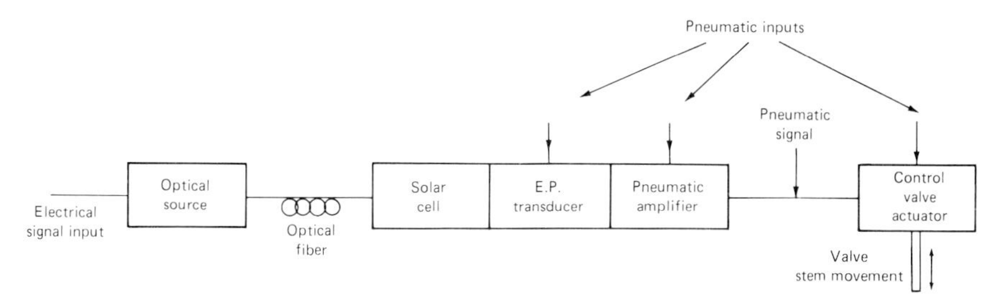


**FIGURE 8.42** Remote actuation of a valve by optical fiber. *(Courtesy GEC Research, Ltd.)*

The initial maximum photovoltaic conversion efficiency is typically 40 percent for silicon and 68 percent for gallium arsenide detectors at room temperature. Thus a 7-mW input to the photovoltaic device from a laser diode and 200-μm fiber core tens of meters long will produce 22 mA in the 185-Ω coil of the electrical/pressure balanced pilot valve. A change of 2 $lb/in^2$ (14 $kN/m^2$) is given at the output, which is then further amplified pneumatically. Subsequent results have been obtained by GEC Research, using lower-power (approximately 1-mW) LEDs but with higher pneumatic amplification. With feedback, it should be possible to proportionally control and sense the position of the valve with an encoder or other optical-fiber sensor.

With higher optical powers, it should be possible to remotely actuate valves using thermal effects,[3] although response times will be slow, and the safety aspects remain a consideration. Nevertheless, the demonstrated ability to pass such powers to remote power storage units means that switching and actuations are possible at very remote locations (over several kilometers distant). Mechanical designs can use discharge and ratcheting techniques to manipulate large bodies. In the thermally stable subsea environment for offshore applications, for example, the conversion efficiencies will be higher at the lower temperatures of the seabed. Potential cost savings involved in rapid dispensing of reduced-volume fiber cables from smaller seagoing or airborne vessels could be appreciable.

### 8.7.3 Safety

The lack of metallic components in optical fibers leads to a reduced spark hazard. However, at higher power levels, the two disadvantages are possible—eye damage and explosion hazard—should the fiber break. The escaping radiation could be sufficiently high to ignite particulates in the beam. The first disadvantage is already covered by international health and safety regulations for laser radiation levels. The mechanisms of the second area of risk—that of explosion or ignition—remain unclear. Several tests by Sira Safety Services have been carried out on explosive effects, where the cause of ignition is postulated as being the absorption of radiation by a small particle. The Sira Safety Services work considered particles in the tens of microns range for ignition.

Most safety work has concentrated on the type of beam likely to emerge from a fractured fiber; e.g., a smooth break causes a divergent beam. The worst case would be a particle of approximately the same diameter as the fiber core, situated within a few diameters' distance from the core. It is estimated that a 50-μm particle, absorbing radiation in the path of a 6-mW beam, may ignite diethyl ether or carbon disulfide. Reducing the particle diameter by a factor of 2 reduces the power required to cause ignition by a factor of 2.5. This confirms that, for several of the applications described in this chapter, the passage of fibers carrying relatively high powers through potentially explosive environments for sensing should be considered carefully.

Applications where relatively high powers are required—for example, for actuation, data transport over buses, and remote sensing (e.g., actuation of some frequency-out transducers)—are not necessarily inhibited by an explosive risk, and installation will depend on the environment in which they are to operate. Provided that adequate precautions (still to be defined) are taken, the other advantages of fiber sensors, e.g., lightness, reduced volume, and flexibility, will remain valid.

## 8.8 APPLICATIONS

Some examples of fiber-optic sensor applications are given in Table 8.17.

### 8.8.1 Defense

Early fiber-sensor developments were driven by defense requirements, where cost has not been an overriding design factor. Multimode systems for fly-by-light concepts using encoders and other point sensors have been successfully implemented and tested. Interferometric devices for hydrophones and gyroscopes are emerging, although the cross-perturbing effects of temperature and other variables have had a major impact on the overall cost and on the packaging designs.

**TABLE 8.17** Fiber-Optic Sensor Application Areas: Some Examples

| Area | Methods (Examples) | Comments |
|---|---|---|
| Medical | Analytical; surface adsorption, e.g., antibody-antigen immobilization. Total internal reflection. Laser doppler. | Throw-away. Small devices for *in vivo* insertion in blood flow. Inert sensors. |
| Defense | Surveillance: acoustic, magnetic. Navigation: magnetic, fiber-optic gyro. Fire, flash, radiation, temperature, fail-safe switches, etc. | All types required, from extended systems to point sensors. |
| Aerospace | Encoders, switches, oil-in-water, particulates in fuels, oil. Pressure displacement. Machinery condition monitoring. | Should have low weight and withstand shock, vibration, and large temperature range. Strain-sensitive systems for structural surveillance. |
| Offshore | Platform gas detection, local and over an area. Fire, smoke. Well head controls. Pollution control with no fire risk. Down-well pressure, temperature, etc. Umbilicals to submersibles, divers. | Sensors are closely linked to developments on cables, which must be hermetic and withstand static fatigue. |
| Utilities | Gas detection: point and over an area. Current and voltage detection, transformer temperatures. Water ingress to feed heater tubes, etc. | Specialized applications. |
| Industrial | Flow, pressure, temperature, etc. Combustion energy control. | Should be cost-competitive, and will usually require multiplexing. Standardization of optical components (e.g., couplers) needed. |

*Source:* Adapted from Ref. 3.

With the advent of stronger fibers, the ability to rapidly dispense them from light aircraft and seaborne vessels will lead to the use of fiber sensors which can be remotely situated and optically powered and interrogated. These will provide arrays for surveillance purposes, where the passive, nearly undetectable nature of the sensor is attractive. Use of interferometric sensors in the form of towed acoustic arrays faces difficulties in design, given the strains and vibrations involved.

### 8.8.2 Industrial

Industrial fiber-optic sensor systems have not yet made a major impact in terms of displacing more conventional sensor techniques, largely because of the cost of specialized optical components. Users have tended to be highly conservative, requiring lengthy trial periods, high reliability, and some form of multiplexing capability. Various attempts at total systems within a plant have been made, but with little evidence of industrial users rushing to embrace fiber-optic sensor technology. Experience has shown that the first successfully engineered solutions have been in the area of specialized sensors. Standardization of optical couplers on a worldwide basis remains a problem. Most requirements will be for low-power LED, multimode-fiber applications, rather than for the more intricate interferometric solutions.

### 8.8.3 Marine

Several marine fiber-optic sensor systems have been installed, largely to overcome interference effects from large machinery in compact areas. On offshore platforms, the reduced volume and weight of fiber-optic cable is attractive, particularly when fiber-optic cable can be laid next to power cables in existing ducting. Use of fibers in hazardous spaces, as well as in cable bulkhead penetrations, is now accepted by some of the relevant classification societies. Wideband cables are used for umbilical connections to submersibles and divers for closed-circuit television and speech transmission.

### 8.8.4 Medical

The "automatic bedside physician," using cheap, throwaway sensors for computer analysis of blood, urine, and other samples, is approaching realization. Surface adsorption, with refractive index changes caused by surface reactions with specialized coatings, has been successfully demonstrated.[72] These techniques can also be applied to control sensing in drug production, for the detection of specific compounds. Once it has been proved that such devices can be made cheaply and reproducibly for one-time, throwaway applications, the technical objection to lack of reversibility of surface-effect sensors presents no further problem. The inertness and small volume of glass fibers are ideal properties for *in vivo* examination. Optical fibers can also be applied in an active mode for localized heating and removal of fatty deposits in arteries, for example. Pressure sensors, using silicon diaphragm sensors at the fiber tip, can be inserted into the blood stream, and methods have been devised for measurement of blood flow by doppler technique.

### 8.8.5 Oil Exploration

Down-hole optical fiber sensors are being increasingly studied for exploration and logging. Larger-bandwidth information is required for greater strata resolution. Pressure sensors and cables must withstand at least 250°C and 20,000 $lb/in^2$ (138 $MN/m^2$) within an extremely corrosive environment. Typically, a pressure measurement in this environment should have an accuracy between 1 part in $10^4$ and 1 part in $10^5$ (e.g., $\pm 1$ $lb/in^2$ in 20,000 $lb/in^2$), and an absolute accuracy of about $\pm 10$ $lb/in^2$ (70 $kN/m^2$). To obtain the accuracies required, temperature compensation is needed (i.e., combined optical fiber pressure and temperature sensors suitably packaged). Directional measurements down-hole, using a small, robust fiber-optic gyro, are a possibility.

### 8.8.6 Utilities

Voltage and current detection schemes have been described, where all-fiber or fiber-birefringent elements are used. Packaging and correct attachment of the sensor to the wire is important. The separate sensor approach (rather than all-fiber devices) has the advantage that local maintenance and replacement are more convenient. Fiber sensors for the detection of radiation in nuclear plants are in use.[73] Defects formed in the fiber over a measurable time give an indication of the type and dosage of radiation. Other process parameters have been measured to advantage because of the sensor's immunity to interference.

## 8.9 CONCLUSIONS

Sensing applications for optical fibers are following closely advances in the use of fibers in telecommunications. However, it is likely that the rate of acceptance will be slow, until fully engineered, robust, and cheap fiber sensor systems become available. It has been emphasized that the mechanical design and systems aspects are important. Controlled laboratory tests are not always relevant to the operation of a sensor system in the field. The engineer should first distinguish which technique will provide advantages over other available technologies, operate in difficult environments and remain robust, with little requirement for maintenance. The field of fiber-optic sensors covers a wide range, varying from highly expensive devices with military application, often requiring specialized and costly components, to simpler process control and medical applications where low cost, safety, and reliability are prime considerations.

Two cautionary points are made:

1. Most fiber-optic sensors are susceptible to thermal effects, e.g., thermal strain in interferometric or polarimetric sensors. In the past, some systems have been proposed for the measurement of several other variables, but in reality,

they have proved to be nothing more than highly expensive temperature sensors.

2. At the other end of the scale, a cheap, robust, and reproducible fiber-optic temperature sensor, which can compete with a thermocouple remains to be developed. At the time of writing, no fiber-optic sensor has been accepted as a standardized temperature-measuring device, submitting to easy replacement and maintenance on a worldwide basis. The cost and degree of standardization of the necessary optical-fiber couplers and other optical components has often been an inhibiting factor.

It has been shown that, with the advent of techniques for transmission of safe optical power on fibers, the control loop for both sensing and actuation has been theoretically closed. This has allowed improved multiplexing methods for optical-fiber sensors. Hybrid fiber-optic and semiconductor sensors show particular promise for high-volume, low-cost applications.

## ACKNOWLEDGMENTS

The author wishes to acknowledge the assistance provided by his former colleagues at STC Technology Ltd., particularly Dr. R. E. Jones, Dr. J. R. Willson, Dr. M. Bone, and Mr. M. M. Ramsay. The information from Dr. D. Wilson, Dr. R. H. Pratt, and Dr. R. J. Chaney at Renishaw Transducer Systems, Ltd., and from that of Dr. D. N. Batchelder at Queen Mary College, London is also gratefully acknowledged. The Directors of STC Technology Ltd. have kindly allowed access to their drawing files, and this support is gratefully acknowledged. Permission has been given by the Institute of Electrical Engineers for reproduction of some drawings from the paper by G. D. Pitt et al., "Optical Fibre Sensors" (*IEE Proceedings,* part J, vol. 132, no. 4, pp. 214–248, 1985).

## REFERENCES

1. T. G. Giallorenzi et al., "Optical Fiber Sensor Technology," *IEEE Journal of Quantum Electronics,* vol. QE-18, no. 4, 1982, pp. 626-664.
2. R. A. Bergh, H. C. Lefevre, and H. J. Shaw, "An Overview of Fiber Optic Gyroscopes," *Journal of Lightwave Technology,* vol. LT-2, no. 2, 1984, pp. 91–107.
3. G. D. Pitt et al., "Optical Fibre Sensors," *IEE Proceedings,* vol. 132, pt. J, no. 4, 1985, pp. 214–248.
4. B. Culshaw, *Optical Fibre Sensing and Signal Processing,* Peter Peregrinus, London, 1984.
5. D.A. Jackson, "Monomode Optical Fibre Interferometers for Precision Measurement," *Journal of Physics E: Scientific Instruments,* vol. 18, 1985, pp. 981–1001.
6. A. J. Rogers, "Distributed Optical-Fibre Sensors," *Journal of Physics D: Applied Physics,* vol. 19, 1986, pp. 2237–2255.
7. *Proceedings of Conference on Optical Fiber Sensors—1,* IEE conference pub. no. 221, London, April 26–28, 1983.
8. *Proceedings of Conference on Optical Fiber Sensors—2,* Stuttgart, September 5–7, 1984, VDE Verlag GmbH.

9. *Proceedings of Conference on Optical Fiber Sensors—3,* San Diego, 1985, OSA/IEEE.
10. *Proceedings of Conference on Optical Fiber Sensors—4,* Tokyo, 1986, Institute of Electrical and Commercial Engineers of Japan, Optoelectronic Industry and Technology Development Association.
11. "Fibre Optics 1985 (SIRA)," *Proceedings of the Society of Photo-Optical Instrumentation Engineers,* vol. 522, London, 1985.
12. *Proceedings of the International Conference on Optical Techniques in Process Control,* June 14–16, 1983, The Hague, Netherlands, BHRA Fluid Engineering Publications, Cranfield, Beds., U.K.
13. *Optical Sensors and Optical Techniques in Instrumentation,* November 12, 1981, London, Institute of Measurement and Control.
14. "Fiber Optics in Laser Doppler Velocimetry," *Lasers and Applications,* July 1986, pp. 71–73.
15. J. Knuhtsen, E. Olldag, and P. Buchhave, "Fibre Optic Laser Doppler Anemometer with Bragg Frequency Shift Utilizing Polarization-Preserving Single-Mode Fibre," *Journal of Physics E: Scientific Instruments,* vol. 15, 1982, pp. 1188–1191.
16. K. Chan, H. Ito, and H. Inaba, "Optical Remote Monitoring of $CH_4$ Gas Using Low-Loss Optical Fiber Link and InGaAsP Light Emitting Diode in 1.33 $\mu$m region," *Applied Physics Letters,* vol. 43, 1983, pp. 634–636.
17. B. E. Jones and R. C. Spooncer, "Two Wavelength Referencing of an Optical Fiber Intensity-Modulated Sensor," *Journal of Physics E: Scientific Instruments,* vol. 16, 1983, pp. 1124–1126.
18. M. C. Hutley, R. F. Stevens, and D. E. Putland, "Wavelength Encoded Optical Fibre Sensors," *Sensor Review,* vol. 5, no. 2, 1985, pp. 64–68.
19. K. Kyama, S. Tai, T. Sanada, and H. Nunoshita, "Fiber Optical Instrument for Temperature Measurement," *IEEE Journal of Quantum Electronics,* vol. QE-18, no. 4, 1982, pp. 676–679.
20. W. B. Spillman and R. A. Soref, "Hybrid Fiber Optic Sensors Using Liquid Crystal Light Modulators and Piezo-Ceramic Transducers," *Applied Optics,* vol. 21, no. 15, 1982, pp. 2696–2701.
21. R. J. Baumbick, "Fiber Optic Sensors for Measuring Angular Position and Rotational Speed," NASA technical memorandum 81454, 1980.
22. A. R. Nelson, D. H. McMahon, and R. L. Gravel, "Passive Multiplexing System for Fiber Optic Sensors," *Applied Optics,* vol. 19, 1980, pp. 2917–2920.
23. B. E. Jones and G. S. Philp, "A Vibrating Wire Strain Sensor," European Conference on Sensors and their Applications, UMIST, Manchester, Technical Notes, 1983, pp. 86–87.
24. S. Venkatesh and B. Culshaw, "Optically Activated Vibrations in a Micromachined Silicon Structure," *Electronics Letters,* vol. 21, 1985, pp. 315–317.
25. G. D. Pitt, "Materials Cross-Fertilization," in E. R. Howells (ed.), *Chemicals and Materials for Electronics,* Ellis Horwood, Chichester, U.K., 1984, pp. 245–265.
26. W. B. Spillman, "Multimode Fiber Optic Pressure Sensor Based on the Photoelastic Effect," *Optics Letters,* vol. 7, no. 8, 1982, pp. 388–390; D. H. McMahon, R. A. Soref, and L. E. Sheppard, "Sensitive, Fieldable Photoelastic Fiber Optic Hydrophone," *Journal of Lightwave Technology,* vol. LT-2, 1984, pp. 469–478.
27. S. Tai, K. Hyuma, and M. Nunashita, "Fiber Optic Acceleration Sensor Based on the Photoelastic Effect," *Applied Optics,* vol. 22, no. 11, 1983, pp. 1771–1774.
28. A. M. Smith, "Polarization and Magneto-Optical Properties of Single Mode Optical Fiber," *Applied Optics,* vol. 17, no. 1, 1978, pp. 52–56.
29. Y. Kuroda, Y. Abe, H. Kuwahara, and K. Yoshinaga, "Field Test of Fiber-Optic Voltage and Current Sensors Applied to Gas Insulated Substation," *Proceedings of the Society of Photo-Optical Instrumentation Engineers,* vol. 586, 1986, pp. 30–37.

30. G. D. Pitt, "Oil on Troubled Waters," *Physics Bulletin,* vol. 28, no. 10, 1977, pp. 459–460.

31. F. K. Kawahara et al., "Development of a Novel Method for Monitoring Oil in Water," *Analytical Chemistry Action,* vol. 151, 1983, pp. 315–327.

32. A. L. Harmer, "Optical Fibre Refractometer Using Attenuation of Cladding Modes," *Proceedings of Conference on Optical Fiber Sensors—1,* IEE conference pub. no. 221, London, April 26–28, 1983.

33. G.B. Hocker, "Fiber Optic Sensing of Pressure and Temperature," *Applied Optics,* vol. 19, 1980, pp. 98–107; G. B. Hocker, "Analysis of Fiber Optic Acoustic Sensors with Composite Structures," *Applied Optics,* vol. 18, 1979, pp. 3679–3683.

34. J. A. Bucaro and T. R. Hickman, "Measurement of Sensitivity of Optical Fibers for Acoustic Detection," *Applied Optics,* vol. 18, 1979, pp. 930–940.

35. N. Lagakos et al., "Optimizing Fiber Coatings for Interferometric Acoustic Sensors," *IEEE Journal of Quantum Electronics,* vol. QE-18, no. 4, 1982, pp. 683–689; N. Lagakos and J. A. Bucaro, "Pressure Desensitization of Optical Fibers," *Applied Optics,* vol. 20, 1981, pp. 2716–2720.

36. D. R. Biswas and S. Raychauduri, "Optomechanical Properties of Long-Length Nickel Coated Fibres," *Proceedings of Conference on Optical Fiber Sensors—3,* San Diego, 1985, OSA/IEEE, p. 124.

37. K. P. Koo, A. Dandridge, A. B. Tveten, and G. H. Sigel, "A Fiber Optic DC Magnetometer," *Journal of Lightwave Technology,* vol. LT-1, 1983, pp. 524–525.

38. H. Talaat, W. J. Moore, J. Jarzyinski, and J. A. Bucaro, "Fourier Transform IR Photoacoustic Study of Piezo-Electric Coatings on Optical Fibers," in *Proceedings of Conference on Optical Fiber Sensors—2,* Stuttgart, September 5–7, 1984, VDE Verlag GmbH, pp. 124, 125.

39. N. Lagakos et al., "Optimization of the Ultrasonic Sensitivity of Single Mode Fibers," in *Proceedings of Conference on Optical Fiber Sensors—2,* Stuttgart, September 5–7, 1984, VDE Verlag GmbH, pp. 134, 135.

40. M. Butler, "Optical Scientist Can Tell by the Swell," *Lasers and Applications,* November 1984, pp. 40–42.

41. J. P. Willson and P. G. Hale, "Limit to Sensitivity of Optical Fiber Magnetometers due to Shape Anisotropy," *Electronics Letters,* vol. 22 no. 11, pp. 567–569.

42. J. A. Bucaro, H. D. Dardy, and E. F. Carome, "Fiber Optic Hydrophone," *Journal of the Acoustical Society of America,* vol. 62, 1977, pp. 1302–1305.

43. A. Dandridge et al., "Single-Mode Diode Phase Noise," *Applied Physics Letters,* vol. 38, 1981, pp. 77–78.

44. K. P. Koo, A. B. Tveten, and A. Dandridge, "Passive Stabilization Scheme for Fiber Interferometers Using (3 × 3) Fiber Directional Couplers," *Applied Physics Letters,* vol. 41, 1982, pp. 616.

45. R. H. Pratt et al., "Optical Fibre Magnetometer Using a Stabilised Semiconductor Laser Source," *Proceedings of Conference on Optical Fiber Sensors—1,* IEE conference pub. no. 221, London, April 26–28, 1983, pp. 45–50.

46. E. W. Saaski et al., "A Family of Fibre Optic Sensors Using Cavity Resonator Microshifts," *Proceedings of Conference on Optical Fiber Sensors—4,* Tokyo, 1986, Institute of Electrical and Commercial Engineers of Japan, Optoelectronic Industry and Technology Development Association, pp. 11–14.

47. J. P. Dakin, C. A. Wade, and M. L. Henning, "Novel Optical Hydrophone Array Using a Single Laser Source and Detector," *Electronics Letters,* vol. 20, 1984, pp. 53–54.

48. R. E. Meyer, S. Ezekiel, D. W. Stowe, and V. J. Tekippe, "Passive Fiber Optic Ring Resonator for Rotation Sensing," *Optics Letters,* vol. 8, no. 12, 1983, pp. 644–646.

49. D. M. Shupe, "Thermally Induced Non-Reciprocity in the Fiber Optic Interferometer, *Applied Optics,* vol. 19, no. 5, 1980, pp. 654–655.

50. M. C. Bone and J. W. Parker, "An Integrated Optic Fiber Gyroscope; Performance and Limitations," *Proceedings of Conference on Optical Fiber Sensors—1,* IEE conference pub. no. 221, London, April 26–28, 1983, pp. 143–146.

51. A. Yariv and H. Winsor, "Proposal for Detection of Magnetostrictive Perturbation of Optical Fibers," *Optics Letters,* vol. 5, 1980, pp. 87–89.

52. A. D. Kersey, M. Clarke, and D. A. Jackson, "Closed-Loop DC Field Fiber Optic Magnetometer," *Proceedings of Conference on Optical Fiber Sensors—2,* Stuttgart, September 5–7, 1984, VDE Verlag GmbH, pp. 51–54.

53. K. P. Koo, F. Bucholtz, and A. Dandridge, "A New Sampling Detection Scheme for High Sensitivity Fiber Optic Magnetometer," *Proceedings of Conference on Optical Fiber Sensors—4,* Tokyo, 1986, Institute of Electrical and Commercial Engineers of Japan, Optoelectronic Industry and Technology Development Association, pp. 77–80.

54. K. P. Koo and G. H. Sigel, "A Fiber Optic Gradiometer," *Journal of Lightwave Technology,* vol. LT-1, 1983, pp. 524–525.

55. S. C. Rashleigh, "Origins and Control of Polarization Effects in Single-Mode Fiber," *Journal of Lightwave Technology,* vol. LT-2, 1983, pp. 312–331.

56. J. P. Dakin, S. Broderick, D. C. Carless, and C. A. Wade, "Operation of a Compensated Polarimetric Sensor with Semiconductor Light Source," *Proceedings of Conference on Optical Fiber Sensors—2,* Stuttgart, September 5–7, 1984, VDE Verlag GmbH, pp. 241–246.

57. N. Lagakos et al., "Microbend Fiber Optic Sensor as Extended Hydrophone," *IEEE Journal of Quantum Electronics,* vol. QE-18, no. 10, 1982, pp. 1633–1638.

58. G. F. Lipscomb, S. K. Yao, and C. K. Asawa, "Stabilization of Single and Multimode Fiber Optical Microbend Sensors," *Proceedings of Conference on Optical Fiber Sensors—1,* IEE conference pub. no. 221, London, April 26–28, 1983, pp. 117–121.

59. Y. Imai, T. Saino, and Y. Ohtuka, "Temperature or Strain Insensitive Sensing Based on Bending-Induced Retardations in a Birefringent Single-Mode Fiber," *Proceedings of Conference on Optical Fiber Sensors—4,* Tokyo, 1986, Institute of Electrical and Commercial Engineers of Japan, Optoelectronic Industry and Technology Development Association, pp. 89–92.

60. G. Meltz and J. R. Dunphy, "Twin-Core Fiber Optic Strained Temperature Sensor," *Proceedings of Conference on Optical Fiber Sensors—3,* San Diego, 1985, OSA/IEEE, pp. 142, 143.

61. A. J. Rogers, "Polarization Optical Time-Domain Reflectometry," *Electronics Letters,* vol. 16, 1980, pp. 489–490.

62. A. H. Hartog and A. P. Leach, "Distributed Temperature Sensing in Solid-Core Fibers," *Electronics Letters,* vol. 21, no. 23, 1985, pp. 1061–1062.

63. M. C. Farries, "Spontaneous Raman Temperature Sensor," *Proceedings of the Society of Photo-Optical Instrumentation Engineers,* vol. 586, 1986, pp. 121–125.

64. J. C. Thevenin, L. R. Allemand, J. Calvet, and J. C. Cavan, "Scintillating and Fluorescent Plastic Optical Fibres for Sensors," *Proceedings of Conference on Optical Fiber Sensors—2,* Stuttgart, September 5–7, 1984, VDE Verlag GmbH, pp. 133–141.

65. K. Inada, "Special Fibers for Sensors," *Proceedings of Conference on Optical Fiber Sensors—4,* Tokyo, 1986, Institute of Electrical and Commercial Engineers of Japan, Optoelectronic Industry and Technology Development Association, pp. 101–108.

66. D. N. Payne, "Optical Fibers for Sensors," *Proceedings of Conference on Optical Fiber Sensors—2,* Stuttgart, September 5–7, 1984, VDE Verlag GmbH, pp. 353–360.

67. D. E. N. Davies, "Signal Processing for Distributed Optical Fibre Sensors," *Proceedings of Conference on Optical Fiber Sensors—2,* Stuttgart, September 5–7, 1984, VDE Verlag GmbH, pp. 285–296.

68. D. E. N. Davies, "Optical Fibre Distributed Sensors and Sensor Networks," *Proceedings of the Society of Photoelectrical Instrumentation Engineers,* vol. 586, 1986, pp. 52–57.

69. I. Sakai, G. Parry, and R. C. Youngquist, "Multiplexing Fiber Optic Sensors by Frequency Modulation-Cross Term Considerations," *Optics Letters,* vol. 11, 1986, pp. 183–185.

70. D. Kreit, R. C. Youngquist, and D. E. N. Davies, "Two Mode Fibre Interferometer/Amplitude Modulator," *Applied Optics,* vol. 25, no. 23, 1986, pp. 4433–4438.

71. K. C. Byron and G. D. Pitt, "Limits to Power Transmission in Optical Fibres," *Electronics Letters,* vol. 21, no. 21, 1985, pp. 850–852.

72. A. M. Scheggi, "Optical Fibre Sensors of Chemical Parameters for Industrial and Medical Applications," *Proceedings of Conference on Optical Fiber Sensors—4,* Tokyo, 1986, Institute of Electrical and Commercial Engineers of Japan, Optoelectronic Industry and Technology Development Association, pp. 117–126.

73. W. Gaebler and D. Bräunig, "Application of Optical Waveguides in Radiation Dosimetry," *Proceedings of Conference on Optical Fiber Sensors—1,* IEE conference pub. no. 221, London, April 26–28, 1983, pp. 185–189.

# CHAPTER 9
# FIBER-OPTIC SYSTEMS DESIGN


**Harish R. D. Sunak**
*University of Rhode Island*


## 9.1 INTRODUCTION

### 9.1.1 The Bandwidth-Length (*BL*) Product

This chapter on fiber-optic systems design provides a methodology by which the three basic components of a fiber-optic communication system (namely, source, fiber, and detector) can be assembled to produce systems meeting various performance specifications. An important parameter in the specification of a fiber-optic system is the so-called *bandwidth-length* ($B \cdot L$) *product* (referred to as $BL$ product hereafter), which will be used throughout this chapter. $B$ is the bit rate transmitted and $L$ is the fiber length between repeaters (or the length of the link if it does not contain repeaters). Fiber-optic systems can be designed to have $BL$ products ranging from a few megabits per second × kilometers [(Mb/s) · km] to many hundreds of gigabits per second × kilometers [(Gb/s) · km]. This chapter will discuss the various options available for the selection of the three basic components so that the desired $BL$ product can be achieved for the system to be designed.

### 9.1.2 Fiber-Optic Systems Descriptions

The typical contemporary fiber-optic communication system using optical fibers is fairly simple. The current stage of development (in terms of commercially available options) of fiber-optic communication systems could be described as elementary, when compared with present-day radio communication systems. The reasons for this evaluation are provided later in the chapter. However, it is pertinent to state also that the fiber-optics industry has made significant technological strides during the 1980s.

Figure 9.1 shows a representative fiber-optic system transmitting a telephone signal, in a very simple form. The electrical signal output from the telephone mouthpiece is first sampled and coded, and a series of binary electrical pulses is produced. These pulses are then used to modulate the light source directly and

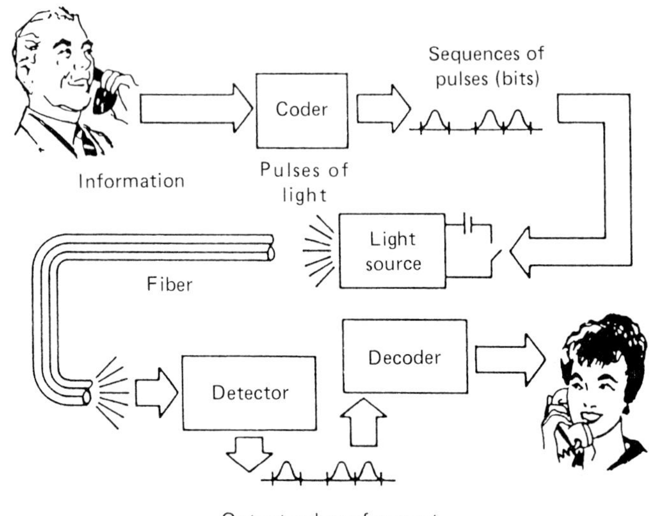


**FIGURE 9.1** Diagram of a fiber-optic telephone system, showing how a telephone call is made. (*From A. H. Cherrin, An Introduction to Optical Fiber, McGraw-Hill, New York, 1983.*)

produce optical pulses corresponding to the electrical pulses. The optical pulses propagate through the optical fiber to the photodetector at the output end, where the optical pulses (now broadened due to dispersion) are converted back to electrical pulses, which are then decoded into sound for the listener. This is an example of a digital communication system, which employs on-off pulse coding. With this system for reference, the basic building blocks of a fiber-optic communication system are illustrated in Fig. 9.2. They are

- The optical source—semiconductor laser or light-emitting diode (LED)
- The optical fiber
- The photodetector—PIN photodiode or avalanche photodiode (APD)

Two other building blocks are also shown in Fig. 9.2, namely:

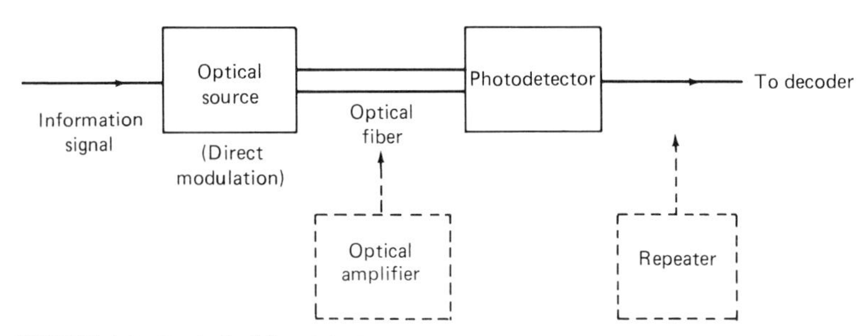


**FIGURE 9.2** Basic building blocks of a fiber-optic communication system. (The photodiode is incorporated in the repeater in long-haul links.)

- A repeater
- An optical amplifier

The former is widely used in long-haul links (intercity and intercontinental) because of limitations imposed by fiber attenuation and dispersion. However, repeaters are fairly expensive and account for nearly one-third of the cost of some communication systems. An alternative to the repeater (which detects an optical signal, amplifies it electronically, and retransmits it optically) would be the optical amplifier. Optical amplifiers, which are much simpler and much less expensive devices (in principle), *optically* amplify the attenuated signal. These devices are still in the research stage, but their potential for practical use in future fiber-optical communication systems is noted. Field evaluation of optical amplifiers installed in demonstration systems had already begun by 1987.

### 9.1.3 Comparison of Microwave and Fiber-Optic Communication Systems

Table 9.1 may be used to illustrate the statement that current fiber-optic communication systems are still at a relatively rudimentary stage of conceptual development,[1] by comparison with contemporary microwave communication systems. The various systems design features may be compared in terms of the six listed parameters, which are discussed below.

***Length of Cable Link.*** Cables for microwave systems have much greater attenuation than fiber-optic cables. Hence, the $L$ for fiber-optic systems can be greater than that for microwave systems by 1 or 2 orders of magnitude. The attenuation for single-mode fiber at 1.55 μm, for example, is only about 0.2 dB/km (which is very near the theoretical minimum). Such low attenuations provide the basis for using very long fiber-optic cable lengths.

***BL Product.*** The bandwidth of single-mode optical fibers is high, since the dispersion can be tailored to be nearly zero in the 1.3- to 1.55-μm region, where the loss of the fiber is at its minimum. Hence, the $BL$ product in fiber-optic systems is 2 to 4 orders of magnitude greater than that for microwave systems.

For $L$ and $BL$ product, it is important to note that fiber-optic systems are far

**TABLE 9.1** Comparison between Microwave and Fiber-Optic Communication Systems*

| Parameter | Microwave system | Fiber-optic system |
|---|---|---|
| Length of cable link $L$ | 2–4 km | 10–300 km |
| $BL$ product | 50 MHz · km | 2–1000 GHz · km |
| Bandwidth relative to carrier frequency | 1 | 0.001 |
| Carrier frequency spread/bandwidth | $10^{-6}$ | ≈240 (laser diode); ≈5000 (LED) |
| Circuit components | Many | Few (and expensive) |
| Multiplexing | Yes | Only simple (few wavelengths) |

*From W. A. Gambling, *Journal of IEE Electronics and Power,* December 1983.

superior to microwave systems principally because of the fiber properties, which are

- Very small attenuation
- Nearly zero dispersion

*Bandwidth Relative to Carrier Frequency.* Microwave systems have a carrier frequency of about $10^{10}$ Hz, while that for fiber-optic systems is about $10^{14}$ Hz—i.e., 4 orders of magnitude higher. Present-day microwave systems use 1 percent of the carrier frequency as the bandwidth, namely about 100 MHz. The bandwidth of the majority of commercial fiber-optic systems is 1 GHz or less, which is only 0.001 percent of the carrier frequency. Research experiments in the laboratory have demonstrated data transmission rates of more than 8 Gb/s.

***Ratio of Carrier Frequency Spread to Bandwidth.*** The ratio of carrier frequency spread to bandwidth, $\Delta f/B$, provides an indirect measure of the spectral purity (or monochromaticity) of the source. For microwave systems, the sources are very monochromatic, having a narrow linewidth corresponding to a carrier frequency spread of about 100 Hz, which results in a $\Delta f/B$ ratio of $10^{-6}$. In contrast, commercially available sources for fiber-optic systems are not very monochromatic. Laser diodes have a linewidth of about 2 nm ($\Delta f \approx 240$ GHz at 1.55 μm), for a $\Delta f/B$ ratio of $2.4 \times 10^2$. LEDs have a linewidth of about 40 nm ($\Delta f \approx 5000$ GHz at 1.55 μm), which results in a $\Delta f/B$ ratio of $5 \times 10^3$. Note that more sophisticated laser designs, called distributed-feedback lasers, can produce fairly narrow spectral linewidths (on the order of tens of megahertz), with a corresponding $\Delta f/B \approx 10^{-3}$, but these devices are not yet considered cost-effective.

Hence typical, commercially available optical sources—i.e., laser diodes and light-emitting diodes—cannot yet compare with microwave or radio-frequency oscillators in terms of stability or coherence, the latter being inversely related to the linewidth of the source. The typical contemporary fiber-optic system uses direct intensity modulation of a noisy light source, where—because the $\Delta f/B$ ratio for optical sources, as illustrated above, is much greater than 1—the carrier frequency spread is much greater than the bandwidth of the modulated signal. Thus, only amplitude modulation (AM) is typically available. Amplitude modulation is accomplished directly by turning the source on and off with the drive current. (The other properties of light which can be modulated with the information signal are frequency, phase, and polarization.) Demodulation is also performed directly, as shown in Figs. 9.1 and 9.2. Hence some contemporary, commercially viable fiber-optic communication systems are at a stage of development in some ways comparable to that of spark transmitters.

***Circuit Components.*** For microwave communication systems, many varieties of circuit components, at very cost-effective prices, are available, whereas for fiber-optic systems, relatively few components are available and these are expensive. One example is the $2 \times 2$ (or 4-port) directional coupler (or power splitter), which is typically several times more expensive for the fiber version than for the microwave version.

***Multiplexing.*** In microwave systems, many thousands of channels can be multiplexed, as is demonstrated in present-day microwave satellite links. Multiplexing and channel selectivity in radio communication systems are also extremely well-developed, as with superheterodyne techniques. For example, portable radio sets used at home can select a radio channel in the shortwave band which is multiplexed 5 kHz away from the next channel. Any one channel can be selected with this resolution, from the many thousands arriving at the radio antenna through the atmosphere. In the frequency-modulation (FM) band, the radio selectivity is about 100 kHz, since a larger channel bandwidth is needed to provide the higher-quality FM channel.

In present-day fiber-optic systems, in contrast, only a few wavelengths can be multiplexed. Receiver selectivity is severely limited by the large linewidth of the sources. Multiplexing of two or more wavelengths at or near 1.3 μm and 1.55 μm has been accomplished, but it is important to note that the frequency spread of these two wavelengths is about 30,000 GHz. It is very instructive to compare this figure with the radio channel selectivities of 5 kHz and 100 kHz mentioned before.

The real multiplexing potential of fiber-optic systems has yet to be exploited. As shown in Fig. 9.3, the available bandwidth of a single-mode fiber in the 1.5- to

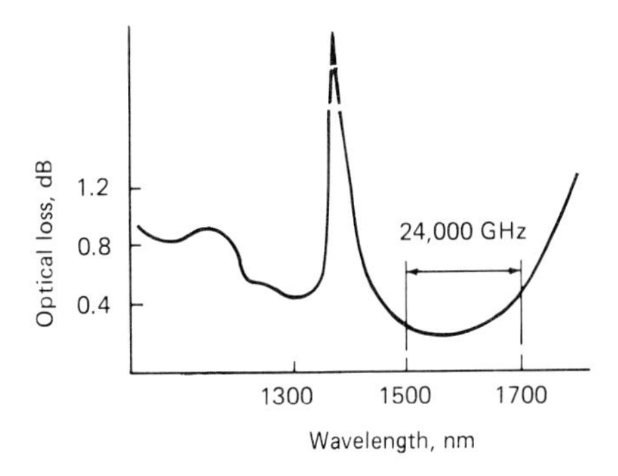


**FIGURE 9.3** Typical loss and available bandwidth of a single-mode optical fiber.

1.7-μm transmission window, where the fiber loss is less than 0.4 dB/km, is about 24,000 GHz (or 24 THz). If optical channels could be multiplexed every 100 MHz and the receiver could select these using an appropriate system design, then it is easy to see that approximately a quarter million channels could be transmitted on one single-mode fiber. Although this calculation cannot be supported by current fiber-optic systems technology, it does illustrate the information-carrying potential of optical fibers.

### 9.1.4 Chapter Perspective

Current commercially available fiber-optic communications systems are relatively primitive in concept, by comparison with radio communications technology. The typical fiber-optic system employs direct intensity modulation of a noisy optical source and direct detection with a photodetector. Because of the limitations of the coherence of commercial optical sources, sophisticated external modulation (frequency, phase, or polarization) of the electromagnetic optical carrier wave or demodulation cannot be accomplished with cost-effective components. Research on coherent fiber-optic techniques is being actively pursued in laboratories all over the world, with the intent of improving fiber-optic system performance substantially. In particular, coherent fiber-optic techniques have been demonstrated to permit substantial improvements in receiver sensitivity and receiver selectivity, with the prospect of exploiting the ultimate potential of fiber-optic systems.

Such research, however, is beyond the scope of this chapter. This chapter is concerned with the design of fiber-optic systems as illustrated in Fig. 9.1 (i.e. , involving direct intensity modulation of an optical source and direct intensity detection with a photodetector). Even though current fiber-optic systems are simple in concept, by comparison with microwave systems, they can be very cost-effective and are therefore being used in many diverse application areas. The discussions and examples in this chapter are intended to help the reader design a system for specific needs by proper selection of the three basic components used in a fiber-optic communication system.

It is *very important* to note that, in present fiber-optic system designs, the extremely high *BL* products achieved are attributable to the *fiber alone* (by virtue of its very low attenuation and nearly zero dispersion) and *not* the elaborate modulation of the source and sophisticated demodulation in the system design.

It is important to mention that the technologies to produce, install, and maintain present-day fiber-optic systems are well-developed, and the progress made, especially since 1980, has been quite dramatic. These technologies support, to name a few areas, fiber cable design, transmitter design and packaging, detector and receiver design, portable splicing machines, demountable connectors, highly reliable optical semiconductor sources, and external modulators.

Many hundreds of direct-detection systems have been installed all over the world, especially in the telecommunications network. Some specific examples are undersea and terrestrial trunk (or long-haul) systems, central-office loops (or junction networks), and optical data links. Long-haul links, especially, are switching over substantially to fiber optics and the thrust of telecommunications efforts now is to support the feeder and subscriber loops (or distribution network) with optical fibers, which is approximately 90 percent of the market. With the continuing evolution in the technological performance of optical semiconductor lasers and other passive and active devices, substantial performance improvements and wider application of fiber-optic communication systems using more sophisticated coherent demodulation techniques can be expected.

## 9.2 FIBER-OPTIC COMMUNICATION SYSTEM DESIGN CONSIDERATIONS

### 9.2.1 Direct-Detection System Components

A typical fiber-optic communication system was outlined, very briefly, in Sec. 9.1.2. This section will provide a more detailed view of the components of a direct-detection fiber-optic communication system, in the context of fiber-optic systems design. Specifically, this section will show how the components discussed in the previous chapters of this handbook can be put together to form working fiber-optic systems for communication and data transmission.

Many of the design considerations for fiber-optic communication systems are somewhat different from those associated with other communication systems. However, the basic undertaking of the systems designer is identical to that for other transmission media, whether waveguide or free space: the system designer has to overcome the following obstacles in order to obtain an efficient, reliable, and cost-effective communication system:

1. The fundamental, random nature of the communication process
2. Nonideal aspects of the transmission medium
3. Nonideal aspects of the various passive and active components in the system

The components of an intensity-modulated, direct-detection fiber-optic communication system are shown in Fig. 9.4. The information signal to be transmit-

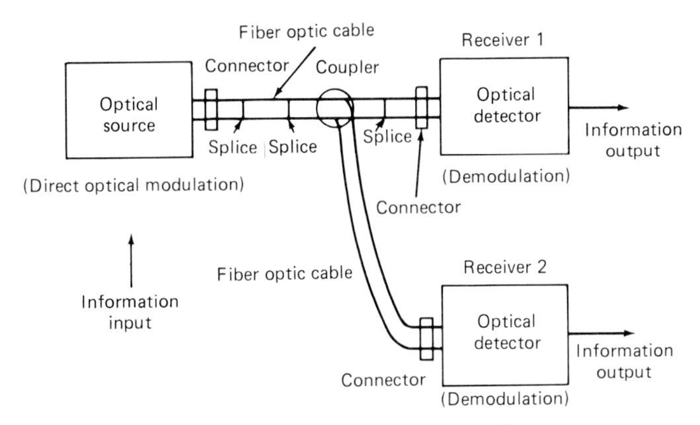


**FIGURE 9.4** Components of a direct-detection fiber-optic communication system.

ted is used for the direct optical modulation (Chap. 7) of the optical semiconductor source (Chap. 5) to produce light which is intensity-modulated optically. Either digital or analog modulation can be used, as will be discussed in more detail in Secs. 9.2.6 and 9.2.7 below. The modulated light is then launched into the optical fiber (Chap. 1), which has been suitably packaged into a cable form (Chap. 2). (The performance of the fiber is characterized using the Test Methods of Chap. 4.) The cable may contain splices and connectors (Chap. 3), depending on the system design requirements but especially when the length of the link is greater than the length of individual cables available. Fiber-optic couplers (Chap. 3) can also be used to split the optical power, if the information is to be transmitted to more than one receiver. Figure 9.4 shows the information being transmitted to two receivers. Couplers can be designed to have a variety of optical properties useful for systems design (more details can be found in Chap. 3).

At the receiver end, the optically modulated signal is converted to an electrical signal by means of a photodetector (Chap. 6) and demodulation (Chap. 7) takes place to recover the original information.

### 9.2.2 System Design Criteria

The many unique features of the optical fiber as a transmission medium give rise directly to many problems in the design of fiber-optic systems. However, in common with other transmission systems, there are two parameters which tend to be the *dominant design criteria* for a specific application. These are

1. Transmission bit rate or frequency $B$, e.g., in units of kb/s, kHz, Mb/s, or MHz
2. Transmission distance or link length $L$ or length between intermediate repeaters in a system, e.g., in meters or kilometers

$B$ is directly linked to the dispersion in the fiber and $L$ to its attenuation. In addition, other basic system parameters must be specified in the design. Several of the more important parameters are

1. Modulation format, e.g., digital or analog
2. Specified system fidelity, which is differently stated depending on the modulation format used:
   *a*. Bit error rate (BER) in digital systems
   *b*. Signal-to-noise ratio (SNR) and signal distortion in analog systems
3. Cost of the system, which includes the cost of the three basic components, the cost of installation, and the life-cycle maintenance costs
4. Reliability of the system (the TAT-8 system, to cite one example, installed in 1988 across the Atlantic Ocean between the United States and the United Kingdom and France, has a specified reliability of three component failures in a 25-year lifetime)
5. The ability to upgrade the system

The ability to upgrade the system is an extremely *important aspect* and will be discussed in Sec. 9.2.4, with optical-fiber considerations. The ability to upgrade the system may be interpreted as the ability to increase the $BL$ product of the system with minimum additional cost. A system might be completely removed and a new one put in to accommodate new $BL$ specifications, but if this is done every 6 months (to accommodate increasing system demands, for example), it is certainly not likely to be a cost-effective solution on a long-term basis. In these circumstances, more care must be taken with the system design at the initial stage, in order to provide for performance upgrades with a minimum of equipment replacement.

In addition to the aforementioned system considerations, two important component considerations apply to the design of fiber-optic systems. These are

1. Technological performance
2. Commercial availability

The technology for the three basic components is still evolving and much progress has been made. Presently, the optical sources (laser diodes and light-emitting diodes) can be identified as imposing the major limitations in the design of fiber-optic systems, as will be discussed in Sec. 9.2.3. If a compact, rugged, stable, and inexpensive (about $10) laser having a single longitudinal mode with a very narrow spectral linewidth (less than 100 kHz) and a very wide wavelength tunability range of about 200 nm (25,000 GHz) at 1.55 μm were available today, it would be possible to design very sophisticated fiber-optic systems with much better performance specifications (e.g., increased receiver sensitivity and selectivity) than is possible today. Hence these two "hidden" variables will always enter into the systems equations and will thereby influence the choice of components for the optimum system. In Secs. 9.2.3 to 9.2.5, the various choices which the systems designer has for selecting the optical sources, the optical fiber, and photodetector will be discussed. Also, the various feature considerations which must be taken into account (so as to optimize the system performance for a particular application) will be discussed.

### 9.2.3 Optical Source Considerations

Various options, materials, and features of the optical source and its associated transmitter module are summarized in Table 9.2. Many of these features have been discussed previously in Chap. 5 but are tabulated here to provide the systems designer with a checklist when considering a particular system. Of course, this list is not exhaustive, and it is hoped that the designer can add to it any other considerations peculiar to the system under consideration and hence be able to select the optimum transmitter module by analyzing the specifications provided by the device manufacturer.

Table 9.3 provides a comparison of light-emitting diodes and laser diodes, in terms of those characteristics which directly affect the system design. From the table, it is obvious that laser diodes are much superior to light-emitting diodes in terms of

**TABLE 9.2** Optical Source Characteristics

| | |
|---|---|
| Device options | Surface light-emitting diode (SLED)<br>Edge light-emitting diode (ELED)<br>Laser diode (LD) |
| Materials | GaAlAs<br>InGaAsP |
| Features | Peak wavelength (850, 1300, 1500 nm)<br>Spectral linewidth<br>Output power and brightness<br>Power coupled into fiber pigtail<br>Linearity<br>Rise and fall time (frequency response)<br>Stability (feedback control)<br>Reliability (lifetime)<br>Cost<br>Upgradability<br>Availability |
| Transmitter module | Extinction ratio<br>Thermal properties<br>Electrical drive requirements<br>Pulse duty cycle<br>Power consumed<br>Modulation format (digital, analog)<br>Line code<br>Electrical interface (RS, TTL, ECL)<br>Mechanical properties<br>Other field considerations |

**TABLE 9.3** Comparison of Optical Source Characteristics

| Parameter | LED | Laser diode |
|---|---|---|
| Threshold current, mA | Not applicable | 5–250 |
| Drive current, mA | 50–300 | 10–300 |
| Voltage drop, V | 1.5–2.5 | 1.5–2.0 |
| Output power, mW | 1–10 | 1–100 |
| Power launched into fiber, mW | 0.0005–0.5 | 0.5–5 |
| Spectral width, root-mean-square (rms) value, nm | 35–50 (at 0.8 μm)<br>70–100 (at 1.3 μm) | 2–3 (at 0.8 μm)<br>3–5 (at 1.3 μm) |
| Radiance, W/($cm^2 \cdot sr$) | 1–10 (SLED)<br>10–1000 (ELED) | $\approx 10^5$ |
| Rise time, ns | 5–50 (SLED)<br>2–20 (ELED) | <1 |
| Frequency response, −3 dB, MHz | < 500 | > 500 |

- Higher output power
- Higher power launched into the fiber
- Lower spectral width
- Higher radiance
- Faster rise time
- Higher frequency response

But taking the other features of Table 9.2 into consideration, lasers *currently* have the following disadvantages:

- Higher chip processing complexity
- Higher packaging complexity
- Higher cost
- The requirement for feedback stabilization (since threshold current is highly temperature-dependent; also lasers can suffer from kinks and nonlinearity)
- Poorer reliability—mean time to failure (MTTF) is still not comparable to that of surface-emitting light-emitting diodes (SLEDs) and edge-emitting light-emitting diodes (ELEDs), especially for devices in the 1.3- to 1.55-μm wavelength region

The design, for example, of a transmitter module with a laser diode and a single-mode fiber pigtail is quite a difficult task, based on the considerations outlined in Table 9.2.

Considering the above problems with lasers, many systems designers are using light-emitting diodes, especially edge-emitting types, for their applications and plan to upgrade their systems with lasers when these are available with higher technological performance at prices comparable to LEDs. This is especially true in subscriber loop systems, with link lengths of less than 10 km, where LEDs and single-mode fiber are being presently considered. This will be discussed further in Sec. 9.3. It is interesting to note that, before 1985, laser diodes were thought to be the only source compatible with single-mode fibers. The very low loss (about 0.2 dB/km at 1.55 μm) of single-mode fibers enables systems to be designed with $L \approx 5$ to 20 km and $B \approx 100$ to 500 Mb/s, even if the power launched from LEDs into single-mode fibers is relatively low—e.g., about 20 to 35 dB less than from laser diodes. The fiber again comes to the rescue, and it is indeed an excellent transmission line.

### 9.2.4 Optical-Fiber Considerations

The various options of fiber type, fiber materials and fiber-optic cable features are listed in Table 9.4. Some of these have been discussed in detail in Chaps. 1 and 2.

As mentioned before, the upgradability of a system to accommodate a higher $BL$ product at minimum additional cost is an extremely important consideration. If this is a requirement for a communications link, then the use of single-mode fiber in the initial installation must be considered by the designer. There are many advantages to single-mode fiber in the feeder and subscriber loop (exchange network) applications area:

1. Single-mode fiber has very large information-carrying capacity, because of its negligible dispersion, which allows the design of systems that use LEDs at hundreds of megabits per second over distances of 5 to 10 km, and which can be upgraded by using laser diodes at many gigabits per second over similar or even longer distances. This ability to upgrade is not possible with multimode fibers. (It is worthwhile mentioning here that signal processing schemes using various forms of pulse-code modulation techniques, which are wasteful in bandwidth but simplify the overall system design and are more cost-effective, should also be considered in the systems design.)

**TABLE 9.4** Optical-Fiber Considerations

| | | | |
|---|---|---|---|
| Fiber Options | Multimode fiber<br>Step-index<br>Graded-index<br>Single-mode fiber<br>Nondispersion-shifted<br>Dispersion-shifted<br>Disperson-flattened | Optical-fiber cable features | Overall cable design considerations<br>Number of fibers in cable<br>Attenuation<br>Dispersion<br>Extrinsic losses (microbending, macrobending, and splicing)<br>Fiber temperature coefficient<br>Strength member<br>Power feeding<br>Ease of installation, operation, maintenance<br>Cost<br>Upgradability<br>Special features for specific applications |
| Fiber materials | Glass<br>Silica<br>Plastic<br>Polymer | | |

2. Single-mode fiber has lower loss and lower cost than multimode fiber, because many millions of kilometers are being mass-produced each year. With further improvement in fabrication processes and increases in the total production, it is anticipated that single-mode fiber will cost as little as a pair of copper wires in the foreseeable future.
3. Single-mode fiber cable will have a one-time installation cost. System upgrading can be done by changing the transmitter and receiver modules only. This is obviously a very cost-effective and desirable aspect of system design with single-mode fiber.
4. System upgrading to higher *BL* products can be done by using higher transmission speeds at one wavelength, since the dispersion of the single-mode fiber is very small.
5. Single-mode fiber does not give rise to the modal noise problems associated with multimode fibers.
6. Single-mode fiber in the exchange network will be very compatible with the long-haul ($L > 20$ km) network, which is totally using single-mode fiber.
7. Single-mode fiber is also compatible with integrated-optic devices planned for future optical communication systems. Integration of various devices and the production of opto-electronic integrated circuits may be critical in future systems, to obtain increased speeds, lower cost, reduced size, and higher reliability.
8. Single-mode fiber is also compatible with future coherent fiber-optic communication systems. Since these coherent systems offer improved receiver sensitivity (5 to 20 dB) and, more important, improved receiver selectivity (100 MHz as compared to 100 to 1000 GHz), the very large bandwidth of single-mode fiber (about 24 THz, as shown in Fig. 9.3) can be used effectively for the first time.
9. The use of single-mode fiber throughout the public network will also facilitate the introduction and eventual standardization of a broadband integrated services digital network (ISDN) of the future with nearly unlimited new service potential.

Given the above advantages of single-mode fiber in the exchange network, the system designer can appreciate the upgradability of this unique transmission line. But the upgradability aspect of system design has to be carefully addressed at the very start, with other system design criteria.

### 9.2.5 Optical Detector and Receiver Considerations

The various device options, materials, features, and receiver design considerations are summarized in Table 9.5. Again, this list is not exhaustive, but it does contain the principal parameters to be considered in designing the receiver module. The sensitivity of the receiver at a given BER or SNR, with the output power of the transmitter (i.e., power available at the output of the fiber pigtail of the transmitter module), determines the effective system dynamic range, i.e., the total loss (number of decibels) that can be incurred by the various components in the system. This will be discussed in more detail in Sec. 9.3.3.

The important points about photodetectors are

**TABLE 9.5** Optical Detector and Receiver Considerations

| | | | |
|---|---|---|---|
| Device options | PIN diode<br>Avalanche photodiode | Receiver design | Preamplifier and main amplifier<br>Equalization and filtering<br>Noise sources and limitation<br>Front-end impedance option<br>High-impedance<br>Transimpedance<br>System fidelity (BER, SNR, etc.)<br>System dynamic range<br>Receiver sensitivity<br>Data format restrictions<br>Acquisition time<br>Power consumed<br>Bit-rate transparency<br>Bit-pattern transparency |
| Materials | Silicon<br>Germanium<br>InGaAs | | |
| Features | Active detector diameter<br>Operating wavelength range<br>Responsivity (quantum efficiency)<br>Wavelength of peak sensitivity<br>Bandwidth (speed of response)<br>Bias voltage<br>Power consumed<br>Avalanche gain and noise<br>Dark current<br>Feedback control<br>Cost, availability, upgradability | | |

1. APDs offer higher sensitivity (5 to 15 dB) than PIN diodes and offer wider receiver dynamic range through control of the avalanche gain.
2. APDs have the following disadvantages:
   *a*. The gain characteristic is temperature-sensitive.
   *b*. High bias supply voltages are required to get the necessary current avalanche gain.
   *c*. They cost more than PIN diodes.
   *d*. The dark current and excess noise of long-wavelength (1.3- to 1.55-$\mu$m) APDs are still high.
3. Although PIN diodes have lower receiver sensitivity and lower dynamic range, they are simple to use and have low cost. Hence, these should be the first choice if the system design allows this. [For short-haul links ($L < 10$ km), PIN diodes can be used very effectively.]
4. For applications in which first window, or short-wavelength (0.8- to 0.9-$\mu$m) optical sources are used, both PIN diodes and APDs in silicon technology have excellent performance. These can therefore be widely used when $BL \approx 1000$ (Mb/s) · km.
5. For applications in the second and third windows, or long-wavelength regime (1.3 to 1.6 $\mu$m), PIN diodes in indium gallium arsenide (InGaAs) technology have matured considerably and offer excellent performance. APDs made with germanium technology have matured as well, but still have high excess noise, high dark current, and limited gain-bandwidth products. InGaAs APDs have better performance than germanium APDs, but are more difficult to fabricate.

Typical characteristics of various photodetectors are summarized in Table 9.6.

**TABLE 9.6** Comparison of Optical Photodetector Typical Characteristics

| | PIN diodes | | | APDs | |
|---|---|---|---|---|---|
| Parameter | Si | Ge | InGaAs | Si | Ge |
| Wavelength range, μm | 0.4–1.1 | 0.5–1.8 | 1.0–1.6 | 0.4–1.1 | 0.5–1.65 |
| Wavelength of peak sensitivity, μm | 0.85 | 1.5 | 1.26 | 0.85 | 1.5 |
| Quantum efficiency, % | 80 | 50 | 70 | 80 | 80 |
| Rise time, ps | 50 | 300 | 100 | 300 | 500 |
| Bias voltage, V | 15 | 6 | 10 | 170 | 40 |
| Responsivity, A/W | 0.5 | 0.7 | 0.4 | 0.7 | 0.6 |
| Avalanche gain | 1.0 | 1.0 | 1.0 | 80–150 | 80–150 |

*Source:* From A. H. Cherrin, *An Introduction to Optical Fibers*, McGraw-Hill, New York, 1983.

### 9.2.6 Modulation Considerations

The modulation techniques employed in fiber-optic communication systems are characterized as being either *analog* or *digital.* The techniques which can be used under the analog category include:

1. Direct intensity modulation of the optical source
2. Electrical subcarrier modulation using one of the following parameters as a variable:
   *a.* Intensity
   *b.* Double sideband
   *c.* Frequency
   *d.* Phase
3. Pulse analog modulation using one of the following pulse parameters as a variable:
   *a.* Amplitude
   *b.* Width
   *c.* Position
   *d.* Frequency

The most widely used digital modulation technique is pulse-code modulation (PCM) with two states, which is also known as on-off keying (OOK). Multilevel pulse modulation can also be used with amplitude-shift keying (ASK) to produce different amplitudes of the optical carrier for different information bits.

*Advantages of Pulse-Code Modulation.* The very extensive use of PCM in fiber-optic systems can be understood in terms of the following discussion of PCM advantages.

1. Digital optical communication systems can tolerate much greater impairments compared to analog systems. In a digital receiver, with binary signals, only a two-level decision has to be accomplished, i.e. whether there is a 1 or a 0

present. Hence the optical pulse propagating in the fiber can suffer large amounts of loss and dispersion without severely degrading the information content, i.e., without impairing the ability to differentiate a 1 from a 0. Therefore, digital fiber-optic systems can, implicitly, operate over greater link lengths than analog fiber-optic systems, and therefore completely dominate the present-day, long-haul ($L > 20$ km), telecommunication systems market.

**2.** Since commercially available optical semiconductor sources (laser diodes and LEDs) can switch very rapidly (the laser diode has a many-gigahertz capability and the LED has a many hundreds of megahertz capability), they can be modulated directly with a binary digital electrical pulse train generated by coding the pulse-amplitude modulated (PAM) signal obtained by sampling the baseband signal at the Nyquist rate. As only two-level signaling is involved (1 and 0 for pulse and no pulse), the optical device nonlinearity, especially with laser diodes, does not affect system performance.

**3.** A digital fiber-optic system using PCM format can handle all types of information signals—voice channels, video, and data. (It is worthwhile mentioning here that with recent advances in analog fiber-optic systems, video, audio, and data channels can also be handled through a star-switched design; see Sec. 9.4 for more details.) Each of these signals can first be coded to produce its own electrical pulse train. The different pulse trains can then be combined to generate the master pulse train which directly drives the optical device to produce optical pulses. Hence a digital PCM fiber-optic system can handle a variety of signals without special techniques for different types of signals or loss of bandwidth of the channel.

**4.** Because a digital receiver usually only differentiates a 1 from a 0, the waveshape of the pulse or its particular format is not usually important. By comparison, an analog system typically requires that a precise waveform be transmitted and maintained, lest the signal-to-noise ratio be degraded.

**5.** The electronics in a PCM terminal can be shared by all the channels. This applies to, for example, the quantizer and coder. Hence, when transmitting many hundreds of channels, the cost per channel of the terminal equipment can be reduced.

*Disadvantages of Pulse-Code Modulation.* PCM also has some disadvantages when applied to the transmission of signals which are originally analog. Some of these disadvantages are discussed below.

**1.** In a typical digital fiber-optic system, the analog signal is converted to a digital signal by first producing a PAM pulse train by sampling at twice the highest frequency component. Then a binary PCM train is produced by coding the PAM train. Hence the bandwidth required to transmit the analog signal digitally is much greater than the bandwidth of the original analog signal. (However, the single-mode fiber provides extremely large bandwidths and hence this disadvantage may be of minor importance in the context of the system advantages discussed above.)

**2.** Analog signals must be coded before they can be transmitted over a digital fiber-optic system.

**3.** The systems constraints in a digital fiber-optic system are very different from those in an analog system because the coding, digital processing, and decoding of analog signals give rise to noise mechanisms which are very different from those encountered in analog systems. One example is degradation of system performance due to quantization noise. These noise mechanisms may be difficult to manage.

**4.** System costs and complexities are increased with digital transmission because of the analog-to-digital and then digital-to-analog conversion terminal equipment. (However, the cost per channel can still be smaller than for analog systems because the number of channels transmitted can be very much greater with multiplexing techniques.)

***Analog Modulation.*** The reasons for choosing an analog modulation format are usually related to economics. Two known applications of analog fiber-optic techniques are (1) individual 4-kHz voice channels and (2) many frequency-division-multiplexed (FDM) 5-MHz video channels. An analog fiber-optic system would be preferred with a larger analog communication network, such as a microwave relay network, since it would be easier to integrate it into the network. Also analog intensity modulation is comparatively easy to apply. However, there are many distinct disadvantages of analog modulation:

**1.** An analog receiver requires a comparatively high SNR (at least 40 dB). Hence the use of analog systems is limited to relatively narrow bandwidths (few megabits) and short-distance applications (approximately 10 to 20 km).

**2.** High end-to-end linearity is required to prevent crosstalk between different frequency components of the analog signal. LEDs have better performance than laser diodes, but the disadvantage in item 1 and the much smaller launched power into the fiber limit the link lengths to about 10 km.

**3.** Intensity-modulated analog signals are very susceptible to harmonic distortion because of nonlinearities in the characteristics of the light source. However, techniques are available to linearize the transmitter input and output. Among these are applying predistortion or using different modulation schemes (other than intensity modulation), such as subcarrier phase modulation, which does not require linear phase.

**4.** Phase-modulated signals can be used to reduce source nonlinearities, but then higher transmission bandwidth is required, as well as more costly and complex terminal equipment. It is better to use PCM, with all the system advantages it offers.

### 9.2.7 Other Considerations

The number of pulse-code formats is large; hence, for a given application, it is important to consider which code format will be best employed. The chosen format will affect transmission speed, stability of clock generation, ease of data detection, delays en route, phase synchronization, and error monitoring, as well as complexity of implementation. The correct choice of a suitable format can save the system designer both time and expense. A simpler code allowing simpler timing extraction can be used, at a penalty of requiring more bandwidth. A more complex code, to conserve bandwidth and optical power, can be used at the expense of the coding electronics and electric power. Other questions which the system designer must address are

1. Is it better to overrate the individual components to obtain the highest technical performance at the expense of reliability?
2. What is the risk of subsequent damage to the cables and other components when new systems are being installed in close proximity to existing systems?
3. What special hazards can render the system useless in the particular application considered? For example, when the TAT-8 undersea cables were tested off the Canary Islands, they were damaged by sharks. The average TAT-8 cable repair is laborious and costs at least $150,000.
4. What is the growth rate on the system route, as this will determine when the system must be upgraded? Should single-mode fiber be selected because of its upgrade potential, as discussed in Sec. 9.2.4, even though its capacity exceeds initial requirements?
5. What other signal processing techniques can be applied, which may require greater bandwidth, but simplify the overall system design and result in more cost-effective systems?

Given the above considerations, options, and features of the system components, it is not surprising to conclude that fiber-optic system design can be highly complex. Many important and delicate judgments have to be made, since the component technology is still in an evolutionary phase. However, by considering the various examples discussed below, the reader will see that the basic system design, in terms of technological performance only, is fairly straightforward and that other, nontechnological, issues need careful analysis and judgment. The options available for the source, fiber, and detector permit a wide variety of cost-effective fiber-optic systems. However, it is important and necessary that the component choices are made, and also the interaction of one component with another considered, so that the system performance is optimized for the particular application.

Optical fiber systems have been shown to have application in a great range of different situations, a number of which will be addressed in the following examples. Although it is impossible to address all component combinations and all possible applications, representative examples have been provided so that the different system groups can be identified. Table 9.7 summarizes the fundamental options available for fiber-optic system design.

## *9.3 DIGITAL FIBER-OPTIC COMMUNICATION SYSTEMS*

### 9.3.1 Introduction

The advantages of digital transmission systems over analog transmission systems have been discussed previously. These systems have been employed extensively in telephone network trunking during the mid 1980s. In this section, the *BL* system design grid will be introduced and explained. The system dynamic range and design procedure will be subsequently discussed and seven different design examples will be given covering all possible system designs on the *BL* design grid. Some novel approaches to system design, which are still in the research phase,

**TABLE 9.7** Fundamental Options for Fiber-Optic System Design

| Parameter | Options |
|---|---|
| Optical source | LED, laser diode |
| Optical fiber | Multimode, single-mode |
| Optical detector | PIN, APD |
| Wavelength windows | 0.8–0.9 μm, 1.3–1.6 μm |
| Modulation | Analog<br>Direct modulation<br>Subcarrier modulation<br>Pulse modulation |
| | Digital<br>Two-level<br>Multilevel |
| Multiplexing | Frequency-division (electrical)<br>Time-division (electrical)<br>Wavelength-division (optical)<br>Space division (optical, multiple fibers) |

will be discussed. This section will conclude with other practical system considerations which need to be addressed when designing digital fiber-optic systems.

### 9.3.2 Digital Bandwidth-Length System Design Grid

The digital *BL* system design grid is shown in Fig. 9.5, wherein link length $L$ is divided into eight convenient sections spanning 1 m to more than 100 km. Similarly, the system transmission rate $B$ has been divided, covering from less than 10 kb/s to more than 1 Gb/s. (Note that the regions at the extremes of the grid, e.g., $B_1L_1$, $B_1L_8$, $B_8L_8$, probably have no commercial system application.)

Once the system *BL* product is known, the designer can identify the corresponding region on the *BL* grid and then proceed to study the reference example provided later in this section. By studying the various options available to change the source, fiber, or detector (and receiver design) the designer can arrive at an optimum system, taking into account the other considerations previously discussed. Figure 9.5 also identifies seven different regions which have the same combination of source and fiber. It is important to point out that the boundaries shown are, of course, only approximate and will depend in more detail on the actual performance of the three basic components available. This will be shown with the examples. Regions I and VI are fairly well defined, as they correspond to two system extremes. Region II requires more careful considerations for optimum system design. All the examples considered are applied to synchronous, digital, point-to-point communication systems. These systems lend themselves to straightforward analysis and, at the same time, provide a basis for examining more complex architectures, such as multiterminal data networks and multichannel wavelength-division multiplexed (WDM) digital fiber-optic communication systems.

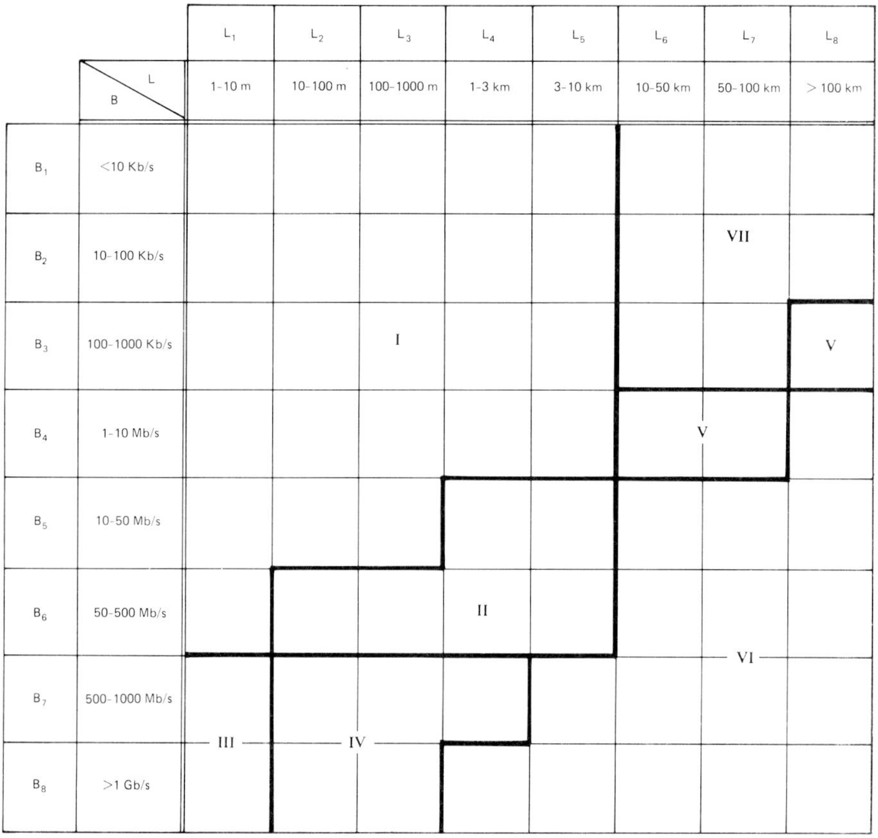

| | B \ L | $L_1$ | $L_2$ | $L_3$ | $L_4$ | $L_5$ | $L_6$ | $L_7$ | $L_8$ |
|---|---|---|---|---|---|---|---|---|---|
| | | 1-10 m | 10-100 m | 100-1000 m | 1-3 km | 3-10 km | 10-50 km | 50-100 km | > 100 km |
| $B_1$ | <10 Kb/s | | | | | | | | |
| $B_2$ | 10-100 Kb/s | | | | | | | VII | |
| $B_3$ | 100-1000 Kb/s | | | I | | | | | V |
| $B_4$ | 1-10 Mb/s | | | | | | | V | |
| $B_5$ | 10-50 Mb/s | | | | | | | | |
| $B_6$ | 50-500 Mb/s | | | | II | | | VI | |
| $B_7$ | 500-1000 Mb/s | III | | IV | | | | | |
| $B_8$ | >1 Gb/s | | | | | | | | |

FIGURE 9.5 The digital $BL$ (bandwidth-length) system design grid (described as: region and optical source/fiber pair). I: SLED with step-index multimode fiber; II: LED or laser diode with step-index or graded-index multimode fiber; III: laser diode or ELED with step-index multimode fiber; IV: laser diode or ELED with graded-index multimode fiber; V: laser diode with graded-index multimode fiber; VI: laser diode with single-mode fiber; VII: laser diode with step-index multimode fiber.

### 9.3.3 Digital System Design Procedure

The system design procedures for different fiber-optic digital systems are very similar even when many options are available for the three basic components. The following steps should be carried out in a typical system design:

**1.** Determine the maximum tolerable bit error rate for the system. BER is the probability that an error will be made in detection of a received bit. This will obviously depend on the desired system performance, but a BER of $10^{-9}$ is commonly used with digital fiber-optic systems. Some computer links require BER $\leq 10^{-12}$.

**2.** Determine the required $BL$ product for the system. The designer's goal is to transmit the desired data rate $B$ over the link length $L$ and achieve the specified BER.

**3.** Determine the appropriate region for the system on the *BL* grid (Fig. 9.5).

**4.** Identify the recommended source, whether laser diode or light-emitting diode, and determine the average transmitted power $P_S$ in dBm (power level in decibels, referred to 1 mW). Typical absolute power levels (Fig. 9.6) for laser diodes are in the range 0.1 to 5 mW (−10 to 7 dBm), while values for LEDs typically range from −10 to −25 dBm.

**5.** Determine the receiver sensitivity $P_R$ to achieve a BER of $10^{-9}$ at the required rate, also in dBm. The receiver sensitivity for PIN FET receivers (1.1 to 1.6 μm) and for silicon APD receivers ($\lambda \leq 1$ μm) is shown in Fig. 9.6 for bit rates between 10 to 1000 Mb/s (which covers most currently used data rates). Typical values of $P_R$ range from −60 to −30 dBm; for further discussion of receiver sensitivity, see Chap. 6 and Ref. 2.

**6.** Determine the maximum allowable loss $\alpha_{max}$ of the system by taking the difference between $P_S$ and $P_R$ (i.e., $\alpha_{max} = P_S - P_R$ dB). If for any reason the system loss is greater than $\alpha_{max}$, the BER will exceed $10^{-9}$ (or other specified BER).

**7.** Determine the total loss $\alpha_s$ of all the splices in the link, in decibels.

**8.** Determine the total loss $\alpha_c$ of all the connectors in the link, in decibels.

**9.** Determine the power margin $\alpha_{mar}$ in decibels for the system. This is always included as a safety factor to allow for unforeseen increases in the total link loss and for system noise parameters. Some of these increases result from aging of components, sensitivity of fiber and components to temperature fluctuations, repairs carried out on a damaged cable, etc. The power margin also allows the system designer to specify and operate at a reduced BER.

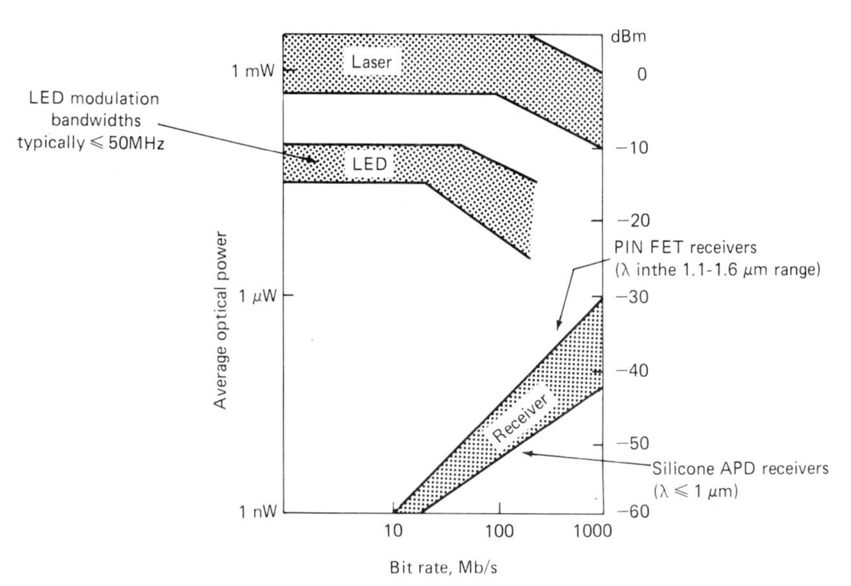


**FIGURE 9.6** Transmission margin vs. bit rate for a digital fiber-optic communication system. *(From D. C. Gloge and T. Li,[2] © 1980 IEEE, with permission.)*

**10.** Determine the loss $\alpha_f$ that can be allowed to the fiber in the link by taking the following difference:

$$\alpha_f = \alpha_{max} - (\alpha_s + \alpha_c + \alpha_{mar}) \qquad \text{dB} \tag{9.1}$$

**11.** Determine the normalized typical loss ($\alpha_n$ dB/km) of the fiber cable to be used at the operating wavelength of the optical source selected for the system.

**12.** Determine the maximum allowable length $L_{max}$ of the link (or the length between repeaters) by carrying out the following division:

$$L_{max} = \frac{\alpha_f}{\alpha_n} \qquad \text{km} \tag{9.2}$$

**13.** If $L_{max} \geq L$, as specified in the initial design in step 3, then the system design is satisfactory and other factors, such as cost, availability, reliability, ability to upgrade, etc., should be addressed. We have assumed so far that the system is loss-limited and not dispersion- or bandwidth-limited. The latter limitation comes about at high data rates, and single-mode fiber must be used to overcome it. The examples to follow will further clarify this point.

**14.** If $L_{max} < L$, then the system design is not satisfactory and the designer has to consider options to increase transmitted source power, increase receiver sensitivity, and reduce fiber normalized attenuation ($\alpha_n$ dB/km) and determine whether $\alpha_s$, $\alpha_c$, and $\alpha_{max}$ can be reduced as well. It is best to carry out this power-budget analysis using worst-case data for the various parameters previously discussed and a reasonable power margin.

### 9.3.4 Digital System Design Examples

In this section, several design examples, covering the major areas marked I through VII on the *BL* grid in Fig. 9.5, are provided. The examples show the implementation of the steps discussed in Sec. 9.3.3. With each example, the various options available to change the system components (and their resultant effect on system performance) will be discussed. Other system-orientated comments will also be provided, as appropriate. The typical maximum bandwidth × length of the various fibers are

Step-index multimode fiber ≈ 100 MHz · km

Graded-index multimode fiber ≈ 1.5 GHz · km

Single-mode fiber > 1 GHz · km

These values are determined by dispersion effects in the fiber and have been discussed in Chap. 1.

**DESIGN EXAMPLE I (Region I of *BL* Grid, Fig. 9.5)**

Required data rate $B$ = 10 Mb/s

Required link length $L$ = 1 km

(This falls in box $B_4L_3$ on the *BL* grid, and was chosen to show that *any* values of $L < 1$ km and $B < 10$ Mb/s can be easily covered by this example.)

Source: SLED, GaAlAs, $\lambda$ = 820 nm (reliable, low cost)

Fiber: step-index fiber (low cost, 100 MHz · km; hence system design is totally loss-limited, since bit rate is much less than fiber bandwidth over link length)
Detector: silicon, PIN, or APD (PIN preferred because it is reliable, has low excess noise, and receiver design is well-developed)

*Loss-Budget Analysis*

$P_S = -12$ dBm; $P_R = -60$ dBm (Fig. 9.6)
Maximum allowable loss $\alpha_{max} = 48$ dB
Total splice losses $\alpha_s = 1$ dB (worst-case)
Total connector loss $\alpha_c = 2$ dB (worst-case)
Power safety margin $\alpha_{mar} = 10$ dB (worst-case)
Total loss available for fiber, $\alpha_f = 48 - (1 + 2 + 10)$ dB $= 35$ dB
Typical fiber attenuation at 0.82 μm, after cabling, $\alpha_n \approx 12$ dB/km
Maximum link length $L_{max} = (\alpha_f/\alpha_n)$ km $= (35/12)$ km
Therefore $L_{max} \approx 3$ km

Initial specified link length $L = 1$ km. Since $L_{max} \gg L$, the system design is satisfactory.

*Options*

As $L_{max} \gg L$, the options for other design considerations are

1. Higher-attenuation fiber with lower cost, e.g., polymer-clad or plastic.
2. Lower-cost SLED with lower coupled power.
3. Different and cheaper SLED at $\lambda = 660$ nm and higher-attenuation fiber at the same wavelength.

*Application Areas*

1. *Industrial Links Requiring Electromagnetic Interference Protection:*
   $B =$ DC to 1 Mb/s
   $L = 50$ m (0.05 km)
   $BL = 0.05$ (Mb/s) · km (maximum)
   Source: SLED (660 nm)
   Fiber: plastic step-index multimode
   Detector: PIN (silicon)
   Interface: transistor-transistor logic (TTL)
2. *RS-232 Extender-Multiplexer:*
   $B = 19.2$ kb/s per channel (1 to 26 channels)
   $L = 3$ km
   $BL = 0.576$ (Mb/s) · km (per channel)
   Source: SLED (820 nm)
   Fiber: glass step-index multimode
   Detector: PIN (silicon)
   Interface: RS-232
3. *Programmable Controller:*
   $B =$ DC to 2 Mb/s
   $L = 2$ km
   $BL = 4$ (Mb/s) · km (maximum)
   Source: SLED (820 nm)
   Fiber: polymer-clad, glass step-index multimode
   Detector: PIN (silicon)
   Interface: TTL

4. *Low-Speed Data Link:*
   $B = 1$ Mb/s
   $L = 2$ km
   BER $= 10^{-7}$
   $BL = 2$ (Mb/s) · km
   Source: LED (820 nm or 1300 nm)
   Fiber: step-index multimode; large core, with loss $< 30$ dB/km
   Detector: PIN (silicon or germanium)
   Interface: TTL
5. *T1 System Extender:*
   $B = 1.544$ Mb/s
   $L = 12$ km
   $BL = 18.53$ (Mb/s) · km
   Source: SLED (820 nm)
   Fiber: step-index multimode (glass, silica)
   Detector: APD (silicon)
   Interface: TTL

*Note:* The $BL$ product for this application is nearly twice the $BL$ product in Design Example I [10 (Mb/s) · km], but the design is satisfactory because of the higher $P_S$ (lower data rate) and lower-attenuation fiber (about 4 dB/km).

6. *Ethernet Extender:*
   $B = 10$ Mb/s
   $L = 4$ km
   $BL = 40$ (Mb/s) · km
   Source: SLED (820 nm)
   Fiber: step-index multimode (glass, silica)
   Detector: APD or PIN (silicon)
   Interface: TTL

*Note:* By using step-index multimode fiber having approximately 8 dB/km attenuation, the goal of 4 km can be achieved, according to Design Example I.

**DESIGN EXAMPLE II (Region II of *BL* Grid, Fig. 9.5)** The maximum $BL$ product in this region can be greater than 100 (Mb/s) · km, hence step-index multimode fiber cannot be used.

Required data rate $B = 100$ Mb/s
Required link length $L = 3$ km
$BL$ product $= 300$ (Mb/s) · km
Source: LED, GaAlAs, 820 nm
Fiber: graded-index multimode (system design is loss-limited)
Detector: silicon, APD

*Loss-Budget Analysis*

$P_S = -12$ dBm; $P_R = -50$ dBm (determined from Fig. 9.6).
$\alpha_{max} = 38$ dB and $(\alpha_s + \alpha_c + \alpha_{mar}) = 13$ dB (as shown in Design Example I)
Use graded-index multimode fiber cable with $\alpha_n = 5$ dB/km attenuation at 820 nm, which gives $L_{max} = 5$ km.
Hence $L_{max} > L$ and the design is satisfactory.

*Options*

1. PIN diode instead of APD for lower cost
2. Higher-attenuation, cheaper fiber of up to about 8 dB/km attenuation
3. Cheaper LED with lower coupled power

*Application Areas*

1. *High-Speed Local-Area Networks (Token Ring) and High-Speed Data Links:*

   $B$ = 40 to 80 Mb/s

   $L$ = 2 km

   $BL$ = 80 to 160 (Mb/s) · km

   Source: SLED (820 nm, 1300 nm)

   Fiber: graded-index multimode (glass, silica)

   Detector: PIN or APD (silicon or germanium)

   Interface: TTL or emitter-coupled logic (ECL)

2. *Wide-Area Network:*

   $B$ = 100 Mb/s

   $L$ = 3 to 10 km

   $BL$ = 300 to 1000 (Mb/s) · km

   Source: ELED (820 nm)

   Fiber: graded-index multimode (silica)

   Detector: PIN or APD (silicon)

   Interface: ECL

*Note:* To achieve $L_{max}$ = 10 km in Design Example II above, it is necessary to use silica fiber with lower attenuation (about 3 dB/km) and higher coupled power from an edge-emitting LED. The other option is to change to an ELED at 1300 nm, where the attenuation of graded-index fiber is much lower (about 1 dB/km). If this is done, then germanium or InGaAs detectors are needed and cost considerations will become important.

3. *Fiber Distributed Data Interface (FDDI):*

   $B$ = 125 to 200 Mb/s

   $L$ = 5 km

   $BL$ = 625 to 1000 (Mb/s) · km

Similar to application 2, wide-area network.

4. *Intercity Telephone Trunk Network (Phase I):*

   $B$ = 44.7 Mb/s or 90 Mb/s

   $L$ = 5 to 15 km

   $BL$ = 224 to 1350 (Mb/s) · km or greater

   BER $< 10^{-9}$

   Source: laser (820 nm, GaAlAs)

   Fiber: graded-index multimode (attenuation < 5.5 dB/km)

5. *Intracity Telephone Loop Distribution Network:*

   $B$ = 44.7 Mb/s

   $L$ = 8 to 30 km

   $BL$ = 58 to 1340 (Mb/s) · km

   BER < than $10^{-9}$

   Source: ELED [1300 nm, indium gallium arsenide phosphide (InGaAsP)]

Fiber: graded-index multimode (attenuation < 2.5 dB/km at 1300 nm)
Detector: PIN or APD (germanium or InGaAs)

Since these two systems have been very widely used, more details are provided in the following on the FT3 system[3] installed by AT&T.

6. *Intracity Trunk Fiber-Optic Communication System (FT3, AT&T)*[3]:

$B$ = 44.7 Mb/s (672 voice channels per fiber)
$L$ = 5 to 10 km
$BL$ = 223.5 to 447 (Mb/s) · km
BER $< 10^{-9}$
Source: laser diode (825 nm, GaAlAs)
Fiber: graded-index multimode (loss < 5 dB/km at 825 nm, $BL$ > 550 MHz · km, 12-mm ribbon cable, 12 ribbons per cable and 12 fibers per ribbon)
Detector: APD (silicon)

*Loss-Budget Analysis*

$P_S = -3$ dBm; $P_R = -48.5$ dBm
$\alpha_{max} = 45.5$ dB
End connector loss $\alpha_c = 3$ dB
Total splice loss $\alpha_s = 7$ dB
Total system margin $\alpha_{mar} = 3$ dB
$\alpha_f = 45.5 - (3 + 7 + 3) = 45.5 - 13 = 32.5$ dB

For graded-index fiber, $\alpha_n = 5$ dB/km; hence the maximum length between repeaters is given by

$$L_{max} = \frac{32.5}{5} = 6.5 \text{ km}$$

*Note:* $L_{max}$ can be increased by using lower-loss fiber. By 1982, more than 40 FT3 systems were in operation in the United States. One example is the Pittsburgh-Greensburg FT3 system[3] with a maximum total distance of 65 km, with 13 regenerator sections. This system is an example of a practical, economically competitive digital fiber-optic communication system which is a complete success with telephone companies across the United States. Similar systems are equally successful in other parts of the world.

**DESIGN EXAMPLE III (Region III on *BL* Grid, Fig. 9.5)**

Required data rate $B$ = 1 Gb/s
Required link length $L$ = 10 m
$BL$ product = 10 (Mb/s) · km
Source: laser diode, GaAlAs, 850 nm (these are commercially available and are necessary for the high data rate; ELEDs can be used if available for operation at these data rates and lower in cost)
Fiber: step-index multimode (loss-limited design)
Detector: APD (silicon)

*Loss-Budget Analysis*

$P_S = -5$ dBm; $P_R = -42$ dBm (Fig. 9.6)
$\alpha_{max} = (42 - 5) = 37$ dB
Power safety margin $\alpha_{mar} = 10$ dB

Coupling loss $\alpha_c = 2$ dB

$\alpha_c + \alpha_{\text{mar}} = 12$ dB (no splicing loss as $L$ is only 10 m, without splices)

$\alpha_f = (37 - 12) = 25$ dB

Using step-index multimode fiber cable with $\alpha_n = 12$ dB/km gives $L_{\max} > 2$ km.

*Options*

1. Inexpensive large-core, high-loss step-index multimode fiber can be used, with a theoretical attenuation of about 2000 dB/km.
2. Change to silicon PIN diode for lower cost. Can be used with higher attenuation fiber.
3. Change to ELED only if speed requirement can be met.

*Application Area: Automotive Applications*

$BL$ product: 1 MHz · km

Source: SLED (600 nm)

Fiber: step-index multimode (plastic, 1000-μm core)

Detector: PIN

Connectors: molded plastic

**DESIGN EXAMPLE IV (Region IV on *BL* Grid, Fig. 9.5)**

Required data rate $B = 1$ Gb/s

Required link length $L = 1$ km

$BL$ product = 1 (Gb/s) · km (step-index multimode fiber cannot be used)

Source: laser diode, GaAlAs, 850 nm (for speed considerations)

Fiber: graded-index multimode, silica (system design is loss-limited)

Detector: APD (silicon)

*Loss-Budget Analysis*

$P_S = -5$ dBm; $P_R = -42$ dBm (Fig. 9.6)

$\alpha_{\max} = (42 - 5) = 37$ dB

$\alpha_{\text{mar}} = 10$ dB

$\alpha_c = 2$ dB

$\alpha_s = 1$ dB

$\alpha_f = 37 - (10 + 2 + 1) = 24$ dB

Using graded-index multimode fiber cable with $\alpha_n = 5$ dB/km at the operating wavelength of 850 nm gives $L_{\max} > 5$ km.

*Options*

1. Use PIN photodetector instead of APD.
2. Use higher-attenuation and cheaper graded-index multimode fiber at 850 nm.

*Application Area: Telephone Subscriber Loops*

Required data rate $B = 565$ Mb/s

Required link length $L < 3$ km

$BL < 1695$ (Mb/s) · km

BER = $10^{-9}$ or less

Source: SLED (1300 nm)

Fiber: graded-index multimode (must be carefully selected for the $BL$ product)

Detector: APD

Option: Use single-mode fiber instead of graded-index multimode fiber.

**DESIGN EXAMPLE V (Region V on *BL* Grid, Fig. 9.5)**

Required data rate $B$ = 10 Mb/s
Required link length $L$ = 30 km or greater
$BL$ product = 300 (Mb/s) · km or greater
Source: laser diode (1300 nm, InGaAsP)
Fiber: graded-index multimode (low loss at 1300 nm)
Detector: PIN (InGaAs with GaAs field-effect transistor)

*Loss-Budget Analysis*

$P_S = 0$ dBm; $P_R = -60$ dBm
$\alpha_{max} = 60$ dB
$\alpha_{mar} = 10$ dB
$\alpha_c = 2$ dB
$\alpha_s = 3$ dB (for 25 splices)
$\alpha_f = 60 - (10 + 2 + 3) = 45$ dB

Using graded-index multimode fiber cable with $\alpha_n = 0.5$ to 1.5 dB/km at 1300 nm gives $L_{max} =$ 30 to 90 km. Hence very large repeater spans are possible.

*Options*

Single-mode fiber could be used. It is cheaper than graded-index fiber, but $\alpha_s$ has to be watched as it could be greater. Also cost of splicing equipment will be important.

**DESIGN EXAMPLE VI (Region VI on *BL* Grid, Fig. 9.5)**

Required data rate $B$ = 140 Mb/s
Required link length $L$ = 150 km
$BL$ product = 21 (Gb/s) · km
Source: laser diode (1550 nm, distributed-feedback laser)
Fiber: single-mode (minimum loss possible at the laser diode wavelength)
Detector: APD (germanium)

*Loss-Budget Analysis*

$P_S = +1.5$ dBm; $P_R = -49$ dBm
$\alpha_{max} = 50.5$ dB
$\alpha_s \approx 4$ dB (37 splices at 0.1 dB/splice)
$\alpha_c = 2$ dB
$\alpha_{mar} = 11.5$ dB (adequate for $L$ = 150 km)
$\alpha_f = 50.5 - (4 + 2 + 11.5) = 33$ dB

Using $\alpha_n = 0.22$ dB/km at 1550 nm gives $L_{max} = 150$ km, which meets the initial system design specifications. The system safety margin can be reduced to increase the link length.

*Application Areas*

1. *Intercity Telephone Trunk Link (Phase II):*
   Required data rate $B$ = 274 Mb/s
   Required link length $L$ = 20 km or greater
   $BL$ = 5.48 (Gb/s) · km or greater
   BER = $10^{-9}$ or less
   Source: laser diode (1300 nm)
   Fiber: single-mode ($\alpha_n < 0.8$ dB/km at 1300 nm)
   Detector: PIN (germanium or InGaAs)
   Interface: ECL

2. *Intercontinental Ocean Cable Link (Undersea Lightwave System*[4] *for TAT-8 Link):*

   Required data rate $B = 295.6$ Mb/s
   Required link length $L = 60$ km (approximate repeater spacing)
   $BL = 17.74$ (Gb/s) · km
   BER $= 10^{-9}$ or less
   Source: laser diode (1300 nm, buried double heterostructure)
   Fiber: single-mode (depressed-cladding design)
   Detector: InGaAs

*Loss-Budget Analysis*

$P_S > -2.0$ dBm; $P_R > -31.0$ dBm
$\alpha_{max} = 29$ dB
$\alpha_{sys} = 1$ dB (added system penalty for dispersion, optical feedback, timing error, and mode partition noise)
$\alpha_{mar} = 5$ dB (added system margin for aging of system components)
$\alpha_f = 23$ dB (including losses for all the splices)

Taking $\alpha_n = 0.45$ dB/km gives $L_{max} > 50$ km.

**DESIGN EXAMPLE VII (Region VII on *BL* Grid, Fig. 9.5)**

Required data rate $B = 1$ Mb/s
Required link length $L = 50$ km
Source: laser diode (1300 nm)
BL = 50 (Mb/s) · km
Fiber: graded-index multimode (to give lower loss at 1300 nm, since $L$ is larger)
Detector: PIN (InGaAs)

*Loss-Budget Analysis*

$P_S = +1$ dBm; $P_R = -60$ dBm (from Fig. 9.6)
$\alpha_{max} = 61$ dB
$\alpha_{mar} = 10$ dB
$\alpha_s = 5$ dB (for 50 splices)
$\alpha_c = 2$ dB
$\alpha_f = 61 - (10 + 5 + 2) = 44$ dB

For $\alpha_n = 0.5$ dB/km, $L_{max} = 88$ km. For a design length of 50 km, the system loss margin increases to 29 dB. A laser at 850 nm cannot be used because the fiber loss is much higher (about 3 dB/km).

*Options*

1. Change laser diode to ELED also at 1300 nm. Coupling efficiency will be lower, but if $\alpha_{mar} = 10$ dB is acceptable, an additional 19-dB launching loss can be tolerated.
2. Change graded-index multimode fiber to single-mode fiber and laser diode to ELED. The launching loss from the ELED into the single-mode fiber will be the important design issue.

### 9.3.5 Digital System Design Evolution

The advantages of single-mode fiber over multimode fiber were discussed in Sec. 9.2. The very low loss (less than 0.5 dB/km), and extremely small dispersion in the 1.3- to 1.6-μm wavelength region, and other relevant aspects were established, especially upgradability. Section 9.2.3 noted that the performance of semiconductor laser diodes has ample room for improvement, whereas LEDs are relatively inexpensive, are readily available, and do not suffer from the many drawbacks of present-day laser diodes. This is why much research in the mid-1980s was focused on LED–single-mode-fiber transmission systems, especially for deployment in the local network and subscriber loop, with full knowledge that upgrading to higher transmission rates with laser diodes would be possible in the future because of the properties of single-mode fiber. Some of the representative results[5] obtained in various laboratories are shown in Table 9.8, using SLEDs, ELEDs, and superluminescent diodes (SLD). It is evident that the APD is the preferred photodetector option to obtain maximum link length $L$ since the launched powers from LEDs into single-mode fiber are relatively poor (−36 dBm for SLEDs and −22 dB to −30 dBm for ELEDs).

These results are also plotted on the $BL$ design grid in Fig 9.7. Only two regions, separated by the curve $BB'$, on the entire $BL$ grid are noted: region I with LED and single-mode fiber and region II with laser diode and single-mode fiber. The horizontal line $AA'$ is drawn at $B$ = 560 Mb/s, on the assumption that this is the maximum for commercially available ELEDs; technological advances can easily improve on this, and then curve $A'B$ can be plotted. It is worth noting that with $B$ = 560 Mb/s and $L$ = 15 km, the $BL$ product is fairly high at 8.4

**TABLE 9.8** Fiber-Optic Communication Experiments Using LEDs and Single-Mode Fibers

| $B$, Mb/s | $L$, km | $BL$, (Gb/s) · km | λ, nm | Detector |
|---|---|---|---|---|
| | | SLED/single-mode fiber | | |
| 140 | 4.5 | 0.63 | 1295 | PIN |
| 280 | 4.5 | 1.26 | | APD |
| 280 | 5 | 1.40 | | APD |
| 560 | 3 | 1.68 | 1295 | APD |
| | | ELED/single-mode fiber | | |
| 34 | 11 | 0.374 | | PIN |
| 90 | 21 | 1.89 | 1278 | PIN |
| 140 | 15 | 2.10 | 1270 | PIN |
| 280 | 7.5 | 2.10 | 1270 | APD |
| 140 | 22.5 | 3.15 | 1278 | APD |
| 140 | 35 | 4.90 | | APD |
| 140 | 35.4 | 4.96 | 1293 | APD |
| 180 | 35 | 6.30 | | APD |
| 560 | 15 | 8.40 | 1290 | APD |
| | | SLD/single-mode fiber | | |
| 16 | 107 | 1.71 | | PIN |
| 140 | 50.1 | 7.01 | 1283 | APD |
| 560 | 25.3 | 14.17 | 1283 | APD |

*Source:* From J. L. Gimlett and N. K. Cheung, *IEEE Journal of Lightwave Technology,* vol. LT-4, 1986.

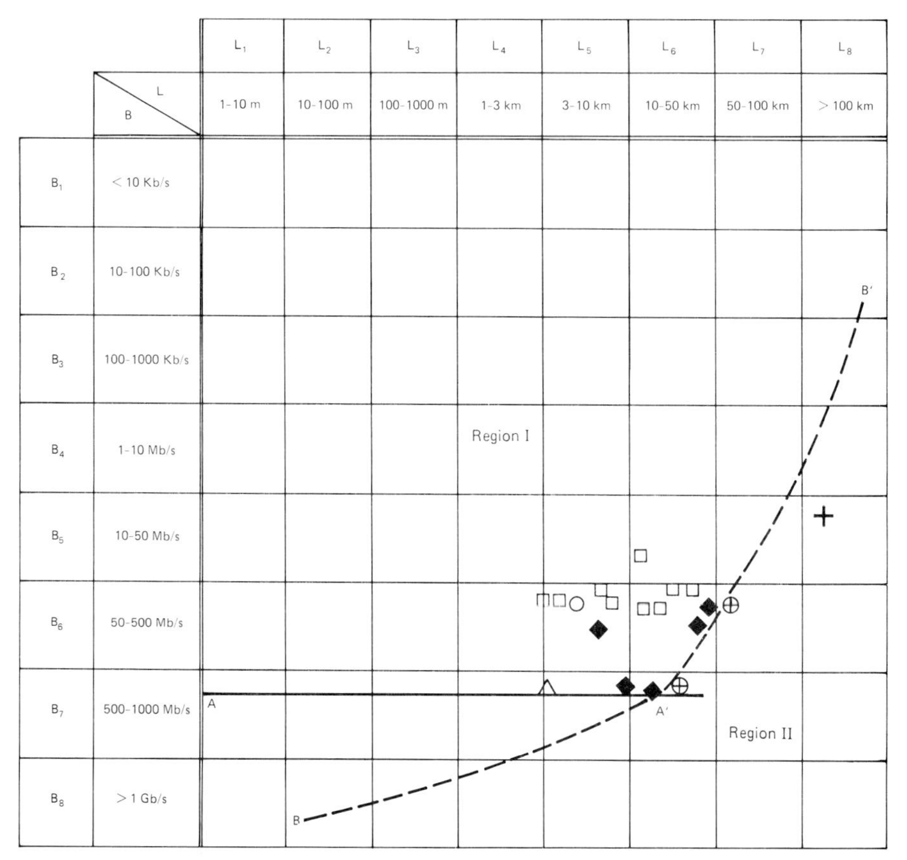


**FIGURE 9.7** *BL* (bandwidth-length) grid for experiments with light-emitting diodes and single-mode fibers, indicated in Region I. Region II would need laser diodes. The various combinations of source and detector are as follows: SLED/PIN ○; ELED/PIN □; SLD/PIN +; SLED/APD △; ELED/APD ♦; SLD/APD ⊕. (*Based on data from Ref. 5.*)

(Gb/s) · km. Since good-quality, single-mode fiber is the cheapest of all fiber types available, it is other considerations which severely limit the use of LED–single-mode-fiber systems for any system having $B$ = 560 Mb/s and $L$ < 15 km. One important consideration in the campus environment, with L ≈ 3 to 5 km, is the large total loss suffered in the system because of the many connectors that have to be used.

It is clear that laser diodes, with higher coupled power into single-mode fiber, would be preferred. Laser diode transmitters, especially for local loop applications at $\lambda$ = 1300 nm, having a coupled power of 100 $\mu$W into single-mode fiber, became commercially available in about 1985. A power level of 100 $\mu$W is −10 dBm and hence is 12 to 20 dB higher than the previously mentioned figures for ELEDs. The modulation performance can easily be in the multi-gigabit-per-second range, limited only by the transmitter package design. The reliability of these transmitters is also high and projections are for an MTTF of 60 years at 25°C and 25 years at 70°C. The cost of these devices in large quantities (greater than 100 to 1000) has become comparable, in some cases, to that for ELEDs at 1300 nm. It is clear that these types of laser diode packages can support the design of cost-effective, single-mode, fiber-optic systems. Assuming an average coupled power increase of 16 dB, if 10 dB is allowed for connectors, then 20 connectors can be incorporated (assuming 0.5 dB loss per connector). The remaining 6 dB would allow an increase in $L$ of about 12 km, assuming a 0.5-dB/km fiber loss.

As the overall performance of laser diodes improves and the cost decreases, and as similar trends occur in the other components and instrumentation needed for use with single-mode fiber, it is plausible to envision the whole $BL$ design grid covered with laser-diode–single-mode-fiber combinations. Two design examples are given below to illustrate potential laser-diode–single-mode-fiber combinations.

**DESIGN EXAMPLE VIII (Region I on *BL* Grid, Fig. 9.7)**

Required bit rate $B$ = 560 Mb/s
Required link length $L$ = 7 km (typical for loop applications)
$BL$ product = 3.92 (Gb/s) · km
Source: ELED (1300 nm)
Fiber: single-mode fiber (non-dispersion-shifted)
Detector: APD (InGaAs)

*Loss-Budget Analysis*

$P_S = -26$ dBm; $P_R = -40$ dBm
$\alpha_{\text{max}} = 14$ dB
$\alpha_{\text{mar}} = 4.0$ dB (safety margin)
Fiber loss $\alpha_f = 3.5$ dB, assuming a typical $\alpha_n = 0.5$ dB/km
$\alpha_d = 3.0$ dB (dispersion penalty; see Ref. 5)
Splice loss $\alpha_s = 1.0$ dB
Connector loss $\alpha_c = 14 - (4.0 + 3.5 + 3.0 + 1.0) = 2.5$ dB

With 0.5-dB loss in each connector, only five connectors are allowable in this typical system "specification" for a local loop application.

**DESIGN EXAMPLE IX (Region II on *BL* Grid, Fig. 9.7)** Specifications exactly as Design Example VIII, but now the source is a laser diode with $P_S = -10$ dBm.

$$\text{Total } \alpha_c = 30 - 11.5 = 18.5 \text{ dB}$$

The number of connectors possible is 37.

## *9.4 ANALOG FIBER-OPTIC COMMUNICATION SYSTEMS*

### 9.4.1 Introduction

Analog fiber-optic communication systems can be very attractive, especially for video transmission over short-haul and medium-haul links, because of their *simplicity* and *cost-effectiveness*. They are not normally used for long-haul links, where digital techniques, with single-mode fiber, are usually superior, even with the expensive terminal equipment for coding, multiplexing, and timing. This is because an analog fiber-optic system requires 20 to 30 dB higher signal-to-noise ratio than is typically required for a comparable digital fiber-optic system.

The need for a higher SNR requirement has the following detrimental consequences for analog fiber-optic systems:

- A higher optical power has to be provided at the detector to ensure the high SNR. This means that, in our loss-budget analysis, the number of decibels available for transmission loss (i.e., fiber loss, connector loss, and splice loss) are reduced considerably. Hence analog systems have applications principally in short-haul and medium-haul links.
- Shot noise at the receiver causes further deterioration of the available transmission loss, because of the higher required power levels being transmitted over the optical fibers in analog systems.

Taking into account the higher sensitivity of digital fiber-optic systems and the very low loss of optical fibers, it can be said, that, as a communications technology, fiber optics is more suited for digital transmission. The transmission of video channels has been the most common application to date of analog fiber-optic systems. It is also one of the more stringent analog applications, because of the requirements for

- A higher SNR
- A frequency response which is accurately controlled
- A highly linear optical power vs. current characteristic of the optical source

Nonetheless, analog fiber-optic technology has remained competitive for some broadband telecommunications applications because of advances made in the last few years. In many ways, some analog systems are superior to digital systems for the integration of voice, data, and video channels, when star-switched topologies are used. The notion that data transmission is best accomplished by digital transmission has been assumed—quite erroneously—by some system designers. Analog systems can be particularly attractive to telephone companies and cable television companies, since the vast embedded base of telephones and television sets operates in the analog mode.

The integration of different types of information on a single channel was previously practical primarily with digital techniques. Advances in analog systems technology, however, now make analog systems a candidate in virtually any situation calling for medium-distance transmission of signals originating and terminating in the analog mode. Now, broadband analog systems deliver 16 or 24 video channels per fiber over a link length (or length between repeaters) of less than 40 km, with an SNR of greater than 60 dB. Single video channels can be transmitted over distances approaching 60 km. The keys to these advances are the use of semiconductor laser diodes at 1300 nm and single-mode optical fibers. This should not be surprising since single-mode fibers, with very low loss at 1300 nm can allow the system designer more efficient use of the loss budget allocated for the transmission loss in the system loss-budget analysis. Further, single-mode fiber has no intermodal dispersion and the chromatic dispersion is nearly zero at 1300 nm in conventional step-index single-mode fiber, so no modal noise phenomena deteriorate the system as is the case with lasers and multimode fibers.

The increased channel capacity with multiplexing and improved performance in analog fiber-optic systems have had a large impact on the *per-channel* cost, which is competitive with coaxial cable and microwave systems in applications ranging from clustered-facility and campus-style networks to point-to-point links stretching across metropolitan regions. The major portion of analog activity, in the mid-1980s, was in point-to-point broadband networks, where the cost advantages are decisive. These systems usually carry a high volume of video information, but analog medium-haul (4 to 45 km) point-to-point fiber systems are applicable where the data at the end locations is in analog form, as in broadband local-area networks (LANs). To link two LANs with an analog fiber-optic system requires only two fiber-optic transmitters and two receivers, one for each data direction; no data conversion equipment is required, and hence this scheme can result in a very cost-effective solution. In video, potential users of modern analog fiber-optic systems are broadcasters, cable television system operators, teleports, providers of security systems, government units, institutions, and the military. Many analog fiber-optic systems use a 70-MHz common interface and hence are directly compatible with satellite and FM microwave links.

Analog fiber-optic systems can be classified into two main types, and these are shown in Fig. 9.8. Type I is a fairly simple system, and the output light intensity of the semiconductor optical source (laser diode or light-emitting diode) is directly changed or modulated by the information-carrying analog signal. This modulated light then travels down the optical waveguide and is demodulated at the receiver with a photodiode, amplifier, and filter. The filter characteristics depend on the nature of the original analog signal; it is a low-pass filter for baseband signals and a bandpass filter for radio-frequency signals. The system performance, especially the SNR, can be evaluated by considering the input optical power to the detector and the noise contributions of the amplifier.

Figure 9.8 also shows the Type II analog fiber-optic system, where a subcarrier, which has been modulated by the information-carrying analog signal, drives the optical source, and hence modulates or changes its output power. This is done in order to trade off two important system considerations, namely, SNR and signal-to-distortion ratio (SDR), with the bandwidth of the optical waveguide.

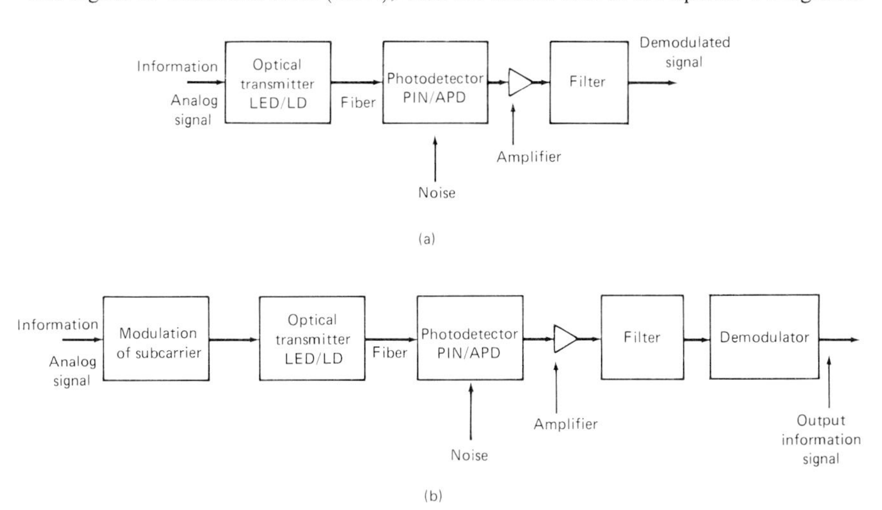


**FIGURE 9.8** Analog fiber-optic communication systems. (*a*) Type I: optical intensity is modulated by the information analog signal; (*b*) type II: a subcarrier is modulated by the information analog signal.

Various forms of subcarrier modulation can be used and these include amplitude modulation, frequency modulation, phase modulation, pulse-frequency modulation (PFM), or pulse-position modulation (PPM). AM and FM are the most widely used modulation formats. The modulated signal then is launched into the optical waveguide and is recovered at the receiver with a photodetector, an amplifier, and a bandpass filter. The final stage consists of a demodulator which returns the signal to its original baseband form (which it had before subcarrier modulation). More details are mentioned below in Sec. 9.4.3. For multichannel analog transmission of video signals, the AM or FM subcarriers can be frequency-division-multiplexed.

### 9.4.2 Analog Transmitter Considerations

The relatively high SNR required by an analog fiber-optic communication system dictates the need for more stringent considerations of the noise and the distortion induced by the optical transmitter and receiver. When considering noise and distortion introduced by a semiconductor laser emitter, it is necessary to take into account the intrinsic distortions and noise which arise if the optical power is fed directly to the photodiode, without having traveled through the transmission line. It has been noted,[6] however, that the effects of any intrinsic aberrations of the optical source are altered considerably by the interaction of the laser light with the optical fiber. Among these are the effects which the mode partition noise exerts on the system, especially when the fiber exhibits significant material dispersion or wavelength-dependent transmission loss. Other important effects include modal-noise-induced phenomena, as well as the effects of the optical power reflected back toward the laser from any fiber discontinuity. When a single-mode fiber is the transmission line, the influence of polarization of the fundamental mode and its fluctuations may also have to be assessed. These are further considered below.

*Intrinsic Distortions and Noise.* The linearity of state-of-the-art semiconductor lasers is good and hence they are well suited for applications where direct analog modulation is employed. For modulation frequencies below 100 MHz, the magnitude of the distortions is determined principally by the nonlinearity in the output power vs. input current characteristic of the laser. However, if the modulation frequencies exceed 100 MHz, then nonlinear distortions due to relaxation oscillations occur. For example, it may be observed that V-groove semiconductor lasers have a smooth transition near the threshold current. This means that they present some curvature in the power-current characteristic. Consequently, such optical sources exhibit somewhat larger second-order harmonic distortions than index-guided semiconductor lasers, which show a sharper transition around threshold (see Chap. 5). Both laser types exhibit very small third-order harmonic distortion. In general, the intrinsic distortion figures obtained for most semiconductor lasers are satisfactory for most analog transmission applications.

In addition to the intrinsic distortions discussed, the inherent noise characteristics of the optical transmitter need to be considered. These are governed by the quantum processes occurring inside the laser diode, and include the following:

1. Shot noise due to the injection current
2. Spontaneous recombination of the carriers within the active layer
3. Stimulated emission
4. Light absorption and scattering in the diode

A parameter which is used to characterize this intrinsic noise in the laser is the relative intensity noise (RIN). An alternative parameter is the DC signal-to-noise ratio, which is the inverse of RIN.

RIN has a maximum value when the laser is operated at a current slightly above the threshold current. It has been shown that this maximum is much higher in index-guided than in gain-guided lasers. The noise maximum, in the neighborhood of the threshold, is related to the smoothness of the transition in the power vs. current characteristic. A low RIN value is observed for a smoother transition, from the spontaneous emission to the stimulated emission region, of the laser power-current curve, thus yielding a smaller value of the noise maximum for the gain-guided laser diode. Therefore, a tradeoff, related to the smoothness of the transition in the power-current curve, exists between intrinsic distortions and RIN, if the laser is to be operated near its threshold current. However, when the injection current is sufficiently higher than the threshold current, both types of laser diodes exhibit tolerable amounts of RIN, with the index-guided laser being slightly less noisy.[6]

***Intermodulation and Cross-Modulation Noise.*** The signal-to-noise ratio with modulation of the source is lower than the DC signal-to-noise ratio. The reduction in the SNR depends on the modulation index of the analog fiber-optic systems. If several channels are to be multiplexed and transmitted over the same waveguide, further deterioration of the SNR will occur, depending on the number of channels to be multiplexed. In such a case, the intermodulation and cross-modulation between channels has to be taken into account as well. The overall transmission quality of the system depends on both the distortion and the noise present. Distortion may be reduced by decreasing the modulation index, whereas increasing the modulation index will result in a higher SNR. Thus a tradeoff of the modulation index has to be performed to optimize system performance. Typically, modulation index lies between 0.5 and 0.7.

*Mode Partition Noise.* The mode partition noise is the fluctuation of the intensity between different longitudinal lasing modes within the total laser emission spectrum, as illustrated in Fig. 9.9. There is experimental evidence[6] of a strong dependence between the amount of partition noise and the emission spectrum of the laser, as indicated in Fig. 9.10. Lasers with a single longitudinal mode present no

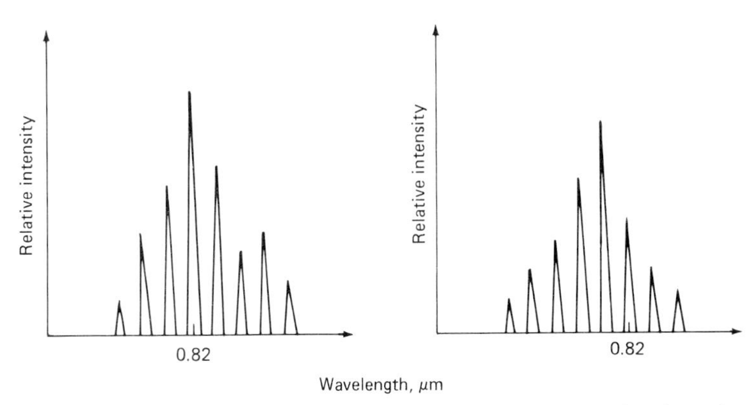


FIGURE 9.9 Partitioning between longitudinal lasing modes in a semiconductor laser (shown for two different times).

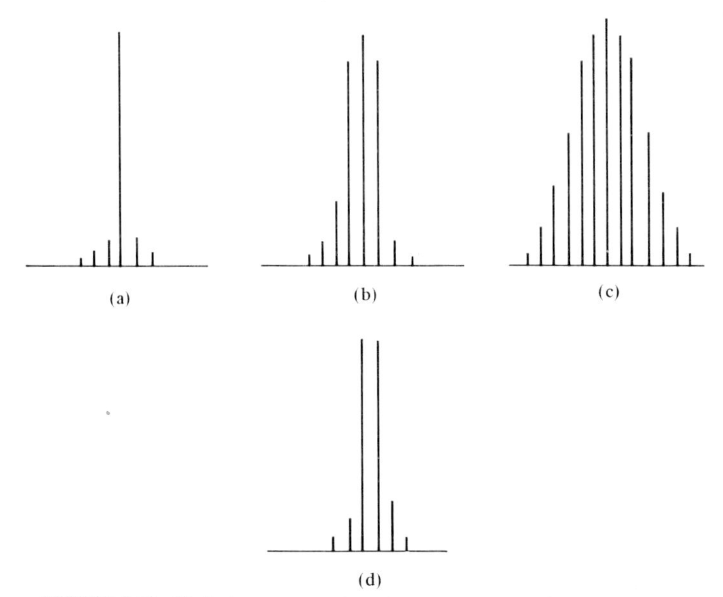


**FIGURE 9.10** Emission spectra of semiconductor lasers. Mode partition noise in them is indicated in parentheses. (*a*) Single-longitudinal-mode laser (low); (*b*) three-longitudinal-mode laser (high); (*c*) multi-longitudinal-mode laser (low); (*d*) mode-hopping longitudinal-mode laser (high). (*From Petermann and Arnold,*[6] *© 1982 IEEE, with permission.*)

partition noise. A semiconductor laser emitting in a large number of modes will also exhibit a relatively low amount of partition noise. This can be explained by the fact that the large number of modes results in a relatively small photon density for each mode, which yields a large amount of spontaneous emission within the lasing modes. This in turn leads to a stabilization of the photon density and damping of the mode partition fluctuations. The worst case occurs when the laser oscillates in about three modes because the number of modes is not large enough to produce the necessary stabilization of the photon density. Very high partition noise is also present in single-longitudinal-mode lasers in the unstable and intermediate mode-hopping state when two lasing modes can oscillate.

One effect of partition noise on the system, when a fiber exists between the laser transmitter and the photodiode, is the deterioration of the SNR because of material dispersion in the optical fiber. Material dispersion causes the different lasing modes to undergo different group-velocity delays. The SNR depends on the uniformity of the transmission of different longitudinal lasing modes; if the nonuniformity of the transmission is sufficiently large because of the fluctuation of the lasing mode power (i.e., mode partition noise), the noise contributions of the different modes do not compensate each other at the receiver, and hence deterioration of the overall SNR occurs.

Partition noise is proportional to signal power, which means that no improvement of the SNR can be achieved by increasing the input signal. Furthermore, because the noise power is also proportional to the square of the nonuniformity in the transmission of the modes, doubling the material dispersion, or (equivalently) the length of the fiber, will result in a 6-dB reduction of the SNR. The problems arising from the presence of partition noise may be reduced, to a great extent, if

(1) the optical fiber is operated near the zero dispersion wavelength and (2) a single-longitudinal-mode laser (e.g., a distributed-feedback laser) is used. However, a single-mode laser, when modulated, may become multimode, resulting in the occurrence of dynamic partition noise effects. Therefore, partition noise can cause significant deterioration of the SNR of the system. This problem can be greatly reduced by using the first option, i.e., operating the fiber at its zero dispersion wavelength.

***Noise and Distortions Caused by Fiber Reflections.*** Another source of noise and distortion encountered in an analog transmitter is the feedback produced when light is reflected back toward the laser by some discontinuity in the fiber or by a fiber connector or splice. To be more specific, with the existence of a waveguide between the laser diode and the photodetector, at least two cavities are formed. One is in the laser itself and the other is between the laser output surface and the fiber discontinuity or connector; the latter forms an external cavity. Noise and distortion are then introduced in the system by the interaction between these two cavities. Because the external cavity is substantially greater in length $L$ than the laser cavity, the frequency spacing $\Delta\nu$ between the laser cavity modes is much wider than that between the external cavity modes, as an inverse relationship exists between $\Delta\nu$ and $L$. A considerable change of the emission spectrum, and hence degradation of the SNR, can then occur because of the interaction between the spectra of these two cavities.

At this point, a distinction should be made between reflections occurring at a distance of a few centimeters from the laser, called *near-end reflections,* and those occurring several kilometers from it, the *far-end reflections.* In the case of near-end reflections, the intensity of the total light output depends on whether the reflected light interferes constructively or destructively with the output light of the laser. This means that any change in the external cavity length—i.e., any change of the external cavity spectrum—will cause a fluctuation of the emitted light intensity. Likewise, when the injection current is modulated, the laser temperature varies with it and shifts the laser cavity modes, and hence fluctuations of the emitted light intensity occur. In most cases, multimode gain-guided lasers present less near-end reflection problems than single-mode index-guided lasers because the interaction of the large number of modes in the multimode laser with those of the external cavity is averaged out.

The noise produced by near-end reflections has a low frequency content with a range of up to a few kilohertz. If the far-end reflections are considered for a single-longitudinal-mode laser, submodes corresponding to the external cavity modes will be observed. If these submodes are sufficiently strong, they may mode-lock, thus causing pulsation of the light intensity, with a frequency up to 100 kHz.

As a general rule, a more coherent laser is more sensitive to fiber reflections. It should also be noted that the use of single-mode fibers carries a higher liability, as far as reflections are concerned, than does the use of multimode fibers. For analog fiber-optic systems, where very low noise levels are required, an optical isolator should be used when a single-longitudinal-mode laser is operated with either a multimode or a single-mode fiber, in order to reduce the amount of reflection-related noise. If, on the other hand, a multi-longitudinal-mode laser is used in conjunction with a multimode fiber, no isolator is necessary.

***Modal Noise.*** When light is launched in a multimode fiber, modal noise should also be considered. The effects of modal noise stem from the interference between the many spatial modes excited in the fiber. Such interference will result in a speckle pattern at the end of the fiber, or at points where the fiber has a discontinuity. This speckle pattern is very sensitive to (1) any temperature changes in the fiber cable, (2) external forces applied to the cable, and (3) minute changes in the emission wavelength. Therefore, if an imperfect fiber connector is present in the system, any changes in the speckle pattern will result in a change in the coupling efficiency at the faulty joint, and consequently noise will be produced. Modal noise is directly related to the degree of coherence of the optical source (coherence is inversely proportional to the laser spectral linewidth); a more coherent source is more susceptible to coupling efficiency fluctuations caused by modal noise than a less coherent source. In fact, for totally incoherent LED sources, having many tens of nanometers spectral linewidth, modal noise is completely absent. Therefore a gain-guided laser which emits in a large number of modes produces less modal-noise-related coupling efficiency fluctuations than an index-guided single-longitudinal-mode laser.

In analog fiber-optic systems, where the linearity of the optical source is particularly important, modal noise creates more concern because nonlinear distortions will also be present as a result of the speckle pattern. As already mentioned, this speckle pattern is very sensitive to even minute changes in the emission wavelength. Since the emission wavelength is inadvertently modulated as the power of the laser is modulated, the coupling efficiency at an imperfect fiber connection will also fluctuate. This puts a serious limitation to the use of index-guided single-longitudinal-mode lasers in connection with multimode fibers in systems where high-quality analog transmission is required.

***Polarization Distortions.*** The modal-noise-related phenomena can be reduced substantially if a single-mode fiber is used as the transmission medium, but nonlinear distortions may occur because of chromatic material dispersion in the fiber, for modulation frequencies of a few gigahertz. These distortions can be reduced if the laser output wavelength matches the zero-material-dispersion wavelength of the fiber. Nevertheless, a single-mode fiber may be considered as a two-mode fiber, the two different modes being the two light-beam orthogonal polarization components propagating in the fiber. In such a case, some mode-related problems, similar to those in the presence of multimode excitation in the fiber, may also affect the performance of the optical source. For example, single-mode fibers with low birefringence have an effective delay which is wavelength-dependent. As mentioned above, an inadvertent wavelength modulation occurs, which, in turn, results in a delay modulation and hence distortion. The distortion eventually limits the bandwidth that can be transmitted over the fiber. If a multi-longitudinal-mode laser is used with a single-mode fiber having 20-ps polarization mode dispersion, the bandwidth will be limited to about 130 MHz. For a single-longitudinal mode laser, the transmission bandwidth[6] may be up to 500 MHz.

***Modulation Bandwidth Considerations.*** The laser structure should have only parasitic capacitance if it is to be modulated at rates close to intrinsic bandwidth of the semiconductor laser. When a communication system is designed, the transmitter must be able to provide a modulation bandwidth compatible with the modulation bandwidth specified for the particular application. Fortunately, most commercially available semiconductor lasers can be modulated at frequencies which are greater than what is required in most analog transmission applications. As an example, consider a multi-longitudinal-mode, constricted mesa laser, operating at 1.3 μm. Such a laser can provide a 3-dB bandwidth of 14.6 GHz when biased at 100 mA. An application where, for example, 60 video channels are frequency-modulated, then frequency-division-multiplexed, will require a bandwidth of approximately 2.5 GHz. Therefore, the response of this laser is quite adequate for the most bandwidth-demanding analog applications. It is more likely that the bandwidth of the system will be limited by the electronic equipment in use, rather than by the laser diode modulation bandwidth. Therefore, the modulation bandwidth of the semiconductor laser is relatively unimportant compared to the design of the electronic equipment, which modulates the signal before it is fed to the laser.

*LED Considerations in Analog Transmitters.* The noise and distortions which are of major concern in semiconductor lasers become less critical if transmitters incorporating LEDs are used. Since LEDs are incoherent sources, modal noise-induced distortions are totally absent. In addition, as the emission spectrum of an LED is at least an order of magnitude greater than that of a laser, the near- and far-end reflection noise is less severe, and the partition-noise-related distortions are averaged out by the presence of a large number of emission modes. Some of the processes responsible for RIN in a laser diode are also present in an LED, but the noise they produce is of lesser concern. On the other hand, systems employing LEDs as optical transmitters present a serious limitation because, in practice, it is very difficult to manufacture LEDs which can be modulated at the desired high frequencies and at the same time provide high output power. Thus the available loss budget of the system is reduced, and consequently the use of LED transmitters is limited to short-haul applications. For medium- or long-haul applications, the use of semiconductor laser diodes is preferable, if not imperative, since they are able to provide more output power at the desired modulation bandwidth.

### 9.4.3 Analog Receiver Considerations

*Nonlinearity in Analog Receivers.* The considerations affecting the selection of a receiver for an analog system, as far as nonlinearities are concerned, are less critical than those for an analog transmitter. In the case of the latter, the nonlinearities of the optical source were the cause of severe limitations in the subsequent design. However, in the case of receivers, nonlinearities penalize the performance of the system less. Optical receivers may employ either PIN photodiodes or APDs. The nonlinearities of PIN photodetectors have been measured and have been found to be negligible.[7] Therefore they can be ignored in system design. As far as APDs are concerned, some nonlinearities may be present because of fluctuations in the avalanche gain with variation of the frequency of the incoming signal. For this reason, APDs should be selected so as to have a flat gain response over the desired bandwidth. Most commercially available APDs provide sufficiently high gain-bandwidth products and the nonlinearities they exhibit can be considered negligible for most analog applications.

*Analog Receiver Noise.* In the design of an analog receiver, care should be taken so that the SNR of the system will be maintained at a very high level. This, in effect, means that more power is needed at the receiver so that the SNR becomes shot-noise-dominated at the receiver. For pn and PIN photodiodes, where no internal gain is provided, shot noise is the sum of quantum noise and dark current noise. If $I_D$ is the dark current of the photodiode and $I_P$ is the photocurrent, then the total shot noise of the detector, $\langle i_s^2 \rangle$ is given by

$$\langle i_s^2 \rangle = 2eB(I_D + I_P) \tag{9.3}$$

where $e$ is the electron charge and $B$ is the electrical bandwidth of the system. Furthermore, the thermal noise $i_t^2$ resulting from the load resistance $R$ of the detector is given by

$$\langle i_t^2 \rangle = \frac{4kTB}{R} \tag{9.4}$$

where $k$ is the Boltzmann constant and $T$ is the absolute temperature of the system. Then the total SNR may be expressed as

$$\text{SNR} = \frac{I_P^2}{2eB(I_D + I_P) + 4kTB/R} \tag{9.5}$$

Note that the noise introduced by the amplifier following the optical detector is not included in the above expression for the SNR. To include this noise, the second term of the denominator has to be multiplied by the noise figure $F_n$ of the amplifier. Moreover, the detector and amplifier each have a capacitance which, when added, imposes a limitation on the bandwidth $B$ of the system, as follows:

$$B < \frac{1}{2\pi RC} \tag{9.6}$$

where $C$ is the total capacitance produced by the combination of the photodiode capacitance and the amplifier input capacitance. The analysis for the SNR in an APD is similar to the one for pn or PIN photodiodes, but special attention should be given to the fact that APDs provide internal amplification of the photocurrent and consequently of the shot noise as well. The SNR for an APD is

$$\text{SNR}_{\text{APD}} = \frac{I_P^2 M^2}{2eB(I_D + I_P)M^x + 4kTBF_n/RM^2} \tag{9.7}$$

where $M$ is the mean avalanche gain and $M^x$ is the excess noise factor of the detector. As in the case of the PIN diode, the bandwidth of the system is again limited by the capacitance of the APD and the input capacitance of the amplifier. It can be observed from the SNR expressions for both APDs and PIN photodiodes that SNR depends on the amplifier noise figure $F_n$. Thus, in order to achieve a high SNR, an amplifier with a very low noise figure should be used. Particularly, if a photodiode without internal gain is to be used as a detector, the thermal noise term becomes dominant and hence the importance of $F_n$ is more critical; hence the use of a very low noise amplifier becomes imperative for high-quality analog systems. In general, APDs are able to provide higher SNRs than PIN and pn photodiodes and therefore offer more sensitivity. Thus, despite the minor nonlinearity problems they exhibit, APDs are preferred for most long-haul applications. For short- and medium-haul applications, the use of PIN photodetectors is more common.

*Other Analog Receiver Considerations.* When many channels are multiplexed and transmitted over the same waveguide, the intermodulation distortion at the receiver has to be examined. The intermodulation distortion levels have been measured for both APDs and PIN photodiodes for the same operating conditions and have been found to be considerably less than those present at the transmitter. Therefore, intermodulation distortion in the receiver is not likely to affect the overall performance of the system.[8]

The reflections from fiber discontinuities can cause noise and distortion and thus degrade the system SNR. A similar effect may result from reflections from the detector facet. One way to tackle this problem would be to position the detector at an angle to the fiber in such a way that the light reflected from the photodiode will not enter the fiber. From experimental measurements it appears that a significant reduction of reflections from the detector can be achieved with tilt angles of less than 20°. However, an angled photodetector loses some optical power. The loss is, in most cases, small and is a reasonable price to pay to minimize the far-end reflection noise. But consideration should be given to whether this approach is needed and is indeed beneficial for the specific system application under consideration.

### 9.4.4 Analog Multiplexing Techniques

Many of the advantages of analog fiber-optic communication systems stem from the fact that such systems are readily compatible with the already well-developed technology of microwave transmission systems. Since this technology utilizes frequency-division multiplexing in the electrical domain as a means of increasing the system channel capacity, such multiplexing techniques become important for analog fiber-optic systems as well. The technique of directly modulating the intensity of the optical source with the already frequency-division-multiplexed electrical signal is called *subcarrier multiplexing* and will be examined in more detail later in this section. This multiplexing scheme may find applications for providing TV conference facilities; local-area networks; point-to-multipoint video distribution, as in community antenna television (CATV) distribution networks; and in earth-station satellite links.

On the other hand, if the individual channels are multiplexed in the optical domain, the technique is known as optical frequency-division multiplexing or wavelength-division multiplexing. In the remainder of this chapter such techniques will be referred to as WDM techniques in order to avoid confusion with subcarrier FDM. Wavelength-division multiplexing is done by having each electrical signal modulate a separate optical source at a different wavelength, and mixing the outputs of all transmitters in a single fiber. This technique is more advantageous when it is used in combination with coherent optical systems than with direct intensity modulation systems. A closer examination of WDM techniques follows.

When multiplexing is used in an analog system, a number of problems arise from the transmission of many signals over the same waveguide. These include:

- Poor modulation depth
- Intermodulation noise
- Crosstalk between different signals

All these considerations will also be dealt with in more detail when each multiplexing technique is examined. As a closure to this brief introduction to multiplexing techniques, the following factors should be considered before deciding on the number of channels to be multiplexed:

- The required quality of the received signal. This will depend on the particular application of the system; for example, whether the system is used in a corporate or medical application; studio, broadcast, CATV, or TV conferencing; monochrome or color video; etc.
- The required or cost-effective distance between repeaters.
- The type of optical fiber to be used as the transmission medium (e.g., single-mode fiber, step-index multimode fiber, graded-index multimode fiber).
- The type of signal processing used before optical modulation (e.g., amplitude-modulated or frequency-modulated subcarrier modulation, signal compression).

*Subcarrier Multiplexing.* This technique takes advantage of the combination of very large bandwidth provided by single-mode optical fibers, semiconductor lasers, and optical detectors and the well-developed technology of microwave electronics, in order to provide an alternative method for transmission of a large number of signals through a single waveguide. The essence of this scheme is as follows: each signal to be transmitted is modulated onto a high-frequency electrical carrier wave. Any modulation method can be used for this purpose; that is, the signal may be FM, AM, pulse-width modulated (PWM), PFM, or PPM. Note that subcarrier multiplexing can be used for digital fiber-optic systems as well, and therefore not all the multiplexed signals have to be of analog nature. The subcarrier-modulated signals are then combined; i.e., they are frequency-division-multiplexed as radio-frequency (RF) signals and the resulting signal is used to modulate the intensity of the optical source. In many cases, in order to avoid the intermodulation noise introduced by the nonlinearities of the current-power characteristics of the transmitter, the combined RF signals are further frequency-modulated to a higher signal frequency before they modulate the intensity of the source. A schematic diagram of a system using subcarrier modulation is shown in Fig. 9.11.

An important parameter, which affects the performance of a subcarrier-multiplexed system to a great extent, is the optical intensity modulation index (OMI). An individual signal, prior to optical modulation, can be expressed as

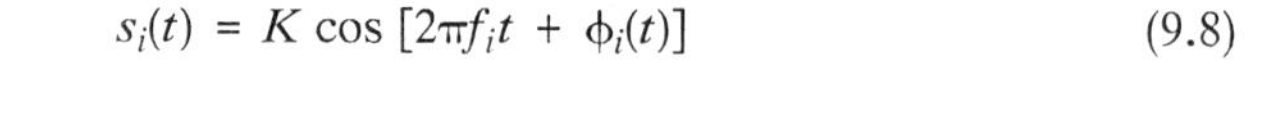

$$s_i(t) = K \cos [2\pi f_i t + \phi_i(t)] \tag{9.8}$$

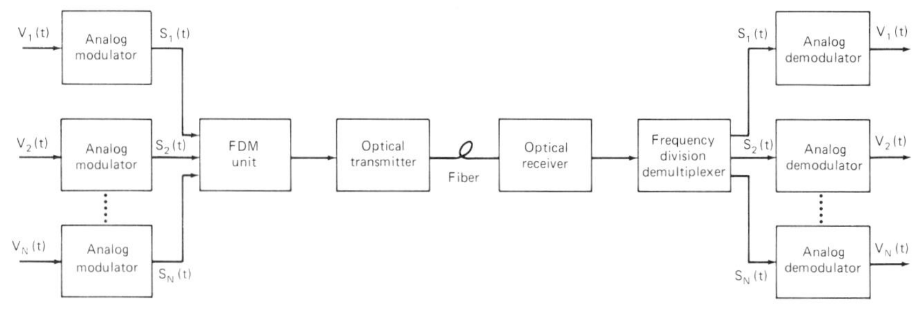


**FIGURE 9.11** Schematic diagram of an analog fiber-optic communication system with subcarrier modulation and frequency division multiplexing.

If $N$ such signals are combined in the electrical domain, the resulting optical power of the received signal will be

$$P_r(t) = P_0 \left[ 1 + m_0 \sum_{i=1}^{N} s_i(t) \right] \tag{9.9}$$

where $P_0$ is the average power of the received light and $m_0$ is the overall OMI, which is given by

$$m_0 = \frac{P_{max} - P_0}{P_0} \tag{9.10}$$

where $P_{max}$ is the peak received light power. Thus a higher OMI will yield a higher carrier-to-noise ratio (CNR) and, consequently, a higher SNR as well. However, if the OMI is too high, unwanted operation of the optical source in the nonlinear region may occur. In these circumstances, intermodulation noise, which is a factor of serious concern in any multichannel system, whether optical, electrical, or microwave, becomes important. The intermodulation noise is added to a channel by interference from other channels in a multichannel system. In analog systems, intermodulation is caused by the nonlinearities of the optical source, that is, by the static nonlinearities of the power-current characteristic and the dynamic nonlinearities due to relaxation oscillation resonance. In general, the intermodulation noise increases as the OMI increases, and thus deterioration of the overall system performance occurs for OMIs beyond a certain value. Further details can be found in Ref. 9.

A comparison between two different subcarrier modulation schemes is now made. The first scheme is AM/FDM subcarrier multiplexing, where the individual signals are first amplitude-modulated and then combined to further intensity-modulate the optical source. This scheme has the following advantages:

1. It is simple to implement and therefore cost-effective.
2. The bandwidth expansion, after frequency-division multiplexing in the electrical domain, is relatively small and means that more channels can be transmitted per fiber.

On the other hand, the first scheme presents the following drawbacks:

1. It requires a high SNR and therefore the repeaterless transmission distance is limited.
2. It is more sensitive to distortion produced by source-fiber interactions.

Note that injection lasers in combination with multimode fibers may not yield acceptable performance for many applications.

The second scheme uses FM/FDM, where the individual signals are first frequency-modulated and then frequency-division-multiplexed into a single signal which, in turn, drives the optical source. The advantages exhibited by such a scheme are

1. It requires a lower SNR than the preceding AM/FDM scheme and therefore may be used for repeaterless transmission over greater distances.
2. Nonlinearities introduced by the source and distortions due to source-fiber interactions have a lesser effect.
3. It provides satellite transmission compatibility; i.e., the signal received from a

satellite can be directly fed to the optical transmitter without signal conversion from the FM to the AM domain.

The FM/FDM scheme, however, has the following disadvantages:

1. It requires a large bandwidth per channel, hence a smaller number of channels can be transmitted per fiber.
2. It is more costly to implement because of the additional cost of frequency-modulation modem equipment.

Despite its drawbacks, the FM/FDM scheme is widely used in short- and medium-haul multichannel analog fiber-optic communication systems. Transmission of 60 FM video channels over 18 km has been achieved.[10] If traveling-wave laser amplifiers are used to enhance the SNR of the system, the capabilities of this technique become even greater. For example, distribution of 90 FM video channels to 2048 terminals has been demonstrated[11] with the use of the FM/FDM scheme and two in-line traveling-wave laser amplifiers.

***Wavelength-Division Multiplexing.*** When frequency-division multiplexing is performed directly in the optical domain, the technique is called *wavelength-division multiplexing.* This name is derived from the fact that each signal is used to modulate a separate optical source, each operating at a different optical wavelength. The outputs of the optical transmitters are combined through passive optical components and then transmitted through a common fiber. The individual signals are separated at the output fiber end by either passive optical filtering or coherent detection techniques. Figure 9.12 illustrates a WDM system with passive optical filters for demultiplexing. If coherent detection is used, frequency selectivity of 3 orders of magnitude better than direct intensity modulation can be achieved because filtering is performed at lower intermediate frequencies in the RF domain. To achieve this, however, use of highly coherent lasers with extremely narrow linewidths (about 100 kHz), good frequency stability, with easy tuning, is imperative. More detailed discussion on coherent systems is beyond the scope of this chapter.

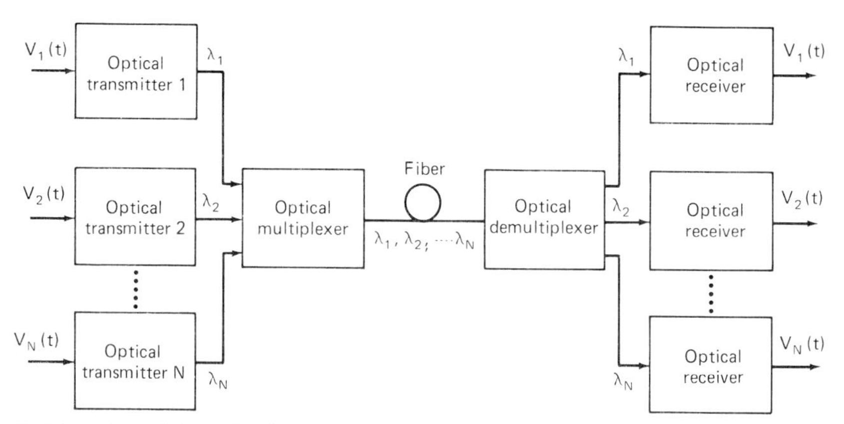


**FIGURE 9.12** Schematic diagram of an analog fiber-optic communication system with wavelength-division multiplexing.

WDM systems may also be used in combination with traveling-wave laser amplifiers to enhance the SNR, but crosstalk between the channels must be considered. Crosstalk is mainly introduced by the gain reduction and resonance frequency shift induced by other channels in the traveling-wave laser amplifier, as well as by the vibration of the electron carrier number at the beat frequency of the incident light. It has been found that the crosstalk induced by such electron vibrations depends to a great degree on the spacing between channels and on the input power to the traveling-wave laser amplifier.

In general, WDM systems have a serious disadvantage with respect to subcarrier modulation techniques because they are not compatible with the highly developed technology of microwave transmission. Furthermore, the implementation of such systems is relatively complicated, compared to that of subcarrier modulation systems, mainly because of the complexity of the optical coherent techniques required for demodulation and demultiplexing and the greater number of optical components (active and passive) required for the operation of such systems. Another drawback of WDM systems is the requirement for a separate loss-budget analysis for each channel, since the optical wavelengths differ. This adds to the complexity of the overall system design. On the other hand, WDM techniques have the advantage of independence between channels; that is, if one of the transmitters fails, the rest of the system will continue to operate. This is not the case in subcarrier modulation systems, where failure of the transmitter signifies the failure of the entire system. In addition, because the channels are independent, the system can be upgraded easily and inexpensively. WDM techniques are most advantageous when coherent detection techniques are utilized. However, coherent techniques are still in the experimental research phase, while subcarrier modulation schemes have been more thoroughly investigated and used.

### 9.4.5 Other Analog System Design Considerations

In multichannel analog systems, it is very important to consider the nonlinearities introduced by the optical components. Even though detector nonlinearities are negligible, nonlinearities in the optical sources may be of serious concern. Information concerning the nonlinearities of LEDs and laser diodes is usually provided by the manufacturers of these devices in terms of the ratios of the second and third harmonics of the fundamental. These ratios are, respectively,[7]

$$M_2 = 10 \log \left( \frac{P_{2f}}{P_f} \right) \tag{9.11a}$$

$$M_3 = 10 \log \left( \frac{P_{3f}}{P_f} \right) \tag{9.11b}$$

where $P_{2f}$ = electric power level of second harmonic
$P_{3f}$ = electric power level of third harmonic
$P_f$ = electric power level of fundamental component

The importance of the nonlinearities is that they give rise to intermodulation and cross-modulation effects in multichannel transmission systems, such as those employing subcarrier modulation techniques.

Intermodulation is generated in multichannel analog systems by the harmonics and the interference of beats between two or more carriers. If the carrier levels of all the signals are the same, then the intermodulation beat power relative to the signal power level can be given[7] (in decibels) by

$$\text{im}_{2P} = M_2 - 20 \log \frac{m_I}{m_c} + C_{2P} + 10 \log N \qquad (9.12a)$$

$$\text{im}_{3P} = M_3 - 40 \log \frac{m_I}{M_c} + C_{3P} + 10 \log N \qquad (9.12b)$$

$$\text{im}_{3V} = M_3 - 40 \log \frac{m_I}{m_c} + C_{3V} + 20 \log N \qquad (9.12c)$$

where $\text{im}_{2P}$ = second-order power-additive intermodulation product
$\text{im}_{3P}$ = third-order power-additive intermodulation product
$\text{im}_{3V}$ = third-order voltage-additive intermodulation product
$N$ = number of repeater sections in system
$m_c$ = modulation index per channel
$m_I$ = modulation index of multiplexed signal

The terms $C_{2P}$, $C_{3P}$, and $C_{3V}$ are coefficients which depend on the frequency of operation of the system.

Cross-modulation is defined as the transfer of modulation from one carrier to another, and is given[7] (in decibels) by

$$\text{xm} = M_3 + 15.5 - 40 \log \frac{m_I}{m_c} + 20 \log (K - 1) \qquad (9.13)$$

where $K$ is the number of multiplexed channels.

### 9.4.6 Analog System Design Procedure and Examples

***System Design Procedure for Single-Channel Transmission.*** The design of any fiber-optic communication system, be it analog or digital, is based upon the tradeoffs between the performance and the cost of the system, where performance is measured in terms of noise, distortion, and available bandwidth. The system designer typically needs to optimize the design by choosing system components which will provide the required performance within specified costs.

The optical performance of a system is determined by transmission distance; the types of optical source, detector, and cable; and the number and type of connectors and splices in the optical path. This has been shown in the previous section with digital technologies. Choosing a source for a particular system has been described previously. A designer should be familiar with the types of available sources, their costs, and their performance specifications, as highlighted in Tables 9.2 and 9.3. Commercially available sources include LEDs and lasers. These are available in three wavelength regions—830 nm, 1300 nm, and 1550 nm—and thus utilize the three transmission windows of silica-based fibers. The first two wavelengths are most commonly used in current applications of analog systems. For short-haul applications (up to 3 km at 830 nm and 7 km at 1300 nm), LEDs are usually chosen because of their cost-effectiveness. For longer transmission lengths, laser diodes are preferred because they yield much better performance. Lasers at 830 nm are recommended for shorter runs (up to 10 km), while for longer runs, transmission at 1300 nm is required. However, the longer wavelength detectors are typically made of germanium and are generally more noisy than the 830-nm detectors, which are usually made of silicon. On the other hand, the 1300-nm wavelength region window exhibits lower attenuation than the 830-nm window in silica fibers; hence longer distances are possible at 1300 nm.

When choosing a fiber cable, the designer has to consider construction, transmission characteristics, and cost as highlighted in Table 9.4. A careful decision should be made so that the cost of the system remains within specifications while, at the same time, upgradability is possible, if needed for future expansion. In choosing a cable, for example, future expansion can be accommodated by providing a sufficient number of spare optical fibers and ensuring that anticipated bandwidth requirements can be met. To provide large available fiber bandwidth, single-mode fiber should be chosen. The cable construction is often specific to the fiber itself (i.e., whether it is step-index multimode, graded-index multimode, or single-mode fiber), and to its operating environment—that is, whether the cable will be used in aerial, buried duct, armored, highly flexible, or underwater applications. Moreover, the number of fibers accommodated in a single cable should be considered, with respect to distribution (i.e., fan-out). Finally, the transmission characteristics of the fiber cable, including its operating wavelengths, optical attenuation, and bandwidth, should be considered. The manufacturer's specifications of the above parameters should be analyzed before deciding on which type of cable to use.

Another consideration is that of connections in the optical path (i.e., the ways in which the fiber cable is joined to sources, detectors, and to other cables). Generally, connections are made by either fusion splices or mechanical connectors. Fusion splices typically exhibit less attenuation, while mechanical connectors prove to be more flexible in the field. More information about connectors and splices can be found in Chap. 3.

A recommended procedure for designing an analog fiber-optic system follows the same pattern as that recommended for the design of a digital system and involves the following steps:

1. Determine the power to be coupled into the fiber from the optical source.
2. Determine the required power at the receiver by specifying the required receiver SNR.
3. Calculate the total available power margin for the optical path by subtracting the required received power from the power coupled into the fiber.
4. Assign to the power margin losses due to fiber attenuation, connector, and splice losses, etc. Note that a part of the loss budget may have to be distributed to source nonlinearities and source-fiber interaction distortion and noise penalties.

A single channel, analog system design example follows.

*Example 1.* This example provides a rationale for the selection of the optical source and fiber for an analog fiber-optic system having a 2.5-km link length, a required SNR at the receiver input of approximately 68 dB (CCIR-weighted), and a 55-MHz bandwidth.

Since the distance is only 2.5 km, cost factors lead to the 830-nm LED as the optical source. Using a laser at full power would saturate the photodetector, while operating it at low power would be too cost-ineffective. Use of a 1300-nm source for such a short distance is also not recommended for cost reasons.

Having decided to use an LED as a source, the type of fiber can be chosen. Because the core diameter of single-mode fiber is about 10 μm, the launching efficiency from the LED is poor. Hence we chose a graded-index multimode fiber, which must have the required bandwidth of 55 MHz over 2.5 km. If, for example, a candidate cable was specified by the manufacturer to have a bandwidth-distance product of 400 MHz · km, it would have an available bandwidth of 400/2.5 = 160 MHz over the 2.5 km of the link. Thus the available bandwidth would be adequate.

Now, for a 68-dB CCIR-weighted SNR, a received power of −29.5 dBm is required at the detector. If the LED launches −16 dBm into the fiber, then the total loss margin for the optical path will be equal to (−16 dBm) − (−29.5 dBm) = 13.5 dB. Assuming that there are two connectors in the path, each with a loss of 1 dB, and that the loss at the fiber-detector coupling is also 1 dB, there will be 10.5 dB of attenuation to be distributed over the length of the fiber cable. That is, the fiber will have to provide a 10.5 dB/2.5 km = 4.2 dB/km attenuation figure (or lower) at 830 nm. Table 9.9 shows the loss budget developed for this example. Provision must also be made to have a safety margin by decreasing the coupling losses and/or using lower-attenuation fiber. For this example any type of detector (PIN or APD) may be used; a PIN detector is preferred because of its cost-effectiveness. Also note that the nonlinearity, RIN, and source-fiber interaction distortion and noise were considered to be negligible.

**TABLE 9.9** Power Budget for Example 1

| | |
|---|---|
| Power coupled into fiber | −16 dBm |
| Power required at the receiver | −29.5 dBm |
| Overall power margin | 13.5 dB |
| Connector losses (2 at 1 dB/connector) | 2 dB |
| Fiber-detector coupling loss | 1 dB |
| Fiber attenuation at 4.2 dB/km (for a length of 2.5 km) | 10.5 dB |

***Multichannel Analog System Design Procedure.*** Before a step-by-step description of the design procedure is presented, it is useful to review the required standards for analog systems in general, insofar as source nonlinearities, cross-modulation and intermodulation, and modulation index in multichannel transmission are concerned. The expressions relating the carrier-to-noise ratio and SNR of an analog system to the transmitter, receiver, and fiber characteristics (e.g., laser power level, avalanche gain of the APD, launching efficiency, and received power level) are provided below.

The reader may wish to refer to Sec. 9.4.5, where the relationships between source nonlinearity, cross-modulation, intermodulation, and modulation index are given. These will be helpful in determining the optimum modulation index once the acceptable levels for intermodulation and cross-modulation are specified.

The carrier-to-noise ratio for an AM/FDM system is expressed by

$$\text{CNR} = \frac{(RMm_{\text{eff}}P_{\text{av}})^2}{2(i_a^2 + 2e[(RP_{\text{av}} + I_{\text{dm}})M^2 + I_{\text{du}}])B} \tag{9.14}$$

where $B$ = signal bandwidth (4 MHz for video transmission)
$m_{\text{eff}}$ = effective modulation index
$M^2$ = mean-square avalanche gain of photodiode
$i_a^2$ = mean-square value of amplifier noise current referred to input, $A^2$/Hz
$e$ = electronic charge
$I_{\text{dm}}$ = primary dark current undergoing multiplication
$I_{\text{du}}$ = unmultiplied portion of dark current
$R$ = responsivity of photodiode
$P_{\text{av}}$ = average received optical power

The nominal modulation index of the optical source, $m_I$, is related to $m_{\text{eff}}$ by the following expression:

$$m_{\text{eff}} = m_I \times 10^{-0.1\alpha} \tag{9.15}$$

where $\alpha$ is expressed as

$$\alpha = -3\left(\frac{f}{f_{\text{mod}}}\right)^2 L^{2z} - 3\left(\frac{f}{f_{\text{mat}}}\right)^2 L^2 - C(f) \tag{9.16}$$

where $C(f)$ = optical-power roll-off characteristics of optical source
$L$ = length of fiber, km
$f$ = frequency of link operation, Hz
$f_{\text{mod}}$ = 3-dB modal dispersion-limited cutoff frequency of fiber
$f_{\text{mat}}$ = 3-dB material dispersion-limited cutoff frequency of fiber

In Eq. (9.16), $0.5 < z < 1$. An estimate of the power level required at the optical receiver can be derived from a prespecified value of the carrier-to-noise ratio in order to achieve the desired performance. This power level is given as an average power $P_{\text{av}}$ by

$$P_{\text{av}} = AM^x + \left(A^2M^{2x} + \frac{C}{M^2}\right)^{1/2} \tag{9.17}$$

where $A = 2eB\rho/Rm_c^2$
$C = 2i_a^2\rho B/m_c^2R^2$
$\rho$ = specified value of CNR

Similarly, the optimum avalanche gain for the photodiode, $M_{\text{opt}}$, can be derived from

$$M_{\text{opt}} = \left[\frac{C}{x(x + 2)A^2}\right]^{1/2x+2} \tag{9.18}$$

where $x$ is 0.35 for a silicon APD and 0.95 for a germanium APD.

For an FM transmission system the SNR is given by

$$\text{SNR} = \frac{12P_c(\Delta F_s)^2}{N_o b_n^3} \tag{9.19}$$

where $P_c$ = carrier power
$N_o$ = noise-power spectral density
$\Delta F_s$ = half of peak-to-peak deviation produced by signal waveform
$b_n$ = noise bandwidth of baseband-filter function with respect to triangular noise

The following step-by-step procedure may be used for the design of analog systems using laser diodes for optical transmission:

1. By definition, the total length of the link is $L_s$ (in kilometers), and the number of repeaters required is $N$.
2. A required value for the SNR is set and the corresponding CNR is computed. For this computation the following expressions may be used. For vestigial sideband (VSB) AM systems,[7]

$$\text{SNR}_{\text{CCIR}} = \text{CNR}_{\text{NCTA}} - 0.2 \text{ dB} \tag{9.20}$$

where $\text{SNR}_{\text{CCIR}}$ = SNR as defined by the International Radio Consultative Committee (CCIR) and $\text{CNR}_{\text{NCTA}}$ = CNR as defined by the National Cable Television Association (NCTA). For FM video systems,

$$\text{SNR} = \text{CNR} + 20 \log \frac{\Delta F_s}{B} + 20.37 \text{ dB} \tag{9.21}$$

where $\Delta F_s$ = half of peak-to-peak deviation produced by the video waveform and $B$ = video waveform bandwidth (usually taken to be 4 MHz).
3. The value of the CNR after the $N$ repeaters, $\text{CNR}_T$, is computed from

$$\text{CNR}_T = 10 \log \text{CNR} - 10 \log \text{N} \tag{9.22}$$

4. The maximum modulation depth is then computed from the equation

$$m_c = \frac{m_I}{Kz} \tag{9.23}$$

where $0.5 < z < 1$ and $K$ is the number of multiplexed carriers.
5. By applying the equation linking $P_{\text{av}}$ to the CNR and $m_c$, just given, the required average power level at the receiver may be calculated.
6. The length $L_R$ between successive repeaters is computed from

$$L_R = \frac{P_B - P_{\text{av}} - \alpha_c - \alpha_m - D_L}{\alpha_f} \tag{9.24}$$

where $P_B$ = average power coupled into the fiber, dBm
$\alpha_c$ = losses due to fiber connectors, dB
$\alpha_m$ = allowed safety margin, dB
$\alpha_f$ = fiber loss, including splice losses, dB/km
$D_L$ = fiber dispersion-equalization penalty, dB

7. The number of repeaters required for the total link is then calculated by evaluating the ratio $L_s/L_R$ and rounding up to the next higher whole number.

When the analog system to be designed will use an LED instead of a laser diode, the design procedure becomes much simpler because the distortions and noise introduced by the source-fiber interaction are absent. In this case, the following step-by-step procedure can be used:

1. The performance objectives of the system (i.e., the required CNR and maximum intermodulation and cross-modulation levels) are calculated.
2. The permissible values of the modulation index $m_c$ for the cross-modulation level specified are calculated from Eq. (9.13):

$$\mathrm{xm} = M_3 + 15.5 - 40 \log \frac{m_I}{m_c} + 20 \log (K - 1) \qquad \mathrm{dB}$$

3. From the values of $m_c$ obtained, the average power level required at the receiver to achieve the specified CNR is calculated.
4. The link loss budget is developed and the permissible repeaterless link length, or the required number of repeaters for a given link length, is estimated.

*Example 2.* This example is based on the design[7] of a multichannel CATV VSB/FDM system, using an LED as a source. The performance objectives for this system are

$\mathrm{CNR} > 43$ dB
$\mathrm{im}_{2P} < -56$ dB
$\mathrm{im}_{3P} < -56$ dB
$\mathrm{im}_{3V} < -56$ dB
$\mathrm{xm} < -50$ dB

Having these objectives in mind, we estimate the maximum repeaterless link length for the system. The permissible values for $m_c$ for $K = 5$ and $K = 1$ are determined from the cross-modulation specification and are found to be

$$m_c = 0.137 \qquad \text{for } K = 5$$

$$m_c = 0.5 \qquad \text{for } K = 1$$

Then the required power level at the receiver is calculated by the appropriate expressions given earlier in this section, so that the specified CNR is achieved, subject to the maximum allowed modulation index. Having estimated the values of the required power at the receiver, we can compute the loss budgets and thus calculate the permissible repeaterless link lengths using any of the available fiber-optic technologies. Numerical examples of the loss budgets for $K = 1$ and $K = 5$ are given in Table 9.10 (taken from Ref. 7).

### 9.4.7 Analog *BL* Product and Systems Examples

As pointed out in the introduction to this chapter, the most important single performance criterion for a fiber-optic communication system is the bandwidth-length product. The *BL* product indicates what the available bandwidth and repeaterless transmission distance are for a system by stating the product of the two. These two quantities, as a product, take into account both the system SNR

**TABLE 9.10** Power Budget for Example 2

| Parameter | 0.8 μm, APD | | 1.3 μm, PIN | |
|---|---|---|---|---|
| | $K = 5$ | $K = 1$ | $K = 5$ | $K = 1$ |
| Average power coupled into fiber,* dBm | −10 | −10 | −15 | −15 |
| Average power level required at the receiver, $P_{av}$, dBm | −17.42 | −22 | −19 | −22 |
| Connector losses, dB | 2 | 2 | 2 | 2 |
| Link margin, dB | 1 | 5.5 | 1.3 | 3 |
| Fiber attenuation, dB/km | 4.5 | 4.5 | 0.7 | 0.7 |
| Maximum link length, km | 0.98 | 1.0 | 2.0 | 2.79 |

*Core diameter = 62.5 μm.
*Source:* From Ref. 7.

and the signal dispersion. In other words, the *BL* product of a system is a quantitative indication of the tradeoff between transmission distance and the dispersion introduced by the system over that distance. It is easy to see that for a fixed *BL,* when the transmission distance is increased, the available bandwidth of the system is decreased, and vice versa.

Furthermore, the *BL* product of a system can be used as an indication of its capabilities; that is, by providing the *BL* product of a particular system, we specify whether it may be used for short-, medium-, or long-haul applications and what the available bandwidth will be for each case. For example, a system using a high-quality laser diode as the optical source, at the zero-dispersion wavelength of 1300 nm, with a single-mode fiber as the transmission medium and a high-quality APD detector in the receiver, will provide a high *BL* figure; i.e., the system may be used for a long-haul, broadband signal application. On the other hand, a system with an LED as its optical source, operating at 830 nm, with a multimode fiber as the transmission medium and a PIN photodetector in the receiver, will provide a relatively low *BL* figure, and hence it can be used only for short-haul, narrowband signal transmission.

The results obtained by a number of research groups[9–13] and fiber-optic communications companies were incorporated in the bandwidth-length grid in Fig. 9.13. The *BL* grid can be used as a quick reference to the capabilities of existing analog systems and various combinations of fiber-optic technologies, such as lasers or LEDs with single-mode or multimode fibers and PIN or APD receivers. The *BL* grid may also be used as an aid for the analog system design procedures. However, it is important to know that in some cases the *BL* product is less significant than other system parameters, for example, when a signal must be transmitted over a short distance and then divided among a number of subscribers. In such a case, the SNR of the system is of more importance than the *BL* product, even though the latter may provide valuable information about the overall transmission quality of the system.

Now we present examples of existing or experimental systems, with information pertinent to the *BL* product and the fiber-optic technology of each system:

*Example 1 (Ref. 6)*

Source type: InGaAsP laser diode having a modulation bandwidth of 11 GHz at 5 mW bias and an RIN of less than 135 dB/Hz

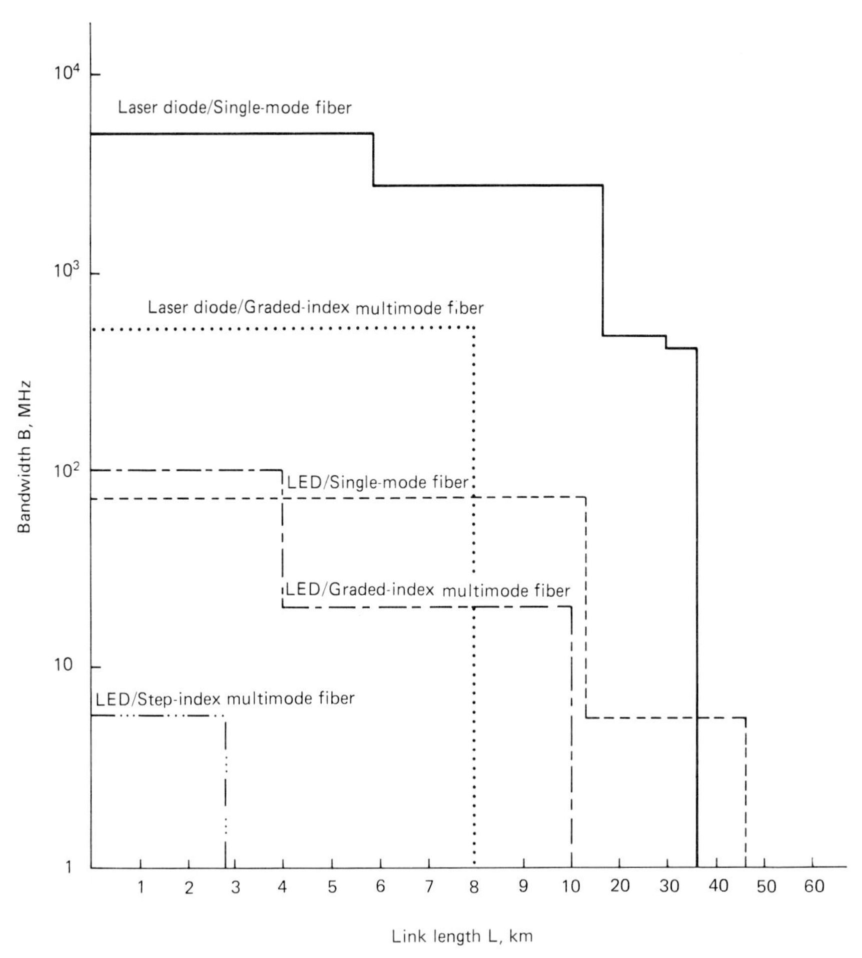


**FIGURE 9.13** *BL* (bandwidth-length) grid for analog fiber-optic communication systems. The source-detector combinations are indicated for each region.

Detector type: InGaAs, PIN diode, modulation bandwidth of 20 GHz, responsivity $R$ = 0.7 A/W

Fiber: single-mode fiber

Transmission wavelength: 1300 nm

Transmission distance: 18 km

Transmitted bandwidth: 2.7 to 5.2 GHz

*BL* product: 45 GHz · km

Application: Transmission, with subcarrier modulation, of 60 FM video channels, possibly for CATV networks

*Example 2 (Ref. 8)*

Source type: edge-emitting LED

Detector type: PIN FET receiver

Fiber: single-mode fiber

Transmission wavelength: 1300 nm

Transmission distance: 11 km

Transmitted bandwidth: about 75 MHz

*BL* product: about 825 MHz · km

Application: transmission, with subcarrier modulation, of two FM video and one DS1 digital channel

*Example 3 (Ref. 9)*

Source type: multilongitudinal-mode InGaAsP laser diode, having a modulation bandwidth of 6 GHz at 15-mA bias

Detector type: InGaAs PIN diode, modulation bandwidth of 6 GHz

Fiber: single-mode fiber

Transmission wavelength: 1300 nm

Transmission distance: 35 km

Transmitted bandwidth: 3.7- to 4.2-GHz band (satellite C band)

*BL* product: 17.5 GHz · km

Application: transmission of C-band satellite signal, possibly for earth-station links

*Example 4 (Ref. 10)*

Source type: LED

Detector type: APD

Fiber: multimode fiber

Transmission wavelength: 0.8 μm

Transmission distance: 1 km

Transmitted bandwidth: 6 MHz (one FM video channel)

*BL* product: 6 MHz · km

Application: transmission of one FM video channel by subcarrier modulation prior to intensity modulation of the LED

*Example 5*

Source type: LED

Detector type: APD (silicon)

Fiber: graded-index multimode fiber

Transmission wavelength: 820 nm

Transmission distance: 1 to 2 km

Transmitter bandwidth: 10 to 100 MHz

*BL* product: 10 to 2000 MHz · km

Application: transmission of 2 FM video channels, with subcarrier modulation

### 9.4.8 Analog System Design Evolution and Novel Approaches

So far, analog fiber-optic communications have been discussed from the viewpoint of established technology and how it can be used for transmission of broadband signals by the various multiplexing techniques described in Sec. 9.4. For example, microwave multiplexed broadband lightwave systems take advantage of the large bandwidth provided by single-mode fibers, lasers, and photodiodes and well-developed microwave electronics technology to provide an attractive alternative to conventional baseband digital systems. Furthermore, microwave multiplexed systems make it possible to transmit both analog and digital signals at the same time, and thus they represent a link between the present predominantly analog video technology and future technology, which may be predominantly digital.

Recent research has been aimed at either enhancing the performance of the existing systems or developing new schemes for the transmission of analog signals. For example, the use of traveling-wave laser amplifiers has been suggested as a way to enhance the SNR of subcarrier multiplexed systems. Traveling-wave laser amplifiers have been used in an experimental subcarrier modulation system[11] to transmit 90 FM video channels over single-mode fiber. The increase in the overall system SNR with traveling-wave laser amplifiers increases significantly the power margin in a subscriber loop, with minimum signal degradation; hence the video signal can be distributed to a large number of subscribers. In the experimental system, the signal was distributed to 2048 terminals, but in practical system applications, with plant allowance for fiber losses, connector and splice losses, etc., distribution ratios may be closer to 1:100 or 1:200. The power budget for the system used in this experiment can be found in Ref. 11. Traveling-wave laser amplifiers can also be incorporated in WDM systems; however, the following should be taken into account: the crosstalk in the traveling-wave laser amplifier resulting from gain reduction, resonance frequency shift induced by other channel light sources, and four-wave mixing.

After recent advances, coherent fiber optics communication systems are also being considered as candidates for improved analog signal transmission. Although coherent systems are significantly more complex in their implementation than direct detection systems, they have the advantages of inherent linearity, improved receiver sensitivity, and orders-of-magnitude of greater selectivity than noncoherent WDM techniques. One coherent optical system[13] takes advantage of the inherent linearity and sensitivity of amplitude modulation in order to transmit broadband analog (such as CATV) signals. This scheme requires significant phase-noise reduction, which is accomplished by using two novel approaches: optical filtering and phase-noise compensation based on linear heterodyne detection. The power budget of this system allows transmission of up to 10 wavelength-division-multiplexed carriers, each carrying up to 30 subcarrier-modulated TV channels.

As mentioned earlier, coherent systems not only have the advantage of inherent linearity and improved receiver sensitivity over direct detection systems, but also have much superior receiver selectivity and hence can exploit the large bandwidth of single-mode fibers (see Fig. 9.3). By taking advantage of the recent advances in tunable laser diode technology, such systems have been experimentally constructed and proved to be very promising. A few field trials have also been successful.

Optical heterodyne detection, one option in coherent receivers, has been aided by the progress in frequency-tunable laser diode technology, using distributed-feedback and distributed Bragg reflector structures and external cavities. The tuning range of frequency-tunable laser diodes in optical heterodyne systems employing WDM seems to be sufficient for the transmission of many channels, each carrying 10 subcarrier-modulated signals. However, frequency-tunable lasers still need to be improved as far as their spectral linewidth is concerned.

### 9.4.9 Commentary

In Sec. 9.4, we have examined the use of analog techniques in fiber-optic communications. The considerations pertaining to the analog transmitter and analog receiver used for such a system have been dealt with and the intermodulation and cross-modulation products have been examined. It has been shown that analog systems require a higher SNR than digital systems, and thus, in order to design systems with high *BL* product, semiconductor lasers should be used, transmitting at 1300 nm, with single-mode fiber as the medium. LEDs may be used in combination with single-mode or multimode fibers, but the *BL* product yielded by such systems is low, and therefore their use is restricted to short-haul applications. The type of detector for an analog system is not so critical, however, and either PIN or APD detectors may be used.

The advantages of analog fiber-optic communications becomes outstanding when multiplexing techniques, such as subcarrier multiplexing or wavelength-division multiplexing, are used. Systems which employ these techniques in combination with state-of-the-art optical components and fiber-optic technology are able to achieve very high *BL* product figures (of up to 45 GHz · km) or distribution of broadband signals to a large number of subscribers. The ability of these systems to provide transmission of broadband analog signals make them a very potent candidate for video signal transmission. Analog fiber-optic systems are at the present being used in CATV distribution applications. Furthermore, such systems provide a bridge between the present analog video technology and future, predominantly digital, video technology.

Considerable research is also taking place in the field of coherent fiber-optic systems and their possible use in combination with analog signal transmission. Such systems, despite the fact that most of them are on the experimental stage, have shown promising capabilities for providing even higher *BL* products and for better frequency utilization of the bandwidth.

The great bandwidth capabilities of single-mode fibers is not fully exploited because of the unavailability of high-speed digital electronics for transmitters, receivers, and multiplexing and demultiplexing circuits. This poses a constraint on the maximum data rates that can be achieved. This fact has shifted attention from developing baseband digital lightwave systems (for transmission of time-division-multiplexed voice and data channels over long-distance and interoffice routes) to extending the fiber-optic network to the subscriber loop and providing a variety of voice, data, and video services to subscribers over lightwave networks. Subcarrier modulation is an attractive way to transmit wideband services over optical fibers, using a complete range of available electronic microwave (analog and digital) techniques, and push to even greater bandwidths is the *natural* progression for exploiting the enormous bandwidth potential of single-mode fiber systems.

## ACKNOWLEDGMENTS

It is a pleasure to thank my graduate students for the many constructive inputs to this chapter and much help in collection of the materials and proof-reading. They are Steven Bastien, Hatem Abdelkader, Antonio Deus, Clark Engert, Il Whan Oh, Jeff Adams, Andreas Xenophontos, George Valliath, Alexis Mendez, and Bob Meindl. Particular thanks are due to Charlotte Knott for typing this chapter.

## REFERENCES

1. W. A. Gambling, "The Development of Optical Communication," *Electronics and Power,* December 1983, pp. 777–780.
2. D. C. Gloge and T. Li, "Multimode Fiber Technology for Digital Transmission," *Proceedings of the IEEE,* vol. 68, 1980, pp. 1269–1275.
3. I. Jacobs and J. R. Stauffer, "FT3—A Metropolitan Trunk Lightwave System," *Proceedings of the IEEE,* vol. 68, 1980, pp. 1286–1290.
4. P. K. Runge and P. R. Trischitta, "The SL Undersea Lightwave System," *Journal of Lightwave Technology,* vol. LT-2, 1984, pp. 744–753.
5. J. L. Gimlett and N. K. Cheung, "Dispersion Penalty Analysis for LED/Single-Mode Fiber Transmission Systems," *Journal of Lightwave Technology,* vol. LT-4, 1986, pp. 1381–1392.
6. K. Petermann and G. Arnold, "Noise and Distortion Characteristics of Semiconductor Lasers in Optical Fiber Communication Systems," *IEEE Journal of Quantum Electronics,* vol. QE-18, 1982, pp. 543-555.
7. M. W. Mesiya, "Design of Multichannel Analog Fiber-Optic Transmission Systems," in E. E. Basch (ed.), *Optical-Fiber Transmission,* Sams, Indianapolis, 1987, pp. 487–501.
8. J. E. Bowers, A. C. Chipaloski, S. Boodaghians, and J. W. Carlin, "Long Distance Fiber-Optic Transmission of C-Band Microwave Signals to and from a Satellite Antenna," *Journal of Lightwave Technology,* vol. LT-5, 1987, pp. 1733–1740.
9. W. I. Way, R. S. Wolff, and M. Krain, "A 1.3-μm 35-km Fiber-Optic Microwave Multicarrier Transmission System for Satellite Earth Stations," *Journal of Lightwave Technology,* vol. LT-5, 1987, pp. 1325–1332.
10. R. Olshansky and V. A. Lanzisera, "60-Channel FM Video Subcarrier Multiplexed Optical Communication System," *Electronics Letters,* vol. 23, 1987, pp. 1196–1198.
11. W. I. Way et al., "90-Channel FM Video Transmission to 2048 Terminals Using Two Inline Traveling-Wave Amplifiers in a 1300 nm Subcarrier Multiplexed Optical System," *Proc. 14th European Conference on Optical Communication, Brighton, U.K.,* 1988.
12. P. S. Venkatesan, P. S. Natarajan, and J. Orost, "Transmission of FDM Wideband Data and Video Channels over a Single-Mode Fibre Using a 1300 nm LED," *Electronics Letters,* vol. 24, 1988, pp. 387–389.
13. A. C. Van Bochove, J. P. Bekooj, and C. M. De Blok, "A Coherent Optical System for Analogue Multichannel Video Transmission," *Proc. 14th European Conference on Optical Communications,* Brighton, U.K., 1988, pp. 541–544.

# INDEX

## ABOUT THE EDITOR

Frederick C. Allard is the Fiber Optics Manager in the Electromagnetic Systems Department of the Naval Underwater Systems Center, New London, Connecticut. Since 1973, he has overseen the application of fiber optics to various naval systems involving sensors and data transfer, often for use in adverse environments. He has presented numerous papers at conferences and holds three optical patents. He has served for many years as a primary U.S. Navy member on national and international fiber optics coordination groups and has been a member of the Optical Society of America since 1967.